Earth as an Evolving Planetary System

D1342149

Kent C. Condie

AMSTERDAM • BOSTON • HEIDELBERG • LONDON
NEW YORK • OXFORD • PARIS • SAN DIEGO
SAN FRANCISCO • SINGAPORE • SYDNEY • TOKYO
Academic Press is an imprint of Elsevier

Elsevier Academic Press
200 Wheeler Road, Burlington, MA 01803, USA
525 B Street, Suite 1900, San Diego, CA 92101-4495, USA
84 Theobald's Road, London WC1X 8RR, UK

This book is printed on acid-free paper

Cover image courtesy of Tasa Graphic Arts, Inc., Taos, NM
http://www.tasagraphicarts.com

Library of Congress: Application submitted

British Library Cataloguing in Publication Data
A catalogue record for this book is available from the British Library

ISBN: 0-12-088392-9

For information on all Academic Press publications
visit our Web site at www.books.elsevier.com

Printed in the United States of America
04 05 06 07 08 09 9 8 7 6 5 4 3 2 1

Preface

Although this book began life in 1976 with the title *Plate Tectonics and Crustal Evolution*, the subject matter has gradually changed focus with subsequent editions, and especially since the third edition in 1989. In the last decade it has become increasingly clear that the various components of Earth act as a single, interrelated system, often referred to as the *Earth System*. One reviewer of the fourth edition pointed out that the title of the book was no longer appropriate, since plate tectonics was not a major focus. This is even more so in this fifth version, and thus, I introduce a new title for the book: *Earth as an Evolving Planetary System*.

Since the first edition in 1976, which appeared on the tail end of the plate tectonics revolution of the 1960s, our scientific database has grown exponentially and continues to grow—in fact, much faster than we can interpret it. If one compares the earlier editions of the book with this edition, a clear trend is apparent. Plate tectonics is now assimilated into geological textbooks and is part of the vernacular. The changing emphasis over the last 30 years is from how one system in our planet works (plate tectonics) to how all systems in our planet work, how they are related, and how they have governed the evolution of the planet. As scientists continue to work together and share information from many disciplines, this trend should continue for many years into the future.

Today, more than at any time in the past, we are beginning to appreciate the fact that to understand the history of our planet requires understanding the various interacting components and how they have changed with time. Although science is made up of specialties, to learn more about how Earth operates requires input from all of these specialties—geology alone is not sufficient. In this new book the various subsystems of the Earth are considered as vital components in the evolution of our planet. Subsystems include such components as the crust, mantle, core, atmosphere, oceans, and life.

As with previous editions, the book is written for advanced undergraduate and graduate students and assumes a basic knowledge of geology, biology, chemistry, and physics that most students in the Earth Sciences acquire during their undergraduate education. It also may serve as a reference book for specialists in the geological sciences who want to keep abreast of scientific advances in this field. I have attempted to synthesize and incorporate data from the fields of oceanography, geophysics, paleoclimatology, geology, planetology, and geochemistry, and to present this information in a systematic manner addressing problems related to the evolution of Earth over the last 4.6 billion years.

The book includes an introduction to some of the new and exciting topics in the Earth Sciences. Among these are results from increased resolution of seismic tomography by which plates can be tracked into the deep mantle. High resolution U-Pb zircon isotopic dating now permits us to better constrain the timing of important events in Earth history. We have detrital zircons with ages up to 4.4 Ga, suggesting the presence of felsic crust and water on the planet by this time. New information on the core provides us with a better understanding of how the inner and outer core interact and how the Earth's magnetic field is generated.

Two expanding areas of knowledge have necessitated new chapters. Exciting work on the origin of life and the possibilities of life beyond the Earth is summarized in a new chapter on Living Systems. Also, the continuing saga of mass extinctions and the role of impacts and mantle plume events have required more coverage. And a contentious new discussion on the Snowball Earth enters the picture. Did the Earth really freeze up some 600-700 million years ago? The episodic nature of crustal production, stable isotope anomalies, black shale deposition, giant dyke swarms and other phenomena have been well documented in the last few years, so much so that a new chapter has been added to cover this subject. Also included in this chapter are new ideas and results from the super-continent cycle and on the possibility of global mantle plume events as the driving force of episodic events.

In addition, a new version of an interactive CD ROM by the author, titled "Plate Tectonics and How the Earth Works (version 1.2)" is available to accompany this book. This CD, with animations and interactive exercises, can be obtained from Tasa Graphic Arts Inc., Taos, NM at http://www.tasagraphicarts.com.

Kent C. Condie
Department of Earth & Environmental Science
New Mexico Tech
Socorro, New Mexico

Contents

Earth Systems

Earth as a Planetary System

A **system** is an entity composed of diverse but interrelated parts that function as a whole (Kump et al., 1999a). The individual parts, often called **components,** interact with each other as the system evolves with time. Components include reservoirs of matter or energy, described by mass or volume, and subsystems, which behave as systems *within* a system. In recent years, the Earth has been considered a complex planetary system that evolved over 4.6 billion years of time. It includes reservoirs, such as the crust, mantle, and core, and subsystems, such as the atmosphere, hydrosphere, and biosphere. Because many of the reservoirs in the Earth interact with each other and with subsystems, such as the atmosphere, there is an increasing tendency to consider most or all of the Earth's reservoirs as subsystems.

The state of a system is characterized by a set of variables at any time during the evolution of the system. For the Earth, temperature, pressure, and various compositional variables are most important. The same thing applies to subsystems within the Earth. A system is at equilibrium when nothing changes as it evolves. If, however, a system is perturbed by changing one or more variables, it responds and adjusts to a new equilibrium state. A **feedback loop** is a self-perpetuating change and a response in a system to a change. If the response of a system amplifies the change, it is known as a **positive feedback loop,** whereas if it diminishes or reverses the effect of the disturbance, it is a **negative feedback loop.** As an example of positive feedback, if volcanism pumps more CO_2 into a CO_2-rich atmosphere of volcanic origin, this should promote greenhouse warming and the temperature of the atmosphere would rise. If the rise in temperature increases weathering rates on the continents, this would drain CO_2 from the atmosphere causing a drop in temperature, an example of negative feedback. Because a single subsystem in the Earth affects other subsystems, many positive and negative feedback loops occur as the Earth attempts to reach a new equilibrium state. These feedback loops may be short lived over hundreds to tens of thousands of years, such as short-term changes in climate, or they may be long lived over millions or tens of millions of years, such as changes in climate related to the dispersal of a supercontinent.

The driving force of planetary evolution is the thermal history of a planet, more fully described in Chapter 10. The methods and rates by which planets cool, either directly or indirectly, control many aspects of planetary evolution. In a silicate-metal planet like Earth, thermal history determines when and if a core will form (Fig. 1.1). It determines whether the core is molten, which in turn determines whether the planet will have a global magnetic field (generated by dynamo-like action in the outer core, as explained in Chapter 5). The magnetic field, in turn, interacts with the solar wind and with cosmic rays, and it traps high-energy particles in magnetic belts around the planet. This affects life because life cannot exist in the presence of intense solar wind or cosmic radiation.

Planetary thermal history also strongly influences tectonic, crustal, and magmatic history (Fig. 1.1). For instance, only planets that recycle lithosphere into the mantle by subduction, as the Earth does, appear capable of generating continental crust and thus having collisional orogens. Widespread felsic and andesitic magmas can only be produced in a plate tectonic regime. In contrast, planets that cool by mantle plumes and lithosphere delamination, as perhaps Venus does today, should have widespread mafic magmas with little felsic to intermediate component. They also appear to have no continents.

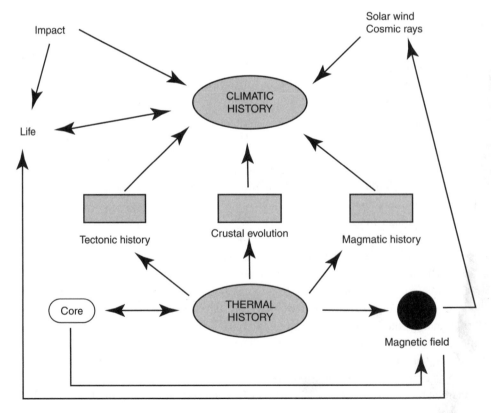

Figure 1.1 Major relationships between thermal and climatic histories of the Earth.

So where does climate come into these interacting histories? Climate reflects complex interactions of the ocean–atmosphere system with tectonic and magmatic components, as well as interactions with the biosphere. In addition, solar energy and asteroid or cometary impacts can have severe effects on climatic evolution (Fig. 1.1). The thermal history of a planet affects directly or indirectly all other systems in the planet, including life. The Earth has two kinds of energy sources: those internal to the planet and those external to the planet. In general, internal energy sources have long-term ($>10^6$ years) effects on planetary evolution, whereas external energy sources have short-term ($<10^6$ years) effects. Gradual increases in solar energy over the last 4.6 Gy have also influenced the Earth's climate on a long timescale. The most important extraterrestrial effects on planetary evolution, and especially on climate and life, are asteroid and cometary impacts, the effects of which usually last less than 10^6 years.

Many examples of interacting terrestrial systems are described in later chapters. However, before describing these systems and their interactions, I first need to review the basic structure of the Earth as determined primarily from seismology.

Structure of the Earth

The internal structure of the Earth is revealed primarily by compressional waves (primary waves, or P-waves) and shear waves (secondary waves, or S-waves) that pass through the planet in response to earthquakes. Seismic-wave velocities vary with pressure (depth), temperature, mineralogy, chemical composition, and degree of partial melting. Although the overall features of seismic-wave velocity distributions have been known for some time, refinement of data has been possible in the last 10 years. Seismic-wave velocities and density increase rapidly in the region between 200 and 700 km deep. Three first-order seismic discontinuities divide the Earth into crust, mantle, and core (Fig. 1.2): the **Mohorovicic discontinuity,** or **Moho,** defining the base of the crust; the **core–mantle interface** at 2900 km; and the **inner-core–outer-core interface** around 5200 km. The core composes about 16% of the Earth's volume and 32% of its mass. These discontinuities reflect changes in composition, phase, or both. Smaller but important velocity changes at 50 to 200 km, 410 km, and 660 km provide a basis for further subdivision of the mantle, as described in Chapter 4.

The major regions of the Earth can be summarized as follows (Fig. 1.2):

1. The **crust** consists of the region above the Moho and ranges in thickness from about 3 km at some oceanic ridges to about 70 km in collisional orogens.
2. The **lithosphere** (50–300 km thick) is the strong outer layer of the Earth—including the crust, which reacts to many stresses as a brittle solid. The **asthenosphere,** extending from the base of the lithosphere to the 660-km discontinuity, is by comparison a weak layer that readily deforms by creep. A region of low seismic-wave velocity and of high attenuation of seismic-wave energy, the **low-velocity zone** (LVZ), occurs at the top of the asthenosphere and is from 50 to 100 km thick. Significant lateral variations in density and in seismic-wave velocity are common at depths of less than 400 km.

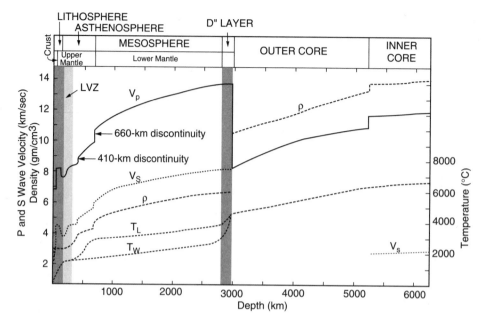

Figure 1.2 The distribution of average compressional-wave, or P-wave (V_p), and shear-wave, or S-wave (V_s), velocities and the average calculated density (ρ) in the Earth. Also shown are temperature distributions for whole-mantle convection (T_W) and layered mantle convection (T_L). LVZ, low-velocity zone.

3. The **upper mantle** extends from the Moho to the 660-km discontinuity and includes the lower part of the lithosphere and the upper part of the asthenosphere. The region from the 410-km to the 660-km discontinuity is known as the **transition zone.** These two discontinuities, further described in Chapter 4, are caused by two important solid-state transformations: from olivine to wadsleyite at 410 km and from spinel to perovskite with the addition of magnesiowustite at 660 km.

4. The **lower mantle** extends from the 660-km to the 2900-km discontinuity at the core–mantle boundary. It is characterized mostly by rather constant increases in velocity and density in response to increasing hydrostatic compression. Between 200 and 250 km above the core–mantle interface, a flattening of velocity and density gradients occurs in a region known as the **D″ layer,** named after the seismic wave used to define the layer. The lower mantle is also referred to as the **mesosphere,** a region that is strong but relatively passive in terms of deformational processes.

5. The **outer core** will not transmit S-waves and is interpreted to be liquid. It extends from the 2900-km to the 5200-km discontinuity.

6. The **inner core,** which extends from the 5200-km discontinuity to the center of the Earth, transmits S-waves—although at very low velocities, suggesting that it is a solid near the melting point.

There are only two layers in the Earth with anomalously low seismic-velocity gradients: the lithosphere and the D″ layer just above the core (Fig. 1.2). These layers coincide with steep temperature gradients; hence, they are thermal boundary layers in the Earth. Both layers play an important role in the cooling of the Earth. Most cooling (>90%) occurs by plate tectonics as plates are subducted deep into the mantle. The D″ layer is important in that steep thermal gradients in this layer may generate mantle plumes, many of which rise to the base of the lithosphere, thus bringing heat to the surface (<10% of the total Earth cooling).

Considerable uncertainty exists regarding the temperature distribution in the Earth. It depends on features of the Earth's history such as the initial temperature distribution in the planet, the amount of heat generated as a function of both depth and time, the nature of mantle convection, and the process of core formation. Most estimates of the temperature distribution in the Earth are based on one or a combination of two approaches: Models of the Earth's thermal history involving various mechanisms for core formation, and models involving redistribution of radioactive heat sources in the Earth by melting and convection processes.

Estimates using various models seem to converge on a temperature at the core–mantle interface of about $4500 \pm 500°$ C and a temperature at the center of the core from 6700 to 7000° C. Two examples of calculated temperature distributions in the Earth are shown in Figure 1.2. Both show significant gradients in temperature in the LVZ and the D″ layer. The layered convection model also shows a large temperature change near the 660-km discontinuity, because this is the boundary between shallow and deep convection systems in this model. The temperature distribution for whole-mantle convection, preferred by most scientists, shows a rather smooth decrease from the top of the D″ layer to the LVZ.

Plate Tectonics

Plate tectonics, which has so profoundly influenced geologic thinking since the early 1970s, provides valuable insight into the mechanisms by which the Earth's crust and mantle have evolved as well as into how the Earth has cooled. **Plate tectonics** is a unifying model that attempts to explain the origin of patterns of deformation in the crust, earthquake distribution, supercontinents, and midocean ridges and that provides a mechanism for the Earth to cool. Two major premises of plate tectonics are as follows:

1. The lithosphere behaves as a strong, rigid substance resting on a weaker asthenosphere.
2. The lithosphere is broken into numerous segments or plates that are in motion with respect to one another and are continually changing shape and size (Fig. 1.3).

The parental theory of plate tectonics, seafloor spreading, states that new lithosphere is formed at ocean ridges and moves from ridge axes with a motion like that of a conveyor belt as new lithosphere fills the resulting crack or rift. The mosaic of plates, which ranges from 50 to more than 200 km thick, is bounded by ocean ridges, subduction zones (partly collisional boundaries), and transform faults (boundaries along which plates slide by each other) (Fig. 1.3, cross-sections). To accommodate the newly created lithosphere, oceanic plates return to the mantle at subduction zones so that the surface area of the Earth remains constant.

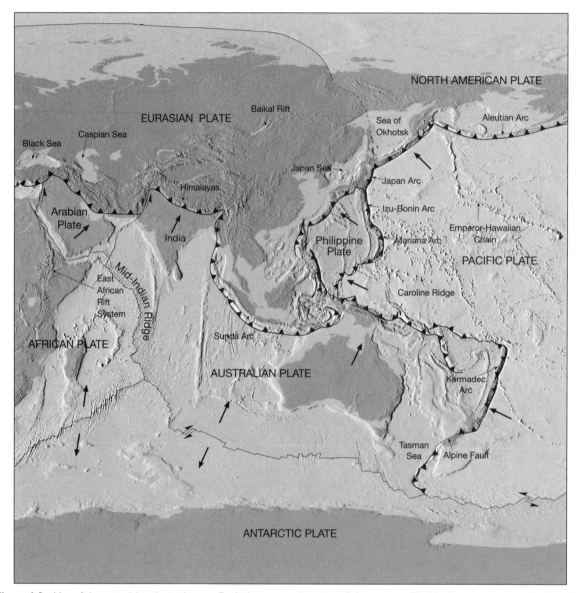

Figure 1.3 Map of the major lithospheric plates on Earth. Arrows are directions of plate motion. Filled barbs are convergent plate boundaries (subduction zones and collisional orogens); single lines are divergent plate boundaries (ocean ridges) and transform faults. Cross-sections show details of typical plate boundaries. Artwork by Dennis Tasa courtesy of Tasa Graphic Arts.

Many scientists consider the widespread acceptance of the plate tectonic model as a revolution in the earth sciences. As pointed out by J. Tuzo Wilson in 1968, scientific disciplines tend to evolve from a stage primarily of data gathering, characterized by transient hypotheses, to a stage at which a new unifying theory or theories are proposed that explain a great deal of the accumulated data. Physics and chemistry underwent such revolutions around the beginning of the twentieth century, whereas earth sciences entered such a revolution in the late 1960s. As with

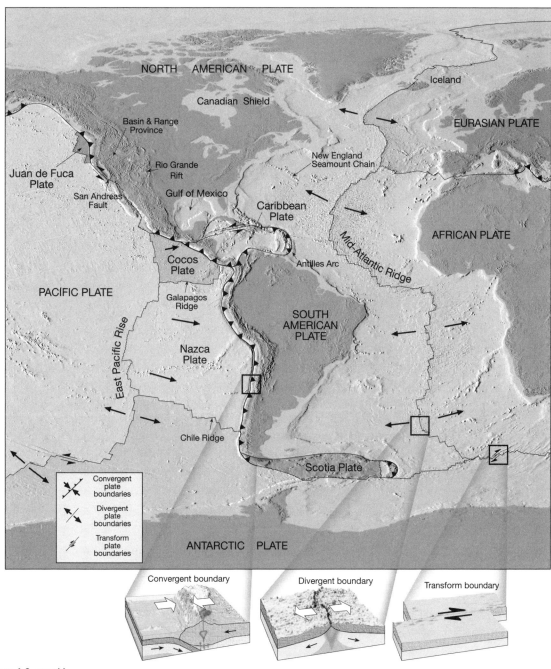

Figure 1.3, cont'd.

scientific revolutions in other fields, new ideas and interpretations do not invalidate earlier observations. On the contrary, the theories of seafloor spreading and plate tectonics offer for the first time a unified explanation for heretofore seemingly unrelated observations in the fields of geology, paleontology, geochemistry, and geophysics.

The origin and evolution of the Earth's crust is a tantalizing question that has stimulated much speculation and debate since the early part of the nineteenth century. Some of the first problems recognized, such as how and when the oceanic and continental crust formed, remain a matter of considerable controversy. Results from the Moon and other planets indicate that the Earth's crust may be a unique feature in the solar system. The rapid accumulation of data in the fields of geophysics, geochemistry, and geology since 1970 has added much to our understanding of the physical and chemical nature of the Earth's crust and of the processes by which it evolved. Evidence favors a source for the materials composing the crust from within the Earth. Partial melting of the mantle produces magma that moves to the surface and forms the crust. The continental crust, being less dense than the underlying mantle, rises isostatically above sea level and is subjected to weathering and erosion. Eroded materials are partly deposited on continental margins and partly returned to the mantle by subduction to be recycled and perhaps again become part of the crust at a later time.

Is the Earth Unique?

There are many features of our planet indicating that it is unique among planets in the Solar System and certainly among planets discovered so far around other stars. Consider, for instance, the following characteristics:

1. The Earth's near-circular orbit results in a more or less constant amount of heat from the Sun. If the orbit were more elliptical, the Earth would freeze over in the winter and roast in the summer. In such a case, higher forms of life could not survive.
2. If the Earth were much larger, the force of gravity would be too strong for higher life forms to exist.
3. If the Earth was much smaller, water and oxygen would escape from the atmosphere and higher life forms could not survive.
4. If the Earth was only 5% closer to the Sun, the oceans would evaporate and greenhouse gases would cause the surface temperature to rise too high for any life to exist (like on Venus today).
5. If the Earth was only 5% farther from the Sun, the oceans would freeze over and photosynthesis would be greatly reduced, leading to a decrease in atmospheric oxygen. Again, higher life forms could not exist.
6. If the Earth didn't have plate tectonics, there would be no continents; thus, large numbers of subaerial higher life forms could not exist.
7. If the Earth did not have a magnetic field of just the right strength, lethal cosmic rays would kill most or all life forms (including humans) on the planet.

8. If the Earth did not have an ozone layer in the atmosphere to filter out harmful ultraviolet radiation from the Sun, higher forms of life could not exist.

9. If the Earth's axial tilt (23.4 degrees) was greater or smaller, surface temperature differences would be too extreme to support life. Without the Moon, the Earth's spin axis would wobble too much to support life.

10. Without the huge gravity field of Jupiter, the Earth would be bombarded with meteorites and comets and higher life forms would not survive on the planet.

11. The massive asteroid collision on the Earth in 65 Ma led to the extinction of dinosaurs and cleared the way for the evolution and diversification of mammals and the eventual appearance of humans.

Are all these features of the Earth that make it suitable for higher life forms simply a coincidence? Although lower life forms, described in Chapter 7, can survive a range of physical and chemical conditions, higher life forms cannot survive such conditions. The Earth's development of "just right" conditions to support higher life forms is sometimes referred to as the Goldilocks problem (Rampino and Caldeira, 1994). As in the Goldilocks story, although Venus is too hot (the papa bear's porridge) and Mars is too cold (the mama bear's porridge), Earth (the baby bear's porridge) is just right! Why is Earth just right? Although some scientists believe that these conditions came about by chance, others argue that the Earth's properties have developed to *prepare* the planet for the origin, evolution, and survival of higher life forms.

Interacting Earth Systems

Earth subsystems (hereafter called *systems*) are not static but have evolved with time, leading the habitable planet we reside on. Current and future interactions of these systems will have a direct effect on life; therefore, it is important to understand how perturbation of one system can affect other systems and how rapidly systems change with time. Short-term climatic cycles are superimposed on and partly controlled by long-term processes in the atmosphere–ocean system, which in turn are affected by even longer processes in the mantle and core. Also affecting Earth systems are asteroid and cometary impacts, both of which appear to have been frequent in the geologic past.

Some of the major pathways of interaction among Earth systems and between Earth and extraterrestrial systems are summarized in Figure 1.4. Although I will consider some of these interactions in detail in later chapters, it may be appropriate to preview some of them now. As an example, the crystallization of metal onto the surface of the inner core may liberate enough heat to generate mantle plumes just above the core–mantle interface. As these plumes rise into the uppermost mantle, spread beneath the lithosphere, and begin to melt, large volumes of basalt may underplate the crust and erupt at the Earth's surface. Such eruptions may pump significant quantities of CO_2 into the atmosphere; because it is a greenhouse gas, it will warm the atmosphere, leading to warmer climates. This, in turn, may affect the continents (increasing weathering and erosion rates), the oceans (increasing the rate of limestone deposition), and life (leading to extinction of those forms not able to adapt to the changing climates). Thus, through a linked sequence of events, processes occurring in the Earth's core could lead to the extinction of life forms at the Earth's surface. Before such changes can affect life, however, negative feedback processes may return the atmosphere–ocean system to an equilibrium level or even reverse these changes. For instance, increased weathering rates caused by increased CO_2 levels in the atmosphere may drain the atmosphere of its excess CO_2, which is then transported by streams to the oceans where it is deposited in limestone. If cooling is sufficient, this could lead to widespread glaciation, which in turn could cause the extinction of some life forms.

Figure 1.4 Diagram showing major subsystems in the Earth. Some of the major paths of interaction are noted with arrows.

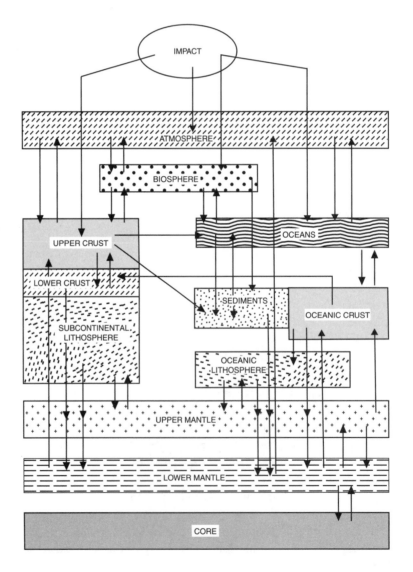

As an example of interactions related to plate tectonics, consider the subduction of oceanic crust into the deep mantle. This crust produces distinct compositional domains in the mantle; if incorporated into mantle plumes, these can rise to the base of the lithosphere, partially melt, and produce basalts that erupt at the Earth's surface. Again, greenhouse gases emitted during the eruptions can lead to climate warming.

To prepare for the continuing survival of living systems on the Earth, it is important to understand the nature and causes of interactions among Earth systems and between Earth and extraterrestrial systems. How fast and how frequently do these interactions occur, and what are the relative rates of forward and reverse reactions? These are important questions that need to be addressed by present and future generations of scientists.

Further Reading

Ernst, W. G., 2000. Earth Systems: Processes and Issues. Cambridge University Press, Cambridge, UK, 576 pp.

Kump, L. R., Kasting, J. F., and Crane, R. G., 1999a. The Earth System. Prentice Hall, Upper Saddle River, NJ, 351 pp.

Lillie, R. J., 1999. Whole Earth Geophysics. Prentice Hall, Upper Saddle River, NJ, 361 pp.

Taylor, S. R., 1998. Destiny or Chance: Our Solar System and Its Place in the Cosmos. Cambridge University Press, Cambridge, UK, 248 pp.

The Crust

2

Introduction

The Earth's crust is the upper rigid part of the lithosphere whose base defined by a prominent seismic discontinuity, the Mohorovicic discontinuity (Moho). There are three crustal divisions—oceanic, transitional, and continental; of these, oceanic and continental crusts dominate (Table 2.1). Typically, oceanic crust ranges from 3 to 15 km thick and comprises 54% of the crust's area and 17% of its volume. Islands, island arcs, and continental margins are examples of transitional crust that have thicknesses of 15 to 30 km. Continental crust ranges from 30 to 70 km thick and comprises 77% of the crust's volume but only 40% of its area. Our knowledge of oceanic crust comes largely from ophiolites, which are thought to represent tectonic fragments of oceanic crust preserved in the continents, and from deep-sea drill cores. Our view of the lower continental crust is based chiefly on a few uplifted slices of this crust in collisional orogens and from xenoliths brought to the surface in young volcanics.

The crust can be further subdivided into **crustal types,** segments of the crust exhibiting similar geological and geophysical characteristics. There are 13 major crustal types; these are listed in Table 2.1 with some of their physical properties. The first two columns of the table summarize the area and volume abundances, and the third column describes tectonic stability in terms of earthquake and volcanic activity and recent deformation.

To better understand the evolution of the Earth, you must understand the origin and evolution of the crust. In this chapter, I will briefly summarize the physical and chemical properties of the crust and will review characteristics of crustal provinces and supercontinents.

					Bouguer	
	Area	**Volume**		**Heat Flow**	**Anomaly**	**Poisson's**
Crustal Type	**(%)**	**(%)**	**Stability**	**(mW/m²)**	**(mgal)**	**Ratio**
Continental						
1. Shield	6	11	S	40	−20 to −30	0.29
2. Platform	18	35	S, I	49	−10 to −50	0.27
3. Orogen						
PA	8	13	S	60	−100 to −200	0.26
MC	6	14	I, U	70	−200 to −300	0.26
4. Continental-margin arc	2	4	I, U	50–70	−50 to −100	0.25
Transitional						
5. Rift	1	1	U	60–80	−200 to −300	0.30
6. Island arc	1	1	I, U	50–75	−50 to +100	0.25
7. Oceanic plateau	3	3	S, I	50–60	−100 to +50	0.30
8. Inland-sea basin	1	1	S	50	0 to +200	0.27
Oceanic						
9. Ocean ridge	10	2	U	100–200	+200 to +250	0.22
10. Ocean basin	38	12	S	50	+250 to +350	0.29
11. Marginal-sea basin	4	2	U, I	50–150	+50 to +100	0.25
12. Volcanic island	<1	<1	I, U	60–80	+250	0.25
13. Trench	2	1	U	45	−100 to −150	0.29
Average Continent	40	77		55	−100	0.27
Average Ocean	54	17		67 (95)*	+250	0.29

Table 2.1 Characteristics of the Earth's Crust

Stability key: I, intermediate stability; S, stable; U, unstable. Poisson's ratio = 0.5 {1−1/[(Vp/Vs)² − 1]}.
MC, Mesozoic–Cenozoic orogen; PA, Paleozoic orogen.
* Calculated oceanic heat flow.

Crustal Types

Oceanic Crust

Ocean Ridges

Ocean ridges are widespread linear rift systems in oceanic crust, where new lithosphere is formed as the flanking oceanic plates move from each other. They are topographic highs on the seafloor and are tectonically unstable. A medial rift valley generally occurs near their crests in which new oceanic crust is produced by intrusion and extrusion of basaltic magmas. The worldwide ocean-ridge system is interconnected from ocean to ocean and is more than 70,000 km long (see Fig. 1.3 in Chapter 1). Ridge crests are cut by numerous transform faults, which may offset ridge segments by thousands of kilometers.

From geophysical and geochemical studies of ocean ridges, it is clear that both structure and composition vary along ridge axes (Solomon and Tooney, 1992). In general, there is a good correlation between the spreading rate and the supply of magma from the upwelling asthenosphere. Within each ridge segment, the characteristics of deformation, magma emplacement, and hydrothermal circulation vary with distance from the magma center. Also, both the forms of segmentation and the seismic crustal structure differ between fast-spreading (≥80 mm/year half rate) and slow-spreading (12–50 mm/year) ridges. Ultra-slow-spreading ridges (≤12 mm/year) have the greatest topography, often with horsts of mantle rock exposed on the sea bottom (Dick et al., 2003).

From the extensive geophysical and petrologic database, the following observations are important in understanding the evolution of slow- and fast-spreading ocean-ridge systems:

1. The lower crust (the oceanic layer; see Fig. 2.2) is thin and poorly developed at slow-spreading ridges. The Moho is a sharp, tectonic boundary and may be a detachment surface (Dilek and Eddy, 1992). In contrast, at fast-spreading ridges, the Moho is a transition zone up to 1 km thick.
2. In general, rough topography occurs on slow-spreading ridges, and smooth topography is more common on fast-spreading ridges. Fast-spreading ridges also commonly lack well-developed axial rifts.
3. Huge, long-lived axial magma chambers capable of producing a thick gabbroic lower crust are confined to fast-spreading ridges.
4. At slow-spreading ridges, like the mid-Atlantic and southwestern Indian ridges, permanent magma chambers are often absent, and only ephemeral intrusions (chiefly dikes and sills) are emplaced in the medial rift (Dick et al., 2003).

Ocean Basins

Ocean basins comprise more of the Earth's surface (38%) than any other crustal type (Table 2.1). Because the oceanic crust is thin, however, they only make up 12% of its volume. They are tectonically stable and are characterized by a thin, sediment cover (approximately 0.3 km thick) and linear magnetic anomalies produced at ocean ridges during reversed and normal polarity intervals. The sediment layer thickens near continents and arcs from which detrital sediments are supplied.

Volcanic Islands

Volcanic islands occur in ocean basins (such as the Hawaiian Islands) or on or near ocean ridges (e.g., St. Paul Rocks and Ascension Island in the Atlantic Ocean; see Fig. 1.3 in Chapter 1). They are large volcanoes erupted on the seafloor whose tops have emerged above sea level. If they are below sea level, they are called *seamounts*. Volcanic islands and seamounts range in tectonic stability from intermediate or unstable in areas where volcanism is active (such as Hawaii and Reunion island) to stable in areas of extinct volcanism (such as Easter Island). Volcanic islands range in size from less than 1 to about 10^4 km². **Guyots** are flat-topped seamounts produced by erosion at sea level followed by submersion, probably because of the sinking of the seafloor. Coral reefs grow on some guyots as they sink, producing atolls. Some volcanic islands may have developed over mantle plumes, which are the magma sources.

Trenches

Oceanic **trenches** mark the beginning of subduction zones and are associated with intense earthquake activity (see Fig. 1.3 in Chapter 1). Trenches parallel arc systems and range in depth from 5 to 8 km, representing the deepest parts of the oceans. They contain relatively small amounts of sediment deposited chiefly by turbidity currents and derived chiefly from nearby arcs or continental areas.

Back-Arc Basins

Back-arc basins are segments of oceanic crust between island arcs (such as the Philippine Sea) or between island arcs and continents (such as the Seas of Japan and Okhotsk). They are abundant in the western Pacific and are characterized by a horst–graben topography (similar to the Basin and Range Province) with major faults subparalleling adjoining arc systems. The thickness of sediment cover is variable, and sediments are derived chiefly from continental or arc areas. Active basins, like the Lau-Havre and Mariana troughs in the southwest Pacific, have thin sedimentary cover, rugged horst-graben topography, and high heat flow. Inactive basins such as the Tasman and West Philippine basins have variable sediment thicknesses and generally low heat flow.

Transitional Crust

Oceanic Plateaus

Oceanic plateaus are large flat-topped plateaus on the seafloor composed largely of mafic volcanic and intrusive rocks (Coffin and Eldholm, 1994). They are generally capped with a thin veneer of deep-sea sediments and typically rise 2 km or more above the seafloor. Next to basalts and associated intrusive rocks produced at ocean ridges, oceanic plateaus are the largest volumes of mafic igneous rocks at the Earth's surface. The magmas in these plateaus, with their continental equivalents known as *flood basalts,* appear to be produced in hotspots caused by mantle plumes. Some of the largest oceanic plateaus— such as the Ontong Java in the South Pacific and the Kerguelen in the southern Indian Ocean, which together cover an area nearly half the size of the conterminous United States—were erupted in the mid-Cretaceous. Most oceanic plateaus are 15 to 30 km thick, although some exceed 30 km.

Arcs

Arcs occur above active subduction zones where one plate dives beneath another. There are two types: **island arcs** that develop on oceanic crust, and **continental-margin arcs**

that develop on continental or transitional crust. Island arcs commonly occur as arcuate chains of volcanic islands, such as the Mariana, Kermadec, and Lesser Antilles arcs (see Fig. 1.3 in Chapter 1). Most large volcanic chains, such as the Andes, Cascades, and Japanese chains, are continental-margin arcs. Some arcs, such as the Aleutian Islands, continue from continental margins into oceanic crust. Modern arcs are characterized by variable but often intense earthquake activity and volcanism and by variable heat flow, gravity, crustal thickness, and other physical properties. Arcs are composed dominantly of young volcanic and plutonic rocks and derivative sediments.

Continental Rifts

Continental rifts are fault-bounded valleys ranging in width from 30 to 75 km and in length from tens to hundreds of kilometers. They are characterized by a tensional tectonic setting in which the rate of extension is less than a few millimeters a year. Shallow magma bodies (<10 km deep) have been detected by seismic studies beneath some rifts such as the Rio Grande rift in New Mexico. The longest modern rift system is the East African system, which extends more than 6500 km from the western part of Asia Minor to south-eastern Africa (see Fig. 1.3 in Chapter 1). The **Basin and Range Province** in western North America is a multiple rift system composed of a complex series of alternating grabens and horsts. **Aulacogens** are rifts that die out toward the interior of continents, and many appear to represent "failed arms" of triple junctions formed during the fragmenta-tion of supercontinents. Young rifts (<30 Ma) are tectonically unstable, and earthquakes, although quite frequent, are generally of low magnitude.

Inland-Sea Basins

Inland-sea basins are partially to completely surrounded by tectonically stable continen-tal crust. Examples include the Caspian and Black Seas in Asia and the Gulf of Mexico in North America (see Fig. 1.3 in Chapter 1). Earthquake activity is negligible or absent. Inland-sea basins contain thick successions (10–20 km) of clastic sediments, and both mud and salt diapirs are common. Some, such as the Caspian and Black Seas, are the remnants of large oceans that closed in the geologic past.

Continental Crust

Precambrian Shields

Precambrian shields are stable parts of the continents composed of Precambrian rocks with little or no sediment cover. Rocks in shields may range in age from 0.5 to more than 3.5 Ga. Metamorphic and plutonic rock types dominate, and pressure–temperature (P–T) regimes recorded in exposed rocks suggest burial depths ranging from 5 km to 40 km or more. Shield areas, in general, exhibit little relief and have remained tectonically stable for long periods. They comprise about 11% of the total crustal volume, with the largest shields occurring in Africa, Canada, and Antarctica.

Platforms

Platforms are stable parts of the crust with little relief. They are composed of Precambrian basement similar to that in shields overlain by 1 to 3 km of relatively undeformed sedimentary rocks. Shields and the Precambrian basement of platforms are collectively called *cratons*. A **craton** is an isostatically positive portion of the continent that is tectonically stable relative to adjacent orogens. As you shall see later, cratons are composed of uplifted, eroded ancient orogens. Sedimentary rocks on platforms range in age from Precambrian to Cenozoic and reach thicknesses of 5 km, such as in the Williston basin in the north-central United States. Platforms comprise most of the crustal volume (35%) and most of the continental crustal area and volume.

Collisional Orogens

Collisional orogens are long, curvilinear belts of compressive deformation produced by the collision of continents. Giant thrust sheets and nappes are found in many orogens. Collisional orogens range from several thousand to tens of thousands of kilometers in length and are composed of a variety of rock types. They are expressed at the Earth's surface as mountain ranges with varying degrees of relief depending on their age. Older collisional orogens, such as the Appalachian orogen in eastern North America and the Variscan orogen in central Europe, are deeply eroded with only moderate relief, whereas young orogens, such the Alps and Himalayas, are among the highest mountain chains on Earth. Tectonic activity decreases with the age of the deformation in orogens. Orogens older than Paleozoic are deeply eroded and are now part of Precambrian cratons. Large plateaus, uplifted crustal blocks that have escaped major deformation, are associated with some orogens, such as the Tibet plateau in the Himalayan orogen.

Continent Size

It has been known for some time that there is an overall positive correlation between the area of continents and microcontinents and their average elevation relative to sea level (Cogley, 1984) (Fig. 2.1). Small continental fragments, such as Agulhas and Seychelles, lie 2 km or more below sea level. Arabia and India are higher than predicted by the area–height relationship, a feature probably related to their tectonic histories. The height of a continent depends on the uplift rate and the rates of erosion and subsidence. Collision between continents resulting in lithosphere thickening is probably the leading cause of continental uplift in response to isostasy. The anomalous heights of India and Arabia undoubtedly reflect the collision of these microcontinents; Asia is between 70 and 50 Ma, and the collision with India is continuing today. The six large continents have undergone numerous collisions in the last 1 Gy, each thickening the lithosphere in collisional zones leading to a greater average continent elevation. Hotspot activity can also elevate continents, and it is probable that the high elevation of Jan Mayen Island may be related to the

900 km

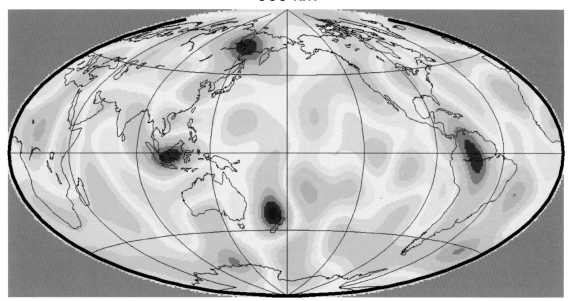

1300 km

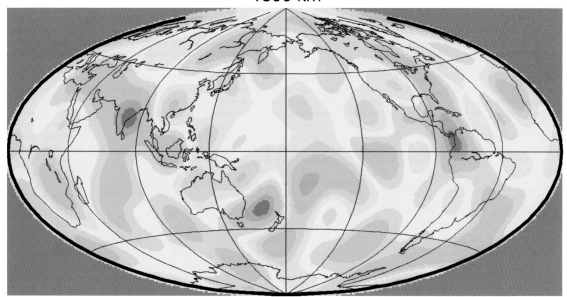

Plate I P-wave tomographic sections of Earth at 900 and 1300 km. Red refers to slow velocities and blue to fast velocities. Model from Boschi and Dziewonski (1999). Reproduced with permission of the American Geophysical Union, courtesy of Lapo Boschi.

Shear Velocity

Density

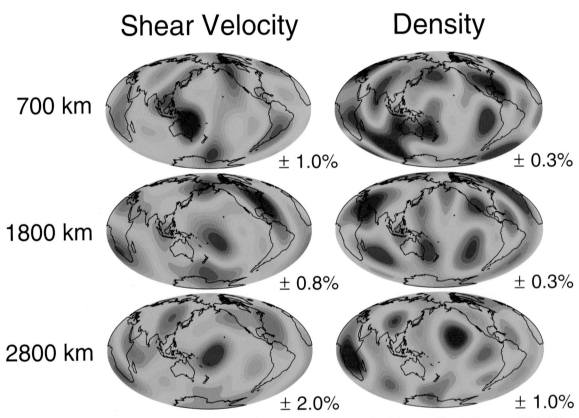

700 km

± 1.0%

± 0.3%

1800 km

± 0.8%

± 0.3%

2800 km

± 2.0%

± 1.0%

Plate 2 S-wave velocity distribution and calculated densities for three depths in the mantle. Colors as in Plate 1. Courtesy of Maiki Ishii.

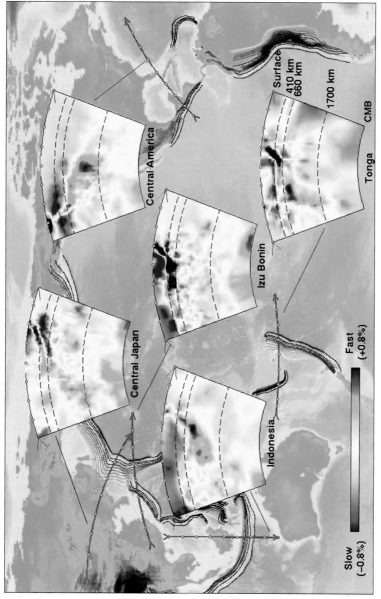

Plate 3a Seismic tomographic cross sections of subduction zones around the Pacific basin. Colors as in Plate I. Although most plates descend into the deep mantle, some are temporarily stalled at the 660-km discontinuity. Courtesy of Rob van der Hilst. Reprinted from Karason et al. (2000) with permission of the American Geophysical Union.

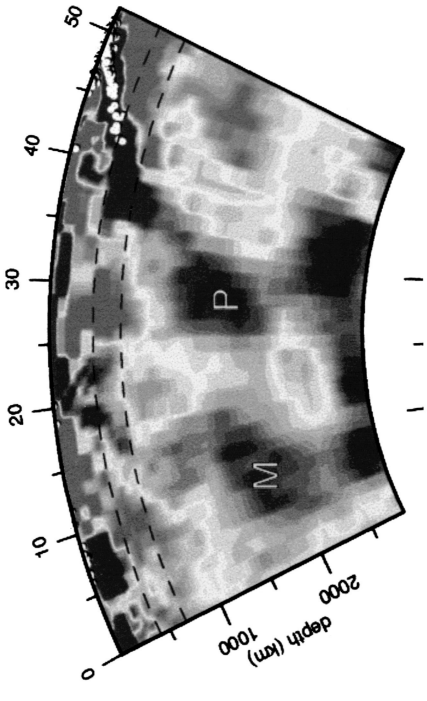

Plate 3b Seismic tomographic section of northeastern Asia. Colors as in Plate 1. Note that the Pacific (P) and Mongol Okhotsk (M) slabs (in blue) extend all the way to the core–mantle boundary. White dots are earthquakes in descending Pacific plate beneath Japan. Courtesy of Rob van der Voo. Reprinted with permission of *Nature*, Copyright © 1999 MacMillan Magazines, Ltd.

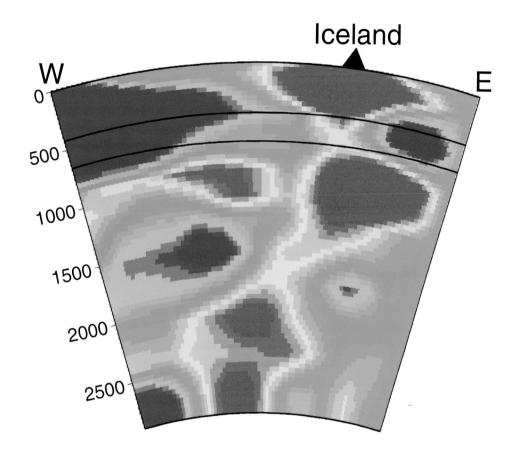

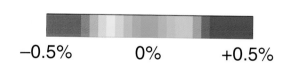

−0.5% 0% +0.5%

Plate 4 P-wave seismic tomographic image of the Iceland plume. Colors as in Plate 1. Courtesy of Dapeng Zhao.

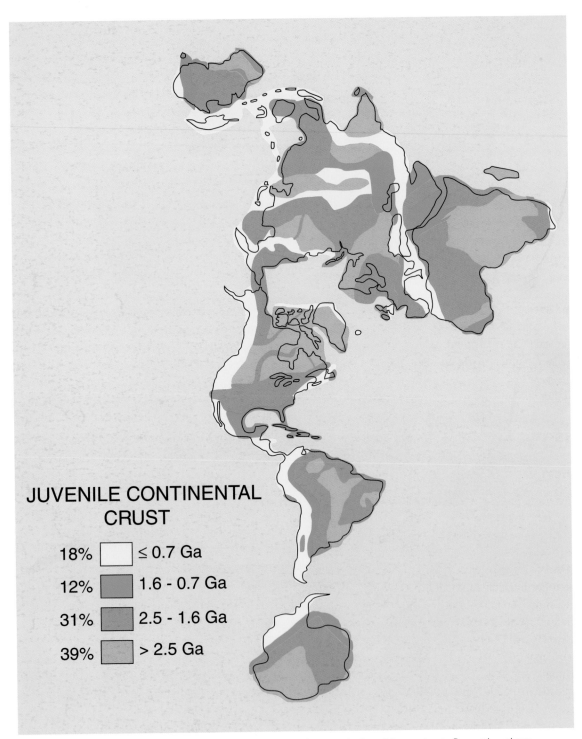

JUVENILE CONTINENTAL CRUST

18% ≤ 0.7 Ga

12% 1.6 - 0.7 Ga

31% 2.5 - 1.6 Ga

39% > 2.5 Ga

Plate 5 Distribution of juvenile continental crust shown on an equal-area projection of the continents. Percent by volume.

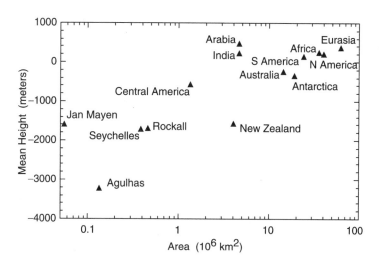

Figure 2.1 Mean continental heights as a function of continental area. Data from Cogley (1984).

Iceland hotspot. Also, Africa, which is higher than the other large continents by 200 m, may reflect the numerous hotspots beneath this continent.

The other five small continents in Figure 2.1 are mostly fragments that have been rifted from supercontinents. Because the breakup of supercontinents leads to a rise in sea level, it is not surprising that these microcontinents are completely flooded. Not until they collide with other continents will they emerge above sea level.

Seismic Crustal Structure

Mohorovicic Discontinuity

The Moho is the outermost seismic discontinuity in the Earth and defines the base of the crust (Jarchow and Thompson, 1989). It ranges in depth from about 3 km at ocean ridges to 70 km in collisional orogens and is marked by a rapid increase in seismic primary-wave (P-wave) velocity from less than 7.6 km/sec to at least 8 km/sec. Because the crust is different in composition from the mantle, the Moho is striking evidence for a differentiated Earth. Detailed seismic refraction and reflection studies indicate that the Moho is not a simple boundary worldwide. In some crust, such as collisional orogens, the Moho is often offset by complex thrust faults. The Himalayan orogen is a superb example in which a 20-km offset in the Moho is recognized beneath the Indus suture (Hirn et al., 1984). This offset was produced as crustal slices were thrust on top of each other during the Himalayan collision. In crust undergoing extension, such as continental rifts, a sharp seismic discontinuity is often missing and seismic velocities change gradually from crustal to mantle values. In some collisional orogens, the Moho may not always represent the base of the crust. In these orogens, thick mafic crustal roots may invert to eclogite (a high-density mafic rock) and seismic velocity may increase to mantle values, yet eclogite may form in the crust (Griffin and O'Reilly, 1987). The petrologic base of the crust where eclogite rests on ultramafic mantle rocks may not show a seismic discontinuity because both rock types have similar velocities. This has produced two types of Mohos: the **seismic Moho** (defined by a jump in seismic velocities) and the **petrologic Moho** (defined by the base of eclogitic lower crust).

Crustal Layers

Crustal models based on seismic data indicate that oceanic crust can be broadly divided into three layers. These are, in order of increasing depth, the **sediment layer** (0–1 km thick), the **basement layer** (0.7–2.0 km thick), and the **oceanic layer** (3–7 km thick) (Fig. 2.2). Models for the continental crust have a greater range in both number and thickness of layers. Although two- or three-layer models for continental crust are most common, one-layer models and models with more than three layers are proposed in some regions (Christensen and Mooney, 1995; Mooney et al., 1998). With the exception of continental borderlands and island arcs, continents range in thickness from about 35 to 40 km (mean = 41 km), and the average thickness of the oceanic crust is only 5 to 7 km.

P-wave velocities in crustal sediment layers range from 2 to 4 km/sec depending on degree of compaction, water content, and rock type. Velocities in the middle oceanic crustal layer are about 5 km/sec, and those in the middle continental crustal layer are about 6.5 km/sec (Table 2.2). Lower crustal layers in both oceans and continents are characterized by P-wave velocities of 6.5 to 6.9 km/sec.

Seismic-wave velocities increase with depth in the continental crust from 6.0 to 6.2 km/sec at depths less than 10 km to 6.6 km/sec at a 25-km depth. Lower crustal velocities range from 6.8 to 7.2 km/sec and in some cases show a bimodal distribution. In some continental crust, there is evidence of a small discontinuity at midcrustal depths, called the **Conrad discontinuity** (Litak and Brown, 1989). When identified, the Conrad discontinuity varies in depth and character from region to region, suggesting that, unlike the Moho, it is not a fundamental property of the continental crust and it is diverse in origin. Seismic reflections also occur at midcrustal depths in some extended crust, such as the Rio Grande rift in New Mexico.

Figure 2.2 Standard cross-section of oceanic crust based on seismic reflection data. Solid lines are examples of major reflectors.

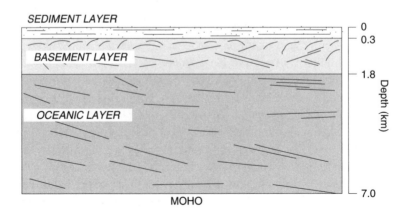

| | | Vp Upper | Vp Middle | Vp Lower | Mean | |
| | Vpn | Crust | Crust | Crust | Thickness | Mean Vp |
Crustal Type	(km/sec)	(km/sec)	(km/sec)	(km/sec)	(km)	(km/sec)
Continental						
1. Shield	8.1	6.2 (13)	6.5 (13)	7.0 (14)	40	6.5
2. Platform	8.1	6.2 (15)	6.5 (13)	7.0 (14)	42	6.5
3. Orogen						
PA	8.1	6.1 (10)	6.4 (12)	6.8 (13)	35	6.4
MC	8.0	6.0 (16)	6.3 (11)	6.7 (25)	52	6.3
4. Continental-margin arc	7.9	6.2 (18)	6.6 (10)	7.0 (10)	38	6.4
Transitional						
5. Rift	7.8	6.0 (11)	6.5 (6)	6.9 (11)	28	6.5
6. Island arc	7.8	6.2 (6)	6.6 (6)	7.2 (7)	19	6.4
7. Oceanic plateau	8.0	6.3 (5)	6.5 (10)	6.8 (10)	25	6.5
8. Inland-sea basin	8.1		2–5 (7)	6.8 (15)	22 [25]	6.0
Oceanic						
9. Ocean ridge	<7.5		5.0 (1)	6.5 (4)	5 [6]	6.4
10. Ocean basin	8.2	2–4 (0.5)	5.1 (1.5)	6.8 (5)	7 [11]	6.4
11. Marginal-sea basin	7.5–8.0	2–4 (1)	5.3 (3)	6.6 (5)	9 [13]	6.3
12. Volcanic island	7.5–8.0		6.6 (6)	7.0 (7)	13	6.5
13. Trench	8.0	2–4 (3)	5.1 (1.5)	6.8 (5)	8 [14]	6.4
Average Continent	8.1	6.2 (13)	6.5 (13)	7.0 (14)	41	6.5
Average Ocean	8.1	2–4 (0.5)	5.1 (1.5)	6.8 (5)	7 [10]	6.4

Table 2.2 Crustal Seismic-Wave Velocities

Reference: Mooney et al. (1998) and references cited therein.
Crustal layer thicknesses (in kilometers) are given in parenthesis. MC, Mesozoic–Cenozoic orogen;
PA, Paleozoic orogen; Vp, compressional velocity; Vpn, mantle velocity at the Moho.
Numbers in square brackets are depths to the Moho below sea level.

Crustal Types

Oceanic Crust

Crustal structure in ocean basins is rather uniform, not deviating greatly in either velocity or layer thickness distribution from that shown in Figure 2.2 (Solomon and Toomey, 1992). Crustal thickness ranges from 6 to 8 km and, unlike the lithosphere, does not thicken with age for faster spreading rates of about 20 mm/year (McClain and Atallah, 1986). Crustal thickness, however, drops rapidly at spreading rates slower than this (Dick et al., 2003). The sediment layer averages about 0.3 km in thickness and exhibits strong seismic reflecting zones with variable orientations (Fig. 2.2), some of which are probably produced by cherty layers, as suggested by cores retrieved by the Ocean Drilling Project. The thickness of the basement layer averages about 1.5 km, seismic-wave velocity increases rapidly with depth, and significant seismic anisotropy has been described in some areas (Solomon and Toomey,

1992). There are also numerous reflective horizons in this layer. The oceanic layer is generally uniform in both thickness (4–6 km) and velocity (6.7–6.9 km/sec).

Beneath ocean ridges, crustal thickness ranges from 3 to 6 km, most of which is accounted for by the oceanic layer (Solomon and Toomey, 1992) (Fig. 2.3). Seismic reflections indicate magma chambers beneath ridges at depths from 1 to 3 km. Unlike other oceanic areas, the velocities in the oceanic layer are quite variable, ranging from 4.4 to 6.9 km/sec. Anomalous mantle (compressional velocity, or Vp, <7.8 km/sec) occurs beneath ridge axes, reflecting high temperatures. Surface-wave data indicate that the lithosphere increases in thickness from less than 10 km beneath ocean ridges to 50 to 65 km at a crustal age of 50 My. Anisotropy in secondary wave (S-wave) velocities in the oceanic mantle lithosphere is often pronounced with the fast wave traveling normal to ocean-ridge axes. Such anisotropy appears to be caused by alignment of olivine c-axes in this direction (Raitt et al., 1969).

The crust of back-arc basins is slightly thicker (10–15 km) than that of ocean basins, principally because of a thicker sediment layer in marginal seas (Fig. 2.3). Crustal thickness in volcanic islands ranges from 10 to 20 km, with upper crustal velocities ranging from 4.7 to 5.3 km/sec and lower crustal velocities from 6.4 to 7.2 km/sec.

Transitional Crust

Oceanic Plateaus. The seismic structure of oceanic plateaus is not well known. On the basis of existing seismic and gravity data, the largest plateaus such as Ontong-Java and

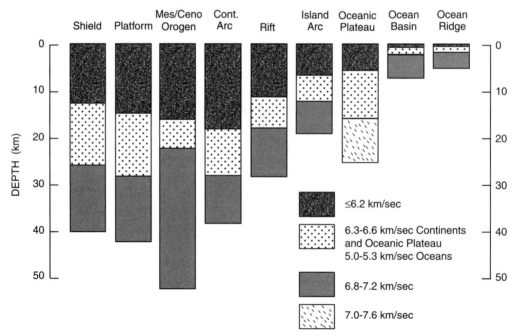

Figure 2.3 Seismic refraction sections of various crustal types. Ceno, Cenozoic; cont., continental-margin; Mes, Mesozoic.

Kerguelen range from 20 to 33 km thick (Condie, 2001) (Fig. 2.3). Lower crustal P-wave velocities are high (6.7–7.0 km/sec), and seismic reflection data from Kerguelen indicate a typical pelagic sediment layer 2 to 3 km thick.

Continental Rifts. Continental rifts have thin crust (typically 20–30 km thick) and low mantle velocities (Vp < 7.8 km/sec). Thinning of the crust in these regions is accomplished by thinning of the lower crustal layer (Fig. 2.3), which ranges from only 4 to 14 km thick. This reflects the ductile behavior of the lower crust during extension. Although most earthquake foci in rifts are less than 20 km deep, some occur as deep as 25 to 30 km. In young crust being extended, such as the Basin and Range Province in the western United States, the lower crust is highly reflective in contrast to a relatively transparent upper crust (Mooney and Meissner, 1992) (Fig. 2.4). Also, the reflection Moho is nearly flat, presumably because of the removal a crustal root during extension. Basin and Range normal faults cannot be traced through the lower crust and do not appear to offset the Moho.

Arcs. Resolution of seismic data is poor in arc systems, and considerable uncertainty exists regarding crustal arc structure. Crustal thickness ranges from 5 km in the Lesser Antilles to 35 km in Japan, averaging about 22 km, and mantle velocities range from normal (8.0–8.2 km/sec) to low (<7.8 km/sec). All values given in Table 2.2 have large standard deviations, and many arcs cannot be modeled as a simple two- or three-layer crust. There is evidence in some island arcs for an intermediate-velocity layer (5.0–6.0 km/sec) of varying thickness (Fig. 2.3).

Seismic reflection profiles in arcs are extremely complex, as shown by a profile across the forearc region of the Aleutian arc in southern Alaska (Fig. 2.5). Although reflections are deformed by steep faults that dip toward the continent (dark lines), they can be traced beneath the forearc basin, where they appear to plunge into the continent. The strong continuous reflection near the middle of the section (R) may be the contact of deformed ocean sediments with metamorphic basement. The sediments appear to be underplated beneath the arc along this boundary. The strong reflectors in the trench represent trench turbidites and oceanic sediments, and those in the forearc basin are chiefly volcanogenic turbidites.

Inland-Sea Basins. Inland-sea basins show a considerable range in crustal thickness and layer distribution. Crustal thickness ranges from about 15 km in the Gulf of

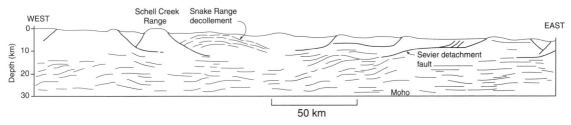

Figure 2.4   East–west cross-section of the Great Basin from Long Valley, Nevada, to the Pavant Range, Utah, showing major seismic reflectors. Data from Gans (1987).

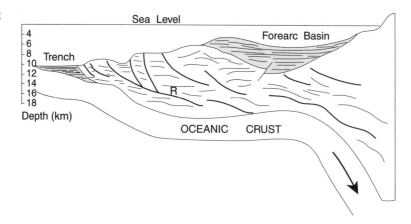

Figure 2.5 Line drawing of a seismic reflection profile across the accretionary prism and forearc basin of the Aleutian arc near Kodiak Island, Alaska. Modified from von Huene and Scholl (1991).

Mexico to 45 km in the Caspian Sea basin. In general, the sedimentary layer or layers (Vp = 2–5 km/sec) rest directly on the lower crust (Vp = 6.3–6.7 km/sec) with little or no upper crust. Differences in crustal thickness among inland-sea basins are accounted for by differences in thickness of both sedimentary and lower crustal layers. Increasing velocities in sediment layers with depth, such as shown by the Gulf of Mexico, appear to reflect an increasing degree of compaction and diagenesis of sediments.

Continental Crust

Shields and Platforms. Shields and platforms have similar upper- and lower-layer thicknesses and velocities (Fig. 2.3). The difference in their mean thickness (Table 2.2) reflects primarily the presence of the sediment layer in the platforms. Upper-layer thicknesses range from about 10 to 25 km and lower-layer thicknesses range from 16 to 30 km. Velocities in both layers are rather uniform, generally ranging from 6.0 to 6.3 km/sec in the upper layer and 6.8 to 7.0 km/sec in the lower layer. Upper mantle velocities are typically in the range from 8.1 to 8.2 km/sec, rarely reaching 8.6 km/sec. Results suggest the existence of a high-velocity layer (~7.2 km/sec) in lower crust of Proterozoic age. Seismic reflection studies show an increase in the number of reflections with depth and generally weak but laterally continuous Moho reflections (Mooney and Meissner, 1992).

Collisional Orogens. Crustal thickness in collisional orogens is extremely variable, ranging from about 30 km in some Precambrian orogens to 70 km beneath the Himalayas. In general, thickness decreases with age. Average layer thicknesses and velocities of the upper two layers of Phanerozoic orogens are similar to platforms (Fig. 2.3; Table 2.2), and average Phanerozoic orogen crustal thickness is about 46 km. In areas with very thick crust, such as the Himalayas, the thickening occurs primarily in the lower crustal layer. The velocity contrast between the lower crust and the upper mantle is commonly smaller beneath young orogenic areas (0.5–1.5 km/sec) than beneath platforms and shields (1–2 km/sec).

Phanerozoic orogens show considerable heterogeneity in seismic reflection profiles. The Caledonian and Variscan orogens in Europe appear to have lost their crustal roots and show well-developed, chiefly subhorizontal reflectors in the lower crust (Mooney and Meissner, 1992). This loss of crustal roots may have occurred during posttectonic extensional collapse. Often, however, major thrust sheets and in some instances sutures are preserved in the reflection profiles (Fig. 2.6). In the southern Appalachians, the reflection data reveal a large decollement at midcrustal depths separating Precambrian basement from overlying allochthonous rocks of the Appalachian orogen (Cook et al., 1981). In very young orogens such as the Alps, lower crustal reflectors dip to the center of a thick root, often with an irregular and offset Moho. At midcrustal depths, however, reflectors typically show a complex pattern suggestive of crustal wedging and of interfingering of nappes (Fig. 2.6).

Heat Flow

Heat-Flow Distribution

Surface heat flow in continental and oceanic crust is controlled by many factors, and for any given segment of crust it decreases with mean crustal age (Nyblade and Pollack, 1993). One of the most important sources of heat loss from the Earth is ocean ridges. Approximately 25% of total global heat flow can be accounted for by hydrothermal

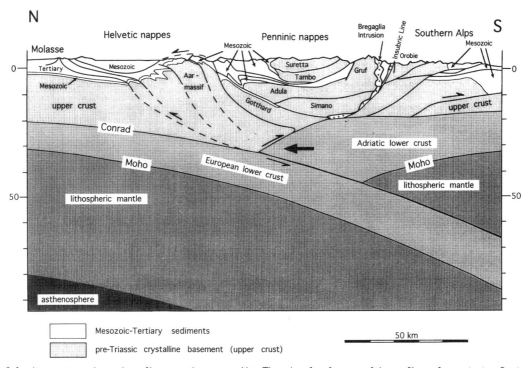

Figure 2.6   A tectonic north–south profile across the eastern Alps. The subsurface features of the profile are from seismic reflection data. Courtesy of Adrian Pfiffner.

transport at ocean ridges (Davies, 1980). Heat flow on continents and islands varies with the age of the last magmatic event, the distribution of heat-producing elements, and erosion level. Considering all sources of heat loss, the total heat loss from the Earth is about 42×10^{12} W, split between 12×10^{12} W from the continents and 30×10^{12} W from the oceans (Sclater et al., 1980). The equivalent heat flows are 55 and 95 mW/m^2, respectively, for a worldwide average heat flow of 81 mW/m^2 (Table 2.1). The difference between the average measured oceanic heat flow (67 mW/m^2) and the calculated value (95 mW/m^2) is caused by heat losses at ocean ridges because of hydrothermal circulation. Models indicate that 88% of the Earth's heat flow is lost from the mantle: 66% is lost at ocean ridges and because of subduction, 10% because of conduction from the subcontinental lithosphere, and 12% from mantle plumes (Davies and Richards, 1992). This leaves 12% lost from the continents because of radioactive decay of heat-producing elements.

Heat flows of major crustal types are tabulated in Table 2.1. Shield areas exhibit the lowest and least variable continental heat-flow values, generally from 35 to 42 mW/m^2 and averaging about 40 mW/m^2. Platforms are more variable, usually falling between 35 and 60 mW/m^2 and averaging about 49 mW/m^2. Young orogens, arcs, continental rifts, and oceanic islands exhibit high and variable heat flow in the range from 50 to 80 mW/m^2. The high heat flow in some arcs and volcanic islands reflects recent volcanic activity in these areas. Heat flow in ocean basins generally falls between 35 and 60 mW/m^2, averaging about 50 mW/m^2. Ocean ridges, on the other hand, are characterized by extremely variable heat flow, ranging from less than 100 to more than 200 mW/m^2, with heat flow decreasing with increasing distance from ocean ridges. Back-arc basins are also characterized by high heat flow (60–80 mW/m^2), whereas inland-sea basins exhibit variable heat flow (30–75 mW/m^2), partly reflecting variable Cenozoic sedimentation rates in the basins.

A **geotherm,** the temperature distribution with depth in the Earth beneath a given surface location, depends on surface and mantle heat flow and the distribution of thermal conductivity and radioactivity with depth. Geotherms in four crustal types are given in Figure 2.7. All of the geotherms have a nearly constant gradient in the upper crust but diverge at depth, resulting in a difference of up to 500° C in crustal types at Moho depths. Geotherms converge again at depths more than 300 km, where convective heat transfer in the mantle dominates.

Heat Production and Heat Flow in the Continents

Heat flow is significantly affected by the age and intensity of the last orogenic event, the distribution of radioactive elements in the crust, and the amount of heat coming from the mantle (Chapman and Furlong, 1992). Continental surface heat flow (q_o) is linearly related to average radiogenic heat production (A_o) of near-surface basement rocks:

$$q_o = q_r + A_o D$$

In this expression, q_r, commonly called the *reduced heat flow,* is the intercept value for rocks with zero heat production; D, the slope of the line relating surface heat flow to radiogenic heat production, has units of depth and is commonly called the *characteristic depth.*

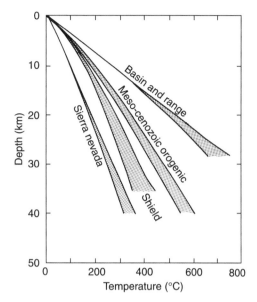

Figure 2.7 Geotherms in four continental crustal types. Shaded areas represent uncertainty ranges for each geotherm. Modified from Blackwell (1971) and Chapman (1986).

The **characteristic depth** is interpreted to reflect the thickness of the upper crustal layer enriched in heat-producing elements, and ranges from about 3 to 14 km. The **reduced heat flow** is the heat coming from below this layer and includes radiogenic heat from both the lower crust and the mantle.

Examples of the linear relationship between surface heat flow and radiogenic heat production for several crustal types are given in Figure 2.8. The Appalachian orogen is typical of pre-Mesozoic orogens and most platform areas. Studies of elevated fragments of lower continental crust, such as the Pikwitonei province in Canada (Fountain et al., 1987), show that heat production and surface heat flow also yield a broadly linear relationship, indicating that the q–A relationship is independent of the crustal level exposed at the surface (Mareschal et al., 1999). The chief factors distinguishing individual crustal types are the reduced heat-flow values and, to a lesser extent, the range of surface heat flow and heat productivity. On average, about 70% of heat flow at the surface is caused by the reduced heat-flow contribution (Artemieva and Mooney, 2001). Thermal modeling indicates that the q–A relationship can best be explained by a decrease in radiogenic heat production with depth in the crust. Uplifted blocks of middle crust have moderate levels of radiogenic elements (U, Th, and K), so most of the drop in radiogenic heat production must occur in the lower crust with a largely mafic composition as inferred from xenolith data (Rudnick and Fountain, 1995).

Reduced heat flow decreases with the average age of the continental crust (Artemieva and Mooney, 2001) (Table 2.3). No relationship, however, appears to exist between the characteristic depth and the mean crustal age, and neither heat flow nor reduced heat flow are related to lithosphere thickness (Jaupart et al., 1998). The reduced heat flow decays between 200 and 300 My to a value between 20 and 30 mW/m², about a factor of three faster than surface heat flow decays. It is probably related to variations in heat production

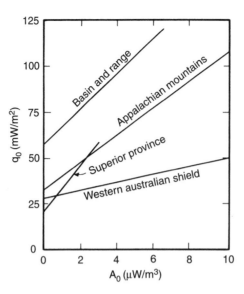

Figure 2.8 Surface heat flow (q_o) versus heat production of near-surface rocks (A_o) for several crustal types.

in the mantle lithosphere (Lenardic, 1997). In contrast, equilibrium heat-flow values for oceans are 30 to 40 mW/m².

Two interpretations of the change in characteristic depth between reduced heat-flow provinces have been proposed (Jessop and Lewis, 1978): differences in the crustal erosion level or differences in the degree of chemical fractionation of the crust. Because the crustal level exposed at the surface, as deduced from metamorphic mineral assemblages, does not correlate with the characteristic depth and the characteristic depth does not correlate with the mean crustal age, the first interpretation seems unlikely. The alternative interpretation necessitates variable intensities of fractionation of U, Th, and K into the upper crust, which is not related to the age of the crust. This could occur by partial melting in the lower crust, producing granitoid magmas (in which U, Th, and K are concentrated) that rise to shallow levels. High-grade metamorphism in the lower crust also liberates fluids in which U, Th, and K can be dissolved and thus transported to higher crustal levels.

Table 2.3 Reduced Heat-Flow Province				
	Age (Ma)	**q_o (mW/m²)**	**q_r (mW/m²)**	**D (km)**
1. Basin and Range provinces (Nevada)	0–65	77	63	9.4
2. Eastern Australia	0.65	72	57	11
3. Appalachian Mountains	400–100	56	34	7.5
4. United Kingdom	1000–300	69	24	16
5. Western Australian shield	>2500	42	30	4.5
6. India shield	>1800	35	22	8
7. Supeior province (Canada)	>2500	42	35	14
8. Baltic shield	>1800	47	25	6.5

D, characteristic depth; q_o, average surface heat flow; q_r, reduced heat flow.

A relatively low heat flow is observed in all Archean cratons (such as the Superior province and Baltic shield) (Table 2.3) compared with post-Archean cratons (Rudnick et al., 1998). This may be caused by either (1) a greater concentration of U, Th, and K in post-Archean cratons or (2) a thick lithospheric root beneath the Archean cratons that effectively insulates the crust from asthenospheric heat (Nyblade and Pollack, 1993; Rudnick et al., 1998). As you shall see in Chapter 4, the second factor is probably the principal cause of the low heat flow from Archean crust.

Age Dependence of Heat Flow

The average surface heat flow decreases with the average age of crustal rocks in both oceanic and continental areas (Sclater et al., 1980; Morgan, 1985). Continental heat flow drops with age to an approximately constant value of between 40 and 50 mW/m^2 in about 1 Gy, and in oceanic areas, similar constant heat-flow values are reached in only 50 to 100 My (Fig. 2.9). Continental heat flow can be considered in terms of three components, all of which decay with time (Vitorello and Pollack, 1980). Radiogenic heat in the upper crust contributes about 40% of the total heat flow in continental crust of all ages. The absolute amount of heat from this source decreases with time in response to erosion of the upper crust. A second component, contributing about 30% to the heat flow in Cenozoic terranes, is residual heat from igneous activity associated with orogeny. These two components decay rapidly in a few hundred million years. The third and generally minor component comes from convective heat from within the mantle. The age of the oceanic crust increases as a function of distance from ocean ridges and as a function of decreasing mean elevation because of seafloor spreading. Decrease in heat flow with distance from an active ocean ridge reflects cooling of new crust formed at the ridge by the injection of magma.

Why should continental and oceanic heat flow decay to similar equilibrium values of between 40 and 50 mW/m^2 when these two types of crust have had such different origins

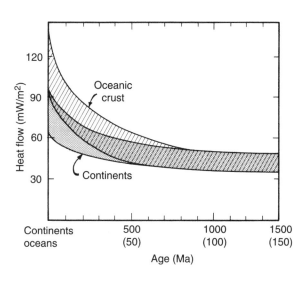

Figure 2.9 Age dependence of average heat flow for continental and oceanic crust. Numbers in parentheses on the horizontal axis refer to the ages in oceanic areas. Modified from Morgan (1985).

Table 2.4 Heat-Flow Model for Ocean Basin and Precambrian Shield

Ocean Basin q_o = 50 mW/m²			Precambrian shield q_o = 38 mW/m²		
Thickness (km)	A (mW/m³)	q (mW/m²)	Thickness (km)	A (mW/m³)	q (mW/m²)
10	0.42	4	35	0.7*	25
90	0.02	2	65	0.02	1
Mantle	0.02	44	Mantle	0.02	12

*$A = A_o\, e^{-(x/D)}$, where the radiogenic heat production (A_o) = 2.5 mW/m³ and the characteristic depth (D) = 10 km. q_o is the reduced heat flow.

and histories? One possibility is summarized in Table 2.4. The model assumes a 10-km-thick oceanic crust and a 40-km-thick continental crust, with an exponential distribution of radiogenic heat sources in the continental crust. Equilibrium-reduced heat-flow values are selected from the range of observed values in both crustal types. Ocean basins and Precambrian shields have similar temperatures at a depth of 100 km, the average thickness of the lithosphere. The results clearly imply that more mantle heat is entering the base of the oceanic lithosphere (44 mW/m²) than of the continental lithosphere (12 mW/m²). Hence, when a plate carrying a Precambrian shield moves over oceanic mantle, the surface heat flow should rise until mantle convective systems readjust so that they are again liberating most heat beneath oceanic areas.

Exhumation and Cratonization

Introduction

To understand how continental crust evolves, you need to understand how collisional orogens evolve because continents are made of collisional orogens of various ages. During continental collisions, large segments of continental crust are highly deformed and thickened; later, they are uplifted and eroded to form cratons. The process of uplift and erosion is known as crustal **exhumation.** Crustal history is imprinted in rocks as pressure–temperature–time trajectories, generally called crustal **P–T–t paths.** Not only collisional orogens but all crustal types record P–T–t paths; to understand continental growth and development, it is important to reconstruct and understand the meanings of these P–T–t paths.

Burial of rocks in the Earth's crust results in progressive metamorphism as the rocks are subjected to increasing pressure and temperature. Given that metamorphic reactions are more rapid with rising than with falling temperature, metamorphic mineral assemblages may record the highest P–T regime to which rocks have been subjected. Later exhumation makes it possible to directly study rocks once buried at various depths in the crust but now exposed at the surface. Such rocks contain metamorphic mineral assemblages "arrested" at some burial depth or metamorphic grade. Minor changes in mineralogy known as *retrograde metamorphism* may occur during uplift, but such changes commonly can be identified without great difficulty by studying textural features of the rocks. Progressive metamorphism is accompanied by losses of water and other volatile constituents, and some high-grade rocks are almost anhydrous.

Zones of increasing metamorphic grade can be classified into **metamorphic facies,** which represent limited ranges of burial depth, temperature, and water content in the crust. Five major facies of regional metamorphism are recognized (Fig. 2.10). The *zeolite facies* is characterized by the development of zeolites in sediments and volcanics and reflects temperatures generally less than 200° C and burial depths up to 5 km. The *greenschist facies* is characterized by the development of chlorite, actinolite, epidote, and albite in mafic volcanics and of muscovite and biotite in pelitic rocks. *Blueschist-facies* assemblages form at high pressures (>800 megapascals, or MPa) yet low temperatures (<400° C) in subduction zones, and typical minerals are glaucophane, lawsonite, and jadeite. The *amphibolite facies* is characterized by kyanite, staurolite, and sillimanite in metapelites and plagioclase and hornblende in mafic rocks. The highest-grade rocks occur in the *granulite facies,* which is characterized by the sparsity or absence of hydrous minerals and the appearance of pyroxenes. Using experimental petrologic and oxygen isotope data, it is possible to estimate the temperature and pressure at which metamorphic mineral assemblages crystallize; from these results, the burial depths of metamorphic terrains exposed at the Earth's surface can be estimated.

Unraveling Pressure–Temperature–Time Histories

To begin to understand complex P–T–t histories, many data sources must be used. It is important to know the sequence in which metamorphic minerals have grown and how their growth is related to deformation (Brown, 1993). This can be established from petrographic studies of metamorphic rocks in which textural relationships between mineral growth and deformational fabric are preserved. Particularly important are mineral inclusions in porhyroblasts, replacement textures of one mineral by another, and mineral zoning, all of which

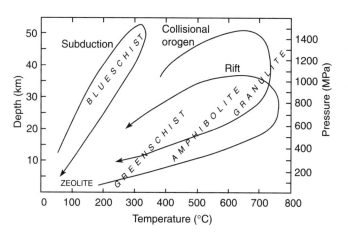

Figure 2.10 Pressure–temperature diagram showing the distribution of metamorphic facies and typical pressure–temperature–time paths.

reflect changing conditions along the P–T–t path. By relating mineral growth relationships to experimentally determined P–T stability fields of metamorphic minerals, it is possible in some instances to pin down the pressure and temperature at which a given mineral assemblage grew. This is known as **geothermobarometry.** In turn, local P–T conditions may be related to broad regional fabrics in crustal rocks, thus relating microscopic growth data to orogenic development.

To add time to these events, it is necessary to isotopically date minerals that form along different segments of the P–T–t path and become isotopically closed at given temperatures. For instance, the $^{40}Ar/^{39}Ar$ method can be used to date hornblende, which closes to argon diffusion around 500° C; muscovite that closes from 350 to 400° C; and K-feldspar that closes from 150 to 350° C. At lower temperatures, fission tracks, which anneal in apatite around 150° C, are useful (McDougall and Harrison, 1988; Crowley and Kuhlman, 1988). To unravel P–T–t histories in the crust, it is necessary to combine results from field with petrographic, experimental, and isotopic studies. Even then it may not be possible to untangle a complex history in which a crustal segment has undergone multiple deformation and metamorphism.

Some Typical Pressure–Temperature–Time Paths

Varying tectonic histories yield different P–T–t paths (Chapman and Furlong, 1992). P–T–t paths of collisional orogens are typically clockwise in P–T–t space, and their general features are reasonably well understood from classic studies in the Appalachians, Caledonides, Alps, and Himalayas (Brown, 1993). This type of P–T–t path results from rapid crustal thickening in which maximum pressure is reached before maximum temperature (Fig. 2.10). Hence, the metamorphic peak generally postdates early deformation in the orogen. This evolutionary path commonly leads to dehydration melting of the lower crust, producing the granitic magmas that abound in collisional orogens.

During the development of a continental rift by crustal extension, the crust is heated from below by mantle upwelling before crustal thickening occurs, and the maximum temperature is reached before the maximum pressure. This produces a counterclockwise P–T–t path (Fig. 2.10). The metamorphic peak usually predates or is synchronous with early deformation in these cases. Heating of the lower crust can lead to dehydration melting as a consequence of the metamorphism, producing felsic magmas. Any crustal environment in which heating precedes thickening of the crust results in counterclockwise P–T–t paths. In addition to rifts, extended margins of platforms and magmatically underplated crust, such as beneath continental flood basalts, have counterclockwise P–T–t paths. In some instances, such as the northern Appalachians, clockwise and counterclockwise P–T–t paths can occur in adjacent segments of the same orogen.

P–T–t paths of subduction zones are very steep, with pressure increasing rapidly as a slab descends into the mantle. Because the slab is relatively cool (<400° C), blueschists and eclogites are stabilized. Just how these dense rocks are returned to the surface is not clear; perhaps they are housed in lower density rocks that rise buoyantly along the slab surface.

Cratonization

Although cratons have long been recognized as an important part of the continental crust, their origin and evolution is still not well understood. Most investigators agree that cratons are the end product of collisional orogenesis; thus, they are the building blocks of continents. Just how orogens evolve into cratons and how long it takes, however, is not well known. Although studies of collisional orogens show that most are characterized by clockwise P–T–t paths (Thompson and Ridley, 1987; Brown, 1993), the uplift–exhumation segments of the P–T paths are poorly constrained (Martignole, 1992). In terms of craton development, the less than 500° C portion of the P–T–t path is most important.

Using a variety of radiogenic isotopic systems and estimated closure temperatures in various minerals, it is possible to track the cooling histories of crustal segments and, when coupled with thermobarometry, the uplift–exhumation histories. Results suggest wide variation in cooling and uplift rates; most orogens having cooling rates <2° C/My, whereas a few (such as southern Brittany) cool at rates <10° C/My (Fig. 2.11). In most cases, it would appear to take a minimum of 300 My to make a craton. Some terranes, such as Enderby Land in Antarctica, have had long, exceedingly complex cooling histories lasting more than 2 Gy. Many orogens, such as the Grenville orogen in eastern Canada, have been exhumed as indicated by unconformably overlying sediments, reheated during subsequent burial, and then reexhumed (Heizler, 1993). In some instances, postorogenic thermal events such as plutonism and metamorphism have thermally overprinted earlier segments of an orogen's cooling history such that only the very early high-temperature history (<500° C) and perhaps the latest exhumation record (<300° C) are preserved. Fission track ages suggest that final uplift and exhumation of some orogens, such as the 1.9-Ga Trans-Hudson orogen in central Canada, may be related to the early stages of supercontinent fragmentation.

An important yet poorly understood aspect of cratonization is how terranes that amalgamate during a continent–continent collision evolve into a craton. Does each terrane maintain its own identity and have its own cooling and uplift history? Or do terranes

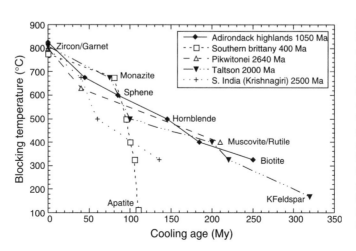

Figure 2.11 Cooling histories of several orogens. Ages of maximum temperatures are given in the explanation and equated to zero age on the cooling age axis. Blocking temperature is the temperature at which the daughter isotope is trapped in a host mineral. Data from Harley and Black (1987), Dallmeyer and Brown (1992), Mezger et. al. (1991) and Kontak and Reynolds (1994).

anneal to each other at an early stage so that the entire orogen cools and is elevated as a unit? What is the effect of widespread posttectonic plutonism? Does it overprint and erase important segments of the orogen cooling history? It is well known that crustal cooling curves are not always equivalent to exhumation curves (Thompson and Ridley, 1987). Some granulite-grade blocks appear to have undergone long periods of isobaric (constant depth) cooling before exhumation. Also, discrete thermal events can completely or partially reset thermochronometers without an obvious geologic rock record, and this can lead to erroneous conclusions regarding average cooling rates (Heizler, 1993).

Posttectonic plutonism, which follows major deformation, or multiple deformation of an orogen can lead to a complex cooling history. Widespread posttectonic plutonism can perturb the cooling curve of a crustal segment, prolonging the cooling history (Fig. 2.12, left panel). In an even more complex scenario, a crustal domain can be exhumed, can be reburied as sediments accumulate in an overlying basin, can age for hundreds of millions of years around the same crustal level, and finally can be reexhumed (A in Fig. 2.12, right panel). In this example, all of the thermal history less than 400° C is lost by overprinting of the final thermal event. A second terrane, B, could be sutured to A during this event (S in Fig. 2.12, right panel), and both domains could be exhumed together. It is clear from these examples that much or all of a complex thermal history can be erased by the last thermal event, producing an apparent gap in the cratonization cooling curve.

Processes in the Continental Crust

Rheology

The behavior of the continental crust under stress depends chiefly on the temperature and the duration of the stresses. The hotter the crust, the more it behaves like a ductile solid deforming by plastic flow. If it is cool, it behaves like an elastic solid deforming by brittle fracture and frictional gliding (Ranalli, 1991; Rutter and Brodie, 1992). The distribution of strength with depth in the crust varies with the tectonic setting, the strain rate, the thickness and composition of the crust, and the geotherm. The **brittle–ductile transition** corresponding to an average surface heat flow of 50 mW/m^2 is around a 20-km depth, which corresponds to the depth limit of most shallow earthquakes. Even in the lower

Figure 2.12 Two possible cooling scenarios in cratons. Left panel: Overprinting of a posttectonic granite intrusion. Right panel: A complex, multiple-event cooling history. S, suturing age of terranes A and B; Te, Ta,b final exhumation age.

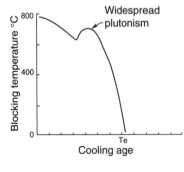

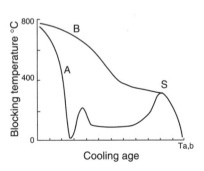

crust, however, if stress is applied rapidly it may deform by fracture; likewise, if pore fluids are present in the upper crust—weakening it—and stresses are applied slowly, the upper crust may deform plastically. In regions of low heat flow, such as shields and platforms, brittle fracture may extend into the lower crust or even into the upper mantle because mafic and ultramafic rocks can be resistant to plastic failure at these depths; thus, brittle faulting is the only way they can deform. Lithologic changes at these depths, the most important of which is at the Moho, are also likely to be rheological discontinuities.

Examples of two rheological profiles of the crust and subcontinental lithosphere are shown in Figure 2.13. The brittle–ductile transition occurs around a 20-km depth in the rift, whereas in the cooler and stronger Proterozoic shield, it occurs around 30 km. In both cases, the strength of the ductile lower crust decreases with increasing depth, reaching a minimum at the Moho. The rapid increase in strength beneath the Moho chiefly reflects the increase in olivine, which is stronger than pyroxenes and feldspars. The rheological base of the lithosphere, generally taken as a strength around 1 MPa, occurs 55 km beneath the rift and 120 km beneath the Proterozoic shield. In general, the brittle–ductile transition occurs at relatively shallow depths in warm and young crust (10–20 km), whereas in cool and old crust, it occurs at greater depths (20–30 km).

Role of Fluids and Crustal Melts

Fluid transport in the crust is an important process affecting both rheology and chemical evolution. Because crustal fluids are mostly inaccessible for direct observation, this process is poorly understood and difficult to study. Studies of fluid inclusions trapped in

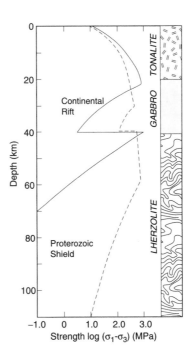

Figure 2.13 Rheological profile of the East African rift and the Proterozoic shield in East Africa. Strength expressed as the difference between maximum and minimum compressive stresses (σ_1 and σ_3, respectively). Diagram modified from Ranalli (1991).

metamorphic and igneous minerals indicate that shallow crustal fluids are chiefly water, whereas deep crustal fluids are mixtures of water and CO_2, and both contain various dissolved species (Bohlen, 1991; Wickham, 1992). Fluids are reactive with silicate melts, and in the lower crust they can promote melting and can change the chemical and isotopic composition of rocks.

In the lower crust, only small amounts of fluid can be generated by the breakdown of hydrous minerals such as biotite and hornblende. Hence, the only major source of fluids in the lower crust is the mantle. Studies of xenoliths suggest that the mantle lithosphere provides a potentially large source for CO_2 in the lower crust, and the principal source for CO_2 may be important in the production of deep crustal granulites.

The formation of granitic melts in the lower crust and their transfer to shallower depths are fundamental processes leading the chemical differentiation of the continents. This is particularly important in arcs and collisional orogens. The melt-producing capacity of a source rock in the lower crust is determined chiefly by its chemical composition but also depends on temperature regime and fluid content (Brown et al., 1995). Orogens that include a large volume of juvenile volcanics and sediments are more fertile (high melt-producing capacity) than those that include chiefly older basement rocks from which fluids and melts have been extracted (Vielzeuf et al., 1990). A fertile lower crust can generate a range of granitic melt compositions and leave behind a residue of granulites. Segregation of melt from source rocks can occur by several processes, and just how much and how fast melt is segregated is not well known. These depend, however, on whether deformation occurs concurrently with melt segregation. Experiments indicate that melt segregation is enhanced by increased fluid pressures and fracturing of surrounding rocks. Modeling suggests that shear-induced compaction can drive melt into veins that transfer it rapidly to shallow crustal levels (Rutter and Neumann, 1995).

Crustal Composition

Approaches

Several approaches have been used to estimate the chemical and mineralogical composition of the crust. One of the earliest methods to estimate the composition of the upper continental crust was based on chemical analysis of glacial clays, which were assumed to be representative of the composition of large portions of the upper continental crust. Estimates of total continental composition were based on mixing average basalt and granite compositions in ratios generally ranging from 1:1 to 1:3 (Taylor and McLennan, 1985) or on weighting the compositions of various igneous, metamorphic, and sedimentary rocks according to their inferred abundances in the crust (Ronov and Yaroshevsky, 1969). Probably the most accurate estimates of the composition of the upper continental crust come from the extensive sampling of rocks exhumed from varying depths in Precambrian shields and from the composition of Phanerozoic shales (Taylor and McLennan, 1985; Condie, 1993). Because the lower continental crust is not accessible for sampling, indirect approaches must be used. These include (1) measuring seismic-wave velocities of crustal rocks in the laboratory at appropriate temperatures and pressures and comparing these with observed velocity distributions in the crust, (2) sampling and analyzing rocks from blocks of continental crust exhumed from middle to lower crustal depths, and (3) analyzing xenoliths of lower crustal rocks brought to the surface during volcanic eruptions. The composition of oceanic crust is estimated from the composition of rocks in ophiolites and from shallow cores into the sediment and basement layers of oceanic crust retrieved by the Ocean Drilling Project. Results are again constrained by seismic velocity distributions in the oceanic crust.

Before describing the chemical composition of the crust, I will review the major sources of data.

Seismic-Wave Velocities

Because seismic-wave velocities are related to rock density and density is related to rock composition, the measurement of these velocities provides an important constraint on the composition of both the oceanic and the continental crust (Rudnick and Fountain, 1995). **Poisson's ratio,** which is the ratio of P-wave to S-wave velocity, is more diagnostic of crustal composition than either P-wave or S-wave data alone (Zandt and Ammon, 1995) (Table 2.1).

Figure 2.14 shows average compressional-wave velocities (at 600 MPa and 300° C) in a variety of crustal rocks. Velocities slower than 6.0 km/sec are limited to serpentinite, metagraywacke, andesite, quartzite and basalt. Many rocks of diverse origins have velocities between 6.0 and 6.5 km/sec, including slates, granites, altered basalts, and felsic granulites. With the exception of marble and anorthosite, which are probably minor components in the crust based on exposed blocks of lower crust and xenoliths, most rocks with velocities from 6.5 to 7.0 km/sec are mafic in composition and include amphibolites and mafic granulites without garnet (Holbrook et al., 1992; Christensen and Mooney, 1995).

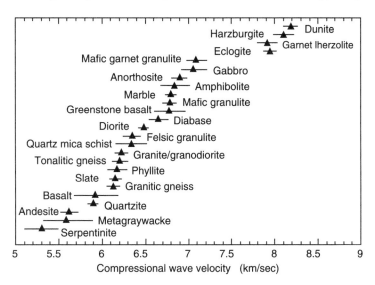

Figure 2.14 Average compressional-wave velocities and standard deviations at 600 MPa (20-km depth equivalent) and 300° C (average heat flow) for major rock types. Data from Christensen and Mooney (1995).

Rocks with average velocities from 7.0 to 7.5 km/sec include gabbro and mafic garnet granulite, and velocities faster than 7.5 km/sec are limited to nonserpentinized ultramafic rocks and eclogite (a high-pressure mafic rock). It is important to note that the order of increasing velocity in Figure 2.14 is not a simple function of increasing metamorphic grade. For instance, low-, medium-, and high-grade metamorphic rocks all fall in the range from 6.0 to 7.5 km/sec.

Although rock types in the upper continental crust are reasonably well known, the distribution of rock types in the lower crust remains uncertain. Platform lower crust, although it has relatively high S-wave velocities, shows similar Poisson's ratios to collisional orogens (Fig. 2.15a; Tables 2.1 and 2.2). The lower crust of continental rifts, however, shows distinctly lower velocities, a feature that would appear to reflect hotter temperatures in the lower crust. Two observations are immediately apparent from the measured rock velocities summarized in Figures 2.14 and 2.15b: (1) the velocity distribution in the lower crust indicates compositional heterogeneity, and (2) metapelitic rocks overlap in velocity with mafic and felsic igneous and metamorphic rocks. It is also interesting that with the exception of rifts, mean lower crustal velocities are strikingly similar to mafic rock velocities. However, because of the overlap in velocities of rocks of different compositions and origins, it is not possible to assign unique rock compositions to the

Figure 2.15
Compressional-wave versus shear-wave velocity diagrams showing lower crust of various crustal types (a) and fields of various crustal and upper mantle rocks (b). Velocities are normalized to a 20-km depth and room temperature. Dashed lines are Poisson's ratio $(0.5\{1-1/[(V_p/V_s)^2-1]\})$. Modified from Rudnick and Fountain (1995).

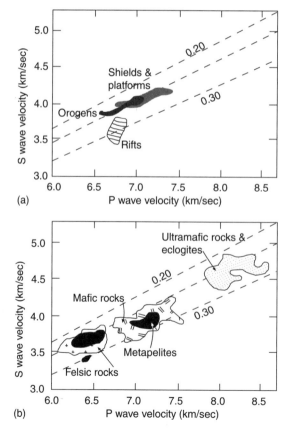

lower crust from seismic velocity data alone. When coupled with xenolith data, however, the seismic velocity distributions suggest that the lower continental crust is composed largely of mafic granulites, gabbros, and amphibolites (50–65%), with up to 10% metapelite, and that the remainder is intermediate to felsic granulite (Rudnick and Fountain, 1995). Based on seismic data, however, the lower crust in the Archean Kaapvaal craton in southern Africa appears to be felsic to intermediate in composition (James et al., 2003).

In common rock types, Poisson's ratio (σ) varies from about 0.20 to 0.35 and is particularly sensitive to composition. Increasing silica content lowers s, and increasing Fe and Mg increases it (Zandt and Ammon, 1995). The average value of s in the continental crust shows a good correlation with crustal type (Fig. 2.16; Table 2.1). Precambrian shield s values are consistently high, averaging 0.29, and platforms average about 0.27. The lower s in platforms and Paleozoic orogens appears to reflect the silica-rich sediments that add 4 to 5 km of crustal thickness to the average shield (Table 2.1). In Meso–Cenozoic orogens, however, s is even lower but more variable, reflecting some combination of lithologic and thermal differences in the young orogenic crust. The high ratios in continental-margin arcs may reflect the importance of mafic rocks in the root zones of these arcs, although again the variation in s is significant.

The origin of the Moho continues to be a subject of widespread interest (Jarchow and Thompson, 1989). Because the oceanic Moho is exposed in many ophiolites, it is better known than the continental Moho. From seismic velocity distributions and from ophiolite studies, the oceanic Moho is probably a complex transition zone from 0 to 3 km thick and between mixed mafic and ultramafic igneous cumulates in the crust and the harzburgites (orthopyroxene–olivine rocks) in the upper mantle. It would appear that large tectonic lenses of differing lithologies occur at the oceanic Moho and that these are the products of ductile deformation along the boundary. The continental Moho is considerably more complex and varies in nature with crustal type and age (Griffin and O'Reilly, 1987). Experimental, geophysical, and xenolith data, however, do not favor a gabbro–eclogite transition to explain the continental Moho. Also, the absence of a correlation between surface heat flow and crustal thickness does not favor a garnet granulite–eclogite phase change

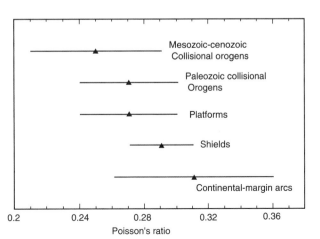

Figure 2.16 Mean values and one standard deviation of Poisson's ratio σ in various crustal types. Data from Zandt and Ammon (1995).

at the Moho. Beneath platforms and shields, the Moho is only weakly (or not at all) reflective, suggesting the existence of a relatively thick transition zone (<3 km) composed of mixed mafic granulites, eclogites, and lherzolites with no strong reflecting surfaces.

Seismic Reflections in the Lower Continental Crust

Many explanations have been suggested for the strong seismic reflectors found in much of the lower continental crust (Nelson, 1991; Mooney and Meissner, 1992) (Fig. 2.17). The most likely causes fall into one of three categories:

1. Layers in which fluids are concentrated
2. Strain banding developed from ductile deformation
3. Lithologic layering

The jury is still out on the role of deep crustal fluids. Some investigators argue that physical conditions in the lower crust allow up to 4% of saline pore waters and that the high electrical conductivity of the lower crust supports such a model. In this model, the seismic reflections are produced by layers with strong porosity contrasts. However, textural and mineralogical evidence from deep crustal rocks exposed at the surface and from xenoliths do not have high porosities, thus contradicting this idea.

Deformation bands hold more promise, at least for some of the lower crustal reflections, in that shear zones exposed at the surface can be traced to known seismic reflectors at depth in shallow crust (Mooney and Meissner, 1992). Some lower crustal reflection patterns in Precambrian cratons preserve structures that date from ancient collisional events, such as in the Trans-Hudson orogen of Paleoproterozoic age in Canada. At least in these cases, it would appear that the reflections are caused by tectonic boundaries or by syntectonic igneous intrusions. In extended crust, such as that found in rifts, ductile shearing in the lower crust may enhance metamorphic or igneous layering.

Figure 2.17 Seismic reflection profile southwest of England. Also shown is a line drawing of the data and a histogram of reflection depths summed over 1-sec travel-time intervals. Modified from Klemperer (1987).

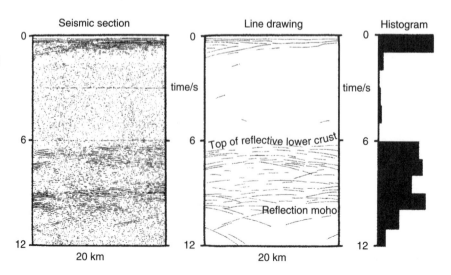

The probable cause of many lower crustal reflections is lithologic layering, caused by mafic sills; compositional layering in mafic intrusions; or metamorphic fabrics. Supporting this conclusion are some shallow reflectors in the crust, which have been traced to the surface, that are caused by such layering (Percival et al., 1989). Furthermore, a bimodal distribution of acoustic impedance in the lower crust favors layered sequences of rocks, especially interlayered mafic and felsic units (Goff et al., 1994). Also, models of reflectivity in the Ivrea zone (a fragment of mafic lower crust faulted to the surface in the Alps) show that lower crustal reflections are expected when mafic rocks are interlayered with felsic rocks (Holliger et al., 1993).

Seismic reflectivity in the lower crust is widespread and occurs in crustal types with differing heat-flow characteristics, favoring a single origin for most reflectors. From studies of exhumed lower crust and lower crustal xenoliths, it would seem that most lower crustal reflections are caused by mafic intrusions and, in some instances, that the reflections have been enhanced by later ductile deformation.

Sampling of Precambrian Shields

Widespread sampling of metamorphic terranes exposed in Precambrian shields and especially in the Canadian shield has provided an extensive sample base to estimate both the chemical and the lithologic composition of the upper part of the Precambrian continental crust (Shaw et al., 1986; Condie, 1993). Both individual and composite samples have been analyzed. Results indicate that although the upper crust is lithologically heterogeneous, granitoids of granodiorite to tonalite composition dominate and the weighted average composition is that of granodiorite.

Use of Fine-Grained Terrigenous Sediments

Fine-grained terrigenous sediments may represent well-mixed samples of the upper continental crust and thus provide a means of estimating upper crustal composition (Taylor and McLennan, 1985). However, to use sediments to estimate crustal composition, it is necessary to evaluate losses and gains of elements during weathering, erosion, deposition, and diagenesis. Elements, such as rare earth elements (REE), Th, and Sc that are relatively insoluble in natural waters and have short residence times in seawater ($<10^3$ y) may be transferred almost totally into terrigenous clastic sediments. The remarkable uniformity of REE in pelites and loess compared with the great variability observed in igneous source rocks attests to the efficiency of mixing during erosion and deposition. Studies of REE and element ratios such as La/Sc, La/Yb, and Cr/Th indicate that they remain relatively unaffected by weathering and diagenesis. REE distributions are especially constant in shales and resemble REE distributions in weighted averages from Precambrian shields. With some notable exceptions (Condie, 1993), estimates of the average composition of the upper continental crust using the composition of shales are in remarkable agreement with the weighted chemical averages determined from rocks exposed in Precambrian shields.

Exhumed Crustal Blocks

Several blocks of middle to lower continental crust have been recognized in Precambrian shields or collisional orogens, the best known of which are the Kapuskasing uplift in southern Canada (Percival et al., 1992; Percival and West, 1994) and the Ivrea Complex in Italy (Sinigoi et al., 1994). Four mechanisms have been suggested to bring these deep crustal sections to the surface: (1) large thrust sheets formed during continent–continent collisions, (2) transpressional faulting, (3) broad tilting of a large segment of crust, and (4) asteroid impacts. However, tectonic settings at the times of formation of rocks within the uplifted blocks appear to be collisional orogens, island arcs, or continental rifts. Common to all studied sections are high-grade metamorphic rocks that formed at depths from 20 to 25 km with a few, such as the Kohistan arc in Pakistan, coming from depths as great as 40 to 50 km. Metamorphic temperatures recorded in the blocks are typically in the range from 700 to 850° C. All blocks consist chiefly of felsic components at shallow structural levels and mixed mafic, intermediate, and felsic components at deeper levels. Commonly, lithologic and metamorphic features in uplifted blocks are persistent over lateral distances greater than 1000 km, as evidenced by three deep crustal exposures in the Superior province in southern Canada (Percival et al., 1992).

Examples of five sections of middle to lower continental crust are given in Figure 2.18. Each section is a schematic illustrating the relative abundances of major rock types, and the base of each section is a major thrust fault. The greatest depths exposed in each section are 25 to 35 km. Each column has a lower granulite zone, with mafic granulites dominating in three sections and felsic granulites in the other two. The sections show considerable

Figure 2.18 Generalized cross-sections of continental crust based on exhumed sections of deep crustal rocks. Modified from Fountain and Salisbury (1981).

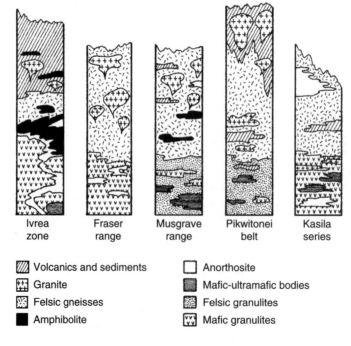

Ivrea zone   Fraser range   Musgrave range   Pikwitonei belt   Kasila series

▨ Volcanics and sediments   ☐ Anorthosite
▦ Granite   ▧ Mafic-ultramafic bodies
▨ Felsic gneisses   ▨ Felsic granulites
■ Amphibolite   ⩗ Mafic granulites

Figure 2.19 Layered mafic granulites from the Ivrea zone in the Alps. Similar rocks may comprise large volumes of the lower continental crust. Courtesy of K. R. Mehnert.

compositional variation at all metamorphic grades, attesting to the heterogeneity of the continental crust at all depths. Mafic and ultramafic bodies and anorthosites occur at deep levels in some sections and probably represent layered igneous sheets intruded into the lower crust (Fig. 2.19). Volcanic and sedimentary rocks are also buried to great depths in some sections. Intermediate and upper crustal levels are characterized by large volumes of granitoids.

More than anything else, the crustal sections indicate considerable variation in lithologic and chemical composition both laterally and vertically in the continental crust. The only large-scale progressive change in the sections is an increase in metamorphic grade with depth. Although there is no evidence for a Conrad discontinuity in the sections, rapid changes in lithology may be responsible for more local seismic discontinuities. Again, it should be emphasized that many uplifted blocks probably do not sample the lower crust but only the middle crust (~25 km). Today, these blocks are underlain by 35 to 40 km of crust, probably largely composed of mafic granulites. The crust in these areas may have thickened during continental collision (to 60–70 km), thus burying upper crustal rocks to granulite grade (35–40 km). Uplift and erosion of this crust brought these felsic granulites to the surface with a possible mafic granulite root still intact. Thus, the differences between the generally felsic to intermediate compositions of uplifted crustal blocks and the mafic compositions of lower crustal xenolith suites (see the next section), may be partly because of different levels of sampling in the crust.

Crustal Xenoliths

Crustal xenoliths are fragments of the crust brought to the Earth's surface by volcanic eruptions. If we can determine the depth from which xenoliths come by thermobarometry and estimate the relative abundances of various xenolith populations in the crust, it should be possible to reconstruct a crustal cross-section (Kay and Kay, 1981). Although metamorphic xenoliths can be broadly ordered in terms of their crustal depths, metamorphic mineral assemblages in many lower crustal xenoliths are not definitive in determining precise depths (Rudnick, 1992). Even more difficult is the problem of estimating the relative abundances of xenolith types in the crust. Some lithologies may be oversampled and others undersampled by ascending volcanic magmas. Hence, it is generally not possible to come up with a unique crustal section from xenolith data alone.

Xenolith-bearing volcanics and kimberlites occur in many tectonic settings, giving a wide lateral sampling of the continents. Lower crustal xenoliths from arc volcanics are chiefly mafic in composition, and xenoliths of sediments are rare to absent. These results suggest that the root zones of modern arcs are composed chiefly of mafic rocks (Condie, 1999). Xenoliths from volcanics on continental crust are compositionally diverse and have complex thermal and deformational histories (Rudnick, 1992). Metasedimentary xenoliths, however, are minor compared with metaigneous xenoliths. In general, xenoliths of mafic granulite are more abundant than those of felsic granulite, suggesting that a mafic lower crust is important in cratons (Rudnick and Taylor, 1987). Most of these xenoliths appear to be basaltic melts and their cumulates that intruded into or underplated beneath the crust. Most granulite-grade xenoliths reflect equilibration depths in the crust of more than 20 km and some of more than 40 km. A few metasedimentary and gneissic xenoliths recording similar depths in many xenolith suites seem to require interlayering of felsic and mafic rocks in the lower crust. When isotopic ages of xenoliths can be estimated, they range from the age of the host crust to considerably younger. For instance, mafic lower crustal xenoliths from the Four Corners volcanic field in the Colorado Plateau appear to be about 1.7 Ga, the same age as the Precambrian basement in this area (Wendlandt et al., 1993).

An Estimate of Crustal Composition

Continental Crust

The average chemical composition of the upper continental crust is reasonably well known from widespread sampling of Precambrian shields, geochemical studies of shales, and exposed crustal sections (Taylor and McLennan, 1985; Condie, 1993). An average composition from Condie (1993) is similar to granodiorite (Table 2.5), although there are

Table 2.5 Average Chemical Composition of Continental and Oceanic Crust

	Continental Crust				Oceanic Crust
	Upper	**Middle**	**Lower**	**Total**	
SiO_2	66.3	60.6	52.3	59.7	50.5
TiO_2	0.7	0.8	0.54	0.68	1.6
Al_2O_3	14.9	15.5	16.6	15.7	15.3
FeOT	4.68	6.4	8.4	6.5	10.4
MgO	2.46	3.4	7.1	4.3	7.6
MnO	0.07	0.1	0.1	0.09	0.2
CaO	3.55	5.1	9.4	6.0	11.3
Na_2O	3.43	3.2	2.6	3.1	2.7
K_2O	2.85	2.0	0.6	1.8	0.2
P_2O_5	0.12	0.1	0.1	0.11	0.2
Rb	87	62	11	53	1
Sr	269	281	348	299	90
Ba	626	402	259	429	7
Th	9.1	6.1	1.2	5.5	0.1
Pb	18	15.3	4.2	13	0.3
U	2.4	1.6	0.2	1.4	0.05
Zr	162	125	68	118	74
Hf	4.4	4.0	1.9	3.4	2.1
Nb	10.3	8	5	7.8	2.3
Ta	0.82	0.6	0.6	0.7	0.13
Y	25	22	16	21	28
La	29	17	8	18	2.5
Ce	59.4	45	20	42	7.5
Sm	4.83	4.4	2.8	4.0	2.6
Eu	1.05	1.5	1.1	1.2	1.0
Yb	2.02	2.3	1.5	1.9	3.1
V	86	118	196	133	275
Cr	112	150	215	159	250
Co	18	25	38	27	47
Ni	60	70	88	73	150

Major elements in weight percentage of the oxide and trace elements in ppm (parts per million).
Lower–middle crust from Rudnick and Fountain (1995), upper crust from Condie (1993), and oceanic crust (NMORB) from Sun and McDonough (1989) and miscellaneous sources.

differences related to the age of the crust, described in Chapter 8. The composition of the lower continental crust is poorly constrained. Uplifted crustal blocks, xenolith populations, seismic velocity, and Poisson's ratio suggest that a large part of the lower crust is mafic in overall composition. I accept the middle and lower crustal estimates of Rudnick and Fountain (1995) based on all of the data sources described previously. If the upper continental crust is felsic in composition and the lower crust is mafic, as most data suggest, how do these two layers form and how do they persist over geologic time? This intriguing question will be returned to in Chapter 8.

The estimate of the total composition of continental crust in Table 2.5 is a mixture of upper, middle, and lower crustal averages in equal amounts. The composition is similar to other published total crustal compositions indicating an overall intermediate composition (Taylor and McLennan, 1985; Wedepohl, 1995; Rudnick and Fountain, 1995). **Incompatible elements,** which are elements strongly partitioned into the liquid phase upon melting, are known to be concentrated chiefly in the continental crust. During melting in the mantle, these elements will be enriched in the magmas and thus transferred upward into the crust as magmas rise. Relative to primitive mantle composition, 35 to 65% of the most incompatible elements (such as Rb, Th, U, K, and Ba) are contained in the continents, whereas continents contain less than 10% of the least incompatible elements (such as Y, Yb, and Ti).

Oceanic Crust

Because fragments of oceanic crust are preserved on the continents as ophiolites, we have direct access to sampling for chemical analysis. The chief problem with equating the composition of ophiolites to average oceanic crust, however, is that some or most ophiolites appear to have formed in back-arc basins and, in varying degrees, to have geochemical signatures of arc systems. Other sources of data for estimating the composition of oceanic crust are dredge samples from the ocean floor and drill cores, retrieved from the Ocean Drilling Project, that have penetrated the basement layer. Studies of ophiolites and P-wave velocity measurements are consistent with basement and oceanic layers being composed largely of mafic rocks metamorphosed to the greenschist or amphibolite facies. The sediment layer is composed of pelagic sediments of variable composition and extent, and it contributes less than 5% to the bulk composition of the oceanic crust.

An estimate of the composition of oceanic crust is given in Table 2.5. It is based on the average composition of normal ocean-ridge basalts, excluding data from back-arc basins. Pelagic sediments are ignored in the estimate. Although ophiolites contain minor amounts of ultramafic rock and felsic rock, they are much less variable in lithologic and chemical composition than crustal sections of continental crust, suggesting that the oceanic crust is rather uniform in composition. Because of the relatively small volume of oceanic crust compared with continental crust (Table 2.1) and because oceanic basalts come from a mantle source depleted of incompatible elements (Chapter 4), the oceanic crust contains little of the Earth's inventory of these elements.

Complementary Compositions of Continental and Oceanic Crust

The average compositions of the continental and oceanic crusts relative to the primitive mantle composition show surprisingly complementary patterns (Fig. 2.20). In continental crust, the maximum concentrations, which reach values 50 to 100 times primitive mantle values, are for the most incompatible elements: K, Rb, Th, and Ba. These same elements reach less than 3 times the primitive mantle values in oceanic crust. The patterns cross at phosphorus (P), and the least incompatible elements—Ti, Yb, and Y—are more enriched in oceanic than in continental crust. The relative depletions in Ta–Nb, P, and Ti are important features of the continental crust and will be explained more fully in Chapter 8. The complementary crustal element patterns can be explained if most of the continental crust is extracted from the upper mantle first, leaving an upper mantle depleted of incompatible elements. The oceanic crust is then continuously produced from this depleted upper mantle throughout geologic time (Hofmann, 1988).

Crustal Provinces and Terranes

Stockwell (1965) suggested that the Canadian shield can be subdivided into structural provinces based on differences in structural trends and style of folding. Structural trends are defined by foliation, fold axes, and bedding and sometimes by geophysical anomalies. Boundaries between the provinces are drawn where one trend crosses another along either unconformities or structural–metamorphic breaks. Large numbers of isotopic dates from the Canadian shield indicate that structural provinces are broadly coincident with age provinces. Similar relationships have been described on other continents and lead to the concept of a crustal province, described later in this section.

Terranes are fault-bounded crustal blocks that have distinct lithologic and stratigraphic successions and that have geologic histories different from neighboring terranes (Schermer et al., 1984). Most terranes have collided with continental crust, along transcurrent faults

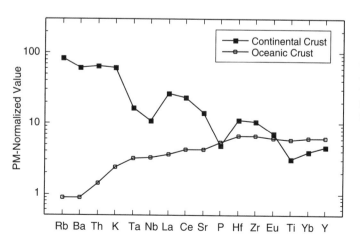

Figure 2.20 Primitive mantle (PM), normalized, incompatible element distributions in continental and oceanic crust. Data from Table 2.5; primitive mantle values from Sun and McDonough (1989).

or at subduction zones, and have been sutured to continents (Maruyama, 1997; von Raumer et al., 2003). Many terranes contain faunal populations and paleomagnetic evidence indicating they were displaced great distances from their sources before continental collision. For instance, Wrangellia, which collided with western North America in the late Cretaceous, traveled thousands of kilometers from what is now the South Pacific. Results suggest that as much as 30% of North America formed by terrane accretion in the last 300 My and that terrane accretion has been an important process in the growth of continents.

Terranes form in a variety of tectonic settings, including island arcs, oceanic plateaus, volcanic islands, and microcontinents (von Raumer et al., 2003). Numerous potential terranes exist in the oceans today and are particularly abundant in the Pacific basin (Fig. 2.21). Continental crust may be fragmented and dispersed by rifting or strike-slip faulting. In western North America, dispersion is occurring along transform faults such as the San Andreas and Fairweather, and in New Zealand movement along the Alpine transform fault is fragmenting the Campbell Plateau from the Lord Howe Rise (see Fig. 1.3 in Chapter 1), Baja California, and California west of the San Andreas fault were rifted from North America about 4 Ma, and today this region is a potential terrane moving northward, perhaps on a collision course with Alaska. Terranes may continue to fragment and disperse after collision with continents, as did Wrangellia, which is now distributed in pieces from Oregon to Alaska. The 1.90-Ga Trans-Hudson orogen in Canada and the 1.75- to 1.65-Ga Yavapai orogen in the southwestern United States are examples of Proterozoic orogens composed of terranes (Karlstrom et al., 2001). The Alps, Himalayas, and American Cordillera are Phanerozoic examples of orogens composed of terranes (Fig. 2.22). Most crustal provinces and orogens are composed of terranes, and in turn, cratons are composed of exhumed orogens. You might consider terranes as the basic building blocks of continents and terrane collision as a major means by which continents grow (Patchett and Gehrels, 1998).

A **crustal province** is an orogen, active or exhumed, composed of terranes; it records a similar range of isotopic ages and exhibits a similar postamalgamation deformational history. Structural trends within provinces range from linear to exceedingly complex swirling patterns reflecting polyphase deformation superimposed on differing terrane structural patterns. Exhumed crustal provinces that have undergone numerous episodes of deformation and metamorphism are old orogens, sometimes called *mobile belts*. Isotopic dating using multiple isotopic systems is critical to defining and unraveling the complex, polydeformational histories of crustal provinces.

The definition of crustal provinces is not always unambiguous. Most crustal provinces contain rocks of a wide age range and record more than one period of deformation, metamorphism, and plutonism. For instance, the Trans-Hudson orogen in North America (Fig. 2.22) includes rocks ranging from about 3.0 to 1.7 Ga and records several periods of complex deformation and regional metamorphism. Likewise, the Grenville province records a polydeformational history with rocks ranging from 2.7 to 1.0 Ga. Some parts of crustal provinces are new mantle-derived crust, known as **juvenile crust,** and other parts represent reworked older crust. **Reworking,** also known as **overprinting** or **reactivation,** describes crust that has been deformed, metamorphosed, and partially melted

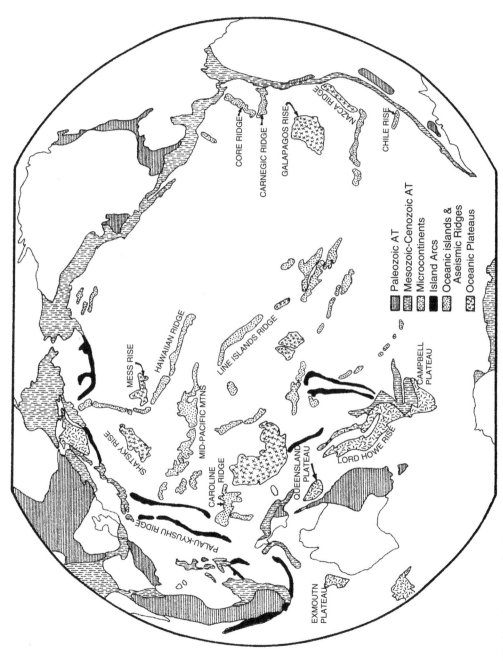

Figure 2.21 Map showing the distribution of accreted (AT) and potential terranes in the Pacific region. Modified from Schermer et al. (1984).

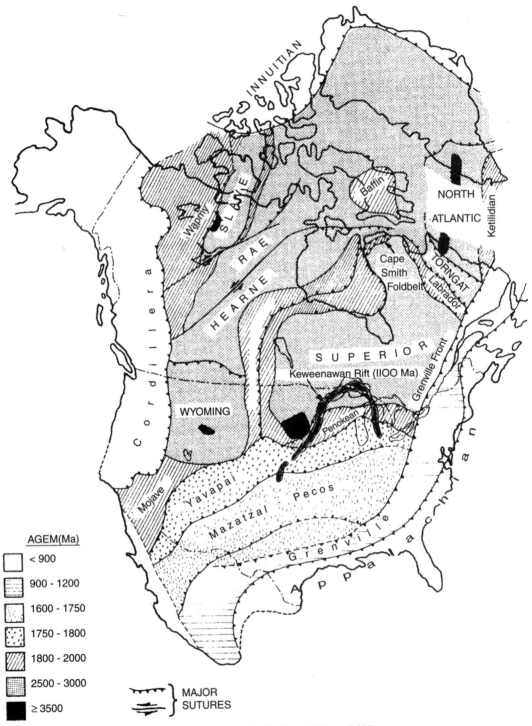

Figure 2.22   Distribution of North American crustal provinces. Modified from Hoffman (1988).

more than once. It is possible, in some instances, to map reworked crust within crustal provinces, and these are sometimes called *relict-age subprovinces*. There is increasing evidence that crustal reworking results from continental collisions, and large segments of Phanerozoic crust appear to have been reactivated by such collisions. For instance, much of central Asia at least as far north as the Baikal rift was affected by the India–Tibet collision beginning about 60 Ma. Widespread faulting and magmatism at present crustal levels suggest that deeper crustal levels may be extensively reactivated. In Phanerozoic collisional orogens where deeper crustal levels are exposed, such as the Appalachian and Variscan orogens, there is isotopic evidence for widespread reactivation.

One of the most important approaches to extracting multiple ages from crustal provinces is dating single zircons by the U-Pb method using an ion probe or a laser probe. Figure 2.23 shows an example of a felsic gneiss from southern Africa. The scatter of data on the Concordia diagram shows complex Pb loss from the zircons and even from within a single zircon. Note, for instance, the complex Pb loss from zircon grain 4. The most concordant domains can be fitted to a discordia line intersecting Concordia at 3505 ± 24 Ma, which is interpreted as the igneous crystallization age of this rock (Kroner et al., 1989). Three spots analyzed on the clear prismatic zircon grain 6 have a near-concordant age of 3453 ± 8 Ma. This records a period of intense deformation and high-grade metamorphism in which new metamorphic zircons formed in the gneiss. Grain 20 has a slightly

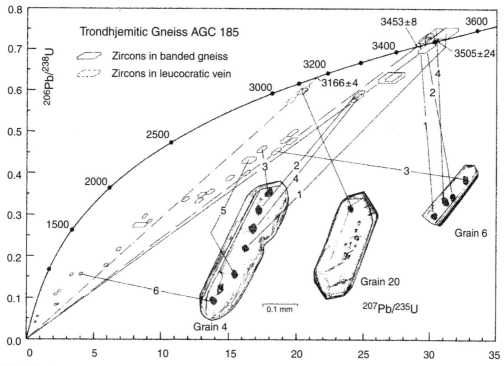

Figure 2.23   U-Pb Concordia diagram showing ion probe analyses of zircons from a trondhjemitic gneiss in northeastern Swaziland, southern Africa. Concordia is the bold solid line defined by concordant $^{206}Pb/^{238}U$ and $^{207}Pb/^{235}U$ ages. From Kroner et al. (1989).

discordant age of 3166 ± 4 Ma and comes from a later granitic vein that crosses the rock. Other discordant data points in Figure 2.23 cannot be fit to regression lines and reflect Pb loss at various times, perhaps some as young as 3 Ga. When combined with other single zircon ages from surrounding gneisses, major orogenic–plutonic events are recorded at 3580, 3500, 3450, 3200, and 3000 Ma in this small geographic area of Swaziland in southern Africa.

Crustal Province and Terrane Boundaries

Contacts between crustal provinces or between terranes are generally major shear zones, only some of which are the actual sutures between formerly colliding crustal blocks. Boundaries between terranes or provinces may be parallel or at steep angles to the structural trends within juxtaposed blocks. Some boundary shear zones exhibit transcurrent motions; others pass from flat to steep structures and may have thrust or transcurrent offsets. Magnetic and gravity anomalies also generally occur at provincial boundaries, reflecting juxtaposition of rocks of differing densities, magnetic susceptibilities, and crustal thicknesses. The Grenville Front, which marks the boundary between the Proterozoic Grenville and Archean Superior provinces in eastern Canada, is an example of a well-known crustal province boundary (Fig. 2.24). Locally, the Grenville Front, which formed about 1 Ga, ranges in width from a few kilometers to nearly 100 km and includes a large amount of reworked Archean crust (Culotta et al., 1990). It also produces a major negative gravity anomaly. Seismic reflection data show that the Grenville Front dips to the east and probably extends to the Moho. K-Ar biotite ages are reset at rather low temperatures (200° C) and gradually decrease from 2.7 Ga to about 1.0 Ga eastward across the Grenville Front. This front, however, is not a suture but is a major foreland thrust associated with the collision of crustal provinces. The suture has not been identified but may be the Carthage-Colton shear zone some 250 km east of the Grenville Front (Fig. 2.24).

Shear zones between crustal provinces are up to tens of kilometers in width, as illustrated by the Cheyenne belt in the Medicine Bow Mountains in southeastern Wyoming (Fig. 2.25). The Cheyenne belt is a near-vertical shear zone separating Archean gneisses of the Wyoming province from Paleoproterozoic juvenile crust of the Yavapai province on the south (Karlstrom and Houston, 1984). The timing of collision along this boundary is

Figure 2.24
Diagrammatic cross-section of the 1-Ga Grenville province in eastern Canada. CCMZ, Carthage-Colton shear zone; CGB, Central gneiss belt; CGT, Central granulite terrane; CMB, Central metasedimentary belt; SP, Archean Superior province.

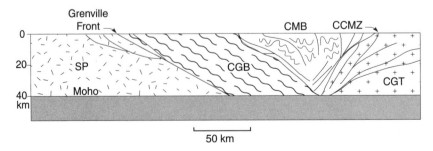

constrained by zircon ages from pre- and posttectonic plutons around 1.75 Ga (Condie, 1992a). This boundary is complicated because deformed metasedimentary rocks (about 2.0 Gy) rest unconformably on the Archean gneisses and are cut by the shear zone. The shear zone, which is up to several kilometers wide, is composed chiefly of mylonitized quartzofeldspathic gneisses.

The United Plates of America

North America provides an example of the birth and growth of a continent through geologic time. Field, geophysical, and Nd and U-Pb isotopic data from the Canadian shield and from borehole samples in platform sediment indicate that North America is an amalgamation of plates, called by Paul Hoffman the "United Plates of America" (Nelson and DePaolo, 1985; Patchett and Arndt, 1986; Hoffman, 1988). The Archean crust includes at least six provinces joined by Paleoproterozoic orogenic belts (Fig. 2.22). The assembly of constituent Archean provinces took only about 100 My between 1950 and 1850 Ma. Apparent from the map is the large amount of crust formed in the Late Archean, comprising at least 30% of the continent. Approximately 35% of the continent appears to have formed in the Early Proterozoic, 9% in the mid-Proterozoic to late Proterozoic, and about 26% in the Phanerozoic.

The systematic asymmetry of stratigraphic sections, structure, metamorphism, and igneous rocks in North American orogens is consistent with an origin by subduction and collision. Such asymmetry is particularly well displayed along the Trans-Hudson, Labrador, and Penokean orogenic belts. In these belts, zones of foreland deformation are dominated

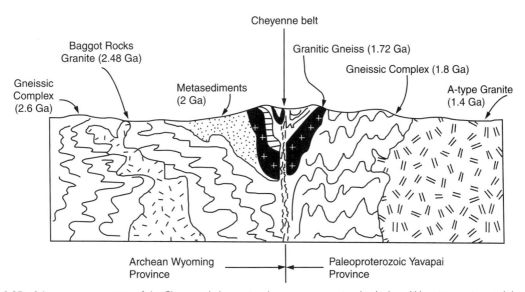

Figure 2.25 Schematic cross-section of the Cheyenne belt, a major shear zone separating the Archean Wyoming province and the Paleoproterozoic Yavapai province in southeastern Wyoming.

by thrusts and recumbent folds, whereas hinterlands typically show transcurrent faults. Both features are characteristic of subduction zones. Some Proterozoic orogens have large accretionary prisms, whereas others do not. For instance, the Rae and Hearne provinces involve only suturing of Archean crust, whereas the Trans-Hudson province is a collisional orogen up to 500 km wide. The Penokean, Yavapai, and Mazatzal provinces are accretionary orogens added to North America at 1.90, 1.75, and 1.65 Ga, respectively, and the Grenville province was added by one or more collisions from 1200 to 1000 Ma. The Cordilleran and Appalachian provinces represent collages of accretionary terranes sutured along transform faults or large thrusts during the Phanerozoic.

Supercontinents

Supercontinents are large continents that include several or all of the existing continents. Matching of continental borders, stratigraphic sections, and fossil assemblages are some of the earliest methods used to reconstruct supercontinents. Wegener (1912) pointed out the close match of opposite coastlines of continents and the regional extent of the Permo-Carboniferous glaciation in the Southern Hemisphere, and DuToit (1937) was first to propose an accurate fit for continents based on geological evidence. Today, in addition to these methods, we have polar-wandering paths, seafloor-spreading directions, hotspot tracks, and correlation of crustal provinces. The use of computers in matching continental borders has resulted in more accurate and objective fits. One of the most definitive matching tools in reconstructing plate positions in a former supercontinent is a piercing point. A **piercing point** is a distinct geologic feature such as a fault or a terrane that strikes at a steep angle to a rifted continental margin, the continuation of which should be found on the continental fragment rifted away.

The youngest supercontinent is **Pangea,** which formed between 450 and 320 Ma and includes most of the existing continents (Fig. 2.26). Pangea began to fragment about 160 Ma and is still dispersing today. **Gondwana** is a Southern Hemisphere supercontinent composed principally of South America, Africa, Arabia, Madagascar, India, Antarctica, and Australia (Fig. 2.27). It formed in the latest Neoproterozoic and was largely completed by the Early Cambrian (750–550 Ma) (Unrug, 1993). Later it became incorporated in Pangea. **Laurentia,** which is also part of Pangea, includes most of North America, Scotland and Ireland north of the Caledonian suture, Greenland, Spitzbergen, and the Chukotsk Peninsula of eastern Siberia. The oldest well-documented supercontinent is **Rodinia,** which formed from 1.3 to 1.0 Ga, fragmented from 750 to 600 Ma, and appears to have included many cratons in a configuration quite different from Pangea (Pisarevsky et al., 2003) (Fig. 2.28). Although the existence of older supercontinents is likely, their configurations are not known. Geologic data strongly suggest the existence of supercontinents in the Early Proterozoic and in the Late Archean (Aspler and Chiarenzelli, 1998; Pesonen et al., 2003). Current thinking is that supercontinents have been episodic, giving rise to the idea of a supercontinent cycle (Nance et al., 1986;

Hoffman, 1991). A **supercontinent cycle** consists of the rifting and breakup of one supercontinent, followed by a stage of reassembly in which dispersed cratons collide to form a new supercontinent with most or all fragments in configurations different from the older supercontinent (Hartnady, 1991).

The supercontinent cycle provides a record of the processes that control the formation and redistribution of continental crust throughout Earth's history. Through magmatism and orogeny associated with supercontinents, the supercontinent cycle influences elemental and isotopic geochemical cycles, climatic distributions, and changing environments that affect the evolution of organisms (Chapter 9).

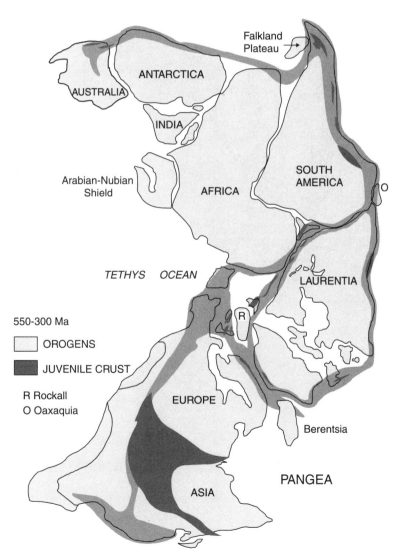

Figure 2.26 Pangea, a supercontinent formed between 450 and 320 Ma and fragmented about 160 Ma. Also shown are major orogens formed as blocks collided to make the supercontinent and the distribution of juvenile crust formed chiefly between 350 and 320 Ma. Juvenile crust is crust extracted from the mantle as the supercontinent formed.

Figure 2.27 Gondwana, a supercontinent formed between 750 and 550 Ma, which became part of Pangea in the late Paleozoic. Other information is given in Fig. 2.26.

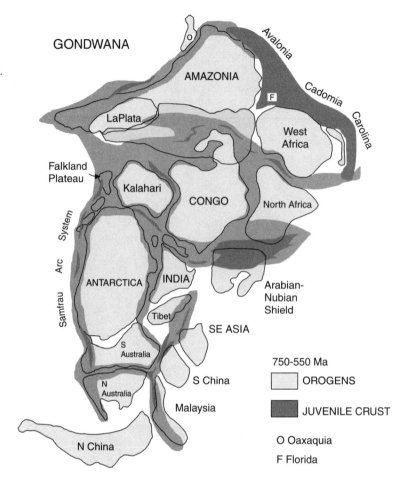

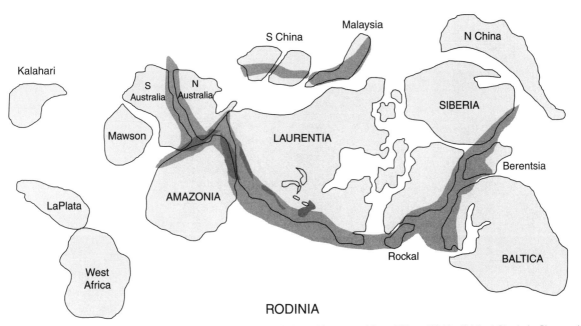

Figure 2.28 Rodinia, a supercontinent formed between 1.3 and 1.0 Ga and fragmented from 750 to 600 Ma. Kalahari, Rio de La Plata, and West Africa cratons were probably never part of Rodinia (Kroner and Cordani, 2003). Reconstruction modified from Pisarevsky et al. (2003) and Tohver et al. (2002). Other information is given in Fig. 2.26.

Further Reading

Beardsmore, G. R., and Cull, J. P., 2001. Crustal Heat Flow. Cambridge University Press, Cambridge, UK, 320 pp.

Brown, M., and Rushmer, T. (eds.), 2003. Evolution and Differentiation of the Continental Crust. Cambridge University Press, Cambridge, UK, 500 pp.

Fountain, D. M., Arculus, R., and Kay, R. W., 1992. Continental Lower Crust. Elsevier, Amsterdam, 486 pp.

Juteau, T., and Maury, R., 1999. The Oceanic Crust: From Accretion to Mantle Recycling. Springer-Verlag, New York, 390 pp.

Kleine, E., 2003. The Ocean Crust. In: R. L. Rudnick (ed.), The Crust: Treatise on Geochemistry, Vol. 3. Elsevier, Amsterdam.

Meissner, R.,1986. The Continental Crust: A Geophysical Approach. Academic Press, New York, 426 pp.

Rudnick, R. L., and Fountain, D. M., 1995. Nature and composition of the continental crust: A lower crustal perspective. Rev. Geophys., 33: 267–309.

Taylor, S. R., and McLennan, S. M., 1985. The Continental Crust: Its Composition and Evolution. Blackwell Scientific Publications, Oxford, 312 pp.

Tectonic Settings

3

Introduction

Rock assemblages that form in modern plate tectonic settings are known as **petrotec-tonic assemblages.** Such assemblages include both supracrustal rocks and intrusive igneous rocks. **Supracrustal rocks** formed at or near the Earth's surface (sediments and volcanics) but lost many of their primary features during metamorphism and deforma-tion. From studying modern plate settings, it is possible to learn what to look for in ancient rocks to identify, or at least constrain, the tectonic setting in which they formed. From the results, it may be possible to evaluate the intriguing question of just how far back in time plate tectonics has operated.

Although we can sample and study sediments forming in modern plate settings, it is more difficult to study young plutonic and metamorphic rocks and deep-seated deforma-tion. One approach is to examine deep canyons in mountain ranges where relatively young (<50 Ma) deep-seated rocks are exposed and where a tectonic setting can be assigned to these rocks with some confidence. Seismic reflection profiles provide a means to study deep-seated deformation. These studies have been particularly useful in modern arc systems. As you saw in the previous chapter, it is also possible to study recently elevated blocks that expose lower crustal rocks and xenolith populations brought up in young volcanoes. Finally, it is possible to simulate pressure–temperature (P–T) conditions that exist in modern plates in laboratory and computer models; from these conditions, constraints can be placed on geologic processes occurring at depth.

In this chapter, I examine some of the processes that leave tectonic imprints in rocks. It is only when results from a variety of studies converge on the same interpretation that ancient tectonic settings can be identified with confidence. Even then, scientists must use care—one or more tectonic settings that no longer exist may have left imprints in the geologic record similar to those of modern plate settings. This is especially true for the Archean period.

Ocean Ridges

Midocean-Ridge Basalts

Experimental results indicate that **midocean-ridge basalts** (MORB) are produced by 10 to 20% partial melting of the upper mantle at depths of 50 to 85 km (Elthon and Scarfe, 1984; Kushiro, 2001). Melting at these depths produces olivine tholeiite magma. Seismic and geochemical results show that this magma collects in shallow chambers (<35 km) where it undergoes fractional crystallization to produce tholeiites and minor amounts of more evolved liquids, including plagiogranites (Forsyth, 1996). A residue of olivine and pyroxenes is left in the mantle and may be represented by the sheared harzburgites preserved in some ophiolites.

The depletion of **large ion lithophile (LIL) elements** (K, Rb, Ba, Th, U, etc.) in **normal MORB** (NMORB) indicates a mantle source that has been depleted in these elements by earlier magmatic events. Low $^{87}Sr/^{86}Sr$ and $^{206}Pb/^{204}Pb$ and high $^{143}Nd/^{144}Nd$ isotopic ratios also demand an NMORB source that is depleted relative to chondrites. Incompatible element contents and isotopic ratios vary in MORB along ocean ridges and from ocean to ocean as illustrated by the distribution of Nb and Zr in modern MORB from the North Atlantic basin (Fig. 3.1). Although some of this variation can reflect differences in the degree of melting of the source, fractional crystallization, or magma

Figure 3.1 Zr and Nb distribution in ocean-ridge basalts from the North Atlantic. Each symbol is for a different geographic location along the mid-Atlantic ridge. Data from Tarney et al. (1979).

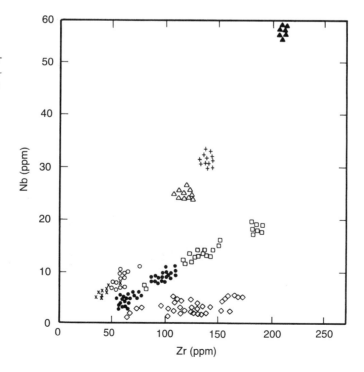

mixing, the large differences among some sites require variation in the composition of the depleted mantle source.

Ophiolites

General Features

Ophiolites are tectonically emplaced successions of mafic and ultramafic rocks considered to represent fragments of oceanic or back-arc basin crust (Coleman, 1977; Moores, 1982). An ideal ophiolite includes, from bottom to top, the following units (Fig. 3.2): (1) ultramafic tectonite (generally harzburgites), (2) layered cumulate gabbros and ultramafic rocks, (3) noncumulate gabbros, diorites and plagiogranites, (4) sheeted diabase dykes, and (5) pillowed basalts. Overlying this succession in many ophiolites are abyssal sediments, pelagic sediments, abyssal–pelagic sediments, or arc-related volcaniclastic sediments. Because of faulting or other causes, the idealized ophiolite succession is rarely found in the geologic record. Instead, one or more of the ophiolite units are missing or they have been dismembered by faulting and occur as blocks in a tectonic melange.

Some ophiolites are in fault contact with underlying shallow marine cratonic sediments, and others occur as tectonic slivers in accretionary prisms with graywackes and other arc-related rocks. These ophiolites appear to be emplaced along passive and active continental margins, respectively. The basal ophiolite melange (Fig. 3.2) consists of a chaotic mixture of diverse rocks in a highly sheared matrix. Clast lithologies include

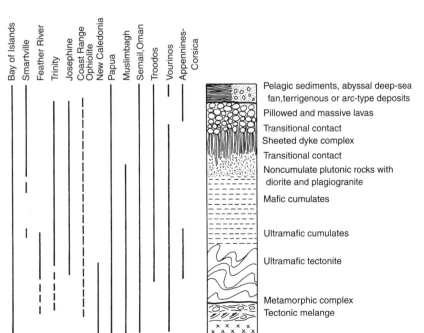

Figure 3.2 An idealized ophiolite succession compared with various exposed ophiolites. Modified from Moores (1982).

ophiolite-derived rocks, pelagic and abyssal sediments, graywackes, and various meta-morphic and volcanic rocks. Matrices are commonly sheared serpentinite. Ophiolite melanges are of tectonic origin formed during ophiolite emplacement.

Sheared and serpentinized ultramafic rock is an important component in lower part of most ophiolites. These tectonites are composed chiefly of harzburgite with pronounced foliation and generally less deformed dunite pods. Lenses of dunite and chromite also occur within the harzburgites. Overlying the tectonites are cumulate ultramafic and gabbroic rocks that formed by fractional crystallization. These rocks have cumulus textures and well-developed compositional banding. Plagiogranite, a minor component in some ophi-olites, is tonalite composed of quartz and sodic plagioclase with minor mafic silicates. These rocks typically have granophyric intergrowths and may be intrusive into layered gabbros (Coleman, 1977).

Above the cumulate unit in an idealized ophiolite is a **sheeted dyke complex** (Baragar et al., 1990) (Fig. 3.2). Dykes may crosscut or be gradational with cumulate rocks. Although dominantly diabase, dykes range from diorite to pyroxenite in composition, and dyke thick-ness is variable, commonly from 1 to 3 m. One-way chilled margins are common in sheeted dykes, a feature generally interpreted to reflect vertical intrusion in an oceanic axial rift zone, where one dyke is intruded in the center of another as the lithosphere spreads. The transition from sheeted dykes to pillow basalts generally occurs over an interval of 50 to 100 m where screens of basalt between dykes become more abundant. The uppermost unit of ophiolites is ocean-ridge basalt occurring as pillowed flows or hyaloclastic breccias. Thickness of this unit varies from a few meters to 2 km and pillows form a honeycomb net-work with individual pillows up to 1 m across. A few dykes cut the pillowed basalt unit.

Many ophiolites are overlain by sediments reflecting pelagic, abyssal, or arc deposi-tional environments. Pelagic sediments include radiolarian cherts, red fossiliferous lime-stones, metalliferous sediments, and abyssal sediments. Abyssal sediments are chiefly pelites and siltstones deposited on abyssal plains and often show evidence of both vol-canic and continental provenance. Modern abyssal plains occur adjacent to passive con-tinental margins and hence have significant input of sediment from continental shelves. Graywackes and volcaniclastic sediments of arc provenance overlie some ophiolites.

Tectonic Setting and Emplacement of Ophiolites

Ophiolites have been described from three tectonic settings: midocean ridges, back-arc basins, and in some instances, such as the Metchosin Complex in British Columbia, immature island arcs (Coleman, 1977; Massey, 1986; Takashima et al., 2002). Incompatible element distributions in many ophiolitic basalts—for example, the Semail ophiolite in Oman and the Troodos Complex in Cyprus—show a **subduction geochem-ical component** (relative depletion in Ta and Nb; Fig. 3.3), suggesting that they are frag-ments of back-arc oceanic crust (Moores et al., 2000). Relatively few ophiolites have basalts showing NMORB element distributions.

Ophiolites are emplaced in arcs or collisional orogens by three major mechanisms (Fig. 3.4): (1) **obduction** or thrusting of oceanic lithosphere onto a passive continental

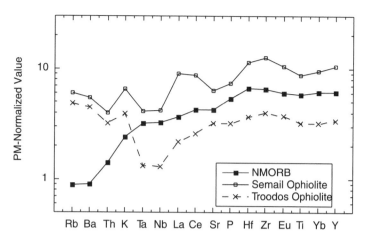

Figure 3.3 Primitive mantle (PM), normalized, incompatible element distributions in ocean-ridge basalts and ophiolites. Primitive mantle values from Sun and McDonough (1989), and other data from Bednarz and Schmincke (1994) and Alabaster et al. (1982).

margin during a continental collision, (2) splitting of the upper part of a descending slab and obduction of a thrust sheet onto a former arc, and (3) addition of a slab of oceanic crust to an accretionary prism in an arc system (Dewey and Kidd, 1977; Cawood and Suhr, 1992).

Compared with seismic sections of oceanic or arc crust, most ophiolites are considerably thinner (chiefly < 5 km compared with 7–10 km for oceanic crust and 10–15 km for arcs) (Fig. 3.5). In some highly deformed ophiolites, thickness may be controlled by tectonic thinning during emplacement. Most ophiolites, however, are not severely deformed and

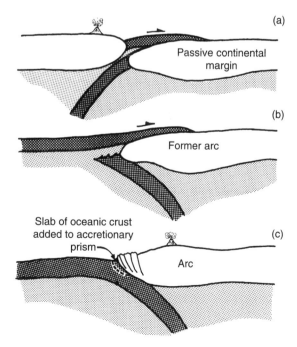

Figure 3.4 Three mechanisms of ophiolite emplacement. Obduction at a passive continental margin (a), obduction of oceanic lithosphere (b), and transfer of a slab of oceanic lithosphere to an accretionary prism (c).

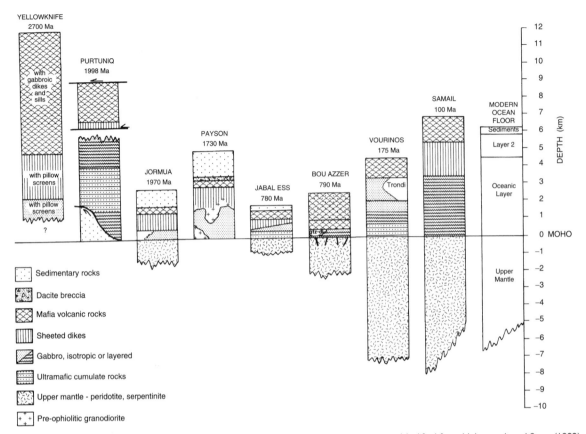

Figure 3.5   Simplified stratigraphic sections of ophiolites compared with average oceanic crust. Modified from Helmstaedt and Scott (1992).

may have been thickened during emplacement (Moores, 1982). So where in the oceanic crust do most ophiolites come from? Perhaps ophiolites are decoupled from the mantle at the asthenosphere–lithosphere boundary near spreading centers where oceanic litho-sphere is thin (Karsten et al., 1996; Bortolotti et al., 2002). In a closing ocean basin, lithosphere decoupling may occur when an ocean ridge enters a subduction zone or during a continent–continent collision. If decoupling occurs only at the asthenosphere–lithosphere boundary, it may not be possible to obduct thicker segments of oceanic lithosphere (up to 150 km thick at the time of subduction), thus accounting for the absence of ophiolites much thicker than 5 km.

Formation of Ophiolites

Although details continue to be controversial, the overall mechanism by which a com-plete ophiolite succession forms is reasonably well understood. As pressure decreases in rising asthenosphere beneath ocean ridges, garnet lherzolite partially melts to produce basaltic magma (Fig. 3.6). These magmas collect in shallow chambers (3–6 km deep) and

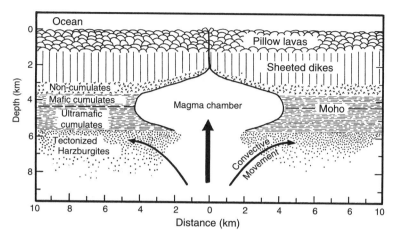

Figure 3.6 Schematic cross-section of oceanic crust near an ocean ridge, showing possible relationships to an ophiolite succession.

undergo fractional crystallization. Layered ultramafic and gabbroic rocks accumulate in the magma chamber, forming the cumulate section of ophiolites. The tectonized harzburgites represent residue left after melting, which is highly deformed and sheared because lateral advective motion of the asthenosphere, deformation during emplacement, or both. Dykes are ejected from the magma chamber, forming the sheeted dyke complex, and many are erupted, forming pillow lavas in the axial rift. Fractional crystallization in the magma chamber also leads locally to diorites and plagiogranites, which are intruded at the top of the layered cumulate series.

Precambrian Ophiolites

Ophiolites and associated deep-sea sediments are first recognized in the geologic record about 2.0 Ga, although a possible ophiolite in China may extend this age to 2.5 Ga (Moores, 2002; Kusky et al., 2001). One of the oldest recognized ophiolites with all of the essential components in the correct stratigraphic order is the 1.9-Ga Jormua Complex in northern Finland (Fig. 3.5). Although older ophiolites have been reported (de Wit et al., 1987; Helmstaedt et al., 1986; Helmstaedt and Scott, 1992), all but the Dongwanzi ophiolite in China lack a convincing sheeted dyke complex and tectonized harzburgites. The thickest, most laterally extensive Paleoproterozoic ophiolite is the Purtuniq ophiolite in the Cape Smith orogen in Canada with an age around 2 Ga (Fig. 3.5) (Scott et al., 1992). Although few well-documented ophiolites have been reported that are older than about 1000 Ma, numerous occurrences in the age range from 1000 to 600 Ma are reported from the Pan-African provinces in Africa and South America and from areas around the North Atlantic. Estimated thicknesses of most Neoproterozoic ophiolites range up to 8 km, but most are less than 5 km. They are bound by thrust faults and appear to have been emplaced by obduction. Although the range in metamorphic grade and in degree of deformation is considerable, many Proterozoic ophiolites preserve primary textures and structures. Most, however, represent only partial ophiolite successions.

As with Phanerozoic ophiolites, many Proterozoic ophiolites carry a weak to strong subduction-zone geochemical imprint. This suggests that they represent fragments of arc-related oceanic crust from either back-arc or intra-arc basins. Why there are few if any occurrences of ophiolite older than 2 Ga is an important question that will be returned to in Chapter 8.

Tectonic Settings Related to Mantle Plumes

Oceanic Plateaus and Aseismic Ridges

Oceanic plateaus, composed chiefly of basalt flows erupted beneath the oceans, are the largest topographic features of the seafloor. **Aseismic ridges** are extinct volcanic ridges on the seafloor. Both are examples of **large igneous provinces,** voluminous eruptions of largely mafic volcanics related to mantle plumes (Condie, 2001). About 10% of the ocean floors are covered by oceanic plateaus and aseismic ridges, and more than 100 are known (Fig. 3.7), many of which are in the western Pacific (Coffin and Eldholm, 1994). These features rise thousands of meters above the seafloor, and some, such as the Seychelles Bank in the Indian Ocean (Fig. 3.7), rise above sea level. Some have granitic basement (such as the Seychelles Bank, the Lord Howe Rise north of New Zealand, and the Agulhas Plateau south of Africa), suggesting that they are rifted fragments of continental crust. Others, such as the Cocos and Galapagos ridges west of South America, are of volcanic origin and related to hotspot activity. Some aseismic ridges, such as the Palau-Kyushu ridge south of Japan, are extinct oceanic arcs. Others, such as the Walvis and Rio Grande ridges in the South Atlantic, are hotspot tracks.

Like flood basalts on the continents, oceanic plateaus are thought to be the products of magmas erupted from mantle plumes (Coffin and Eldholm, 1994; Carlson, 1991). The largest oceanic plateau, which straddles the equator in the western Pacific, is the Ontong-Java plateau erupted from a mantle plume about 120 Ma in the South Pacific. This plateau is capped by seamounts and is covered by a veneer of pelagic sediments (limestone, chert, and radiolarite), in places more than 1 km thick (Berger et al., 1992). Because only the tops of oceanic plateaus are available for sampling, to learn more about their rock assemblages, we must go to older plateaus preserved on the continents as terranes. Studies of the Wrangellia terrane, composed largely an oceanic plateau accreted to western North America in the Cretaceous, show that pillow basalts and hyaloclastic breccias comprise most of the succession with only a thin capping of pelagic limestone, shale, and chert. Some oceanic plateaus (like Kerguelen in the southern Indian Ocean) emerge above sea level and are capped with subaerial basalt flows and associated pyroclastic volcanics. Oceanic-plateau basalts are largely tholeiites with only minor amounts of alkali basalt, and most show incompatible elements only slightly enriched compared with NMORB (Fig. 3.8).

Continental Flood Basalts

Flood basalts are thick successions of basalt erupted on the continents over short periods of time, like the Columbia River basalts in the northwestern United States (Fig. 3.7)

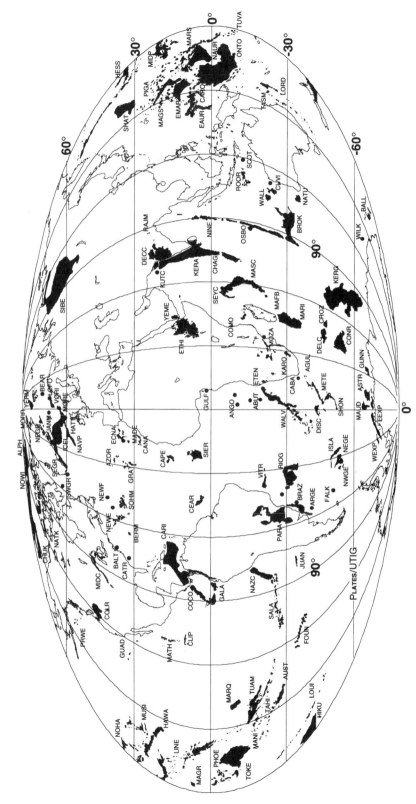

Figure 3.7 Distribution of Phanerozoic oceanic plateaus, aseismic ridges, and flood basalts. AGUL, Agulhas Plateau; COCO, Cocos ridge; DECC, Deccan Traps; GALA, Galapagos ridge; LORD, Lord Howe Rise; RIOG, Rio Grande ridge; SEYC, Seychelles Bank; WALV, Walvis ridge. Modified from Coffin and Eldholm (1994).

Figure 3.8 Primitive
mantle (PM), normalized,
incompatible element distributions in various basalts.
Primitive mantle values and
NMORB from Sun and
McDonough (1989).

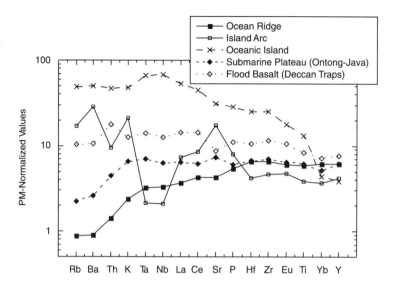

and the Deccan Traps in India. They are composed chiefly of the tholeiitic basalt flows, and like oceanic-plateau basalts, they appear to be derived from mantle plumes or, in some cases, from melting of the subcontinental lithosphere caused by mantle plumes (Carlson, 1991; Condie, 2001). Thick successions of flood basalts comprise important plateaus on the continents. One of the characteristics of both oceanic-plateau basalts and flood basalts is the rapidity with which large volumes of basaltic magma are erupted. The Deccan Traps, which were erupted at the Cretaceous–Tertiary boundary (65 Ma), include a preserved volume of basalt about 1.5×10^6 km^3 that erupted in before or at 1 My; probably more than 80% of these flows are younger than 500,000 years. The Columbia River basalts formed from 17.5 to 6.0 Ma, at which time 0.17×10^6 km^3 of magma were erupted. Eruption rates of flood basalts and oceanic-plateau basalts range from about 0.5 to less than 1 km^3/year, considerably greater than rates typical of ocean ridges or volcanic islands such as Hawaii (with rates of 0.02–0.05 km^3/year) (Carlson, 1991; White and McKenzie, 1995).

Continental flood basalts tend to be lower in Mg and Fe and other compatible elements than MORB or island basalts. They typically show Fe-enrichment trends, but in some instances have lower Ti than other oceanic basalts. Incompatible element distributions vary widely in flood basalts, often within the same volcanic field. Those with relatively high contents of LIL elements, such as the Deccan Traps (Fig. 3.8), are either contaminated by continental crust or come from enriched sources in the subcontinental lithosphere. In both cases, a subduction geochemical component, as shown by the Deccan basalts, may be transferred to the magmas by magma contamination or, in the case of the lithosphere, directly from the source.

Hotspot Volcanic Islands

The hotspot model (Wilson, 1963), which suggests that linear volcanic chains and ridges on the seafloor form as oceanic crust moves over relatively stationary magma sources, has been widely accepted in the geological community. Hotspots are generally thought to form in response to mantle plumes, which rise like salt domes in sediments through the mantle to the base of the lithosphere (Duncan and Richards, 1991). Partial melting of plumes in the upper mantle leads to large volumes of magma partly erupted (or intruded) at the Earth's surface. Hotspots also may be important in the breakup of supercontinents.

Hotspots are characterized by the following features:

1. In ocean basins, hotspots form topographic highs of 500 to 1200 m with typical widths of 1000 to 1500 km. These highs are probably indirect manifestations of ascending mantle plumes.
2. Many hotspots are capped by active or recently active volcanoes. Examples are Hawaii and Yellowstone National Park in the western United States.
3. Most oceanic hotspots are characterized by gravity highs reflecting the rise of denser mantle material.
4. One or two aseismic ridges of mostly extinct volcanic chains lead from many oceanic hotspots. Similarly, in continental areas, the age of magmatism and deformation may increase with distance from a hotspot. These features are known as **hotspot tracks** (Fig. 3.9).
5. Most hotspots have high heat flow, probably reflecting a mantle plume at depth.

Chains of seamounts and volcanic islands are common in the Pacific basin and include such well-known island chains as the Hawaiian-Emperor, Line, Society, and Austral, all of which are subparallel to either the Emperor or the Hawaiian chains and approximately perpendicular to the axis of the East Pacific rise (Fig. 3.9). Closely spaced volcanoes form aseismic ridges such as the Ninetyeast ridge in the Indian Ocean and the Walvis and Rio Grande ridges in the South Atlantic. Isotopic dates demonstrate that the focus of volcanism in the Hawaiian chain has migrated to the southeast at a linear rate of about 10 cm/year for the last 30 My. The bend in the Hawaiian-Emperor chain around 43 Ma was long assumed to reflect a change in the spreading direction of the Pacific plate. However, paleomagnetic results show that before 43 Ma the hotspot track was caused by rapid movement of the hotspot rather than by a change in plate motion (Tarduno et al., 2003). Linear decreases in the age of volcanism that occurred in the Society and Austral hotspot tracks in the South Pacific during the last 30 My (Fig. 3.9) have inferred rates of plate migration of 11 cm/year, similar to the Hawaiian track. The Pratt-Welker seamount chain in the Gulf of Alaska, however, records a rate of plate motion of about 4 cm/year.

Hotspot tracks also occur in the continents, although they are less well defined than in ocean basins because of the thicker lithosphere. For example, North America moved northwest over the Great Meteor hotspot in the Atlantic basin between 125 and 80 Ma (Van Fossen and Kent, 1992). The trajectory of the hotspot is defined by the New England seamount chain in the North Atlantic (see Fig. 1.3 in Chapter 1) and by the Cretaceous kimberlites and alkalic complexes in New England and Quebec. Dated igneous rocks fell near the calculated position of the hotspot track at the time they formed. The calculated trajectory of the Trindade hotspot east of Brazil matches the locations of

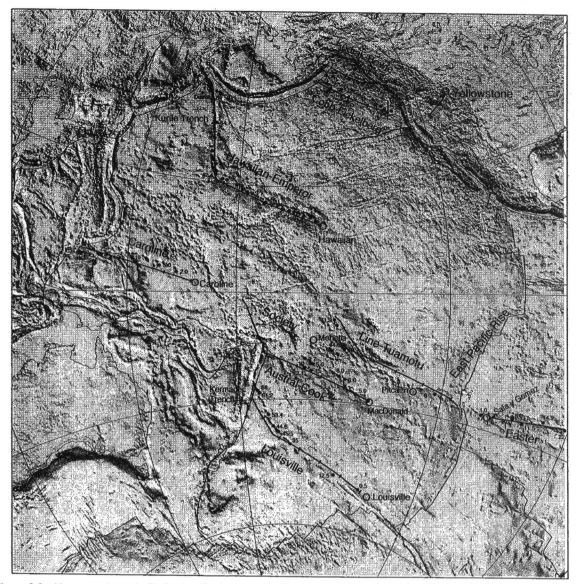

Figure 3.9 Hotspot tracks on the Pacific plate. Tracks assume fixed hotspots and show 10-My tick marks. Modified from Steinberger and O'Connell (1998), courtesy of Bernhard Steinberger.

three dated kimberlites from Brazil and roughly coincides with the distribution of alluvial diamond deposits (derived from nearby kimberlites). High heat flow, low seismic-wave velocities, low densities at shallow depth, and high electric conductivity at shallow depth beneath Yellowstone National Park in Wyoming reflect a mantle hotspot at this locality (Smith and Braile, 1994) (Fig. 3.9). A 600-meter-high topographic bulge is centered on the Yellowstone caldera and extends across an area 600 km in diameter.

Direct evidence of a mantle plume at depth is manifest by anomalously low P-wave velocities that extend to depths of 200 km. The movement of the North American plate over this hotspot during the past 16 My has been accompanied by the development of the Snake River volcanic plain, with the oldest hotspot-related volcanics found in south-eastern Oregon.

As exemplified by Hawaii, hotspot volcanic islands are some of the largest mountains on Earth, often rising many kilometers above the seafloor. During the subaerial stages of eruption, volcanoes typically evolve from a shield-building stage (like Mauna Loa and Kilauea) through a caldera-filling stage (like Mauna Kea) to terminate in a highly eroded shield volcano with small, often alkaline magma eruptions (like most of the extinct volcanoes in Hawaii). Hotspot islands are composed chiefly of tholeiitic basalts, with only minor amounts of alkali basalts and their derivatives erupted during terminal volcanism. Dredging of seamounts and oceanic slopes of islands reveals a dominance of hyaloclastic volcanics in contrast to the abundance of flows found along oceanic ridges (Bonatti, 1967). This difference indicates that island–seamount magmas are considerably more viscous than MORB magmas and readily fragment upon eruption into seawater (Kokelaar, 1986). In striking contrast to subduction-related basalts, many island basalts show relative enrichment in Nb and Ta (Fig. 3.8), reflecting a mantle plume source enriched in these elements.

Giant Mafic Dyke Swarms

Vast swarms of mafic dykes have been intruded into the continents, covering areas that are tens to hundreds of thousands of square kilometers (Ernst et al., 1995; Ernst and Buchan, 2001). These swarms, which are up to 500 km in width and more than 3000 km in length, occur on all Precambrian shields and include thousands of dykes. Individual dykes range from 10 to 50 m in width, with some up to 200 m. Most dykes have rather consistent strikes and dips, and individual dykes have been traced up to 1000 km. Dyke widths tend to be greater in dykes with the least depth of erosion, suggesting that dykes widen in the upper crust; dyke spacing ranges from 0.5 to 3.0 km, with dykes branching vertically and along the strike. Major swarms appear to have been intruded episodically, with important ages of intrusion at 2500, 2390 to 2370, 2150 to 2100, 1900 to 1850, 1270 to 1250, and 1150 to 1100 Ma. Structural studies indicate that most swarms are intruded at right angles to the minimum compression direction and, except near the source, most are intruded laterally, not vertically. Some swarms, such as the giant Mackenzie swarm in Canada (Fig. 3.10), appear to radiate from a point commonly interpreted as a plume head (Baragar et al., 1996). The Mackenzie swarm is also an example of a swarm associated with flood basalts (Coppermine River basalts; Fig. 3.10) and a layered intrusion (Muskox intrusion). U-Pb dating of baddeleyite shows that most swarms are emplaced in short periods, often 2 to 3 My (Tarney, 1992). Giant dyke swarms are typically composed of gabbros, although norite swarms are also important in the Proterozoic.

Like the Mackenzie swarm, many giant dyke swarms appear to be associated with mantle plumes, and their emplacement is accompanied by significant extension

Figure 3.10 The giant Mackenzie dyke swarm in northern Canada. This dyke swarm was intruded at 1267 Ma, probably in response to a mantle plume north of the Coronation Gulf (CG). The Muskox intrusion (M) and the Coppermine River flood basalts (Cb) are part of the same event. GBL, Great Bear Lake; GSL, Great Slave Lake. Courtesy of A. N. LeCheminant.

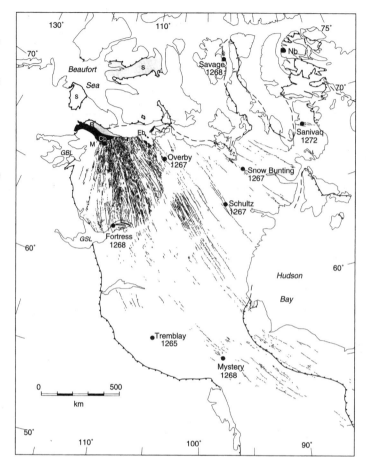

(more than 30%) of the continental crust (Tarney, 1992; Baragar et al., 1996). During the early stages of plume activity and rifting, dykes may develop radial patterns from an underlying plume source (Fig. 3.11, stage 1). As with the Red Sea and Gulf of Aden rifts associated with the Afar hotspot, an ocean basin may open between two of the radiating dyke swarms, leaving the third swarm as part of an aulacogen (Fig. 3.11, stage 2). Later, during closure of the ocean basin and continental collision, dykes are consumed or highly deformed. The swarms intruded in the aulacogen, however, are more likely to be preserved if they are on the descending plate (Fig. 3.11, stage 3). Hence, major dyke swarms that radiate from a point, such as the Mackenzie swarm, may indicate the presence of a former ocean basin. Some giant dyke swarms may have been emplaced during supercontinent fragmentation (Chapter 9); thus, their ages and distributions can be helpful in reconstructing supercontinents.

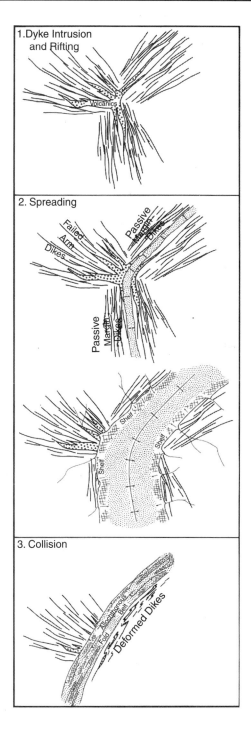

Figure 3.11 Three stages in the tectonic evolution of mafic dyke swarms. From Fahrig (1987).

Continental Rifts

General Features

Continental rifts are fault-bounded basins produced by extension of continental crust. They may be single, like the rifts in East Africa, or multiple, like the Basin and Range Province in the western United States. Also included in this category are **aulacogens,** which are failed or less-active arms of triple junctions such as the Ethiopian and Benue rifts in Africa. Spreading began in the Red Sea and Gulf of Aden from 25 to 15 Ma and may be beginning today in the Ethiopian aulacogen. The Benue rift became a failed arm in the late Cretaceous as the other two rift segments opened as part of the Atlantic basin. Rifts are of different origins and occur in different regional tectonic settings. Although the immediate stress environment of rifts is extensional, the regional stress environment may be compressional, extensional, or nearly neutral. Rifts that form in cratons, such as the East African rift system, are commonly associated with domal uplifts, although the timing of doming relative to rifting may vary (Mohr, 1982). Geophysical data indicate that both the crust and the lithosphere are thinned beneath rifts and that most or all of the crustal thinning occurs in the ductile lower crust over a much broader area than represented by the surface expression of the rift (Thompson and Gibson, 1994). Extension in rifts ranges from 10 km in the Baikal rift to more than 50 km in the Rio Grande rift. Magma chambers have been described in rifts based on seismic data, of which the best-documented cases are magma chambers in central New Mexico beneath the Rio Grande rift and in Iceland (Sanford and Einarsson, 1982). Although in some instances tectonic and volcanic activity in rifts are related in space and time, in many cases they are not. Zones of major eruption seldom coincide with the main rift faults.

Rock Assemblages

Continental rifts (hereafter called *rifts*) are characterized by immature terrigenous clastic sediments and bimodal volcanics (Wilson, 1993). **Bimodal volcanics** are basalts (tholeiites) and felsic volcanics, as found in the Rio Grande rift, or alkali basalts and phonolites, as occur in the East African rift. Few, if any, igneous rocks of intermediate composition occur in rifts. Felsic volcanics are emplaced usually as ash-flow tuffs or glass domes. In older uplifted rifts, the plutonic equivalents of these rocks are exposed, as in the Oslo rift in Norway. Granitic rocks range from granite to monzodiorite in composition with granites and syenites usually dominating (Williams, 1982). The Kenya and Oslo rifts show a decrease in magma alkalinity with time, a feature that probably reflects a secular decrease in the depth of magma generation. Geochemical and isotopic studies of rift basalts show that they are derived from mantle plumes, from the subcontinental lithosphere, or both (Bradshaw et al., 1993; Thompson and Gibson, 1994; Peate, 1997).

    Layered igneous intrusions are common in exhumed rifts. These intrusions are large mafic to ultramafic bodies that exhibit internal layering formed by the accumulation of crystals during fractional crystallization of basaltic or komatiitic parent magmas. Cyclic layering in large mafic intrusions is generally interpreted to reflect episodic injections of new magma into fractionating magma chambers. The largest well-studied intrusion is the vast Bushveld Complex in South Africa, which is more than 8 km thick, covers a minimum area 66,000 km^2, and was intruded at 2 Ga (Fig. 3.12). Large layered intrusions are major sources of metals, such as Cr, Ni, Cu, Fe, and, in the case of the Bushveld Complex, Pt.

Rift sediments are chiefly arkoses, feldspathic sandstones, and conglomerates derived from rapidly uplifted fault blocks in which granitoids are important components. Evaporites also are deposited in many rifts. If a rift is inundated with seawater, as exemplified by the Rhine graben in Germany, marine sandstones, shales, and carbonates may also be deposited.

Rift Development and Evolution

As a starting point for understanding rift development and evolution, rifts can be classified into two categories depending on the mechanisms of rifting (Sengor and Burke, 1978; Ruppel, 1995). **Active rifts** are produced by doming and cracking of the lithosphere, where doming results from upwelling asthenosphere or rising mantle plumes (Fig. 3.13). **Passive rifts** are produced by stresses in moving lithospheric plates or by drag at the base of the lithosphere. Active rifts are represented by ocean ridges, continental rifts, aulacogens, and back-arc basins. As continental rifts and back-arc rifts continue to open, they can evolve into ocean ridges. Active rifts contain relatively large volumes of volcanic rock; in passive rifts, immature clastic sediments exceed volcanics in abundance. Active rifts are characterized by early uplift and basement stripping resulting from crustal expansion because of a deep heat source. In general, uplift in passive rifts is confined to the stretched and faulted near-surface region and to the shoulder of the rift zone, whereas in active rifts, uplift commonly extends hundreds of kilometers beyond the rift zone. Lithospheric thinning is laterally confined to the rift zone in passive rifts; in active rifts, the zone of thinning is several times wider than the rift width (Thompson and Gibson, 1994).

Although active rifts may form in several tectonic settings, they commonly exhibit similar overall patterns of development. The major stages are as follows (modified from Mohr, 1982):

1. A broad, shallow depression develops before doming or volcanism.
2. Diapiric asthenosphere or a mantle plume is forcefully injected into the base of the lithosphere. During injection, diapirs undergo adiabatic decompression, leading to partial melting and the onset of basaltic magmatism.

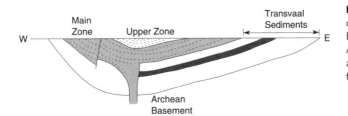

Figure 3.12 Idealized cross-section of the 2-Ga Bushveld Complex in South Africa. Length of section about 150 km. Modified from Hunter (1975).

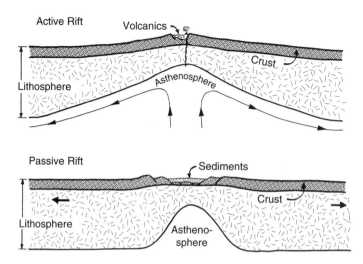

Figure 3.13
Diagrammatic cross-sections of active and passive continental rifts.

3. Buoyant isostatic uplift of heated lithosphere leads to doming.
4. Doming and extensional forces cause crustal attenuation and thinning of the lithosphere.
5. Episodic dyke injection and volcanism alternate with faulting, and the duration of episodes and volumes of magma erupted decrease with time.
6. Rift valleys develop and may be associated with voluminous felsic magmatism, both extrusive and intrusive.
7. An active rift may be aborted at an early stage of its development, such as the Mesoproterozoic Keweenawan rift in the north–central United States, or it may continue to open and evolve into an ocean basin, such as the Red Sea.

Passive rifts develop along faulted continental margins, in zones of continental collision, and in arc systems (Ingersoll, 1988). Examples of rifts associated with faulted continental margins are those found adjacent to the San Andreas and related faults in California and the rifts in western Turkey associated with the Anatolian transform fault. Rifts can be generated along collision boundaries by irregularities in continental margins and by transcurrent and normal faulting caused by nonperpendicular collision. The Rhine graben in Germany is an example of a rift that developed at a steep angle to a collisional boundary, and the term *impactogen* has been applied to this type of rift. The Rhine graben appears to have been produced as Africa collided with an irregularity along the European continental margin. An aborted attempt to subduct the irregularity resulted in depression and rifting. Numerous studies confirm that stresses associated with the Tibet–India collision were transmitted as transcurrent and normal faulting into the Eurasian plate, forming such rifts as the Baikal rift in Siberia and the Shanshi graben system in China.

As rifts evolve, their rock assemblages change. Oceanic rifts are represented by ophiolites. Arc-related rifts contain arc volcanics and graywackes, and moderate to large back-arc basins are characterized by mixed assemblages including some combination of ophiolites and deep-sea sediments, arc-derived graywackes, and cratonic sediments. Continental rifts and aulacogens contain arkoses, feldspathic sandstones, conglomerates, and bimodal volcanics. Passive rifts contain a variety of immature sediments and, in some instances, minor volcanic rocks.

Cratons and Passive Margins

Rock assemblages deposited in cratonic basins and passive margins are mature clastic sediments, chiefly quartz arenites and shales, and shallow marine carbonates. In late Archean–Paleoproterozoic successions, banded iron formation may also be important (Klein and Beukes, 1992; Eriksson and Fedo, 1994). Because passive continental margins began life as continental rifts, rift assemblages generally underlie passive continental margin successions. When back-arc basins develop between passive margins and arcs, like the Sea of Japan, cratonic sediments may interfinger with arc sediments in these basins. Cratonic sandstones are relatively pure quartz sands reflecting intense weathering, low relief in source areas, and prolonged transport across subdued continental surfaces. Commonly associated marine carbonates are deposited as blankets and as reefs around the basin margins. Transgression and regression successions in large cratonic basins reflect the rise and the fall of sea level, respectively.

Depositional systems in cratonic and passive margin basins vary depending on the relative roles of fluvial, eolian, deltaic, wave, storm, and tidal processes. Spatial and temporal distribution of sediments is controlled by regional uplift, the amount of continent covered by shallow seas, and climate (Klein, 1982). If tectonic uplift is important during deposition, continental shelves are narrow and sedimentation is dominated by wave and storm systems. However, if uplift is confined chiefly to craton margins, sediment yield increases into the craton and fluvial and deltaic systems may dominate. For transgressive marine clastic sequences, shallow seas are extensive and subtidal, and storm-dominated and wave-dominated environments are important. During regression, fluvial and eolian depositional systems become dominant.

The rates of subsidence and uplift in cratons are a function of the time interval over which they are measured. Current rates are a few centimeters per year, whereas data from older successions suggest rates 1 to 2 orders of magnitude slower. In general, Phanerozoic rates of uplift appear to have been 0.1 to 1.0 cm/year over periods of 10^4 to 10^5 years and over areas of 10^4 to 10^6 km^2. One of the most significant observations in cratonic basins is that they exhibit the same exponential subsidence as ocean basins (Sleep et al., 1980). Their subsidence can be considered in two stages: in the first stage, the subsidence rate varies greatly, whereas the second stage subsidence is widespread. After about 50 My, the depth of subsidence decreases exponentially to a constant value.

Several models have been suggested to explain cratonic subsidence (Bott, 1979; Sleep et al., 1980). Sediment loading, lithosphere stretching, and thermal doming followed by contraction are the most widely cited mechanisms. Although the accumulation of sediments in a depression loads the lithosphere and causes further subsidence, calculations indicate that the contribution of sediment loading to subsidence must be minor compared with other effects. Subsidence at passive margins may result from thinning of continental crust by progressive creep of the ductile lower crust toward the sub-oceanic upper mantle. As the crust thins, sediments accumulate in overlying basins. Alternatively, the lithosphere may be domed by the upwelling asthenosphere or by a mantle plume, which erodes the uplifted crust. Thermal contraction following doming results in platform basins or a series of marginal basins around an opening ocean, which fill with sediments. Most investigators now agree that subsidence along passive continental margins is caused by the combined effects of thermal contraction of continental and adjacent oceanic lithosphere and of sediment loading.

Igneous rocks are rare in cratonic and passive margin basins; when found, they are small intrusive bodies, dykes, sills, or volcanic pipes, generally of alkaline compositions. **Kimberlites** are ultramafic breccias found in cratonic areas. They are significant in that they contain xenoliths of ultramafic rocks from the upper mantle and diamonds and other high-pressure minerals that indicate depths of origin for kimberlitic magmas of at least 300 km (Pasteris, 1984). They range in age from Archean to Tertiary, and most occur as pipes or dykes less than 1 km^2 in cross-sectional area. Kimberlite magmas appear to have been produced by partial melting of enriched sources in the subcontinental lithosphere.

Arc Systems

Subduction-Related Rock Assemblages

Numerous geologic environments are associated with subduction. In an idealized arc system, three zones are recognized: the arc–trench gap, the arc, and the arc-rear area (Hamilton, 1988; Stern, 2002). From the ocean side landward, a continental-margin arc is characterized by a trench, an accretionary prism with overlying forearc basins, a volcanic arc with intra-arc basins, a fold–thrust belt, and a retroarc foreland basin (Fig. 3.14a). An oceanic arc (Fig. 3.14b) differs from a continental-margin arc primarily in the arc-rear area, where it includes some combination of active and inactive back-arc basins and, in some instances, remnant arcs. Each of these arc environments is characterized by different rock assemblages described in the sections that follow.

Trenches

Trenches are formed where lithospheric slabs begin to descend into the mantle. Trench sediments are dominantly fine-grained graywacke turbidites with minor pelagic components. Turbidity currents generally enter trenches at oceanic canyons and flow along trench axes. Sediments can be transported along trenches for up to 3000 km—as in the Sunda trench south of Sumatra, where detritus from the Himalayas enters the trench on the north from the Bengal oceanic fan (Moore et al., 1982). Although most detrital sediment is clay or silt, sand and coarser sediments may be deposited in proximal facies. Down-axis changes in facies suggest that trenches are filled by a succession

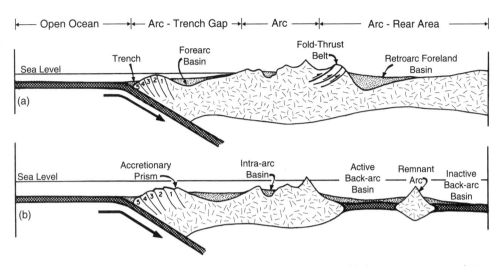

Figure 3.14   Diagrammatic cross-sections of a continental margin arc (a) and an island arc (b) showing major tectonic divisions. Numbers in accretionary prism indicate the relative age of slabs (one the oldest).

of radiating oceanic fans and that sediments are derived from local arc volcanic and plutonic sources.

Accretionary Prisms

An **accretionary prism** (or subduction-zone complex) consists of a series of steeply inclined, fault-bounded wedges of sediment and volcanic rocks above a descending slab. These wedges represent oceanic crust and trench sediments that have been accreted to the front of the arc. Individual wedges in the accretionary prism decrease in age as the trench is approached (Fig. 3.14). Accretionary prisms are intensely deformed, producing **melange,** a mappable body of rock characterized by the lack of continuous bedding and the inclusion of fragments of rocks of all sizes (to more than a kilometer across) contained in a fine-grained, deformed matrix. Both sedimentary and tectonic processes can give rise to melange. **Olistostromes** are melanges produced by gravitational sliding and accumulate as semifluid bodies that do not have bedding but include associated turbidites. Melange clasts may be exotic (derived from another environment) or native (reworked from the immediate environment) and are generally matrix supported (Fig. 3.15). Clast lithologies include graywacke, mudstone, chert, basalt (greenstone) and other ophiolite lithologies, arc volcanics, and rare granitoids. Melanges are commonly folded and may contain more than one cleavage or foliation. Sheared matrices are usually composed of serpentine and fine-grained rock and mineral fragments.

   Although melanges are typical of accretionary prisms, they are formed by several processes and occur in different tectonic environments (Cowan, 1986). Tectonic melanges are produced by compressive forces along the upper part of descending slabs at shallow depths. Fragmentation and mixing of largely lithified rocks may occur along a migrating

Figure 3.15 Franciscan melange near San Simeon, California. Large fragments of greenstone (metabasalt) (left) and graywacke (center) are enclosed in a sheared matrix of serpentine and chlorite. Courtesy of Darrel Cowan.

shear zone subparallel to a subducting slab. Fragments of oceanic crust and trench sediment are scraped off the descending plate and accreted to the overriding plate. Gravitational slumping may produce olistostromes or debris flows on oversteepened trench walls or along the margins of a forearc basin. Debris flows, in which clay minerals and water form a single fluid possessing cohesion, are probably the most important transport mechanism of olistostromes.

Forearc Basins

Forearc basins are marine depositional basins on the trench side of arcs (Fig. 3.14), and they vary in size and abundance with the evolutionary stage of an arc. In continental-margin arcs, such as the Sunda arc in Indonesia, forearc basins can be up to 700 km in strike length. They overlie the accretionary prism, which may be exposed as oceanic hills within and between forearc basins. Sediments in forearc basins, which are chiefly turbidites with sources in the adjacent arc system, can be many kilometers in thickness. Hemipelagic sediments are also of importance in some basins, such as in the Mariana arc. Olistostromes can form in forearc basins by sliding and slumping from locally steepened slopes. Forearc-basin clastic sediments may record progressive unroofing of adjoining arcs, as shown by the Great Valley Sequence (Jurassic–Cretaceous) in California (Dickinson and Seely, 1986). Early sediments in this sequence are chiefly volcanic detritus from active volcanics, and later sediments reflect progressive unroofing of the Sierra Nevada batholith. Volcanism is rare in modern forearc regions, and neither volcanic nor intrusive rocks are common in older forearc successions.

Arcs

Volcanic arcs range from entirely subaerial, such as the Andean and Middle America arcs, to mostly or completely oceanic, such as many of the immature oceanic arcs in the southwest Pacific. Other arcs, such as the Aleutians, change from subaerial to partly oceanic along the strike. Subaerial arcs include flows and associated pyroclastic rocks, which often occur in large stratovolcanoes. Oceanic arcs are built of pillowed basalt flows and large volumes of hyaloclastic tuff and breccia. Volcanism begins rather abruptly in arc systems at a volcanic front. Both tholeiitic and calc-alkaline magmas characterize arcs, with basalts and basaltic andesites dominating in oceanic arcs and andesites and dacites often dominating in continental margin arcs. Felsic magmas are generally emplaced as batholiths, although felsic volcanics are common in most continental-margin arcs.

Back-Arc Basins

Active back-arc basins occur over descending slabs behind arc systems (Fig. 3.14) and commonly have high heat flow, relatively thin lithosphere, and in many instances, an active ocean ridge enlarging the size of the basin (Jolivet et al., 1989; Fryer, 1996). Sediments are varied depending on basin size and nearness to an arc. Near arcs and remnant arcs, volcaniclastic sediments generally dominate, whereas in more distal regions, pelagic, hemipelagic, and biogenic sediments are widespread (Klein, 1986). During the early stages of basin opening, thick epiclastic deposits largely representing gravity flows are important. With continued opening of a back-arc basin, these deposits pass laterally into turbidites, which are succeeded distally by pelagic and biogenic sediments (Leitch, 1984). Discrete layers of air-fall tuff may be widely distributed in back-arc basins. Early stages of basin opening are accompanied by diverse magmatic activity, including felsic volcanism, whereas later evolutionary stages are characterized by an active ocean ridge. As previously mentioned, many ophiolites carry a subduction-zone geochemical signature and thus appear to have formed in back-arc basins.

Subaqueous ash flows may erupt or flow into back-arc basins and form in three principal ways (Fisher, 1984). The occurrence of felsic, welded ash-flow tuffs in some ancient back-arc successions suggests that hot ash flows enter water without mixing and retain enough heat to weld (Fig. 3.16a). Alternatively, oceanic eruptions may eject large amounts of ash into the sea, which falls onto the seafloor and forms a dense, water-rich debris flow (Fig. 3.16b). In addition to direct eruption, slumping of unstable slopes composed of pyroclastic debris can produce ash turbidites (Fig. 3.16c).

Because of the highly varied nature of modern back-arc sediments and the lack of a direct link between sediment type and tectonic setting, scientists cannot assign a distinct sediment assemblage to these basins. It is only when a relatively complete stratigraphic succession is preserved and detailed sedimentologic and geochemical data are available that ancient back-arc successions can be identified. Inactive back-arc basins, such as the western part of the Philippine plate, have a thick pelagic sediment blanket and lack evidence for recent seafloor spreading.

Remnant Arcs

Remnant arcs are oceanic aseismic ridges that are extinct portions of arcs rifted away by the opening of a back-arc basin (Fig. 3.14b) (Fryer, 1996). They are composed chiefly of subaqueous mafic volcanic rocks similar to those formed in oceanic arcs. Once isolated

Figure 3.16 Mechanisms for the origin of subaqueous ash flows (from Fisher, 1984). (a) Hot ash flow erupted on land flowing into water. (b) Ash flow forms from column collapse. (c) Ash turbidites develop from slumping of hyaloclastic debris.

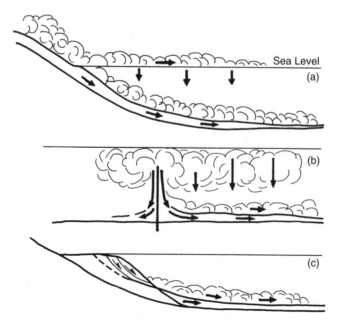

by rifting, remnant arcs subside and are blanketed by progressive deepwater pelagic and biogenic deposits and distal ash showers.

Retroarc Foreland Basins

Retroarc foreland basins form behind continental-margin arc systems (Fig. 3.14a), and they are filled largely with clastic terrigenous sediments derived from a fold–thrust belt behind the arc. A key element in foreland basin development is the syntectonic character of the sediments (Graham et al., 1986). The greatest thickness of foreland basin sediments borders the fold–thrust belt, reflecting enhanced subsidence caused by thrust-sheet loading and deposition of sediments. Another characteristic of retroarc foreland basins is that the proximal basin margin progressively becomes involved with the propagating fold–thrust belt (Fig. 3.17). Sediments shed from the rising fold–thrust belt are eroded and redeposited in the foreland basin only to be recycled again with basinward propagation of this belt. Coarse, arkosic alluvial-fan sediments characterize proximal regions of foreland basins and distal facies by fine-grained sediments and variable amounts of marine carbonates. Progressive unroofing in the fold–thrust belt should lead to an "inverse" stratigraphic sampling of the source in foreland basin sediments (Fig. 3.17). Such a pattern is well developed in the Cretaceous foreland basin deposits in eastern Utah (Lawton, 1986). In this basin, early stages of uplift and erosion deposited Paleozoic carbonate-rich, clastic sediments followed by quartz- and feldspar-rich detritus from the elevated Precambrian basement. Foreland basin successions also typically show upward coarsening and thickening terrigenous sediments, a feature that reflects progressive propagation of the fold–thrust belt into the basin.

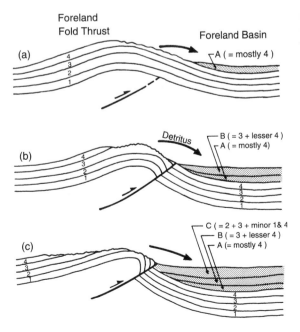

Figure 3.17 Progressive unroofing of an advancing foreland thrust sheet. Modified from Graham et al. (1986).

High- and Low-Stress Subduction Zones

Uyeda (1983) suggested that subduction zones are of two major types, each representing an end member in a continuum of types (Fig. 3.18). The relatively high-stress type, exemplified by the Peru–Chile arc, is characterized by a pronounced bulge in the descending slab, a large accretionary prism, relatively large shallow earthquakes, buoyant subduction (producing a shallow dipping slab), a relatively young descending slab, and a wide range in composition of calc-alkaline and tholeiitic igneous rocks (Fig. 3.18a). The low-stress type, of which the Mariana arc is an example, has little or no accretionary prism, few large earthquakes, a steep dip of the descending plate that is relatively old, a dominance of basaltic igneous rocks, and a back-arc basin (Fig. 3.18b). In the high-stress type, the descending and overriding plates are more strongly coupled than in the low-stress type, explaining the importance of large earthquakes and the growth of the accretionary prism. This stronger coupling, in turn, appears to result from buoyant subduction. In the low-stress type, the overriding plate is retreating from the descending plate, opening a back-arc basin (Scholz and Campos, 1995). In the high-stress type, however, the overriding plate is either retreating slowly compared with the descending plate or perhaps converging against the descending plate. Thus, the two major factors contributing to differences in subduction zones appear to be (1) relative motions of descending and overriding plates and (2) the age and temperature of the descending plate.

Figure 3.18 Idealized cross-sections of high- and low-stress subduction zones. Modified from Uyeda (1983).

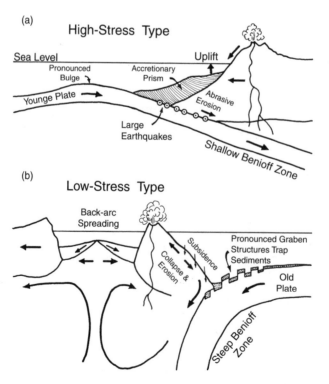

Arc Processes

Seismic reflection profiling and geological studies of uplifted and eroded arc systems have led to a greater understanding of arc evolution and of accretionary prism and fore-arc basin development (Stern, 2002). Widths of modern arc–trench gaps (75–250 km) are proportional to the ages of the oldest igneous rocks exposed in adjacent arcs (Dickinson, 1973). As examples, the arc–trench gap width in the Solomon Islands is about 50 km, with the oldest igneous rocks about 25 Ma, and the arc–trench gap width in northern Japan (Honshu) is about 225 km, with the oldest igneous rocks about 125 Ma. The correlation suggests progressive growth in the width of arc–trench gaps with time. Such growth appears to reflect some combination of outward migration of the subduction zone by accretionary processes and inward migration of the zone of maximum magmatic activity. **Subduction-zone accretion** involves the addition of sediments and volcanics to the margin of an arc in the accretionary prism (von Huene and Scholl, 1991). Seismic profiling suggests that accretionary prisms are composed of sediment wedges separated by high-angle thrust faults produced by the offscraping of oceanic sediment (Fig. 3.19). During accretion, oceanic sediments and fragments of oceanic crust and mantle are scraped off and added to the accretionary prism (Scholl et al., 1980). This offscraping results in outward growth of the prism and controls the location and evolutionary patterns of overlying forearc basins. Approximately half of modern arcs are growing because of offscraping accretion. Reflection profiles suggest deformational patterns are considerably more complex than simple thrust wedges. Deformation may include large-scale structural mixing and infolding of forearc basin sediments. Geological evidence for such mixing comes from exposed accretionary prisms, exemplified by parts of the Franciscan Complex in California. Fluids also play an important role in facilitating mixing and metasomatism in accretionary prisms (Tarney et al., 1991).

In addition to accretion to the landward side of the trench, material can be underplated beneath the arc by a process known as duplex accretion. A duplex is an imbricate package of isolated thrust slices bounded on top by a thrust and below by a low-angle detachment fault (Sample and Fisher, 1986). During transfer of displacement from an upper to a lower detachment horizon, slices of the footwall are accreted to the hanging wall (accretionary prism) and rotated by bending the frontal ramp. Observations from seismic reflection profiles, as well as exposed accretionary prisms, indicate that duplex accretion occurs at greater depths than offscraping accretion. Although some arcs, such as the Middle America and Sunda arcs, appear to have grown by accretionary processes, others, such as the New Hebrides arc, have little if any accretionary prism. In these latter arcs, either little sediment is deposited in the trench or most of the sediment is subducted. One way to subduct sediments is in grabens in the descending slab, a mechanism supported by the distribution of seismic reflectors in descending plates. Interestingly, if sediments are subducted in large amounts beneath arcs, they cannot contribute substantially to arc magma production as constrained by isotopic and trace element distributions in modern arc volcanics.

Subduction erosion is another process proposed for arcs with insignificant accretionary prisms. It involves mechanical plucking and abrasion along the top of a descending slab,

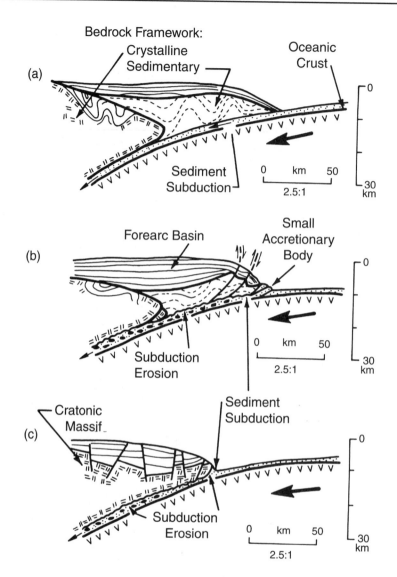

Figure 3.19   Sediment subduction and sediment erosion at a convergent plate boundary. Modified from Scholl et al. (1980).

which causes a trench's landward slope to retreat shoreward (Fig. 3.19). Subduction erosion may occur either along the top of the descending slab (Fig. 3.19b) or at the leading edge of the overriding plate (Fig. 3.19c). Evidence commonly cited for subduction erosion includes (1) an inland shift of the volcanic front, as occurred in the Andes in the last 100 My; (2) truncated seaward trends and seismic reflectors in accretionary prisms and forearc basins; (3) not enough sediment in trenches to account for the amount delivered by rivers; and (4) evidence for crustal thinning such as the tilting of unconformities toward the trench, most easily accounted for by subsidence of the accretionary prism. All of these can be explained by erosion along the top of the descending slab. Subduction erosion rates have been estimated along parts of the Japan and Chile trenches at 25 to 50 km^3/My for each kilometer of shoreline (Scholl et al., 1980).

Accretion, mixing, subduction erosion, and sediment subduction are all potentially important processes in subduction zones, and any of them may dominate at a given place and evolutionary stage. Studies of modern arcs indicate that about half of the ocean-floor sediment arriving at trenches is subducted and does not contribute to growth of accretionary prisms either by offscraping or by duplex accretion (von Huene and Scholl, 1991). At arcs with significant accretionary prisms, 70 to 80% of incoming sediment is subducted, and at arcs without accretionary prisms, all of the sediment is subducted. The combined average rates of subduction erosion (0.9 km^3/year) and sediment subduction (0.7 km^3/year) suggest that, on average, 1.6 km^3 of sediment are subducted each year.

High-Pressure Metamorphism

Blueschist-facies metamorphism is important in subduction zones, where high-pressure, relatively low-temperature mineral assemblages form. Glaucophane and lawsonite, both of which have a bluish color, are common minerals in this setting. In subduction zones, crustal fragments can be carried to great depths (>50 km) yet remain at rather low temperatures, usually less than 400° C (Fig. 2.10). A major unanswered question is how these rocks return to the surface. One possibility is by continual underplating of the accretionary prism with low-density sediments, resulting in fast, buoyant uplift during which high-density pieces of the slab are dragged to the surface (Cloos, 1993). Another possibility is that blueschists are thrust upward during later collisional tectonics.

One of the most intriguing fields of research at present examines how far crustal fragments are subducted before returning to the surface. Discoveries of coesite (high-pressure silica phase) and diamond inclusions in pyroxenes and garnet from eclogites from high-pressure metamorphic rocks in eastern China record astounding pressures of 4.3 gigapascals (GPa, about 150-km burial depth) at 740° C (Schreyer, 1995). Several other localities have reported coesite-bearing assemblages recording pressures from 2.5 to 3.0 GPa. Also, several new, high-pressure hydrous minerals have been identified in these assemblages, indicating that some water is recycled into the mantle and that not all water is lost by dehydration to the mantle wedge. Perhaps the most exciting aspect of these findings is that for the first time we have direct evidence that crustal rocks (both felsic and mafic) can be recycled into the mantle.

Igneous Rocks

The close relationship between active volcanism in arcs and descending plates implies a genetic connection between the two. Subduction-zone-related volcanism starts abruptly at the **volcanic front,** which roughly parallels oceanic trenches and begins 200 to 300 km inland from trench axes adjacent to the arc–trench gap (Fig. 3.20). It occurs where the subduction zone is 125 to 150 km deep, and the volume of magma erupted decreases in the direction of subduction-zone dip. The onset of volcanism at the volcanic front probably reflects the onset of melting above the descending slab, and the decrease in volume of erupted magma behind the volcanic front may be caused by either a longer vertical distance

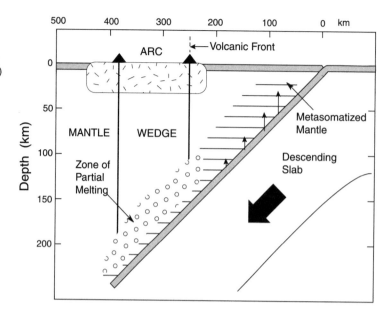

Figure 3.20
Cross-section of a subduction zone showing shallow devolatilization of descending slab (short vertical lines) and magma production in the mantle wedge.

for magmas to travel or a decrease in the amount of water liberated from the slab as a function of depth.

The common volcanic rocks in most island arcs are basalts and basaltic andesites; andesites and more felsic volcanics also become important in continental-margin arcs. Whereas basalts and andesites are erupted chiefly as flows, felsic magmas are commonly plinian eruptions in which much of the ejecta are ash and dust. These eruptions produce ash flows and associated pyroclastic (or hyaloclastic) deposits. Arc volcanic rocks are generally porphyritic, containing up to 50% phenocrysts in which plagioclase dominates.

Arc volcanoes are typically steep-sided stratovolcanoes composed of varying proportions of lavas and fragmental materials (Fig. 3.21). Their eruptions range from mildly explosive to violently explosive and contrast strikingly to the eruptions of oceanic-island and continental-rift volcanoes. Large amounts of water are given off during eruptions. Rapid removal of magmas may result in structural collapse of the walls of stratovolcanoes, producing calderas such as Crater Lake in Oregon. The final stages of eruption in some volcanic centers are characterized by the eruption of felsic ash flows that may travel great distances. Seismic shadow-zone studies indicate that modern magma reservoirs in subduction-zone areas are commonly 50 to 100 km deep. The migration of earthquake hypocenters from depths up to 200 km over periods of a few months before eruption reflects the ascent of magmas at rates of 1 to 2 km/day.

The cores of arc systems comprise granitic batholiths as shown in deeply eroded arcs. Such batholiths, composed of numerous plutons, range in composition from diorite to granite with granodiorite often dominating.

In contrast to oceanic basalts, arc basalts are commonly quartz normative, with high Al_2O_3 (16–20%) and low TiO_2 (<1%) contents. Igneous rocks of the tholeiite and calc-alkaline series are typical of both island arcs and continental-margin arcs. $^{87}Sr/^{86}Sr$ ratios in

Figure 3.21 Eruption of Mount St. Helens, southwest Washington, May 1980.

volcanics from island arcs are low (0.702–0.705), and those from continental-margin arcs are variable, reflecting variable contributions of continental crust to the magmas. Arc basalts also exhibit a subduction-zone component (Hawkesworth et al., 1994; Pearce and Peate, 1995) (depleted Nb and Ta relative to neighboring incompatible elements on a primitive mantle, normalized graph; Fig. 3.8). Arc granitoids are chiefly I-types, typically meta-aluminous, with tonalite or granodiorite dominating.

Compositional Variation of Arc Magmas

Both experimental and geochemical data show that most arc basalts are produced by partial melting of the mantle wedge in response to the introduction of volatiles (principally water) from the breakdown of hydrous minerals in descending slabs (Pearce and Peate, 1995; Poli and Schmidt, 1995) (Fig. 3.20). Other processes such as fractional crystallization, assimilation of crust, and contamination by subducted sediment, however, affect magma composition. Trace element and isotope distributions cannot distinguish between a subducted sediment contribution to arc magmas and a continental assimilation.

The very high Sr and Pb isotope ratios and low Nd isotope ratios in some felsic volcanics and granitic batholiths from continental-margin arcs, such as those in the Andes, suggest that these magmas were produced either by partial melting of older continental crust or by significant contamination with this crust.

One unsolved problem is that of how the subduction-zone geochemical component is acquired by the mantle wedge. Whatever process is involved, it requires decoupling of Ta-Nb, and in some cases Ti, from LIL elements and rare earth elements (REEs) (McCulloch and Gamble, 1991). Liberation of saline aqueous fluids from oceanic crust may carry LIL elements, which are soluble in such fluids, into the overlying mantle wedge, metasomatizing the wedge and leaving Nb and Ta behind (Saunders et al., 1980; Keppler, 1996). The net result is relative enrichment in Nb-Ta in the descending slab and corresponding depletion in the mantle wedge. Thus, magmas produced in the mantle wedge should inherit this subduction-zone signature. A potential problem with this model is that hydrous secondary minerals in descending oceanic crust (e.g., chlorite, biotite, amphiboles, and talc) break down and liberate water at or above 125 km (Fig. 3.20). Only phlogopite may persist to greater depths. Yet the volcanic front appears at subduction-zone depths of 125 to 150 km. Devolatilization of descending slabs by 125 km should add a subduction component only to the mantle wedge above this segment of the slab, which is shallower than the depth of partial melting. How then, do magmas acquire a subduction-zone component in the mantle wedge? Two possibilities have been suggested, although neither has been fully evaluated:

1. Asthenosphere convection extends into the "corner" of the mantle wedge and carries ultramafic rocks with a subduction component to greater depths (Fig. 3.20).
2. Frictional drag of the descending plate drags mantle with a subduction component to greater depths.

Orogens

Two Types of Orogens

Two types of orogens are recognized in the continental crust (Windley, 1992). **Collisional orogens** form during the collision of two or more cratons. When the direction of colliding plates is orthogonal, the crust is greatly thickened and thrusting, metamorphism, and partial melting may rework the colliding cratons. A relatively minor amount of juvenile crust is produced or tectonically "captured" during collisional orogeny. In contrast, **accretionary orogens** involve collision and the suturing of largely juvenile crustal blocks (ophiolites, island arcs, oceanic plateaus, etc.) to continental crust. Accretionary orogens contain relatively little reworked older crust.

Collisional Orogens

During continental collisions, major thrusts and nappes are directed toward the converging plate as the crust in a collisional zone thickens by ductile deformation and perhaps by underplating with mafic magmas. In some instances, sheet-like slabs commonly referred to as **allochthons** or flakes may be sheared from the top of the converging plate and thrust over the overriding plate (Fig. 3.22). In the eastern Alps, for instance, Paleozoic metamorphic rocks have been thrust more than 100 km north over the Bohemian Massif, and a zone of highly sheared Mesozoic metasediments lies within the thrust zone (Pfiffner, 1992). The allochthon is less than 12 km thick and appears to represent part of the Carnics plate detached during the mid-Tertiary. Seismicity along continent–continent collision boundaries suggests partial subduction of continental crust. Thickening of crust in collisional zones partially melts the lower crust, producing felsic magmas chiefly intruded as plutons with some surface eruptions. Fractional crystallization of basalts may produce anorthosites in the lower crust, and losses of fluids from the thickened lower crust may leave behind granulite-facies mineral assemblages. Isostatic recovery of colli-sional orogens is marked by the development of continental rifting with fluvial sedimen-tation and bimodal volcanism, as evident in the Himalayas and Tibet (Dewey, 1988).

To some extent, each collisional belt has its own character. In some instances, plates lock together with relatively little horizontal transport, such as along the Caledonian suture in Scotland or the Kohistan suture in Pakistan. In other orogens, such as the Alps and Himalayas, allochthons are thrust considerable distances and stacked one upon another. Hundreds of kilometers of shortening may occur during such a collision. In a few collisional belts, such as the Neoproterozoic Damara belt in Namibia, a considerable amount of deformation and thickening may occur in the overthrust plate. In the Caledonides and parts of the Himalayan belt, ophiolite obduction precedes continent–continent collision and does not seem to be an integral part of the collision process. In some cases, the collisional zone is oblique, with movement occurring along one or more transform faults. In other examples, one continent will indent another, thrusting thickened crust over the indented block. Thickened crust tends to spread by gravitational forces, and spreading directions need not parallel the regional plate movement. Which block has the greatest effects of deformation and metamorphism depends on such factors as the age of

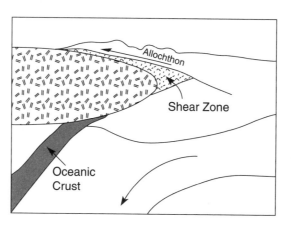

Figure 3.22 Schematic cross-section showing the emplacement of an allochthon during a continent–continent collision.

the crust, its thermal regime, the crustal anisotropy, and the nature of the subcrustal litho-sphere. Old lithosphere generally has greater strength than young lithosphere.

Accretionary Orogens

Accretionary orogens, such as the western Cordillera in Alaska and western Canada, develop as oceanic terranes such as island arcs, oceanic plateaus, and ophiolites collide with a continental margin. Collisions may occur between oceanic terranes, producing a superterrane, before collision with a continent. In some instances, older continental blocks may be involved in the collisions, such as those found in the 1.9-Ga Trans-Hudson orogen in eastern Canada and in the Ordovician Taconian orogen in the eastern United States. Seismic reflection profiles in western Canada show west-dipping seismic reflectors that are probably major thrusts, suggesting that accretionary orogens comprise stacks of thrusted terranes. Because accretionary orogens are chiefly made of juvenile crust, they represent new crust added to the continents during collisions and, as you shall see in Chapter 8 represent one of the major processes of continental growth. As an example, during the Paleoproterozoic, an aggregate area of new crust up to 1500 km wide and more than 5000 km long was added to southern margin the Baltica–Laurentia supercontinent (see Fig. 2.22 in Chapter 2) (Hoffman, 1988; Karlstrom et al., 2001). One of the major questions I will address in later chapters is how this accreted mafic crust changes into continental crust.

Orogenic Rock Assemblages

It is difficult to assign any particular rock assemblage to collisional and accretionary orogens, because rock assemblages change with time and space as collision progresses. Also, scientists are faced with sorting out a myriad of older rock assemblages contained in the colliding blocks and representing every conceivable tectonic setting. Sediments accumulate in peripheral foreland and hinterland basins, which develop in response to uplift and erosion of a collisional zone. These basins and the sediments therein evolve in a manner similar to retroarc foreland basins. Classic examples of peripheral foreland basins developed adjacent to the Alps and the Himalayas during the Alpine–Himalayan collisions in the Tertiary (Fig. 3.23). During the Alpine collision, up to 6 km of alluvial fan deposits were left in foreland basins (Homewood et al., 1986). Individual alluvial fans up to 1 km thick and 40 km wide have been recognized in the Alps. Coarsening upward cycles and intraformational unconformities characterize collisional deposits, both of which reflect uplift of an orogen and propagation of thrusts and nappes into foreland basins.

At deeper exposure levels (10–20 km) in collisional orogens, granitoids are common and appear to be produced by partial melting of the lower crust during a collision. Thickening of continental crust, in descending and overriding plates, produces granulites at depths greater than 20 km as fluids escape upward. Anorthosites may form as cumulates from fractional crystallization of basalt in the lower and middle crust.

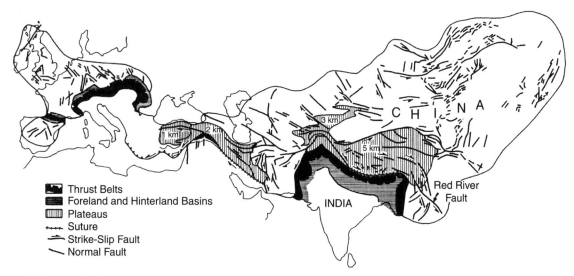

Figure 3.23   Simplified tectonic map of the Alpine–Himalayan orogen. Collisional plateau elevations are given in kilometers. 1 km, Anatolia Plateau; 5 km, Tibet Plateau. Modified from Dewey et al. (1986).

Basaltic magma also may underplate the crust and occurs as gabbro or mafic granulites in uplifted crustal sections. Collisional granites include precollisional, syncollisional, and postcollisional types (Pearce et al., 1984). Pre- and postcollisional granitoids are chiefly I-type granites, whereas syncollisional granites are commonly leucogranites (in collisional orogens) and many exhibit features of S-type granites (i.e., derived by partial melting of sediments). Silica contents of leucogranites exceed 70% and most show a subduction geochemical component inherited from their lower crustal source. Supporting a crustal source for many collisional granitoids are their high $^{87}Sr/^{86}Sr$ ratios (>0.725) and high $\delta^{18}O$ values (Vidal et al., 1984). Postcollisional granites, which generally postdate collision by 40 to 50 My, are sharply crosscutting and are dominantly tonalite to granodiorite in composition.

Tectonic Elements of a Collisional Orogen

Continental collision involves progressive compression of buoyant terranes within subduction zones. These terranes may vary in scale from seamounts or island arcs to large continents. The scale of colliding terranes dictates the style, duration, intensity, and sequence of strain systems (Dewey et al., 1986). If colliding continental margins are irregular, the strain sequences are variable along great strike lengths. Prior to terminal collision, one or both continental margins may have had a long, complex history of terrane assembly. Continental collisional boundaries are wide, complicated structural zones, where plate displacements are converted into complex and variable strains (Fig. 3.24).

Collisional orogens can be considered in terms of five tectonic components (Figs. 3.24 and 3.25): thrust belts, foreland flexures, plateaus, widespread foreland–hinterland

Figure 3.24 Schematic cross-sections of Alpine (a) and Himalayan (b) collisional orogens. Modified from Dewey et al. (1986). M, Mohorovicic discontinuity.

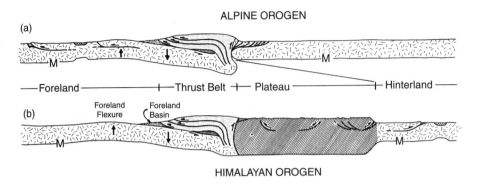

deformational zones, and zones of orogenic collapse. **Foreland** and **hinterland** refer to regions beyond major overthrust belts in the direction and from the direction of principal orogenic vergence, respectively. Thrust belts develop where thinned continental crust is progressively restacked and thickened toward the foreland. If detachment occurs along a foreland thrust, rocks in the allochthon can shorten significantly independent of shortening in the basement. The innermost nappes and the suture zone are usually steepened and overturned during advanced stages of collision. The upper crust in collisional orogens is a high-strength layer that may be thrust hundreds of kilometers as relatively thin, stacked sheets that merge along a decollement surface. For instance, in the southern Appalachians,

Figure 3.25 Schematic cross-sections of Tibet showing three tectonic models for the uplift of the Tibet Plateau.

(a) Underthrusting of Continental Lithosphere

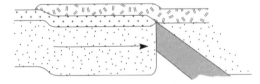

(b) Compressional Thickening of the Lithosphere

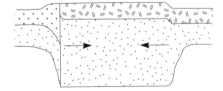

(c) Delamination of the Lithosphere

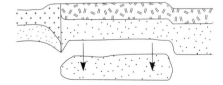

a decollement has moved westward over the foreland for at least 300 km. Foreland flexures are upward warps in the foreland lithosphere caused by progressive bending downward of the lithosphere by advancing thrust sheets.

Collisional plateaus, such as the Tibet and Anatolia plateaus (Fig. 3.24), form in the near-hinterland area adjacent to the suture and may rise to between 1 and 5 km above sea level. Three models have been suggested to explain collisional plateaus (Harrison et al., 1992) (Fig. 3.25): (1) underthrusting of continental crust and lithosphere, which buoyantly elevates the plateau; (2) compressive shortening and thickening of both the crust and lithosphere, followed by isostatic rebound to form the plateau; and (3) either model 1 or model 2 followed by delamination of the thickened lithosphere. Neither model 1 nor model 2 adequately explains why the delayed uplift of Tibet began in 20 Ma yet the India–Tibet collision began about 55 Ma. However, if mantle lithosphere was detached some 30 My after collision and sank into the mantle, it would be replaced with hotter asthenosphere, which would immediately cause isostatic uplift, elevating Tibet.

The widespread foreland–hinterland deformational zones are evidence that stresses generated by collision can affect continental areas thousands of kilometers from the thrust belt (Fig. 3.23). For instance, the Tien Shan Range in central Asia appears to have formed in response to stresses transmitted across the continent from continued postcollisional indentation of India into Eurasia. The preferential hinterland deformation in Asia associated with the India–Tibet collision probably reflects warm, thin Tibetan lithosphere produced by precollisional subduction and accretion. Extensional collapse zones, such as the Aegean basin, are generally of local occurrence in foreland areas and may develop in response to rollback of a partially subducted slab.

Sutures

Sutures are ductile shear zones produced by thrusting along converging plate boundaries, and they range from a few hundred meters to tens of kilometers wide (Coward et al., 1986). Rocks on the hinterland plate adjacent to sutures are chiefly arc-related volcanics and sediments from a former arc system on the overriding plate, whereas those on the foreland plate are commonly passive-margin sediments. Fragments of rocks from both plates and ophiolites occur in suture melanges. These fragments are tens to thousands of meters in size and are randomly mixed in a sheared, often serpentine-rich, matrix.

One of the problems in recognizing Precambrian collisional orogens is the absence of well-defined suture zones. However, at the crustal depths exposed in most reactivated Precambrian terranes, suture zones are difficult to distinguish from other shear zones. The classic example of a Cenozoic suture is the Indus suture in the Himalayas. Yet in the Nanga Parbat area where deep levels of this suture are exposed, it is difficult to identify the suture because it looks like any other shear zone (Coward et al. 1982). Rocks on both sides of the suture are complexly deformed amphibolites and gneisses indistinguishable from each other. If you compare equivalent crustal levels of the Indus suture with those of exhumed Precambrian orogens, there are striking similarities, and deep-seated shear zones in these orogens may or may not be sutures. Without precise geochronology and detailed geologic mapping on both sides of shear zones, it is not possible to correctly identify which shear zones are sutures and which are not.

Foreland and Hinterland Basins

Peripheral foreland and hinterland basins are like retroarc foreland basins in terms of sediment provenance and tectonic evolution. Major peripheral foreland and hinterland basins developed in the Tertiary in response to the Alpine–Himalayan collisions (Allen and Homewood, 1986) (Fig. 3.23). These basins exhibit similar stages of development that have been explained by thermal–mechanical models (Stockwell et al., 1986). In the case of foreland basins, during the first stage, a passive-margin sedimentary assemblage is deposited on a stretched and rifted continental margin. Collision begins as a terrane is thrust against and over continental crust on the descending slab. Continental convergence thickens this terrane and partially melts the root zones to form syntectonic granite magmas. Topographic relief develops at this stage as the thrust belt rises above sea level and sediments derived from erosion of the highland begin to fill a foreland basin. The first sediments come from distal low-relief terranes and are largely fine-grained, giving rise to deepwater marine shales and siltstones exposed in the lowest stratigraphic levels of foreland basin successions. In contrast, in hinterland basins, the early sediments are commonly alluvial fan deposits shed from the rising mountain range. Continued convergence causes thin-skinned thrust sheets to propagate into foreland basins, and relief increases rapidly. Erosion rates increase as do grain size and feldspar content of derivative sediments in response to increased rates of tectonic uplift in the thrust belt. During this stage of development, thick alluvial fans also may be deposited in foreland basins.

The Himalayas

As an example of a young collisional mountain range, none can surpass the High Himalayas. The Himalayan story began some 80 Ma when India fragmented from Gondwana and started on its collision course with eastern Asia. Collision began about 55 Ma and is still going on today. Prior to collision, Tibet was a continental-margin arc system with voluminous andesites and felsic ash-flow tuffs, and northern India was a passive continental margin with a marine shelf-facies on the south, passing into a deepwater Tethyan facies on the north. As the collision began, folds and thrusts moved southward onto the Indian plate (Searle et al., 1987). This resulted in thickening of the crust, high-pressure metamorphism, and partial melting of the root zones to produce migmatites and leucogranites. Thrusting continued on both sides of the Indus suture as India continued to converge on Tibet. By 40 Ma, deformation had progressed southward across both the Lower and Higher Himalayan zones (Fig. 3.26). Collapse of the Lower and Sub-Himalayas during the Miocene juxtaposed lower Paleozoic shelf sediments north of the Main Boundary thrust against metamorphic rocks and leucogranites south of the thrust. Continued convergence of the two continental plates led to oversteepening of structures in the Indus suture, and finally to backthrusting on to the Tibet plate and continued

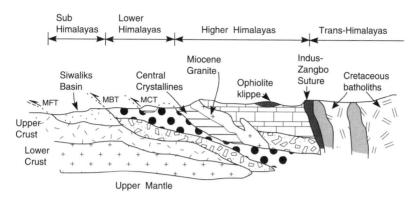

Figure 3.26 Schematic cross-section of the Himalayas in Nepal and Tibet. MBT, Main Boundary thrust; MCT, Main Central thrust; MFT, Main Frontal thrust. Modified from Windley (1983).

southward-directed thrusting. The Siwalik foreland basin continued to move to the south in the Sub-Himalayas as thrust sheets advanced from the north.

The amount of crustal shortening recorded across the Himalayan orogen is almost 2500 km for a time-averaged compression rate of about 5 cm/year (Searle et al., 1987). However, the crust of Tibet is only about 70 km thick, which can account for only about 1000 km of shortening. The remainder appears to have been taken up by transcurrent faulting north of the collision zone (Tapponnier et al., 1986; Windley, 1995). The Tertiary geological record in Southeast Asia required 1000 to 1500 km of cumulative strike-slip offset in which India has successively pushed Southeast Asia followed by Tibet and China in an east–southeast direction. Most of the mid-Tertiary displacement has occurred along the left-lateral Red River fault zone (Fig. 3.23), accompanying the opening of the South China Sea.

Most models for the India–Tibet collision involve buoyant subduction of continental crust. Earthquake data from the Himalayas suggest a shallow (~3 degrees) northward-dipping detachment zone extending beneath the foreland basins on the Indian plate (Fig. 3.26). This detachment surface is generally interpreted as the top of the descending Indian plate and may have a surface expression as the Main Frontal thrust. Convergence since the early Miocene has been taken up chiefly along the Main Central thrust and its splays by counterclockwise rotation of India beneath Tibet.

Uncertain Tectonic Settings

Anorogenic Granites

General Features

A wide belt of Proterozoic granites and associated anorthosites extends from southwestern North America to Labrador, across southern Greenland and into the Baltic shield in Scandinavia and Russia. The granites are massive and relatively undeformed, thus the name *anorogenic* granite (Anderson and Morrison, 1992; Windley, 1993). Many have rapakivi textures and other features of A-type granitoids. Two important field observations for anorogenic granites are that (1) they occur chiefly in accretionary (juvenile) orogens and (2) often there is a close spatial and temporal relation between granite magmatism and crustal extension. Proterozoic anorogenic granites in the Laurentia–Baltica belt range in age from about 1.0 to 1.8 Ga. Most of those in North America are 1.5 to 1.4 Ga and tend to increase from 1.44 to 1.43 Ga in the southwestern United States to 1.48 to 1.46 Ga in the midcontinent area (Fig. 3.27). Major subprovinces of 1.40 to 1.34 Ga and 1.50 to 1.42 Ga granites occur in the midcontinent region. The largest and oldest anorogenic granites in this Proterozoic belt occur in Finland and Russia and date from 1.80 to 1.65 Ga (Haapala and Ramo, 1990). Large anorthosite bodies are associated with some anorogenic granites, and most occur in the Grenville province and adjacent areas in eastern Canada (Fig. 3.27). Although Mesoproterozoic anorogenic granites have received the most attention, granites with similar field and geochemical characteristics are known in the Archean and Phanerozoic, some of the youngest of which are in the American Cordillera.

Proterozoic anorogenic granites are A-type granites enriched in K and Fe and depleted in Ca, Mg, and Sr relative to I- and S-type granitoids. They are subalkalic to marginally peraluminous and plot near the minimum in the Q-Ab-Or system at 5 to 10 kilo bars (kb) of pressure (Anderson, 1983). Anorogenic granites are typically enriched in REE, Zr, and Hf and have striking depletions in Sr, P, and Ti compared with most other granites. In addition, they appear to have been emplaced under relatively dry conditions at temperatures from 650 to 800°C and depths of chiefly less than 15 km. Rapakivi textures

Figure 3.27 The Proterozoic anorogenic granite–anorthosite belt in North America. Modified from Anderson (1983).

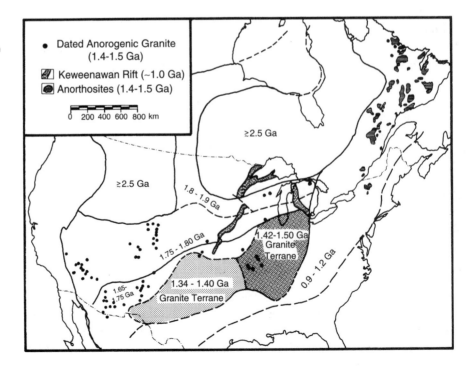

may have developed by volatile losses at shallow depths during emplacement. Another characteristic feature of anorogenic granites is that they crystallized over 3 orders of magnitude of oxygen fugacity as reflected by their Fe-Ti oxide mineralogy. Their relatively high initial $^{87}Sr/^{86}Sr$ ratios (0.705 ± 0.003) and negative or near zero E_{Nd} values are consistent with a lower crustal source, as are incompatible-element and oxygen isotope distributions.

Associated Anorthosites

Associated anorthosites, composed of more than 90% plagioclase (An_{45-55}), are interlayered with gabbros and norites and exhibit cumulus textures and rhythmic layering. Most bodies range from 10^2 to 10^4 km^2 in surface area. Gravity studies indicate that most anorthosites are from 2 to 4 km thick and are sheet-like in shape, suggesting that they represent portions of layered igneous intrusions. The close association of granites and anorthosites suggests a genetic relationship. Geochemical and isotopic studies, however, indicate that the anorthosites and granites are not derived from the same parent magma by fractional crystallization or from the same source by partial melting. Data are compatible with an origin for the anorthosites as cumulates from fractional crystallization of high-Al_2O_3 tholeiitic magmas produced in the upper mantle (Emslie, 1978). The granitic magmas, on the other hand, appear to be the products of partial melting of lower crustal rocks of intermediate or mafic composition (Anderson and Morrison, 1992) or the products of fractional crystallization of basaltic magmas (Frost and Frost, 1997).

Tectonic Setting

The tectonic setting of anorogenic granites continues to baffle geologists. Unlike most other rock assemblages, young counterparts of anorogenic granites have not been recognized. The general lack of preservation of supracrustal rocks further hinders identifying the tectonic setting of these granites. The only well-documented outcrops of coeval supracrustal rocks and anorogenic granite are in the St. Francois Mountains of Missouri, where felsic ash-flow tuffs and calderas appear to represent surface expressions of granitic magmas. Both continental rift and convergent-margin models have been proposed for anorogenic granites, and both models have problems. The incompatible element distributions in most anorogenic granites are suggestive of a tectonic setting within plates. If an extensive granite–rhyolite province of the late Paleozoic to Jurassic age in Argentina represents an anorogenic granite province (Kay et al., 1989), this would favor a back-arc continental setting. The intrusion of anorogenic granites in the Mesoproterozoic belt in Laurentia–Baltica described previously usually follows the main deformational events in this belt by 60 to 100 My. This may reflect the time it takes to heat the lower crust to the point at which it begins to melt. Although the current database seems to favor an extensional regime for anorogenic granites ($\pm$ anorthosites), we cannot draw a strict parallel to modern continental rifts. It would appear that an event that transcended time from about 1.9 to 1.0 Ga is responsible for the Proterozoic anorogenic granite belt in Laurentia–Baltica.

Perhaps this represents the movement of a supercontinent over one or more mantle plumes that heated the lower crust. In any case, it seems clear that Proterozoic-style anorogenic magmatism is not a one-time event but that it has occurred in the late Archean and perhaps many times in the post-Archean.

Archean Greenstones

Although at one time it was thought that greenstones were an Archean phenomenon, it is now clear that they have formed throughout geologic time (Condie, 1994). It is equally clear that all greenstones do not represent the same tectonic setting, nor do the proportions of preserved greenstones of a given age and tectonic setting necessarily reflect the original proportions of that tectonic setting. The term *greenstone* has been used rather loosely in the literature, so herein I will define **greenstone belt** as a supracrustal succession in which the combined mafic volcanic and volcaniclastic sediment component exceeds 50%. Thus, from a modern perspective, greenstones are volcanic-dominated successions that have formed in arcs, oceanic plateaus, volcanic islands, and oceanic crust. It is now known that greenstones contain various packages of supracrustal rocks separated by unconformities or faults (Thurston and Chivers, 1990; Williams et al., 1992). Greenstone belts are linear- to irregular-shaped, volcanic-rich successions that average 20 to 100 km wide and extend several hundred kilometers. They contain several or many greenstone assemblages or domains, and in this sense, you might equate a greenstone belt to a terrane or more specifically to an oceanic terrane. As an example, the largest preserved greenstone belt, the late Archean Abitibi belt in eastern Canada, contains several greenstone domains and hence can be considered a terrane or, more accurately, a superterrane amalgamated around 2.7 Ga (Fig. 3.28). Subprovinces in the Archean Superior province, such as the Wawa-Abitibi and Wabigoon subprovinces (Fig. 3.29), can be considered superterranes and represent amalgamations of greenstone terranes of various oceanic settings. High-precision U-Pb zircon isotopic dating of Archean greenstone terranes indicates that they formed in short periods of time, generally less than 50 My. In some areas, more than one volcanic–plutonic cycle may be recorded for a cumulative history of 200 to 300 My. Although Archean greenstones can be described in terms of terranes, their tectonic settings continue to be a subject of lively debate among Archean investigators. Let me review some of the principal features of Archean greenstones that are useful in constraining tectonic settings.

General Features

Although field evidence clearly indicates that most greenstones are intruded by surrounding granitoids, there are some areas in which greenstone successions lie unconformably on older granitic basement (Condie, 1981; de Wit and Ashwal, 1995). In a few greenstone belts, such as Kambalda in southwest Australia, volcanic rocks contain zircon xenocrysts from gneissic basement at least 700 My older than the host rocks. Thus, although many Archean greenstones are clearly juvenile oceanic terranes, some were erupted on or close to continental crust and may be contaminated by this crust.

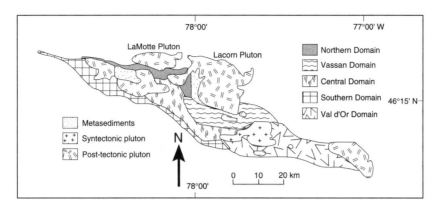

Figure 3.28 Schematic map of part of the late Archean Abitibi greenstone belt in southeast Ontario, Canada, showing tectonic domains. Modified from Descrochers et al. (1993).

In Canadian Archean greenstones, four lithologic associations are recognized (Thurston and Chivers, 1990; Thurston, 1994). Most widespread are the basalt–komatiite (mafic plain) and mafic to felsic volcanic cycle associations comprising most of the major greenstone belts in Canada. These two associations, which are also the most common associations on other continents, appear to represent, respectively, oceanic plateau and volcanic arc settings (Condie, 1994). Of more local importance are the calc-alkaline volcanic and fluvial sediment association and the carbonate–quartz arenite association, the latter of which is volumetrically insignificant. In the Archean Superior province in eastern Canada, greenstone subprovinces alternate with metasedimentary subprovinces (Fig. 3.29). Granitoids are more abundant than volcanic and sedimentary rocks in both subprovinces with gneisses and migmatites most abundant in the metasedimentary belts. U-Pb zircon ages indicate younging in volcanism and plutonism from the northwest (Sachigo subprovince) to the southeast (Wawa-Abitibi subprovinces). The oldest magmatic events in the northwest occurred at 3.00, 2.90 to 2.80, and 2.75 to 2.70 Ga followed by major deformation, metamorphism, and plutonism around 2.70 Ga. In the south, magmatism occurred chiefly between 2.75 and 2.70 Ga. The near contemporaneity of magmatic and deformational events along the lengths of the volcanic subprovinces, coupled with structural and geochemical evidence, supports a subduction-dominated tectonic regime in which oceanic terranes were successively accreted from northwest to southeast.

Greenstone Volcanics

Archean greenstones are structurally and stratigraphically complex. Although stratigraphic thicknesses up to 20 km have been reported, because of previously unrecognized tectonic duplication, it is unlikely that any sections exceed 5 km (Condie, 1994). Most Archean greenstones are composed chiefly of subaqueous basalts (Fig. 3.30) and komatiites (ultramafic volcanics) with minor amounts of felsic tuff and layered chert. Many Archean volcanics are highly altered and silicified, probably from oceanic hydrothermal fluids in a manner similar to that characteristic of modern oceanic arcs and ocean ridges.

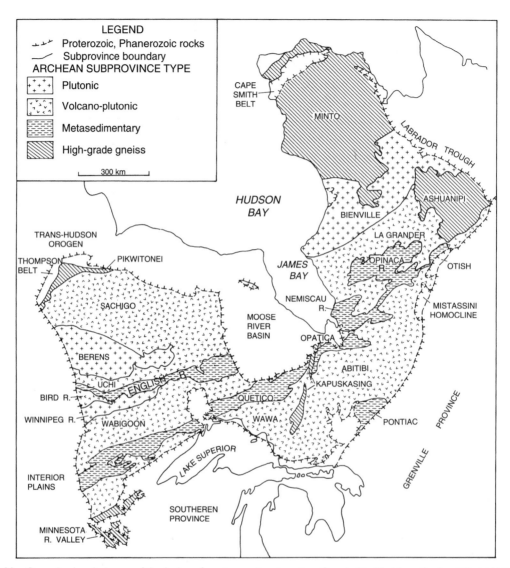

Figure 3.29 Generalized geologic map of the Archean Superior province in eastern Canada. Modified from Card and Ciesielski (1986).

Three general trends observed with increasing stratigraphic height in some late Archean greenstone successions are (1) a decrease in the amount of komatiite, (2) an increase in the ratio of volcaniclastics to flows, and (3) an increase in the relative abundance of andesitic and felsic volcanics. These changes reflect an evolution from voluminous oceanic eruptions of basalt and komatiite, commonly referred to as a **mafic plain** (oceanic plateau), to more localized calc-alkaline and tholeiitic stratovolcanoes (volcanic arc) that may emerge with time and to intervening sedimentary basins.

Archean volcanoes were in some respects similar to modern oceanic volcanoes in arc systems (Ayes and Thurston, 1985). Similarities include (1) a general upward change

Figure 3.30   Pillows in Archean basalts from the Abitibi greenstone belt in southern Ontario.

from basalts to calc-alkaline and tholeiitic volcanics, (2) the eventual emergence of sub-aqueous volcanoes to form islands, (3) a linear alignment of volcanoes, and (4) flanking sedimentary aprons leading into basins between volcanoes. Differences between modern arc volcanoes and Archean volcanoes include (1) the occurrence of komatiites in many Archean volcanoes; (2) the bimodal character of some Archean volcanics, especially in older greenstone successions; (3) the paucity of Archean shoshonitic volcanics; and (4) the thick, laterally continuous oceanic flows that collectively form large oceanic mafic plains upon which Archean volcanoes grew.

Perhaps the most distinctive volcanic rock of the Archean is komatiite (Arndt, 1991). **Komatiites** are ultramafic lava flows (or hypabyssal intrusives) that exhibit a quench tex-ture known as spinifex texture and contain more than 18% MgO. Although komatiites are common in mafic plain Archean greenstone successions, they are uncommon in the Proterozoic and rare in the Phanerozoic. Spinifex texture is commonly preserved in the upper parts of komatiite flows. This texture is characterized by randomly oriented skeletal crystals of olivine or pyroxene (Fig. 3.31) and forms by rapid cooling in the near-absence

Figure 3.31 Spinifex texture in an Archean komatiite from the Barberton greenstone in South Africa (x 6). Radiating crystals are serpentine pseudomorphs after olivine.

of crystal nuclei. Archean komatiite and basalt flows range from about 2 m to more than 200 m thick, are commonly pillowed, and may be associated with sills or volcaniclastic rocks of similar composition. Intrusive mafic and ultramafic igneous rocks are also found in most greenstone successions, and geochemical studies indicate that they are closely related to enclosing volcanic rocks. Archean komatiitic and mafic volcanics are depleted in LIL elements and show variable amounts of light-REE depletion. Although Archean mafic plain basalts have incompatible element distributions similar to modern oceanic plateaus, Archean calc-alkaline basalts (associated with andesites and felsic volcanics) show a distinct subduction-zone geochemical component (Condie, 1989; Condie, 1994).

Andesites and felsic volcanic rocks in greenstone successions occur chiefly as volcaniclastic rocks and minor flows. Tuffs, breccias, and agglomerates are common, and their distribution can be used to define volcanic centers. Archean andesites are similar to modern andesites from volcanic arcs in incompatible element distributions. Alkaline volcanic rocks are rare in Archean greenstone successions and, when found, occur chiefly as volcaniclastics and associated hypabyssal intrusives.

Greenstone Sediments

Four types of sediments are recognized in Archean greenstones (Mueller, 1991; Lowe, 1994b). In order of decreasing importance, they are volcaniclastic, chemical, biochemical, and terrigenous sediments. Volcaniclastic sediments (including graywackes) range from mafic to felsic in composition and are common as turbidites. Chemical and biochemical sediments, principally chert, banded iron formation, and carbonate, are of minor importance but of widespread distribution, and terrigenous sediments such as shale, quartzite, arkose, and conglomerate are only of local significance. Although plutonic sources are of local importance in some graywackes, most greenstone detrital sediments are clearly derived from nearby volcanic sources. Layered chert and banded iron formation are the most important nonclastic sediments in greenstones. The earliest evidence of life on Earth occurs in the form of microstructures in Archean cherts. Relict detrital textures in some cherts, however, indicate that they are chertified clastic sediments or volcanics. Few, if any, Archean cherts appear to represent pelagic sediments. Most are local and are probably hydrothermal vent deposits. Low $\delta^{18}O$ values for many Archean cherts support a volcanogenic origin. Some cherts in early Archean greenstones contain casts of gypsum crystals (and in some instances, halite), suggesting a shallow-water evaporite origin. Carbonates and barite are minor components in some greenstone belts. The barites appear to represent hydrothermal deposits and the carbonates, either hydrothermal or evaporite-related deposits.

Five sedimentary environments are recognized in Archean greenstones (Lowe, 1994a). In some greenstones, such as the northern volcanic zone in the Abitibi belt, more than one sedimentary environment occurs in the same succession (Fig. 3.28; 3.32). The most widespread environment, especially in early Archean greenstones, is the mafic plain environment. In this setting, large volumes of basalt and komatiite were erupted to form a mafic plain, which may have risen to shallow depths beneath sea level or even became emergent. Sediments in this setting include hyaloclastic sediments, chert, banded iron formation, carbonate, and—locally—shallow-water evaporites and barite. These rocks, which commonly preserve

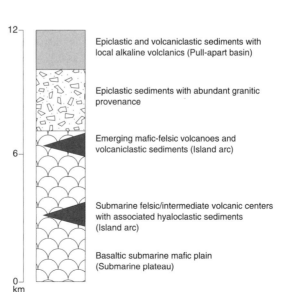

Figure 3.32 Generalized stratigraphic section of the Northern Volcanic Zone in the Abitibi greenstone belt, Quebec, showing the major lithologic assemblages. Modified from Lowe (1994a).

primary textures such as mud cracks, oolites, and gypsum casts, indicate shallow-water deposition. A second environment is a deepwater, nonvolcanic environment in which chemical and biochemical chert, banded iron formation, and carbonate are deposited. The third association, a graywacke–volcanic association widespread in late Archean greenstones, and often stratigraphically on top of the mafic plain succession, is composed chiefly of graywackes and interbedded calc-alkaline volcanics and may have been deposited in or near island arcs. Fluvial and shallow-marine detrital sediments probably deposited in pull-apart basins and mature sediments (quartz arenites, etc.) deposited in continental rifts are the last two environments. These two environments are generally not volumetrically important in most Archean greenstones, although the pull-apart basin setting is widespread.

Granitoids

Granitoids associated with Archean greenstones fall into one of three categories: gneissic complexes and batholiths, diapiric (syntectonic) plutons of variable composition, and late discordant granite plutons (Martin, 1995; Sylvester, 1994). Gneissic complexes and batholiths, which are dominantly tonalite, trondhjemite, and granodiorite, the so-called **TTG suite,** comprise most of the preserved Archean crust. They contain large, infolded remnants of supracrustal rocks and numerous inclusions of surrounding greenstone belts. Contacts between TTG complexes and greenstones are generally intrusive and usually strongly deformed. Granitic plutons range from foliated to massive and discordant to concordant; some are porphyritic. Geophysical studies indicate that most Archean plutons extend to depths less than 15 km. Diapiric plutons have well-developed concordant foliation around their margins and appear to have been deformed during forceful injection. Others, which are posttectonic, are usually granite *(sensu strictu)* in composition and have discordant contacts and massive interiors. Most of these are posttectonic (possibly anorogenic) plutons with A-type characteristics.

All but the posttectonic granites are geochemically similar to felsic volcanics in greenstones. With their common syntectonic mode of emplacement, this fact suggests that most granitoids and felsic volcanics are genetically related. In areas where detailed U-Pb zircon chronology is available, such as in parts of the Superior Province, entire cycles of volcanism, sedimentation, deformation, and plutonism occur in less than 50 My, and late discordant granites are generally emplaced during the following 50 My. In some early Archean greenstones (Pilbara, Barberton) plutonism can continue up to 500 My after major volcanism.

Geochemically, the TTG suite and most diapiric plutons are I-type granitoids, and they are distinct from most of their post-Archean counterparts in that they are extremely depleted in heavy REE (Martin, 1995). Most show a subduction geochemical component and no significant Eu anomalies. Geochemical and experimental data favor an origin for these rocks as partial melts of garnet amphibolites or eclogites. Archean posttectonic and anorogenic granites are calc-alkaline, strongly peraluminous, and commonly have A- or S-type affinities (Sylvester, 1995). They appear to be the products of partial melting of igneous and sedimentary rocks in the lower crust.

Mineral and Energy Deposits

Plate tectonics provides a basis for understanding the distribution and origin of mineral and energy resources in space and time (Rona, 1977; Sawkins, 1990). The occurrence of energy and mineral deposits can be related to plate tectonics in three ways:

1. Geological processes driven by energy liberated at plate boundaries control the formation of energy and mineral deposits.
2. Deposits form in specific tectonic settings, which are controlled by plate tectonics.
3. Reconstruction of fragmented supercontinents can be used in the exploration for new mineral and energy deposits.

In attempting to relate mineral and energy deposits to plate tectonics, it is important to know the relationship between the deposits and their host rocks. If deposits are syngenetic with host rocks, then they formed in the same tectonic setting. If, however, mineral and energy deposits are secondary in origin and entered the host rocks later (such as by hydrothermal fluids or migration of oil), they may have formed in a tectonic setting different from the host rocks. Major *in situ* mineral and energy deposits in various tectonic settings are summarized in Table 3.1.

Mineral Deposits

Ocean Ridges

Employing deep-sea photography and using small submersibles, numerous hydrothermal fields have been identified on ocean ridges (Rona et al., 1986; Von Damm, 1990). The submersibles allow direct observations and measurements on the seafloor. On the Galapagos and Juan De Fuca ridges, hydrothermal veins ranging from 500 to 1600 m^2 are spread along fissure systems for 500 to 2500 m. The TAG hydrothermal field on the mid-Atlantic ridge (26° N) occurs along a fault zone on the east wall of the medial rift valley, and vent waters have temperatures up to 300° C. Animal communities living near the vents appear to totally depend on energy derived from seawater–rock reactions and sulfur-oxidizing bacteria. Seafloor observations also indicate that hydrothermal vents are locally developed in areas of the youngest seafloor and occur along both fast- and slow-spreading ridge segments. Each hydrothermal field has multiple discharge sites, with sulfide chimneys rising up to 10 m above the seafloor. Minerals recovered from chimneys are chiefly Fe, Cu, and Au sulfides with minor amounts of anhydrite and amorphous silica (Humphris et al., 1995). Hydrothermal waters are acid, are rich in H_2S, and are major sources of Mn, Li, Ca, Ba, Si, Fe, Cu, and Au. These metals are deposited in the chimneys or with sediments near the vents.

The hydrothermal water erupted along ocean ridges is seawater that has circulated through hot oceanic crust with thermal gradients of more than 150° C/km. The seawater circulates by convection through the upper part of the permeable crust (1–2 km deep) and is heated at depth and discharges along active fissures in the medial rift.

Table 3.1 Summary of Major Mineral and Energy Deposits by Tectonic Setting

Tectonic Setting	Mineral Deposit	Energy Deposit
Oceanic Settings		
1. Ophiolite	Cyprus-type Cu-Fe massive sulfides Podiform chromite	
2. Ocean ridge		Geothermal
3. Back-arc basin		Hydrocarbons
Subduction Zone		
1. Arc	Hydrothermal: Au, Ag, Cu, Mo, Pb, Sb, Hg, Sn, W Porphyies: Cu, Mo, Sn Massive sulfides: Cu, Pb, Zn	Geothermal Hydrocarbons
2. Foreland basin	Redbed U, V, Cu	Hydrocarbons Coal
3. Forearc basin		Hydrocarbons
Orogens		
1. Highlands	Sn-W granites Gemstones Deposits from older tectonic settings	
2. Foreland–hinterland basin	Redbed U, V, Cu Stratiform Pb-Zn-Ag	Hydrocarbons Oil shale Coal
Continental Rifts	Stratiform Pb-Zn REE, Nb, U, Th, P, Cl, F, Ba, Sr associated with alkaline intrusives Sn granites Stratiform Cu Evaporites	Geothermal
Cratons, Passive Margins	Diamonds (kimberlites) Bauxite Ni laterite Evaporites Clays	Hydrocarbons Coal
Archean Greenstones	Cu-Zn massive sulfides Ni-Cu sulfides Au quartz veins	

The discharging water leaches metals from the oceanic crust and alters the crust by additions of water and elements such as Mg. Upon erupting on the seafloor, the hydrothermal waters are rapidly cooled and an increase in pH deposits sulfides and sulfates.

In considering ancient oceanic crust, two major mineral deposits are formed in ophiolites: Cyprus-type Cu-Fe massive sulfides and podiform chromite. Cyprus-type ores occur as stratiform deposits in pillowed basalt layers. These ores are exhalative deposits formed by hydrothermal vents along ocean ridges as described previously. Podiform chromite deposits are formed in ultramafic cumulates, and relict textures suggest that they are the products of fractional crystallization.

Arc Systems

Metalliferous mineral deposits are important in both continental-margin arcs and island arcs. Base-metals (Zn, Cu, Mo, and Pb), precious-metals (Ag and Au), and other metals (Sn, W, Sb, and Hg) are found in hydrothermal veins and lodes formed in arc systems. Following erosion, placer deposits of these metals are important in some geographic areas. Veins and lodes are commonly associated with volcanics or granitic plutons, where they represent late-stage fluids derived from differentiated arc magmas. Cu, Mo, and Sn porphyry deposits are also formed in arc systems. These large-volume, low-grade, disseminated deposits occur in altered porphyritic granites and are important sources of Cu and Mo in the southwestern United States and in the Andes. Kuroko-type massive sulfides (Cu, Pb, and Zn) are important in intra-arc and back-arc basin successions where they have formed as exhalatives on the seafloor. Redbed U, V, and Cu deposits occur in some retroarc foreland basins such as in the Mesozoic redbeds of the Colorado Plateau.

Zonation of metallic mineral deposits has been reported in late Tertiary rocks from the Andes (Sillitoe, 1976). In the direction of the dipping lithospheric slab, the major metallic mineral zones encountered are contact metasomatic Fe deposits; Cu, Au, and Ag veins; porphyry Cu-Mo deposits; Pb-Zn-Ag vein and contact metasomatic deposits; and Sn and Mo vein and porphyry deposits. Zonation is believed to result from progressive liberation of metals from the descending slab, with Sn coming from an extreme depth of about 300 km. Geochemical and isotopic data support the general concept that metal deposits associated with subduction are derived from some combination of the descending slab and the overlying mantle wedge. Metals move upward in magmas or fluids and are concentrated in late hydrothermal or magmatic phases.

Orogens

Metalliferous deposits are abundant at collisional boundaries where a variety of tectonic settings exist, depending on location and stage of evolution. In addition, older mineral deposits associated with ophiolites, arc, craton, and continental rift assemblages occur in collisional zones. Sn and W deposits are associated with collisional leucogranites in the Himalayas and in the Variscan orogen; at deeper crustal levels, Fe and Ti deposits occur associated with anorthosites. Many gemstones (e.g., ruby and sapphire) also are found in high-grade metamorphic rocks or in syntectonic nepheline syenites from collisional orogens. In peripheral foreland basins, stratiform Pb-Zn-Cu sulfides and redbed U-V-Cu deposits may be of economic importance.

Continental Rifts

Pb-Zn-Ag stratiform deposits occur in ancient continental rift sediments (Sawkins, 1990). These deposits, which are not associated with igneous rocks, occur in marine carbonates and are probably deposited from brines, which migrate to the edge of rift basins. REE, Nb, U, Th, Ba, P, Sr, and halogens are concentrated in carbonatites and other alkaline igneous rocks that occur in some continental rifts. Granites intruded during the late stages of rifting are often associated with Sr and fluorite. Stratiform Cu deposits occur in rift-related shales and sandstones as exemplified by deposits in the Zambian Copper Belt. The Cu appears to be derived from associated basalts. Evaporites are important non-metallic deposits found in some rifts.

Some of the major occurrences of Cr, Ni, Cu, and Pt are found in Proterozoic, layered, igneous intrusions. Chromite occurs as primary cumulates within the ultramafic parts of these intrusions, and Cu and Ni generally occur as late-stage hydrothermal replacements. Pt occurs in a variety of cumulus minerals, the most famous occurrence of which is the Merensky Reef in the Bushveld Complex in South Africa. In addition, some layered intrusions have magmatic ore deposits of Sn (in late granites) and Ti- or V-rich magnetite.

Cratons and Passive Continental Margins

Few if any metallic deposits are known to form on modern cratons or passive continental margins. Among the nonmetallic deposits formed in cratonic areas are diamonds from kimberlite pipes and associated placer deposits, bauxite, Ni-laterite deposits, and evaporites.

Important mineral deposits are the placer deposits of Au and U that occur in quartz arenites and conglomerates in cratonic successions. The largest deposits are in the Witwatersrand (2900 Ma) and Huronian (2300 Ma) Supergroups (Pretorius, 1976). Detrital Au and uraninite appear to have been concentrated by fluvial and deltaic processes in shallow-water, high-energy environments. Sources for the Au and U are older greenstone–granite terranes. Other important Proterozoic sedimentary mineral deposits include banded iron formation and manganese-rich sediments. Banded iron formation, described more fully in Chapter 6, reached the peak of its development around 2500 Ma. It occurs as alternating quartz-rich and magnetite-rich (or hematite-rich) laminae, and some was deposited in basins that are hundreds of kilometers across. Most banded iron formation is not associated with volcanic rocks and appears to have been deposited in shallow cratonic basins. Manganese-rich sediments also occur in cratonic successions associated with carbonates.

Archean Greenstones

Some of the world's major reserves of Cu, Zn, and Au occur in Archean greenstone belts (Groves and Barley, 1994). Those greenstones at least 3.5 Ga contain only minor mineral deposits, including Cu-Mo porphyry and stockwork deposits and small occurrences of barite and banded iron formation. Late Archean greenstones contain major Cu-Zn massive sulfides associated with oceanic felsic volcanics and Ni-Cu sulfides associated with komatiites. The latter sulfides appear to have formed as cumulates by fractional crystallization of immiscible sulfide melts associated with komatiite magmas. Minor occurrences of banded iron formation occur in most late Archean greenstones. One of the most important deposits in greenstones is Au, which occurs in quartz veins and in late disseminated sulfides commonly associated with hydrothermal chert or carbonate deposits.

Energy Deposits

Several requirements must be met in any tectonic setting for the production and accumulation of hydrocarbons such as oil or natural gas. First, the preservation of organic matter requires restricted seawater circulation to inhibit oxidation and decomposition. High geothermal gradients are needed to convert organic matter into oil and gas, and finally, tectonic conditions must create traps for the hydrocarbons to accumulate. Several tectonic settings potentially meet these requirements (Table 3.1).

Both oil and gas are formed in forearc and back-arc basins, which can trap and preserve organic matter and geothermal heat facilitates conversion of organic matter into hydrocarbons. Later deformation, generally accompanying continental collisions, creates a variety of structural and stratigraphic traps in which hydrocarbons can accumulate. An important site of hydrocarbon formation is in foreland basins. The immense accumulations in the Persian Gulf area formed in a peripheral foreland basin associated with the Arabia–Iran collision in the Tertiary.

Most oil and gas reserves in the world have formed in either intracratonic basins or passive continental margins (Table 3.2). During the early stages of opening of a continental rift, seawater can move into the rift valley, and if evaporation exceeds inflow, evaporites are deposited. This environment is also characterized by restricted water circulation in which organic matter is preserved. As the rift continues to open, water circulation becomes unrestricted and accumulation of organic matter and evaporite deposition ceases. High geothermal gradients beneath the opening rift and increasing pressure because of the burial of sediments facilitate the conversion of organic matter into oil and gas. At a later stage of opening, salt in the evaporite succession, because of its gravitational instability, may rise as salt domes and trap oil and gas. Oil and gas may also be

Table 3.2 Oil Reserves in Devonian and Younger Reservoirs			
	Intracratonic Basin (%)	**Passive Margin (%)**	**Foreland Basin (%)**
Cenozoic	5	8	18*
Mesozoic	43	5	6
Paleozoic	13	1	1
Total	**61**	**14**	**25**

*14% in the Persian Gulf and 4% elsewhere.

trapped in structural or stratigraphic traps as they move upward in response to increasing pressures and temperatures at depth. Supporting this model are data from wells in the Red Sea, which represents an early stage of ocean development. These wells encounter hydrocarbons associated with high geotherms and with rock salt up to 5 km thick. Also, around the Atlantic basin there is a close geographic and stratigraphic relationship between hydrocarbons and evaporite accumulation.

Hydrocarbon production in intracratonic basins may also be related to plate tectonic processes (Rona, 1977). Increases in seafloor-spreading rates and ocean-ridge lengths may cause marine transgression, which results in deposition in intracratonic basins. Decreasing spreading rates causes regression, which results in basins with limited circulation; hence, organic matter and evaporites accumulate. Unconformities also develop during this stage. Burial and heating of organic matter in intracratonic basins facilitates hydrocarbon production, and salt domes and unconformities may provide major traps for accumulation.

Coal is formed in two major tectonic settings: cratonic basins and foreland basins. For coal to form, plant remains must be rapidly buried before they decay. Such rapid burial occurs in swamps with high plant productivity. Widespread transgression in the Cretaceous was particularly suitable for coal swamps in cratonic areas. Swamps in foreland basins are generally part of large lake basins, and the widespread late Paleozoic coals of central Europe appear to have formed in such environments. Oil shales can accumulate under similar conditions as exemplified by the Tertiary oil shales in the early Tertiary foreland basins of eastern Utah.

Geothermal energy sources are associated with hotspots (such as Iceland and Yellowstone National Park), arc systems (such as the Taupo area in New Zealand), and continental rifts (such as the Jemez site in New Mexico).

Plate Tectonics with Time

As data continue to accumulate, it becomes more certain that plate tectonics in some form is the principal mechanism by which the Earth has cooled for the last 4 Gy. One way of tracking plate tectonics with time is with the petrotectonic assemblages summarized in this chapter. How far back in time do we see modern petrotectonic assemblages, and are their time–space relationships, tectonic histories, and chemical compositions similar to modern assemblages? Except for ophiolites, the greenstone and TTG assemblages are recognized throughout the geologic record from the oldest known rocks between 4.0 and 3.6 Ga to the present (Fig. 3.33). The oldest well-preserved cratonic–passive margin sediments are in the Moodies Group in South Africa, deposited at 3.2 Ga, and such sediments are minor yet widespread in the rock record by 3.0 Ga. Thus, it would appear that cratons, although probably small, were in existence by 3.2 to 3.0 Ga. Although the oldest isotopically dated mafic dyke swarm is the Ameralik swarm in southwestern Greenland, intruded around 3.25 Ga, deformed remnants of dykes in TTG complexes indicate earlier swarms, perhaps as early as 4.0 Ga in the Acasta gneisses. The oldest dated anorogenic

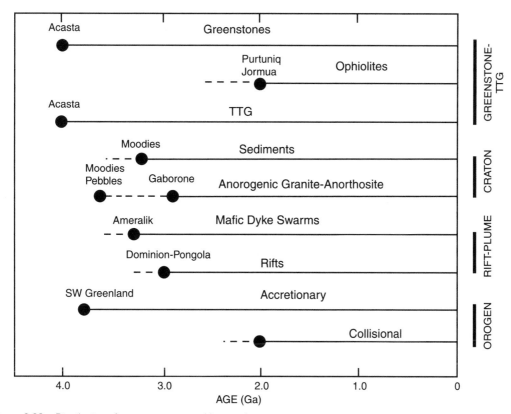

Figure 3.33 Distribution of petrotectonic assemblages with time.

granite is the Gaborone granite in Botswana emplaced at 2875 Ma. However, clasts of granite with anorogenic characters in conglomerates of the Moodies Group have igneous zircons with U-Pb isotopic ages of 3.6 Ga, indicating that highly fractionated granites formed in some early Archean crust. The oldest known continental rift assemblages are in parts of the Dominion and Pongola Supergroups in South Africa, which were deposited on the Kaapvaal craton around 3.0 Ga. The oldest accretionary orogens are the Acasta gneisses (4.0 Ga) and the Amitsoq gneisses (3.9 Ga) in northwestern Canada and southwestern Greenland, respectively. Although the oldest well-documented collisional orogens are Paleoproterozoic in age (such as the Wopmay orogen in northwestern Canada and the Capricorn orogen in Western Australia), it is likely that late Archean collisional orogens with reworked older crust exist in the granulite terranes of east Antarctica and southern India.

That greenstones, TTG, anorogenic granites, mafic dyke swarms, and accretionary orogens all appear in the earliest vestiges of our preserved geologic record from 4.0 to 3.5 Ga strongly supports some sort of plate tectonics operating on the Earth by this time. By 3.2 to 3.0 Ga, cratonic–passive margin sediments and continental rifts appeared, recording the development of the earliest continental cratons. Although plate tectonics appears to have been with us since at least 4.0 Ga, there are differences between some Archean and some post-Archean rocks that indicate Archean tectonic regimes must have differed in some respects from modern ones; these are described in Chapter 8. These differences have led to the concept of having our cake and eating it, too—in other words, plate tectonics operated in the Archean but differed in some ways from modern plate tectonics. Scientists are now faced with the question of how and to what degree Archean plate tectonics differed from modern plate tectonics and what these differences mean in terms of the evolution of the Earth.

Further Reading

Condie, K. C. (ed.), 1994. Archean Crustal Evolution. Elsevier, Amsterdam, 528 pp.

de Wit, M. J., and Ashwal, L. D., 1997. Greenstone Belts. Oxford University Press, Oxford, UK, 840 pp.

Gass, I. G., Lippard, S. J., and Shelton, A. W. (eds.), 1984. Ophiolites and Oceanic Lithosphere. Blackwell Scientific Publications, Oxford, 413 pp.

Leitch, E. C., and Scheibner, E. (eds.), 1987. Terrane Accretion and Orogenic Belts. Geological Society of America-American Geophysical Union, Geodynamic Series, Vol. 19, 354 pp.

Moores, E. M., and Twiss, R. J., 1995. Tectonics. W. H. Freeman, New York, 415 pp.

Olsen, K. H. (ed.), 1995. Continental Rifts: Evolution, Structure, Tectonics. Elsevier, Amsterdam, 520 pp.

Stern, R. J., 2002. Subduction zones. Rev. Geophys., 40 (4): 1 to 38.

Smellie, J. L., 1994. Volcanism Associated with Extension at Consuming Plate Margins. Geological Society of London, Special Publication No. 81, 272 pp.

Treloar, P. J., 1993. Himalayan Tectonics. Geological Society of London, Special Publication No. 74, 640 pp.

Windley, B. F., 1995. The Evolving Continents, Third Edition. John Wiley & Sons, New York, 526 pp.

The Mantle

4

Introduction

The Earth's mantle plays an important role in the evolution of the crust and provides the thermal and mechanical driving forces for plate tectonics. Heat liberated by the core is transferred into the mantle where most of it (>90%) is convected through the mantle to the base of the lithosphere. The remainder is transferred upward by mantle plumes generated in the core–mantle boundary layer. The mantle is also the graveyard for descending lithospheric slabs, and the fate of these slabs in the mantle is a subject of ongoing discussion and controversy. Do they partially collect at the 660-km discontinuity in the upper mantle, or do they descend to the bottom of the mantle? What happens to these slabs that control convection in the mantle? If they do not penetrate the 660-km discontinuity, the upper mantle may convect separately from the lower mantle, whereas if they sink to the base of the mantle, whole-mantle convection is probable. A related question is as follows: Do mantle plumes exist? If they exist, how and where they are generated, and what role do they play in mantle–crust evolution?

Another exciting mantle topic is that of the origin and growth of the lithosphere and whether its role in plate tectonics has changed with time. The Archean continental lithosphere, for instance, is considerably thicker than post-Archean continental lithosphere, and geochemical data from xenoliths suggest it had quite a different origin. Another hot topic is the origin of isotopic differences in basalts, which reflect different compositions and ages of mantle sources. How did these sources form and survive for billions of years in a convecting mantle? These are some of the questions I address in this chapter.

Seismic Structure of the Mantle

Upper Mantle

From studies of spectral amplitudes and travel times of body waves, it is possible to refine details of the structure of the mantle. The primary-wave (P-wave, or compressional-wave) and secondary-wave (S-wave, or shear-wave) velocity structure beneath several crustal types is shown in Figures 4.1 and 4.2. Cratons are typically underlain by a high-velocity lid (Vp > 7.9 km/sec) overlying, in turn, the low-velocity zone (LVZ) at a depth as shallow as 100 km to a depth of 300 km. A prominent minimum measure of the attenuation of seismic-wave energy (Q) coincides with the LVZ (Fig. 4.1). S-wave LVZs may extend to 400 km (Fig. 4.2). The top of the LVZ is generally assumed to mark the base of the lithosphere and averages a 50- to 100-km depth beneath oceans. Beneath ocean ridges and most continental rifts, the LVZ may extend nearly to the Mohorovicic discontinuity at depths as shallow as 15 km. The LVZ is minor or not detected beneath most Precambrian shields and may be nonexistent beneath Archean shields. The thickest lithosphere also occurs beneath shields (100–200 km), and beneath Archean shields, it may be more than 300 km thick.

At depths less than 200 km, seismic-wave velocities vary with crustal type, whereas at greater depths, the velocities are rather uniform regardless of crustal type. From the base of the LVZ to the 410-km discontinuity and from the 410-km discontinuity to the 660-km discontinuity, P-wave velocities increase only slightly (Fig. 4.1). Unlike P-wave velocities, S-wave velocities show significant lateral variation in the upper mantle, indicating compositional heterogeneity, anisotropy, or both. S-wave velocities beneath ocean ridges are strongly correlated with spreading rates at shallow depths (<100 km), and slow

Figure 4.1 (a) P-wave velocity distribution in the mantle (modified from Walck, 1985). (b) Average distribution of the specific attenuation factor (Q) in the mantle (excluding regions beneath Precambrian shields). Q varies inversely with the amount of seismic-wave attenuation; thus low values of Q show the greatest attenuation.

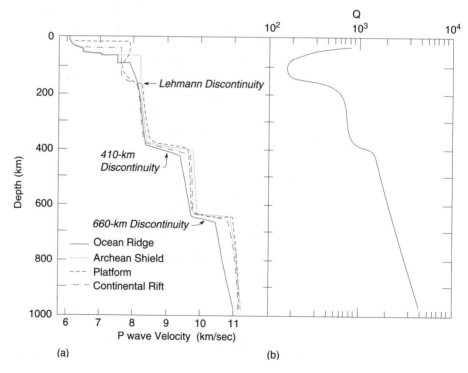

(a)   (b)

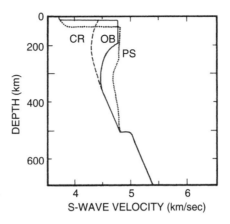

Figure 4.2 S-wave velocity distribution in the upper mantle beneath three crustal types. (CR, continental rift; OB, ocean basin; PS, Proterozoic shield or platform. Modified from Grand and Helmberger (1984).

velocities are limited to depths less than 50 km (Zhang and Tanimoto, 1993). Also, some hotspots are associated with slow S-wave velocities.

Lower Mantle

In the lower mantle, seismic-wave velocities continue to increase with depth and no further discontinuities are recognized until the D″ zone at the base of the mantle (see Fig. 1.2 in Chapter 1). However, seismic tomographic studies have shown that various "velocity domains" exist in both the upper and the lower mantle. **Seismic tomography,** like its medical analogue, combines information from numerous crisscrossing waves to construct three-dimensional images of the interior of the Earth. Relatively hot and cool regions of the Earth's interior can be mapped because seismic velocities vary inversely with temperature. Orientation of minerals in convecting mantle can also increase seismic velocities parallel to "fast" crystallographic axes of minerals. The accumulation of digital data from global seismic networks makes it possible to construct reliable three-dimensional models of mantle structure, and such information provides important constraints on the style of convection in the Earth (Hager and Clayton, 1989; Boschi and Dziewonski, 1999).

Combining data from surface and body waves, it is possible to compare the seismic structure of the upper and lower mantles. Two models of seismic-wave velocity distribution in the deep mantle are shown in Plate 1. The results clearly show that large inhomogeneities exist in the mantle on the scale of thousands of kilometers and that they extend to great depths. LVZs (red, Plate 1) underlie ocean ridges to several hundred kilometers, especially evident beneath the East Pacific rise. Also major low-velocity regions underlie the South Pacific, the West Coast of North America, and southern Africa. High-velocity zones (blue, Plate 1) underlie regions around the Pacific basin and reflect recently subducted slabs. The high-velocity zone extending from eastern North America into northern South America probably represents the descending Farallon plate, an oceanic plate that is almost subducted. Although the origin and the significance of all of the anomalies shown on these sections are not yet understood, the fact that many anomalies cross the 660-km discontinuity favors whole-mantle rather than layered-mantle convection.

Mantle Upwellings and Geoid Anomalies

To balance the subducted slabs that sink into the mantle, there must be an upward return flow. Because subducted plates are relatively cool, they decrease the temperature of nearby mantle, leaving relatively warm mantle in the regions between subduction zones. These broad, warm regions of mantle, known as **mantle upwellings** (or sometimes as superplumes), are relatively buoyant and rise, providing the return flow. Today, there are two large upwellings most investigators recognize in the mantle: one beneath the African plate and one beneath the Pacific plate (Maruyama, 1994; Ni and Helmberger, 2003). Anomalously low P- and S-wave velocity distributions in the mantle are consistent with the existence of these two large upwellings (red, Plates 1 and 2). Mantle upwellings elevate the Earth's surface a few hundred meters, producing *superswells*. Because they elevate the temperature of the uppermost mantle, widespread small degrees of melting occur near the tops of upwellings, giving rise to volcanism and to mafic underplating of the crust. Another characteristic of upwellings is that they contain most of the modern hotspots and, hence, most of the modern mantle plumes. A possible explanation for this is that deep flow in a mantle upwelling, just above the core–mantle boundary, sweeps material upward, thickening the D″ layer. This promotes instability in D″, which produces new plumes that rise within the upwelling. Consistent with this idea, plumes with the highest buoyancy flux are concentrated in the Pacific upwelling (Ribe and de Valpine, 1994. Davaille (1999) has shown experimentally that upwellings with constituent plumes (hotspots) can form with only a small amount of density layering (~1%) in the mantle.

The **geoid** is the gravitational equipotential surface of the Earth that coincides with sea level in oceanic areas. Because gravitational potential decreases inversely with distance from source mass, whereas gravitational acceleration decreases inversely with the square of the distance, the geoid provides a long-range probe into the Earth. Deviations of the geoid from an idealized hydrostatic ellipsoid are known as **geoid anomalies.**

If the mantle were rigid, a positive mass anomaly in the upper mantle would produce a positive geoid anomaly. However, as illustrated in Figure 4.3a, the mantle is not rigid but deforms as a ductile solid, so a positive mass anomaly causes a downward flow in the mantle, which in turn causes a depression on the surface of the planet (Hager et al., 1985). Because the depression results in air or seawater replacing rock, it can be considered a mass deficiency relative to a laterally uniform Earth. This results in a negative contribution to the geoid. If the mass anomaly is near the surface, then the mass deficiency caused by the surface depression is opposite but larger than the mass anomaly; thus, the net geoid anomaly will be negative and much smaller than either of its contributors. The mass anomaly will also cause a small deflection at the bottom of the mantle, which makes a

(b) NEGATIVE MASS ANOMALY

...ositive (a) and negative (b) mass anomalies on the geoid.

...a general case for deflection of top and
...ies will balance the mass excess and the net
...the Earth, subducted plates sinking into the

...ou are dealing with a negative mass anom-
...ow raises both the surface of the Earth and
...itive geoid anomaly because air or water is
...d African upwellings are both characterized
...be contoured much like a topographic map
...the upward flow in the mantle should raise
...ntle interface and produce a positive geoid
...erved if the effect of sinking slabs is sub-
...odel, whole-mantle convection is assumed
...hat of the upper mantle (Hager et al., 1985).
...oid anomalies at low spherical harmonic
...er beneath Africa. These coincide with and
...ntle upwellings from the core–mantle interface.
Supporting this interpretation is the correlation of hotspots and seismic-velocity anomalies
with the geoid highs (Richards et al., 1988).

The core–mantle boundary is also raised beneath the upwellings as predicted by the
model (Fig. 4.4b): that is, the upper and lower surfaces of the mantle (+crust) are
deformed in the same direction and in the same geographic locations. About 3 km of
relief occurs on the core–mantle anomalies. These dynamically induced deformations at
both the top and the bottom of the convecting mantle are caused by substantial mass
anomalies.

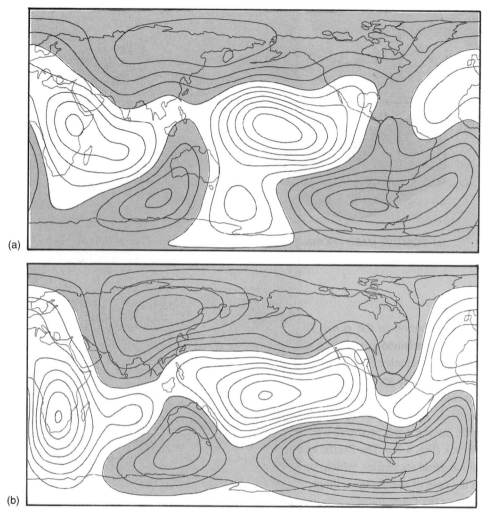

Figure 4.4   (a) The calculated geoid for a convecting Earth. Density contrast is assumed equal to 2.3 gm/cm^3, the contour interval is 200 m, and the degrees are 1 to 6. Geoid lows are shaded. (b) Calculated topography at the core–mantle boundary. Density contrast across the boundary is assumed to be 4.5 gm/cm^3, the contour interval is 400 m, and the degrees are 1 to 6. Geoid lows are shaded. Modified from Hager and Clayton (1987).

Temperature Distribution in the Mantle

Estimates of temperatures in the lower crust and mantle can be made from measurements of surface-heat flow, models of heat production, and thermal conductivity distributions with depth (Artemieva and Mooney, 2001). Convection is the dominant mode of heat transfer in the asthenosphere and mesosphere where an adiabatic gradient is maintained, and thus temperature increases at a slow rate with depth. On the other hand, conduction is the main way heat is lost from the lithosphere, and temperatures change rapidly with depth and with tectonic setting (Fig. 4.5). In this respect, the lithosphere is both a mechanical and a thermal boundary layer in the Earth.

Although major differences in temperature distribution exist in the upper mantle, it is necessary that all geotherms converge at depths of no greater than 400 km; otherwise, large, unobserved gravity anomalies would exist between continental and oceanic areas. Heat-flow distribution, heat-production calculations, and seafloor-spreading rates suggest that most geotherms to about 50 km deep range from 10 to 30° C/km. With the exception of beneath ocean-ridge axes, geotherms must decrease rapidly between 50 and 200 km to avoid a large amount of melting in the upper mantle, which is not allowed by seismic or gravity data. Oceanic geotherms intersect the mantle adiabat at depths less than 200 km, whereas continental geotherms intersect the adiabat at greater depths. Beneath ocean ridges (line A, Fig. 4.5) geotherms are steep and intersect the **mantle solidus** (i.e., the temperature at which the mantle begins to melt) at depths generally less than 50 km. Thus, ocean-ridge basaltic magma is produced at shallow depths beneath ridges. With increasing distance from a ridge axis as the lithosphere cools, geotherms decrease at a rate inversely proportional to lithosphere-spreading rates (Bottinga and Allegre, 1973). The decrease results in progressively greater depths of intersection of the geotherm with the mantle solidus and hence in a deepening of the lower thermal boundary of the oceanic lithosphere as it ages. After about 100 My, the thermal lithosphere is about 100 km thick, in agreement with the thicknesses estimated from surface-wave studies. On average, neutral buoyancy of the oceanic lithosphere is reached around 20 My, and further cooling leads to negative buoyancy and to subduction, where the mean age is about 120 My. Rarely are plates more than 100 km thick when they subduct.

Geotherms beneath Archean shields are not steep enough to intersect the mantle solidus, whereas most other continental geotherms intersect a slightly hydrous mantle solidus at depths from 150 to 200 km (Fig. 4.5). Subduction geotherms vary with the age

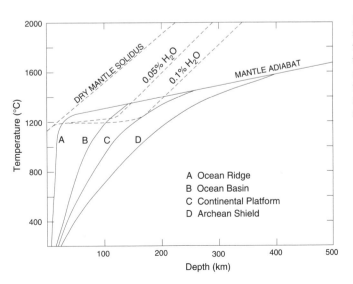

Figure 4.5 Typical lithosphere geotherms. Note that geotherms intersect the mantle adiabat at variable depths. Also shown is the dry mantle solidus and mantle solidus with 0.10 and 0.05% water.

of the onset of subduction. In most cases, however, they are not steep enough for the descending slab to melt. The temperature distribution beneath a descending slab 10 My after the onset of subduction is shown in Figure 4.6. Note that isotherms are bent downward as the cool slab descends and that the phase changes at 150 km (gabbro-eclogite) and 410 km (olivine–wadsleyite) occur at shallower depths in the slab than in surrounding mantle in response to cooler temperatures in the slab. As slabs descend into the mantle, they warm by heat transfer from surrounding mantle, adiabatic compression, frictional heating along the upper surface of the slab, and exothermic phase changes in the slab (Toksoz et al., 1971).

The Lithosphere

Understanding the structure and evolution of the lithosphere is critical to understanding the origin and evolution of both the oceanic and the continental crust. Although the lithosphere can be loosely thought of as the outer rigid layer of the Earth, more precise definitions in terms of thermal and mechanical characteristics are useful. For the oceanic lithosphere, where cooling controls thickness, it is the outer shell of the Earth with a conductive temperature gradient overlying the convecting adiabatic interior (White, 1988; Jaupart and Mareschal, 1999). This is known as the **thermal lithosphere.** Asthenosphere can be converted to oceanic lithosphere simply by cooling. The progressive thickening of the oceanic lithosphere continues until about 70 My; afterward, it remains relatively constant in thickness until subduction. The thickness of the rigid part of the outer layer

Figure 4.6 Temperature distribution in a descending slab 10 My after initiation of subduction for a spreading rate of 8 cm/year. Shaded areas are the gabbro–eclogite, olivine–wadsleyite, and 660-km discontinuity phase changes in order of increasing depth. Modified from Toksoz et al. (1971).

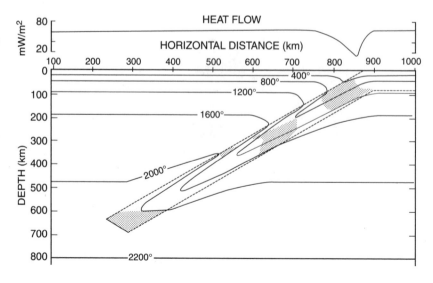

of the Earth that readily bends under a load, known as the **elastic lithosphere,** is less than that of the thermal lithosphere. The base of the oceanic elastic lithosphere varies with composition and temperature increasing from about 2 km at ocean ridges to 50 km just before subduction. It corresponds roughly to the 500 to 600° C isotherm. In continental lithosphere, the elastic thickness is less than crustal thickness, often by 10 to 15 km.

The mineralogical and chemical composition of the mantle lithosphere can be approximated from the combined results of seismic-velocity distributions, high-pressure–temperature (P–T) experimental studies, and geochemical and isotopic studies of mantle xenoliths. It is now possible to attain static pressures in the laboratory to more than 50 GPa, which allows direct investigation of the physical and chemical properties of minerals that may occur throughout most of the Earth.

Several lines of evidence indicate that ultramafic rocks compose large parts of the upper mantle. Geochemical, isotopic, and seismic studies all agree that the mantle is heterogeneous. The distribution of radiogenic isotopes in basalts clearly indicates distinct mantle reservoirs in existence for more than 10^9 years. The sizes, shapes, and locations of these reservoirs, however, are not well constrained by the isotopic data, and it is only through increased resolution of seismic-velocity distributions that these reservoirs can be more accurately characterized.

Composition

Seismic-Velocity Constraints

Laboratory measurements of P-wave velocity (Vp) at pressures up to 5 GPa provide valuable data regarding mineral assemblages in the upper mantle (Christensen, 1966; Christensen and Mooney, 1995). Dunite, pyroxenite, harzburgite, various lherzolites, eclogite, or some mixture of these rock types are consistent with observed P-wave velocities in the upper mantle. Higher than normal temperatures, some combination of garnet granulite or serpentinite, or both may account for regions with anomalously low velocities (Vp < 8.0 km/sec). Measured velocities of highly serpentinized ultramafic rocks (5.1–6.5 km/sec), however, are too low for most of the upper mantle. The high temperatures characteristic of the upper mantle also exceed the stability of serpentine, thus greatly limiting the extent of even slightly serpentinized lherzolite in the mantle.

Mantle Xenoliths

Mantle xenoliths are fragments of the mantle brought to the Earth's surface during volcanic eruptions. They are common in alkali basalts and in kimberlite pipes, and they provide an important constraint on mineral assemblages in the upper mantle. Experimental studies of coexisting minerals in mantle xenoliths provide a means to constrain the temperature and the pressure at which xenoliths last crystallized in the mantle (Ryan et al., 1996; O'Reilly et al., 1997). Hence, clinopyroxene compositions in ultramafic rocks provide information on minimum depths for mantle xenoliths. Results sometimes define a P–T curve, which may be a **paleogeotherm** through the lithosphere (O'Reilly and Griffin,

1985; Boyd, 1989). Supporting such an interpretation, these P–T paths are similar in shape to calculated modern geotherms.

From measured seismic velocities, mantle xenoliths, and high-pressure experimental data, possible upper-mantle mineral assemblages are shown in Figure 4.7. Also shown are the mantle solidus and ocean-ridge, ocean-basin, and Archean shield geotherms. Xenoliths from kimberlites indicate sources with depths greater than 100 km, and ultramafic xenoliths from alkali basalts record depths of 50 to 100 km. The data are consistent with an upper mantle lithosphere beneath both the oceans and the post-Archean continents composed chiefly of spinel lherzolite (olivine-cpx-opx-spinel) and lherzolite (olivine-cpx-opx), whereas in the thick Archean lithosphere and in the asthenosphere, garnet lherzolite (olivine-cpx-opx-garnet) is the dominant rock type. Although measured seismic-wave velocities also allow dunite (all olivine), pyroxenite (all pyroxenes), eclogite (garnet-cpx), and harzburgite (olivine-opx) in the upper mantle, the first three of these rocks are not abundant in xenolith populations or ophiolites, suggesting that they are not important constituents in the upper mantle. Harzburgites, however, are widespread in most ophiolites and may be the most important rock type in the shallow oceanic upper mantle. As suggested by the steep ocean-ridge geotherm in Figure 4.7, some ophiolite ultramafics are dominantly plagioclase lherzolite (olivine-cpx-opx-plag). Ultramafic rocks from ophiolites record source depths from 10 to 75 km and verify the interpretation that they are slices of the oceanic upper mantle. Ophiolites interpreted as fast-spreading ocean ridges (>1 cm/year) commonly contain harzburgites from which less than 20% basaltic melt has been extracted. Ophiolites from slow-spreading ridges (<1 cm/year) record less melt extraction (15%) and a residual mineral assemblage in which plagioclase lherzolite dominates.

One of the exciting applications of paleogeotherm information is to "map" the distribution of mantle mineral assemblages in the lithosphere with increasing depth. With multiple

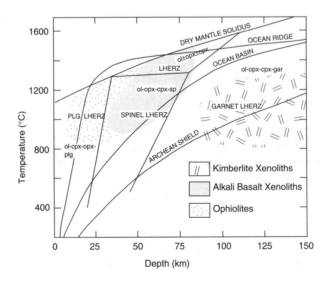

Figure 4.7   Stability fields of mantle mineral assemblages in pressure–temperature space. Also shown are lithosphere geotherms and the dry mantle solidus. (cpx, clinopyroxene; gar, garnet; LHERZ, lherzolite; ol, olivine; opx, orthopyroxene; plg, plagioclase; sp, spinel).

sample locations, mantle lithosphere stratigraphy can be mapped and followed laterally, producing three-dimensional images of the lithosphere that can be correlated with geophysical data and surface geology (Griffin et al., 1999; Poudjom Djomani et al., 2001) (Fig. 4.8). Lithosphere domains often correlate with crustal age provinces, implying that crustal province boundaries are lithospheric boundaries and that each province carried its own lithospheric root during the assembly of cratons and supercontinents. In the Slave province in northwest Canada, lithosphere mapping suggests two layers, with the upper layer (100–150 km) consisting of depleted harzburgite and the lower layer (150–220 km) of less depleted and metasomatized lherzolites (Fig. 4.8). Although the lithosphere is thinner, a similar pattern is shown by Proterozoic lithosphere in Siberia. In the Archean Siberian lithosphere, however, the metasomatized layer extends to much shallower depths.

Chemical Composition

Using the compositions of mantle xenoliths with results of model calculations, estimates of the average composition of primitive mantle, post-Archean lithosphere, Archean lithosphere, and depleted mantle are given in Table 4.1. The four estimates have a combined total of more than 90% of MgO, SiO_2, and FeO, and no other oxide exceeds 4%. In a plot of modal olivine versus Mg number of olivine, ophiolite ultramafics and spinel lherzolite xenoliths from post-Archean lithosphere define a distinct trend, and those from the Archean lithosphere define a population with a relatively high Mg number but no apparent trend (Fig. 4.9). The trends for oceanic and post-Archean suites are readily interpreted as restite trends, where **restite** is the material remaining in the mantle after extraction of

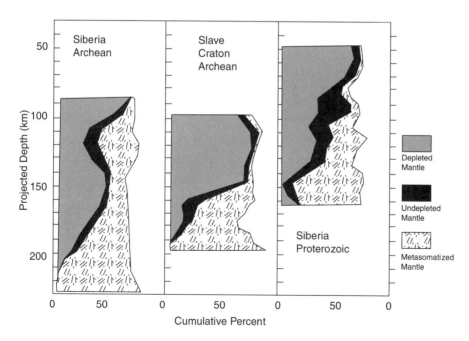

Figure 4.8 Chemical cross-sections of the sub-continental lithosphere based on geochemical data from mantle xenolith and xenocryst populations. Modified from Griffin et al. (1999).

Table 4.1 Average Chemical Composition of the Mantle				
	Primitive Mantle	Post-Archean Lithosphere	Archean Lithosphere	Depleted Mantle
SiO_2	46.0	44.1	46.6	43.6
TiO_2	0.18	0.09	0.04	0.134
Al_2O_3	4.06	2.20	1.46	1.18
FeOT	7.54	8.19	6.24	8.22
MgO	37.8	41.2	44.1	45.2
CaO	3.21	2.20	0.79	1.13
Na_2O	0.33	0.21	0.09	0.02
K_2O	0.03	0.028	0.08	0.008
P_2O_5	0.02	0.03	0.04	0.015
Mg/Mg + Fe	90	90	93	91
Rb	0.64	0.38	1.5	0.12
Sr	21	20	27	13.8
Ba	7.0	17	25	1.4
Th	0.085	0.22	0.27	0.018
U	0.02	0.04	0.05	0.003
Zr	11.2	8.0	7.3	9.4
Hf	0.31	0.17	0.17	0.26
Nb	0.71	2.7	1.9	0.33
Ta	0.04	0.23	0.10	0.014
Y	4.6	3.1	0.63	2.7
La	0.69	0.77	3.0	0.33
Ce	1.78	2.08	6.3	0.83
Eu	0.17	0.10	0.11	0.11
Yb	0.49	0.27	0.062	0.30
Co	104	111	115	87
Ni	2080	2140	2120	1730

Major elements are in weight percentage of the oxide, and trace elements are in parts per million. Mg number = Mg/Mg + Fe mole ratio × 100. Depleted mantle was calculated from NMORB, Archean lithosphere from garnet lherzolite xenoliths, and post-Archean lithosphere from spinel lherzolite xenoliths. Data from Hofmann (1988), Sun and McDonough (1989), McDonough (1990), Boyd (1989), and miscellaneous sources.

variable amounts of basaltic magma (Boyd, 1989). In the case of ophiolites, they represent the restite remaining after the extraction of ocean-ridge basalts, and the xenoliths from post-Archean lithosphere may be restites from the extraction of plume-related basalts (flood basalts). The high Mg numbers of olivines from Archean xenoliths suggest that they are restites from the extraction of komatiite magmas. Because garnet, the densest mineral in the upper mantle, is preferentially removed compared with olivine and orthopyroxene during melting, the restites in all cases are less dense than a primitive mantle of garnet lherzolite; thus, they would tend to rise and remain part of the lithosphere.

The Earth's **primitive mantle** composition is calculated from geochemical modeling of mantle–crust evolution and represents the average composition of the silicate part of

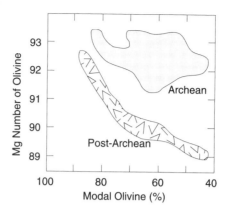

Figure 4.9 Mg number versus modal olivine in post-Archean mantle xenoliths and ophiolite ultramafics and in Archean lithosphere xenoliths from South Africa. Mg number = Mg/Mg + Fe molecular ratio. Modified from Boyd (1989).

the Earth just after planetary accretion (Sun and McDonough, 1989). Compared with primitive mantle, incompatible element distributions in the mantle lithosphere and depleted mantle are distinct (Table 4.1). The striking depletion in the most incompatible elements (Rb, Ba, Th, etc.) in depleted mantle (Fig. 4.10), as represented by ophiolite ultramafics and mantle compositions calculated from ocean-ridge basalts, reflects the removal of basaltic liquids enriched in these elements, perhaps early in the Earth's history (Hofmann, 1988). In striking contrast to depleted mantle, lithosphere mantle shows prominent enrichment in the most incompatible elements. Spinel lherzolites from post-Archean lithosphere show a positive Nb-Ta anomaly, suggesting that they represent plume material plastered on the bottom of the lithosphere (McDonough, 1990). Archean garnet lherzolites commonly show textural and mineralogical evidence for metasomatism (i.e., modal metasomatism), such as veinlets of amphibole, micas, and other secondary minerals (Waters and Erlank, 1988). The high content of the most incompatible elements in modally metasomatized xenoliths (Fig. 4.10) probably records metasomatic additions

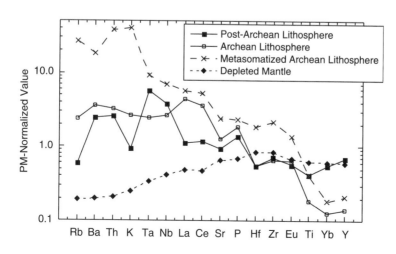

Figure 4.10 Primitive mantle (PM), normalized, incompatible element distributions in subcontinental lithosphere and depleted mantle. Primitive mantle values from Sun and McDonough (1989); data from Nixon et al. (1981), Hawkesworth et al. (1990), Wood (1979), and miscellaneous sources.

of these elements to the lithosphere. In some cases, such as in the xenoliths from Kimberley in South Africa (Hawkesworth et al., 1990), these additions occurred after the Archean.

Thickness of Continental Lithosphere

The continental lithosphere varies considerably in thickness depending on its age and mechanism of formation. S-wave tomographic studies of the upper mantle have been most definitive in estimating the thickness of continental lithosphere (Grand, 1987; Polet and Anderson, 1995). Most post-Archean lithosphere is 100 to 200 km thick, and lithosphere beneath Archean shields is commonly more than 300 km thick. Rheological models suggest thicknesses in these same ranges (Ranalli, 1991). In an S-wave tomographic cross-section around the globe, high-velocity roots underlie Archean crust, as in northern Canada, central and southern Africa, and Antarctica (Fig. 4.11). The base of the lithosphere in these and other areas overlain by Archean crust may nearly reach the 410-km discontinuity. Under Proterozoic shields, however, lithospheric thicknesses rarely exceed 200 km. Consideration of elongation directions of Archean cratons relative to directions of modern plate motions suggests that the thick Archean lithosphere does not aid or hinder plate motions (Stoddard and Abbott, 1996). As expected, hotspots (plumes), such as Hawaii and Iceland, are associated with slow velocities between 50 and 200 km deep (Fig. 4.11). Thermal and geochemical modeling has shown that the lithosphere can be thinned by as much as 50 km by extension over mantle plumes (White and McKenzie, 1995).

Isotopic and geochemical data from mantle xenoliths indicate that the mantle lithosphere beneath Archean shields formed during the Archean and that it is chemically distinct from post-Archean lithosphere. Because of the buoyant nature of the depleted

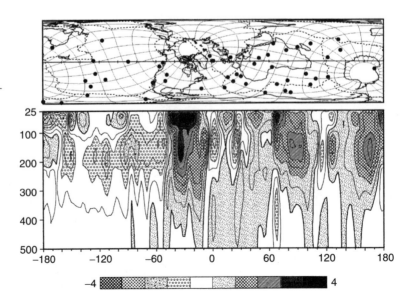

Figure 4.11   S-wave velocity distribution in the upper mantle along a great circle passing through Hawaii and Iceland. Darker shades indicate faster velocities. The map shows the location of the great circle and major hotspots (black dots). From Zhang and Tanimoto (1993).

Archean lithosphere, it tends to ride high compared with adjacent Proterozoic lithosphere, as shown by the extensive platform sediment cover on Proterozoic cratons compared with Archean cratons (Hoffman, 1990). The thick roots of Archean lithosphere often survive later tectonic events and thermal events, such as continental collisions and supercontinent rifting. However, mantle plumes or extensive later reactivation can remove the thick lithosphere keels, as for instance is the case with the Archean Wyoming province in North America and the north China craton.

Seismic Anisotropy

P-wave velocity measurements at various orientations to rock fabric show that differences in mineral alignment can produce significant anisotropy (Ave'Lallemant and Carter, 1970; Kumazawa et al., 1971). Vp differences of more than 15% occur in some ultramafic samples and are related primarily to the orientation of olivine grains. Seismic-wave anisotropy in the mantle lithosphere beneath ocean basins may be produced by recrystallization of olivine and pyroxene accompanying seafloor spreading with the [100] axes of olivine and [001] axes of orthopyroxene oriented normal to ridge axes (the higher velocity direction) (Estey and Douglas, 1986). Supporting evidence for alignment of these minerals comes from studies of ophiolites and upper mantle xenoliths, and flow patterns in the oceanic upper mantle can be studied by structural mapping of olivine orientations (Nicholas, 1986). The mechanism of mineral alignment requires upper mantle shear flow, which aligns minerals through dislocation glide. The crystallographic glide systems have a threshold temperature necessary for recrystallization of about 900° C, which yields a thermally defined lithosphere depth similar to that deduced from seismic data (~100 km). Creep actively maintains mineral alignment below this boundary in the LVZ, and it is preserved in a fossil state in the overlying lithosphere.

The subcontinental lithosphere also exhibits seismic anisotropy of S-waves parallel to the surface of the Earth. This is evidenced by **S-wave splitting,** where the incident wave is polarized into two orthogonal directions traveling at different velocities (Silver and Chan, 1991). As with the oceanic lithosphere, this seismic anisotropy appears to be caused by the strain-induced preferred orientation of anisotropic crystals such as olivine. Seismic and thermal modeling indicate that the continental anisotropy occurs within the lithosphere at depths from 150 to 400 km. The major problem in the subcontinental lithosphere has been to determine how and when such alignment occurred in tectonically stable cratons. Was it produced during assembly of the craton in the Precambrian, or is it a recent feature caused by deformation of the base of the lithosphere as it moves about?

In most continental sites, the azimuth of the fast S-wave has been closely aligned with the direction of absolute plate motion for the last 100 My (Silver and Chan, 1991; Vinnik et al., 1995) (Fig. 4.12). This coincidence suggests that the anisotropy is not a Precambrian feature but results from resistive drag along the base of the lithosphere. Supporting this interpretation, seismic anisotropy does not correlate with single terranes in Precambrian crustal provinces. These provinces were assembled in the Precambrian by terrane collisions, and if anisotropy was acquired at this time, the azimuths should vary from terrane to terrane

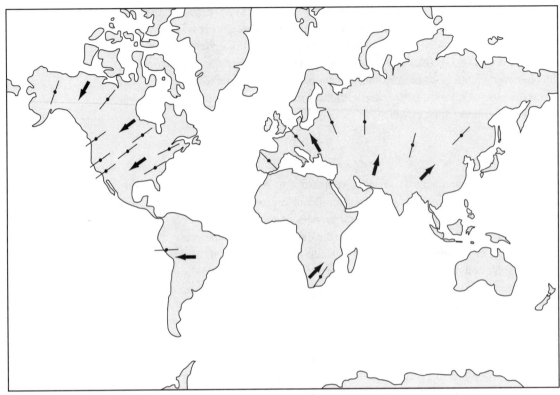

Figure 4.12 Fast S-wave velocity directions in the subcontinental lithosphere compared with motion directions of modern plates (bold arrows). Modified from Silver and Chan (1991).

according to their preassembly deformational histories. Instead, the seismic anisotropies show a uniform direction across cratons, aligned parallel to modern plate motions (Fig. 4.12).

Thermal Structure of Precambrian Continental Lithosphere

It has long been known that heat flow from Archean cratons is less than that from Proterozoic cratons (Fig. 4.13). Two explanations for this relationship have been proposed (Nyblade and Pollack, 1993; Jaupart and Mareschal, 1999):

1. There is greater heat production in Proterozoic crust than in Archean crust.
2. A thick lithospheric root beneath the Archean lithosphere is depleted in radiogenic elements.

The Proterozoic upper continental crust appears to be enriched in K, U, and Th, whose isotopes produce most of the heat in the Earth, relative to its Archean counterpart. Using estimates of the concentration of these elements in the crust (Condie, 1993), only part of the difference in heat flow from Proterozoic and Archean lithosphere can be explained.

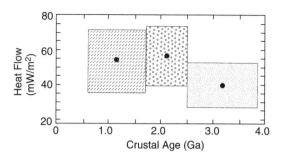

Figure 4.13 Heat flow versus age of Precambrian lithosphere. The width of each box shows the age range, and the height is one standard deviation of the mean heat flow. From Nyblade and Pollack (1993).

Thus, it would appear that the thick root beneath the Archean cratons must also be depleted in radiogenic elements and contribute to the difference in heat flows.

Age of Subcontinental Lithosphere

It is important in terms of crust–mantle evolution to know whether the thick lithospheric roots beneath Archean cratons formed in the Archean in association with the overlying crust or whether they were added later by underplating. Although in theory scientists can use mantle xenoliths to isotopically date the lithosphere, because later deformation and metasomatism may reset isotopic clocks, ages obtained from xenoliths are generally too young. Some xenoliths give the isotopic age of eruption of the host magma. What scientists really need to determine the original age of the subcontinental lithosphere are minerals that did not recrystallize during later events or an isotopic system that was not affected by later events. At this point, diamonds and Os isotopes enter the picture. Diamonds, which form at depths greater than 150 km, are resistant to recrystallization at lithosphere temperatures. Sometimes they trap silicate phases as they grow, shielding these minerals from later recrystallization (Richardson, 1990). Pyroxene and garnet inclusions in diamonds, which range from about 50 to 300 microns in size, have been successfully dated by the Sm-Nd isotopic method and appear to record the age of the original ultramafic rock. Often more than one age is recorded by diamond inclusions from the same kimberlite pipe, as with the Premier pipe in South Africa. Diamonds in this pipe with garnet-opx inclusions have Nd and Sr mineral isochron ages older than 3 Gy, suggesting that the mantle lithosphere formed in the early Archean when the overlying crust formed (Richardson et al., 1993). Those diamonds with garnet-cpx-opx inclusions record an age of 1.93 Ga, and those with garnet-cpx (eclogitic) record an age of 1.15 Ga, only slightly older than kimberlite emplacement. The younger ages clearly indicate multiple events in the South African lithosphere. Diamond inclusion ages from lithosphere xenoliths from Archean cratons in South Africa, Siberia, and Western Australia indicate that the lithosphere in these regions is also Archean.

The Re-Os isotopic system differs from the Sm-Nd and Rb-Sr systems in that Re is incompatible in the mantle but Os is compatible. In contrast, in most other isotopic systems, both parent and daughter elements are incompatible. Hence, during the early magmatic event that left the Archean mantle lithosphere as a restite, Re was completely or largely extracted from the rock and Os was unaffected (Carlson et al., 1994; Pearson et al., 2002). When Re was extracted from the rock, the Os isotopic composition was "frozen" into the system; hence, by analyzing a mantle xenolith later brought to the surface, scientists can date the Re depletion event. Calculated Os isotopic ages of xenoliths from two kimberlite pipes in South Africa are shown in Figure 4.14, plotted at depths of origin inferred from thermobarometry. In the Premier and North Lesotho pipes, ages range from 3.3 to 2.2 Ga, similar to the ranges found in xenoliths from pipes in the Archean Siberian craton. Results support the diamond inclusion ages, indicating an Archean age for the thick Archean mantle keels. It is not yet clear whether the range in ages from a given pipe records the range in formation ages of the lithosphere or a series of metasomatic remobilization events of lithosphere that occurred about 3.0 Ga.

The maximum isotopic ages obtained for mantle keels in the Siberia and South African cratons are similar to the oldest isotopic ages obtained from the overlying crust (Carlson et al., 1994; Pearson et al. 1995; 2002). This suggests that substantial portions of the mantle keels beneath the continents formed at the same time as the overlying crust and that they have remained firmly attached to the crust.

The Low-Velocity Zone

The LVZ in the upper mantle is characterized by low seismic-wave velocities, high seismic-energy attenuation, and high electric conductivity. The bottom of the LVZ, sometimes called the **Lehmann discontinuity,** has been identified from the study of surface-wave and S-wave data in some continental areas (Gaherty and Jordan, 1995) (Fig. 4.1). This discontinuity, which occurs at depths from 180 to 220 km, appears to be thermally

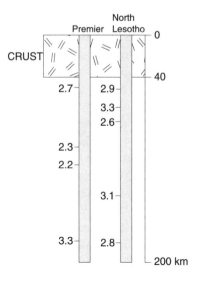

Figure 4.14   Idealized cross-section of the Archean lithosphere in South Africa constructed from mantle xenoliths from two kimberlite pipes. Ages are Re-depletion model ages (in Ga) from Carlson et al. (1994).

controlled and to at least partly reflect a change from an anisotropic lithosphere to an isotropic asthenosphere. The LVZ plays a major role in plate tectonics, providing a relatively low-viscosity region upon which lithospheric plates can slide with little friction.

Because of the dramatic drop in S-wave velocity and the increase in attenuation of seismic energy, it would appear that partial melting must contribute to LVZ production. The probable importance of incipient melting is attested to by the high surface-heat flow observed when the LVZ reaches shallow depths, such as beneath ocean ridges and in continental rifts. Experimental results show that incipient melting in the LVZ requires a minor amount of water to depress silicate melting points (Wyllie, 1971). With only 0.05 to 0.10% water in the mantle, partial melting of garnet lherzolite occurs in the appropriate depth range for the LVZ, as shown by the geotherm–mantle solidus intersections in Figure 4.5. The source of water in the upper mantle may be from the breakdown of minor phases that contain water such as amphibole, mica, titanoclinohumite, or other hydrated silicates. The theory of elastic-wave velocities in two-phase materials indicates that only 1% melt is required to produce the lowest S-wave velocities measured in the LVZ (Anderson et al., 1971). If, however, melt fractions are interconnected by a network of tubes along grain boundaries, the amount of melting may exceed 5% (Marko, 1980). The downward termination of the LVZ appears to reflect the depth at which geotherms pass below the mantle solidus (Fig. 4.5). Also possibly contributing to the base of the LVZ is a rapid decrease in the amount of water available (perhaps free water enters high-pressure silicate phases at this depth). The width or even the existence of the LVZ depends on the steepness of the geotherms. With steep geotherms such as those characteristic of ocean ridges and continental rifts, the range of penetration of the mantle solidus is large; hence the LVZ should be relatively wide (lines A and B, Fig. 4.5). The gentle geotherms in continental platforms, which show a narrow range of intersection with the hydrated mantle solidus, produce a thin or poorly defined LVZ (line C, Fig. 4.5). Beneath Archean shields, geotherms do not intersect the mantle solidus; hence there is no LVZ (line D, Fig. 4.5).

The Transition Zone

The 410-km Discontinuity

The transition zone is that part of the upper mantle in which two major seismic discontinuities occur: one at 410 km and the other at 660 km (Fig. 4.1). High-pressure experimental studies document the breakdown of Mg-rich olivine to a high-pressure phase known as **wadsleyite** (beta phase) around 14 GPa, which is equivalent to a 410-km depth in the Earth (Fig. 4.15). There is no change in chemical composition accompanying this phase change or other phase changes described in this section. Mantle olivine (Fo_{90}) transforms to wadsleyite at pressures less than 300 MPa at appropriate temperatures for the 410-km discontinuity (~1000° C) (Ita and Stixrude, 1992; Helffrich and Wood, 2001). This pressure range agrees with the less than 10-km width of the 410-km discontinuity deduced from seismic data (Vidale et al., 1995). In some places, the discontinuity is

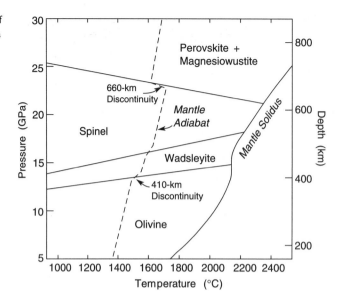

Figure 4.15 Summary of phase relations for Mg₂SiO₄ in the mantle from high-pressure and high-temperature experimental studies. The dashed line is the mantle adiabat. Modified from Christensen (1995).

broader than normal (20–25 km), a feature that may reflect water incorporated into the wadsleyite crystal structure (van der Meijde et al., 2003). If olivine composes 40 to 60% of the rock, as it does in garnet lherzolite, the olivine–wadsleyite phase change may account for the approximately 6% increase in density observed at this discontinuity (Table 4.2). Measurements of elastic moduli of olivine at high pressures suggest that 40% olivine explains the velocity contrast better than 60% olivine (Duffy et al., 1995). Because garnet lherzolites typically have 50 to 60% olivine, modal olivine may decrease with depth in the upper mantle to meet this constraint.

Experimental data indicate that wadsleyite should transform to a more densely packed spinel-structured phase (gamma phase) at equivalent burial depths of 500 to 550 km. This mineral, hereafter called **spinel,** has the same composition as Mg-rich olivine but the crystallographic structure of spinel. The small density change (~2%) associated with this transition, however, does not generally produce a resolvable seismic discontinuity.

High-pressure experimental data also indicate that at depths from 350 to 450 km, both clino- and orthopyroxene are transformed into a garnet-structured mineral known as **majorite garnet,** involving a density increase of about 6% (Christensen, 1995). This transition has been petrographically observed as pyroxene exsolution laminae in garnet in mantle xenoliths derived from the Archean lithosphere at depths from 300 to 400 km (Haggerty and Sautter, 1990). It is probable that an increase in velocity gradient sometimes observed from 350 km to the 410-km discontinuity is caused by these pyroxene transformations. At a slightly higher temperature, Ca-garnet begins to transform to Ca-perovskite (a mineral with Ca-garnet composition but perovskite structure). All of the preceding phase changes have positive slopes in P–T space; thus the reactions are exothermic (Table 4.2).

Table 4.2   Summary of Mantle Mineral Assemblages for Average Garnet Lherzolite from High-Pressure Studies

	Depth (km)	Mineral Assemblage (minerals in vol. %)	Density Contrast (%)	Slope of Reaction (MPa/°C)
	<410	Olivine 58 Opx 11 Cpx 18 Garnet 13		
	350–450	Opx-cpx → Majorite garnet	6	+1.5
410-km discontinuity	410	Olivine (α phase) → Wadsleyite (β phase)	6	+5.5
	410–550	Wadsleyite 58 Majorite garnet 30 Cpx 9 Opx 3		
	500–550	Wadsleyite → Spinel (γ phase)	2	+3.0
	550–660	Spinel 58 Majorite Garnet 37 Ca-perovskite 5 Ca-garnet → Ca-perovskite		
660-km discontinuity	660	Spinel → Perovskite + magnesiowustite	5	−2.5 to −2.8
	650–680	Majorite garnet → perovskite		+1.5 to +2.5
	680–2900	Perovskite 77 Magnesiowustite 15 Ca-perovskite 8 Silica (?)		

Data from Ita and Stixrude (1992), Christensen (1995), Mambole and Fleitout (2002), and Hirose (2002).

The 660-km Discontinuity

One of the most important questions related to the style of mantle convection in the Earth is the nature of the 660-km discontinuity (Fig. 4.1). If descending slabs cannot readily penetrate this boundary or if the boundary represents a compositional change, two-layer mantle convection is favored, with the 660-km discontinuity representing the base of the upper layer. Large increases in both seismic-wave velocity (5–7%) and density (5%) occur at this boundary. High-frequency seismic waves reflected at the boundary suggest that it has a width of only about 5 km but has up to 20 km of relief over hundreds to thousands of kilometers (Wood, 1995).

As with the 410-km discontinuity, it appears that a phase change in Mg_2SiO_4 is responsible for the 660-km discontinuity (Christensen, 1995; Helffrich and Wood, 2001). High-pressure experimental results indicate that spinel transforms to a mixture of perovskite and magnesiowustite at a pressure of about 23 GPa and can account for both the seismic velocity and the density increases at this boundary if the rock contains 50 to 60% spinel:

$$(Mg,Fe)_2 \, SiO_4 \rightarrow (Mg,Fe)SiO_3 + (Mg,Fe)O$$
$$spinel \rightarrow perovskite + magnesiowustite$$

Mg-**perovskite** and **magnesiowustite** are extremely high-density minerals and appear to comprise most of the lower mantle. High-pressure experimental studies show that small amounts of water may be carried as deep as the 660-km discontinuity in hydrous phases stable to 23 GPa (Ohtani et al., 1995).

Unlike the shallower phase transitions, the spinel–perovskite transition has a negative slope in P–T space (–2.5 to –2.8 MPa/°C) (Fig. 4.15; Table 4.2); thus the reaction is endothermic and may impede slabs from sinking into the deep mantle or impede plumes from rising into the upper mantle. The latent heat associated with phase transitions in descending slabs and rising plumes can deflect phase transitions to shallower depths for exothermic (positive P–T slope) reactions and to greater depths for endothermic (negative P–T slope) reactions (Liu, 1994). For a descending slab in an exothermic case, such as the olivine–wadsleyite transition, the elevated region of the denser phase exerts a strong downward pull on the slab or an upward pull on a plume, helping drive convection. In contrast, for an endothermic reaction, such as the spinel–perovskite transition, the low-density phase is depressed, enhancing a slab's buoyancy and resisting further sinking of the slab. This same reaction may retard a rising plume.

Around the same depth (650–680 km), majorite garnet transforms to perovskite (Table 4.2), but unlike the spinel transition, the garnet transition is gradual and does not produce a seismic discontinuity. This transition is sensitive to temperature and the Al content of the system. Unlike the spinel–perovskite reaction, the garnet–perovskite reaction has a positive slope (1.5 to 2.5 MPa/°C) (Hirose, 2002; Mambole and Fleitout, 2002).

Computer models by Davies (1995) suggest that stiff slabs can penetrate the boundary more readily than plume heads and that plume tails are the least able to penetrate it. Some investigators have suggested that slabs may locally accumulate at the 660-km discontinuity, culminating in occasional "avalanches" of slabs into the lower mantle. The fate of the oceanic crust in descending plates may be different from that of the suboceanic lithosphere because of their different compositions. Irifune and Ringwood (1993) suggested that the 660-km discontinuity may be a density "filter," which causes the crust to separate from the mantle in descending slabs. At depths less than 720 km, basaltic crust has a greater density than surrounding mantle, which could cause separation of the two components when they intersect the discontinuity. At a depth of about 700 km, however, the crust becomes less dense than surrounding mantle because of the majorite–perovskite transition (Hirose et al., 1999). Hence, if the mafic parts of the slab accumulate to sufficient thickness (>60 km) just beneath the 660-km discontinuity, the density of the basaltic component will increase and may drive the slabs into the deep mantle.

The Lower Mantle

General Features

High-pressure experimental studies clearly suggest that Mg-perovskite is the dominant phase in the lower mantle (Table 4.2). However, it is still not clear whether the seismic properties of the lower mantle necessitate a change in major element composition (Wang et al., 1994). Results allow, but do not require, the Fe/Mg ratio of the lower mantle to be greater than that of the upper mantle. If this were the case, it would greatly limit the mass flux across the 660-km discontinuity to maintain such a chemical difference and thus favor layered convection. It is also possible that free silica could exist in the lower mantle. Stishovite (a high-P phase of silica) inverts to an even denser silica polymorph with a $CaCl_2$ structure around 50 GPa (Kingma et al., 1995), and it is possible that this silica phase exists in the Earth at depths greater than 1200 km.

New interpretations of the isostatic rebound of continents following Pleistocene glaciation and new views of gravity data indicate that the viscosity of the mantle increases with depth by 2 orders of magnitude, with the largest jump at the 660-km discontinuity. This conclusion agrees with other geophysical and geochemical observations. For instance, although mantle plumes move upward relatively quickly, it would be impossible for them to survive convective currents in the upper mantle unless they were anchored in a "stiff" lower mantle. Also, only a mantle of relatively high viscosity at depth can account for the small number (two today) of mantle upwellings. Geochemical domains that appear to have remained isolated from each other for billions of years in the lower mantle (as described later in this chapter) can also be accounted for in a stiff lower mantle that resists mixing.

Descending Slabs

Although there is still considerable variation in the seismic-wave tomography sections of the mantle among different investigators, some of the general features of the deep mantle are becoming well established (Plates 1 and 2). S-wave anomalies continue over distances of thousands of kilometers with apparent widths as small as several hundred kilometers (van der Hilst et al., 1997; Grand et al., 1997). Fast anomalies can be tracked thousands of kilometers into the mantle and appear to represent descending oceanic plates. One prominent anomaly extending from about 30° S to 50° N in North and South America can be traced into the deep mantle and, as previously mentioned, is interpreted as the Farallon plate. High-velocity anomalies also extend beneath Asia, probably to the base of the mantle (Plates 2 and 3). Note also the low-velocity anomalies beneath Africa and the South Pacific, which reflect the two large mantle upwellings previously described.

Data suggest that the degree of slab penetration into the mantle is related to slab dip and migration of converging margins (Zhong and Gurnis, 1995; Li and Yuan, 2003). Slabs with steep dip angles and relatively stationary trenches, such as the Mariana and Tonga slabs (van der Hilst, 1995), are more likely to descend into the lower mantle (Plate 3a). In contrast, those with shallow dip angles, such as the Izu-Bonin and Japan slabs, are commonly associated with rapid retrograde trench migration and these have greater difficulty in penetrating the 660-km discontinuity. Although some slabs may be delayed at the 660-km discontinuity, tomographic images suggest that all or most modern slabs eventually sink into the lower mantle. Thus, there is little evidence for layered convection in the Earth in terms of slab distributions in the mantle.

The high-velocity anomaly beneath Asia that extends to the lower mantle appears to represent one or more ancient oceanic slabs (the Mongol-Okhotsk plate) subducted into the mantle during closure of several ocean basins as microcontinents collided to form Asia in the Mesozoic (M in Plate 3b) (Van der Voo et al., 1999). Subduction ended about 135 Ma, at which time the subducted slabs became detached from the lithosphere and continued to sink into the mantle. These slabs have sunk about 1 cm/year into the mantle. The Pacific plate, which subducts beneath Japan, is also apparent in the tomographic section, and it appears to merge with the Mongol-Okhotsk plate in the deep mantle.

One interesting yet controversial feature of some tomographic sections is that many, if not most, descending plates seem to break up below about 1700 km (Plate 3a). The pattern of high-velocity anomalies beneath this depth is irregular and dispersed. This suggests to some investigators that the mantle beneath 1700 km may be convectively isolated from the upper part of the mantle (van der Hilst and Karason, 1999; Kaneshima and Helffrich, 1999).

The D″ Layer

The D″ layer is a region of the mantle just above the core in which seismic-velocity gradients are anomalously low (Young and Lay, 1987; Loper and Lay, 1995; Helffrich and Wood, 2001) (Fig. 4.16). Estimates of the thickness of the D″ layer suggest that it ranges from 100 to about 400 km. Calculations indicate that only a relatively small temperature gradient (1–3° C/km) is necessary to conduct heat from the core into the D″ layer. Because of diffraction of seismic waves by the core, the resolution in this layer is not as good as at shallower mantle depths; thus, details of its structure are not well known. However, results clearly indicate that D″ is a complex region that is both vertically and laterally heterogeneous and that it is layered on a kilometer scale in the lower 50 km (Kendall and Silver, 1996; Sidorin et al., 1999; Thybo et al., 2003). Data also indicate the presence of a major solid–solid phase transition about 200 km above the core–mantle interface (Sidorin et al., 1999) (Fig. 4.16). Seismic results confirm the existence of a 5 to 50 km thick ultra-low-velocity layer just above the core–mantle boundary with S-wave velocity decreases of 10 to 50%, consistent with more than 15% melt in this layer (Thybo et al., 2003). As an example of lateral heterogeneity in D″, regions beneath circum-Pacific subduction zones have anomalously fast P- and S-waves, interpreted by many to represent lithospheric

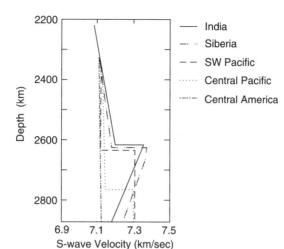

Figure 4.16 S-wave velocity distribution for various geographic regions in the lower mantle. Although most regions show a sharp upper boundary at the top of the D″ layer, the velocity profiles show a great deal of lateral heterogeneity in this region. Data modified from Knittle and Jeanloz (1991).

slabs that have sunk to the base of the mantle. It is noteworthy that slow velocities in D″ occur beneath the central Pacific and correlate with the surface and core–mantle boundary geoid anomalies (Fig. 4.4) and a concentration of hotspots.

There are three possible contributions to the complex seismic structures seen in D″: temperature variations, compositional changes, and mineralogical phase changes. Temperature variations appear to be caused chiefly by slabs sinking into D″ (a cooling effect that produces relatively fast velocities) and heat released from the core (causing slow velocities). Mixing of molten iron from the core with high-pressure silicates can lead to compositional changes with corresponding velocity changes. Experiments have shown, for instance, that when liquid iron comes in contact with silicate perovskite at high pressures, these substances vigorously react to produce a mixture of Mg-perovskite, a high-pressure silica polymorph, wustite (FeO), and Fe silicide (FeSi) (Knittle and Jeanloz, 1991; Dubrovinsky et al., 2003). These experiments also suggest that liquid iron in the outer core will seep into D″ by capillary action to hundreds of meters above the core–mantle boundary and that the reactions will occur on timescales of less than 10^6 years. The S-wave velocity discontinuity around 2600 km (Fig. 4.16) may be caused by a phase change. This may be because of the breakdown of perovskite to magnesiowustite and silica, a reaction that has a positive Clapeyron slope of about 6 MPa/°C (Sidorin et al., 1999). Although the solidus in the deep mantle is more than 1500° C above the average present-day geotherm, experimental data suggest that at the core–mantle boundary the solidus is near the core temperature and thus partial melting of the silicate mantle is possible near the boundary (Zerr et al., 1998).

Because the low seismic-wave velocities in D″ reflect high temperatures, and thus lower mantle viscosity, this layer is commonly thought to be the source of mantle plumes. The lower viscosity will also enhance the flow of material into the base of newly forming plumes, and the lateral flow into plumes will be balanced by slow subsidence of the overlying mantle. The results of Davies and Richards (1992) suggest that a plume could be fed for 100 My from a volume of D″ only tens of kilometers thick and 500 to 1000 km in diameter. These results are important for mantle dynamics as they suggest that plumes are fed from the lowest mantle, whereas ocean ridges are fed from the uppermost mantle.

The heterogeneous nature of the D″ layer is consistent with the presence of denser material, commonly called *dregs*. Slow upward convection of the mantle may pull dense phases such as wustite and FeSi upward from the core–mantle boundary. Eventually, they begin to sink because of their greater density, and these may form dregs that accumulate near the base of D″. Modeling suggests that dregs should pile up in regions of mantle upwelling (the slow regions in Plate 2, 2800 km) and thin in regions of downwelling, with the possibility that parts of D″ could be swept clean of dregs beneath downwellings. This means that the dregs must be continually supplied by reactions and upwelling from the core–mantle boundary. Lateral variations in the thickness of D″ caused by lateral dreg movements could account for the large-scale seismic-wave velocity variations and the variations in the thickness of D″. Small-scale compositional heterogeneity in D″ or convection, including the dregs in this layer, could account for small-scale variations and for scattering of seismic waves.

Plate-Driving Forces

Although the question of what drives the Earth's plates has stirred a lot of controversy in the past, investigators now seem to be converging on an answer. Most investigators agree that plate motions must be related to thermal convection in the mantle, although a generally accepted model relating the two processes remains elusive. The shapes and sizes of plates and their velocities exhibit large variations and do not show simple geometric relationships to convective flow patterns. Most computer models, however, indicate that plates move in response chiefly to slab-pull forces as plates descend into the mantle at subduction zones, and that ocean-ridge push or stresses transmitted from the asthenosphere to the lithosphere are small (Vigny et al., 1991; Lithgow-Bertelloni and Richards, 1995; Conrad and Lithgow-Bertelloni, 2002). In effect, stress distributions are consistent with the idea that at least oceanic plates are decoupled from underlying asthenosphere (Wiens and Stein, 1985). Ridge-push forces are caused by two factors: (1) horizontal density contrasts resulting from cooling and thickening of the oceanic lithosphere as it moves from ridges and (2) the elevation of the ocean ridge above the surrounding seafloor (Spence, 1987). The slab-pull forces in subduction zones reflect the cooling and negative buoyancy of the oceanic lithosphere as it ages. The gabbro–eclogite and the other high-pressure phase transitions that occur in descending slabs also contribute to slab-pull by increasing the density of the slab.

Using an analytic torque balance method, which accounts for interactions between plates by viscous coupling to a convecting mantle, Lithgow-Bertelloni and Richards (1995) show that the slab-pull forces amount to about 95% of the net driving forces of plates. Ridge-push and drag forces at the base of the plates are no more than 5% of the total. Computer models using other approaches and assumptions seem to agree that slab-pull forces dominate (Vigny et al., 1991; Carlson, 1995a). Although slab-pull cannot initiate subduction, once a slab begins to sink, the slab-pull force rapidly becomes the dominant force for continued subduction.

Mantle Plumes

Introduction

A **mantle plume** is a buoyant mass of material in the mantle that, because of its buoyancy, rises. The existence of mantle plumes in the Earth was first suggested by Wilson (1963) as an explanation of oceanic-island chains, such as the Hawaiian-Emperor chain, which change progressively in age along the chain. Wilson proposed that as a lithospheric plate moves across a fixed hotspot (the mantle plume), volcanism is recorded as a linear array of volcanic seamounts and islands parallel to the direction the plate is moving. However, we know today that plumes can also move, so this simplified model will not hold for all hotspot tracks. There even are a few investigators who question whether mantle plumes exist.

Laboratory experiments show that plume viscosity has an important effect on the shape of a plume. If a plume has a viscosity greater than its surroundings, it rises as a finger, whereas if it has a lower viscosity, it rises in a mushroom with a distinct head and tail (Fig. 4.17). The tail contains a hot fluid that "feeds" the head as it buoyantly rises. Griffiths and Campbell (1990) were the first to confirm by experiment and theory the existence of thermal plume heads and tails and to distinguish between thermal and compositional plumes. In thermal plume heads, the boundary layer around the plume is heated by conduction, becomes buoyant, and rises. This results in a plume head of 1000 km or more. This size is consistent with many large igneous provinces, which may be 500 to 3000 km in diameter. Because of the thermal buoyancy of the head, it entrains material from its surroundings as it grows; depending on the rate of ascent, it may entrain up to 90% of its starting mass. Streamlines in computer models show that most of the entrained fraction should come from the lower mantle (Hauri et al., 1994).

Although many details of plumes and their effects are still controversial and debated, the basic theory of mantle plumes is well established and there is considerable observational evidence to support the plume concept. Only recently, however, has the resolution of seismic tomography improved sufficiently that at least some plumes in the upper mantle may be detected seismically (Li et al., 2000; Rhodes and Davies, 2001; Ritsema and Allen, 2003) (Plate 4).

Hotspots

As explained in Chapter 3, hotspots are generally thought to form in response to mantle plumes that reach the base of the lithosphere (Duncan and Richards, 1991). Partial melting of plumes in the upper mantle leads to large volumes of magma, which are partly erupted (or intruded) at the Earth's surface.

Figure 4.17   Photograph of a starting plume in a laboratory experiment showing the large head and narrow tail. The plume was produced by continuously injecting hotter, lower viscosity, dyed fluid into the base of a cool, higher viscosity layer of the same fluid. Light regions in the head are entrained from surrounding undyed fluid. Courtesy of Ian Campbell.

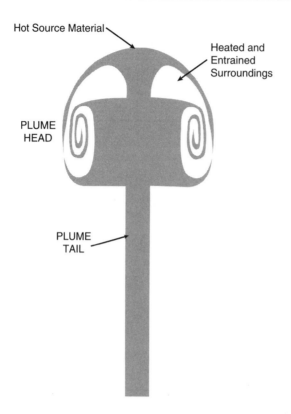

Somewhere between 40 and 150 active hotspots have been described on the Earth. The best-documented hotspots have a rather irregular distribution occurring in both oceanic and continental areas (Fig. 4.18). Some occur on or near ocean ridges, such as Iceland, St. Helena, and Tristan in the Atlantic basin, and others occur near the centers of plates (such as Hawaii). How long do hotspots last? One of the oldest hotspots is the Kerguelen hotspot in the southern Indian Ocean, which began to produce basalts about 117 Ma. Most modern hotspots, however, are less than 100 My. The life spans of hotspots depend on such parameters as the plume size and the tectonic environment into which a plume is emplaced. On the Pacific plate, three volcanic chains were generated by hotspots between 70 and 25 Ma, whereas 12 chains have been generated in the last 25 My. The large number of hotspots in and around Africa and in the Pacific basin corresponds to the two major geoid highs (Stefanick and Jurdy, 1984) (Fig. 4.18). The geoid highs appear to reflect mantle upwellings, supporting the idea that hotspots are caused by mantle plumes rising from the deep mantle.

A major problem that has stimulated controversy is whether hotspots remain fixed relative to plates (Duncan and Richards, 1991). If they have remained fixed, they provide a means of determining absolute plate velocities. The magnitude of interplume motion can be assessed by comparing the geometry and age distribution of volcanism along hotspot tracks with reconstructions of past plate movements based on paleomagnetic data.

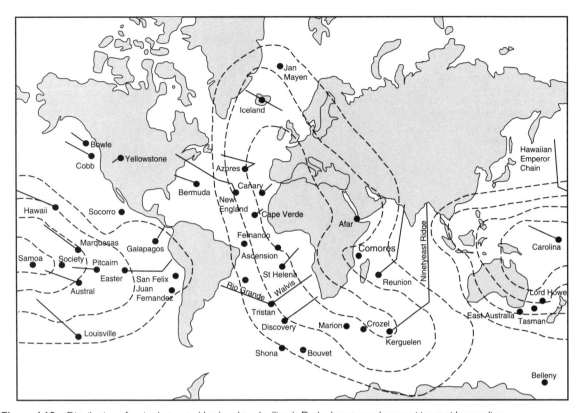

Figure 4.18 Distribution of major hotspots (dots) and tracks (lines). Dashed contours show positive geoid anomalies.

If hotspots are stationary beneath two plates, then the calculated motions of both plates should follow hotspot tracks observed on them. The close correspondence between observed and modeled tracks on the Australian and African plates (Fig. 4.19) supports the idea that these hotspots are fixed relative to each other and are maintained by deeply rooted mantle plumes. Also, the almost perfect fit of volcanic chains in the Pacific plate with a pole of rotation at 70° N 101° W and a rate of rotation about this pole of about 1 deg/My for the last 10 My suggest that Pacific hotspots have remained fixed relative to each other over this short period (Steinberger and O'Connell, 1998).

   If all hotspots have remained fixed with respect to each other, it should be possible to superimpose the same hotspots in their present positions on their predicted positions at other times in the last 150 to 200 My. Except for hotspots near each other, however, it is not possible to do this, suggesting that hotspots move in the upper mantle (Duncan and Richards, 1991; Van Fossen and Kent, 1992). As mentioned in Chapter 3, there is now strong paleomagnetic evidence for movement of the Hawaiian hotspot more than 43 Ma (Tarduno et al., 2003). Also, in comparing Atlantic with Pacific hotspots, there are significant differences between calculated and observed hotspot tracks (Molnar and Stock, 1987). Using paleolatitudes deduced from seamounts, Tarduno and Gee (1995) show that Pacific hotspots have moved relative to Atlantic hotspots at a rate of only 30 mm/year.

Figure 4.19 Computer-generated hotspot tracks for the Indian Ocean and South Atlantic basins, showing calculated (heavy lines) and observed hotspot trajectories. Current hotspots are indicated by black dots, and past ages of hotspot basalts are given in millions of years. Modified from Duncan (1991).

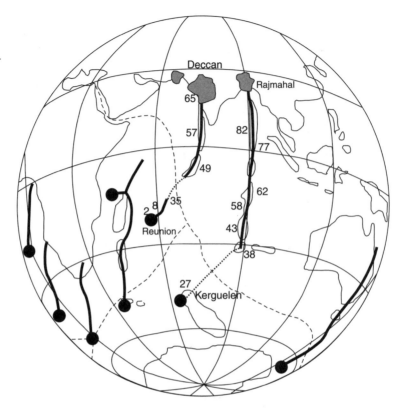

Plume Characteristics

Theoretical and laboratory models have clarified the dynamics of plumes and have suggested numerous ways that plumes may interact with the mantle. Although the bases of both the upper and the lower mantle have been suggested as sites of mantle-plume generation, five lines of evidence suggest that plumes are produced just above the core–mantle boundary in the D″ layer (Campbell and Griffiths, 1992a; Davies and Richards, 1992):

1. Computer modeling indicates that plume heads can achieve a size (~1000 km in diameter) required to form large volumes of flood basalt and large submarine plateaus, such as the Ontong-Java plateau, only if they come from the deep mantle.

2. The approximate fixed position of hotspots relative to each other in the same geographic region is difficult to explain if plumes originate in the upper mantle but is consistent with a lower mantle source.

3. The amount of heat transferred to the base of the lithosphere by plumes, no more than 12% of the Earth's total heat flux, is comparable with the amount of heat estimated to be emerging from the core as it cools.

4. Periods of increased plume activity in the past seem to correlate with normal polarity epochs and decreased pole reversal activity in plume-related basalts (Larson, 1991a). Correlation of plume activity with magnetic reversals implies that heat transfer across the core–mantle boundary starts a mantle plume in D″ and changes the pattern of convective flow in the outer core, which in turn affects the Earth's magnetic field.

5. By analogy with iron meteorites, the Earth's core should be enriched in Os with a high $^{187}Os/^{188}Os$ ratio compared with the mantle (Walker et al., 1995). Thus, plumes produced in the D″ layer could be contaminated with radiogenic Os from the core. Some plume-derived basalts have $^{187}Os/^{188}Os$ ratios up to 20% higher than primitive or depleted mantle, suggesting core contamination and thus a source near the core–mantle interface.

Numerical models of Davies (1999) and others of temperature-dependent viscosity plumes confirm and expand upon the laboratory experiments. A sequence of snapshots of a mantle plume as it ascends and spreads over a 175-My period is shown in Figure 4.20. The plume ascends rapidly in the first 100 My and then more slowly as it begins to flatten against the lithosphere. As the hot fluid in the plume tail reaches the top of the head, it flattens against the lithosphere and the head becomes thin and increases significantly in radius (176 Ma). Heat liberated from the plume partly escapes and partly heats material entrained in the plume head so that the head has a temperature intermediate between the tail and the surrounding mantle. Such features were also recognized in the experimental models of Griffiths and Campbell (1990). As the head continues to grow and cool, it rises more slowly than the tail, which continues to feed the head.

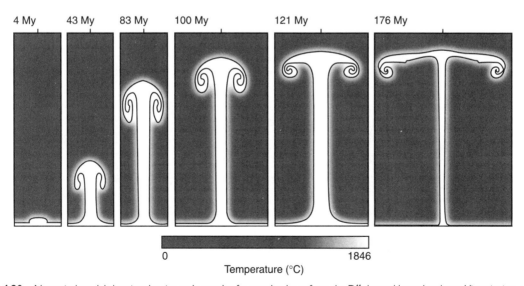

Figure 4.20 Numerical model showing the rise and growth of a mantle plume from the D″ thermal boundary layer. Viscosity is a strong function of temperature, and ambient viscosity is 10^{22} Pa sec. The bottom boundary temperature is 430° C above the interior plume temperature, and plume viscosity here is about 1% of the plume interior. Courtesy of Geoff Davies.

Although convection should not affect plumes, what about the endothermic phase change at the 660-km discontinuity? Davies (1995) has shown by computer models that for the temperature- and depth-dependent viscosity in the Earth, most plumes should pass through the spinel–perovskite phase boundary unimpeded. Supporting this model is that the majorite–perovskite transition around the same depth may partially or completely offset the spinel–perovskite effect because of the opposite Clapeyron slopes previously described.

As a plume head approaches the lithosphere, it begins to flatten and eventually intersects the mantle solidus, producing large volumes of basaltic magma in relatively short times (White and McKenzie, 1995). Plumes with flattened heads no less than 1000 km in diameter are plausible sources for flood basalts and large oceanic plateaus, as explained in Chapter 3.

Mantle Geochemical Components

Introduction

Radiogenic isotopes can be used as geochemical tracers to track the geographic and age distribution of crustal or mantle reservoirs from which the elements that house the parent and daughter isotopes have been fractionated from each other. For instance, any process that fractionates U from Pb or Sm from Nd will result in a changing rate of growth of the daughter isotopes ^{207}Pb and ^{206}Pb (from ^{237}U and ^{235}U, respectively) and ^{143}Nd (from ^{147}Sm). If the element fractionation occurred long ago geologically, the change in parent/daughter ratio will result in a measurable difference in the isotopic composition of the melt and its source. This is the basis for the isotopic dating of rocks. On the other hand, if the chemical change occurs during a recent event, such as partial melting in the mantle just before magma eruption, there is not enough time for the erupted lava to evolve a new isotopic signature distinct from the source material reflecting its new parent/daughter ratio. Hence, young basalts derived from the mantle carry with them the isotopic composition of their mantle sources, and this is the basis of the **geochemical tracer** method.

When isotopic ratios from young oceanic basalts are plotted on isochron diagrams, the data often fall close to an isochron with a Precambrian age; these are sometimes called *mantle isochrons* (Brooks et al., 1976). The interpretation of these isochrons, however, is ambiguous. They could represent true ages of major fractionation events in the mantle, or they may be mixing lines between end member components in the mantle. However, most ages calculated from U-Pb and Rb-Sr isotopic data from young oceanic basalts fall between 1.8 and 1.6 Ga and most Sm-Nd results yield ages from 2.0 to 1.8 Ga, which seem to favor their interpretation as ages of mantle events. Even if the linear arrays on isochron diagrams are mixing lines, at least one of the end members has to be Paleoproterozoic.

Identifying Mantle Components

Summary

At least four and perhaps as many as six isotopic end members may exist in the mantle from results available from oceanic basalts (Hart, 1988; Hart et al., 1992; Helffrich and

Wood, 2001) (Figs. 4.21 and 4.22). These are **depleted mantle,** the source of normal midocean-ridge basalts (NMORB); **HIMU,** distinguished by its high $^{206}Pb/^{204}Pb$ ratio, which reflects a high U/Pb ratio ($\mu = {}^{238}U/^{204}Pb$) in the source; and two enriched mantle sources **(EM1 and EM2),** which reflect long-term enrichment in light rare earth elements in the source. A fifth component, **primitive mantle,** may be preserved in parts of the mantle.

The existence of at least four mantle end members is well documented. What remains to be verified is the origin and location of each end member and their mixing relations. As summarized in the next sections, much progress has been made on these questions using rare gas isotopic data and trace element ratios. Also, the hierarchy of mixing of components in basalts from single islands or island chains can provide useful information about the location of the components in the mantle.

Depleted Mantle

Depleted mantle has undergone one or more periods of fractionation involving extraction of basaltic magmas. Depleted mantle is known to underlie ocean ridges and probably extends beneath ocean basins, although it is not the source of oceanic-island magmas. The depleted isotopic character (low $^{87}Sr/^{86}Sr$, $^{206}Pb/^{204}Pb$, and high $^{143}Nd/^{144}Nd$) and low LIL element contents of NMORB require the existence in the Earth of a widespread depleted mantle reservoir. Rare gas isotopic compositions also require that this reservoir be highly depleted in rare gases compared with other mantle components. Although most of the geochemical variation within NMORB can be explained by magmatic processes such as fractional crystallization, variations in isotopic ratios demand that the depleted mantle reservoir is heterogeneous, at least on scales from 10^2 to 10^3 km. This heterogeneity may be caused by small amounts of mixing with enriched mantle components.

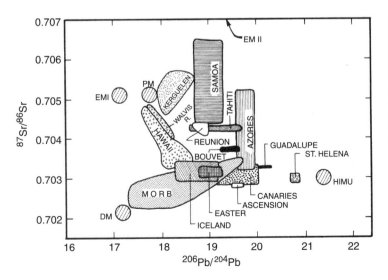

Figure 4.21 Sr and Pb isotope distributions in MORB and oceanic island basalts. Modified from Zindler and Hart (1986). DM, depleted mantle; EM1 and EM2, enriched mantle components; HIMU, high U/Pb ratio; PM, primitive mantle.

Figure 4.22 Nd and Pb
isotope distributions in
MORB and oceanic island
basalts. Modified from
Zindler and Hart (1986).
DM, depleted mantle; EM1
and EM2, enriched mantle
components; HIMU, high
U/Pb ratio; PM, primitive
mantle.

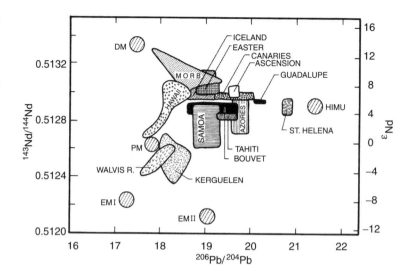

HIMU

The extreme enrichment in ^{206}Pb and ^{208}Pb in some oceanic-island basalts (such as those from St. Helena, Figs. 4.21 and 4.22) requires the existence of a mantle source enriched in U+Th relative to Pb, and mantle isochrons suggest an age for this HIMU source of 2.0 to 1.5 Ga. This reservoir is also enriched in radiogenic ^{187}Os (Hauri and Hart, 1993). Because HIMU has ^{87}Sr/^{86}Sr ratios similar to NMORB, however, it has been suggested that it represents subducted oceanic crust in which the U+Th/Pb ratio was increased by preferential loss of Pb in volatiles escaping upward from descending slabs during sub-duction. Supporting a recycled oceanic crust origin for HIMU are relative enrichments in Ta and Nb in many oceanic-island basalts. As explained in Chapter 3, devolatilized descending slabs should be relatively enriched in these elements because neighboring incompatible elements such as Th, U, K, and Ba have been lost to the mantle wedge. Thus, the residual mafic part of the slab that sinks into the lower mantle and becomes incorporated in mantle plumes should be relatively enriched in Ta and Nb. Although an origin for HIMU as recycled oceanic crust is widely agreed upon, Kamber and Collerson (1999) have proposed that in terms of helium and Pb isotopes it can best be explained as recycled oceanic lithosphere that has been metasomatized.

Enriched Mantle

Enriched mantle components are mantle reservoirs enriched in incompatible elements such as Rb, Sm, U, and Th. At least two enriched components are required to explain the isotopic and trace element distributions in the sources of oceanic basalts (Zindler and Hart, 1986): EM1 with moderate ^{87}Sr/^{86}Sr ratios and low ^{206}Pb/^{204}Pb ratios and EM2 with high ^{87}Sr/^{86}Sr ratios and moderate ^{206}Pb/^{204}Pb ratios. Both have low ^{143}Nd/^{144}Nd ratios. Among the numerous candidates proposed for EM1 are old oceanic mantle lithosphere

(±sediments) that has been recycled back into the mantle (Hart et al., 1992; Hauri et al., 1994) and metasomatized lower mantle (Kamber and Collerson, 1999). Also, studies of Hf isotope distributions in Hawaiian basalts have been interpreted to indicate the presence of old pelagic sediments in their plume sources (Blichert-Toft et al., 1999). At some localities (the Walvis ridge, the southwest Indian ridge, and the Marion hotspot), low but variable $^{206}Pb/^{204}Pb$ ratios occur in the volcanics, and a large range of $^{207}Pb/^{204}Pb$ and $^{208}Pb/^{204}Pb$ ratios are found relative to the $^{206}Pb/^{204}Pb$ ratio (e.g., Mahoney et al., 1995; Douglass et al., 1999). These variations clearly indicate that EM1 cannot be considered one mantle component but must represent several components. All of these components must be depleted in the more highly incompatible element U relative to Pb on a time-integrated basis compared with the average depleted mantle from which NMORB comes, and they must have a time-integrated Rb/Sr similar to that of primitive mantle.

EM2, if there is a single EM2 component, has isotopic ratios closer to average upper continental crust or to modern subducted continental sediments (i.e., $^{87}Sr/^{86}Sr$ >0.71 and $^{143}Nd/^{144}Nd$ ~0.5121). Subducted continental sediments are favored by some investigators for this end member because EM2 commonly contributes to island arc volcanics in which continental sediments have been subducted, such as the Lesser Antilles and the Sunda arc (Hauri et al., 1994). However, based on helium and Pb isotope distributions in Hawaiian lavas, Kamber and Collerson (1999) propose that EM2 is recycled subcontinental lithosphere and that it has a variable composition; Workman et al. (2003) suggest that it may be recycled oceanic lithosphere with trapped melt fractions.

Helium Isotopes

One of the most important observations in oceanic basalts is that their helium isotope ratios differ according to tectonic setting. There are two isotopes of helium: ^3He, a primordial isotope incorporated in the Earth as it accreted, and ^4He, an isotope produced by radioactive decay of U and Th isotopes. Plume-related basalts in oceanic areas have relatively high $^3He/^4He$ ratios, often more than 20 times that of air ($R/R_A \geq 20$), whereas midocean-ridge basalts (MORB) generally has R/R_A values of 7 to 9 (Hanan and Graham, 1996). In some Pacific MORB, R/R_A values increase with increasing $^{206}Pb/^{204}Pb$ ratios, whereas in the Atlantic, the opposite trend is observed. Also, in the Atlantic and Indian Oceans, helium isotope ratios tend to be higher in basalts with EM1 characteristics. These high ratios, which apply to both MORB and OIB, may be sampling two EM1 reservoirs, both with low U/Pb and high Th/Pb ratios.

Two classes of models have been suggested to explain the origin of the high $^3He/^4He$ reservoir in the mantle. Some investigators interpret the high ratios to reflect recycled oceanic lithosphere in the deep mantle (Albarede, 1998). Such lithosphere should have a high $^3He/^4He$ ratio because partial melting at ocean ridges extracts almost all of the U and Th from the mantle source. Because these elements are responsible for the accumulation of ^4He over time, causing the $^3He/^4He$ ratio to decrease in the source, without U and Th, the depleted oceanic lithosphere would acquire a high $^3He/^4He$ ratio. Alternative models call upon primitive, unfractionated sources deep in the mantle that retain their original high ^{3}He/^{4}He ratios (Helffrich and Wood, 2001). So far, the interpretation of helium isotope distributions in oceanic basalts remains ambiguous.

The Dupal Anomaly

With some important exceptions, most oceanic islands showing enriched mantle components occur in the Southern Hemisphere (Hart, 1988). Also, NMORB in the South Atlantic and Indian Oceans exhibit definite contributions from enriched mantle sources, unlike NMORB from other ocean-ridge systems. This band of mantle enrichment in the Southern Hemisphere is known as the Dupal anomaly (Fig. 4.23). Pb isotopic data indicate that the anomaly may have existed for billions of years, although not always as an intact geographic feature in the Southern Hemisphere (Hart, 1984; Mahoney et al., 1998). Except a few samples from the Arctic ridge, an intriguing feature of this anomaly is the apparent lack of enriched mantle signatures in hotspot basalts from polar regions. Both shallow and deep origins have been proposed for the Dupal anomaly. Favoring a deep source is the correlation of the Dupal anomaly with body-wave anomalies in the mantle.

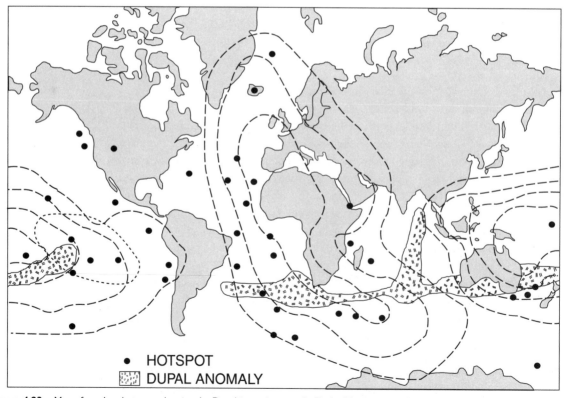

Figure 4.23 Map of modern hotspots showing the Dupal isotopic anomaly. Dashed lines are geoid anomalies.

Some of the Dupal maxima occur over the slowest seismic velocities in the lower mantle, suggesting that the anomaly is an expression of material rising in plumes from the core–mantle boundary region (Hart, 1988).

On the other hand, a shallow origin for Dupal is supported by enriched mantle components in South Atlantic and Indian Ocean NMORB and in island arc basalts in the southwest Pacific (Mahoney et al., 1992). The best way out of this dilemma is for Dupal to have both deep and shallow sources. If EM2, which dominates the Dupal anomaly, is caused by modern continental sediments, the plume source could be from sediments that have been recycled through the deep mantle with sinking slabs. In contrast, the shallow source could be produced by subducted sediments that pass their geochemical signature to mantle wedges, to the subcontinental lithosphere, or both during plate devolatilization. Basalts along the Discovery section of the Mid-Atlantic ridge (47°30′ S) show a Dupal-type geochemical signature but have relatively low $^{206}Pb/^{204}Pb$ and $^{3}He/^{4}He$ ratios, a feature interpreted to represent recycled oceanic crust or delaminated subcontinental lithosphere (Sarda et al., 2000).

Mixing Regimes in the Mantle

When viewed in Nd-Sr-Pb isotopic space, arrays defined by various oceanic islands tend to cluster along two and three component mixing lines with depleted mantle, HIMU, EM1, and EM2 at the corners of a tetrahedron (Fig. 4.24). There is, however, a notable lack of mixing arrays joining EM1, EM2, and HIMU, which seems to rule out random mixing in the mantle (Hart et al., 1992; Hauri et al., 1994). The mixing arrays are systematically oriented originating from points along the EM1–HIMU and EM2–HIMU joins and converging in a region within the lower part of the tetrahedron but not at depleted mantle. This convergence suggests the existence of another mantle component that appears in oceanic mantle plumes. This component, called FOZO (or C) after the focal zone from the converging arrays, is a depleted mantle component (Hart et al., 1992) (Fig. 4. 24). Helium isotope ratios also progressively increase toward FOZO, and because ^{3}He is not formed by radiogenic decay but was incorporated in the Earth during planetary accretion, the high $^{3}He/^{4}He$ ratios are commonly interpreted to mean that the FOZO component is relatively primitive and probably resides deep in the mantle. Alternatively, as mentioned previously, the high $^{3}He/^{4}He$ ratios in FOZO may represent a recycled oceanic lithosphere in this source.

Mixing arrays do not emanate from the EM1–EM2 join or from the EM1–EM2–HIMU face of the tetrahedron, indicating that there was some process that caused the juxtaposition of enriched mantle components and HIMU that predates the mixing with FOZO. The probability that most or all hotspot sources do not mix with the shallow depleted mantle is consistent with most plume entrainment occurring in the lower mantle.

The isotopic mixing arrays do not tell investigators which components occur in the plume sources and which have been entrained in the mantle. A high $^{3}He/^{4}He$ ratio in FOZO in contrast to a low $^{3}He/^{4}He$ ratio in depleted mantle may indicate that FOZO is in the lower mantle and does not mix with depleted mantle in the upper mantle. It is not

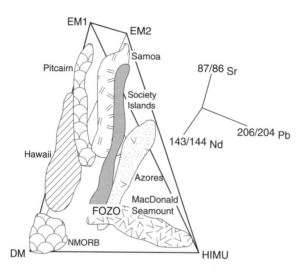

Figure 4.24 Three-dimensional plot of Sr, Nd, and Pb isotopic ratios from intraplate ocean basalts. The tetrahedron is defined by the mantle end members. DM, depleted mantle; EM1 and EM2, enriched mantle components; FOZO, focal zone; HIMU, high U/Pb ratio; NMORB, normal midocean-ridge basalts. Modified from Hauri et al. (1994).

possible from isotopic data alone to determine whether the enriched mantle–HIMU components are the plume sources in the D″ layer, and FOZO is entrained in the lower mantle as the plumes rise, or the opposite is true (FOZO is the plume and enriched mantle–HIMU are entrained). When coupled with previously described geophysical data, however, the former model seems most acceptable.

Although isotopic, rare gas, and trace element distributions in oceanic basalts demand a heterogeneous mantle, the nature, scale, preservation, and history of mantle heterogeneities remain problematic. Experimental studies of diffusion rates of cations in mantle minerals provide a basis to constrain the minimum scale of mantle heterogeneities. Results suggest that even in the presence of a melt in which diffusion rates are high, heterogeneities larger than 1 km can persist for several billion years (Zindler and Hart, 1986; Kellogg, 1992). Geochemical heterogeneities in ultramafic portions of ophiolites occur on scales of centimeters to kilometers, indicating that small-scale heterogeneities survive in the upper mantle. Geochemical and isotopic variations in basalts from a single volcano require heterogeneity on a scale of several kilometers, and variations of MORB along ocean ridges or of arc basalts along single volcanic arcs demand heterogeneities on scales from 10^2 to 10^3 km. At the large end of the scale is the Dupal anomaly (Fig. 4.23), which requires global-scale anomalies that have survived for billions of years. Numerical stirring models show that if the viscosity of the lower mantle is 1 to 2 orders of magnitude greater than the upper mantle, mantle heterogeneities should survive for billions of years as isotopic data from oceanic-island basalts suggest (Davies and Richards, 1992). These models, furthermore, allow heterogeneities of a variety of scales to survive for long periods (Ogawa, 2003).

Because the mantle was hotter in the Archean, and hence probably convected more rapidly and chaotically, it is of interest to see whether mantle heterogeneities were preserved in the Archean mantle. Nd isotopic ratios and Sm/Nd ratios provide a means to compare the Archean mantle with the modern mantle by analyzing for these quantities in greenstone basalts. The mantle array on an E_{Nd}-$^{147}Sm/^{144}Nd$ plot varies significantly from the modern arrays (Fig. 4.25). For a similar overall range in Sm/Nd ratio, the Archean array varies by only 1 epsilon (E) unit compared with at least 8E units in modern arrays (Blichert-Toft and Albarede, 1994). These results demonstrate that both long-term depletion and enriched mantle domains developed in the mantle by the late Archean. The absence of a larger range of E_{Nd} values in the Archean mantle must mean that heterogeneities with variable Sm/Nd ratios did not survive long enough to be recorded in the $^{143}Nd/^{144}Nd$ isotopic ratio. This, in turn, suggests faster mixing in the Archean mantle.

Summary

Perhaps the most significant chemical signatures recorded by the mantle are the complementary relationships between the incompatible element enrichment in the continental crust (Fig. 2.20)—and probable siderophile element (Fe, Ni, Co, Cr, etc.) enrichment of the core—and the corresponding depletion of these elements in the depleted mantle reservoir (Carlson, 1994). It would seem that the key events controlling compositional variations in the Earth are early core formation followed by gradual or episodic extraction of continental crust to leave at least part of the mantle depleted in incompatible and siderophile elements. The depleted mantle and FOZO components may represent depleted mantle remaining from these events. The other mantle components would appear to reflect subducted inhomogeneities that were not sufficiently remixed into the

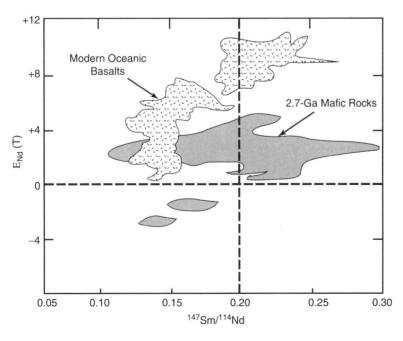

Figure 4.25 Distribution of late Archean and modern basalts on an E_{Nd} (T) versus $^{147}Sm/^{144}Nd$ graph. T (age) = 0 Ma for modern basalts and 2.7 Ga for Archean rocks. Modified from Blichert-Toft and Albarede (1994).

convecting mantle to lose their geochemical and isotopic signatures. Many of these components, with depleted mantle, appear to have been "fossilized" in the subcontinental lithosphere by accretion of spent plume material to the base of the lithosphere throughout geologic time (Menzies, 1990). Geochemical evidence for layering in the mantle is equivocal and controversial. Although the data do not exclude chemical layering, the nonrandom occurrence of geochemical components in plume sources would seem to require exchange between the upper and the lower mantle.

Convection in the Mantle

Nature of Convection

It is generally agreed that convection in the mantle is responsible for driving plate tectonics (Schubert et al., 2001). Convection arises because of buoyancy differences with lighter material rising and denser material sinking. In terms of quantitative laboratory models, **Rayleigh-Bernard convection** is best understood. This type of convection arises because of heating at the base and cooling at the surface of fluid. Lord Rayleigh was the first to show that convective behavior of a substance is dependent on a dimensionless number, now known as the *Rayleigh number.* For a simple homogeneous liquid to convect as it is heated at the base, the Rayleigh number must exceed 2000, and for the convection to be vigorous, it must be the order of 10^5. Irregular turbulent convection begins when the Rayleigh number reaches about 10^6, and such convection probably exists in the Earth when the Rayleigh number is less than 10^6. Another factor contributing to mantle convection is lateral motion of subducted slabs. Absolute plate motions indicate that slab migration is generally opposite to the direction of subduction at rates from 10 to 25 mm/year. As a result, the downward motions of slabs are generally steeper than their dips (Garfunkel et al., 1986). Calculations indicate that the mass flux in the upper mantle caused by lateral slab motions may be an important contribution to large-scale mantle convection.

A voluminous literature exists on models for convection in the Earth. In general, they fall into two categories: (1) layered convection and (2) whole-mantle convection. In the **layered convection** models, convection occurs separately below and above the 660-km discontinuity; **whole-mantle convection** involves the entire mantle, although parts of the deep mantle may be isolated from convection. It is important to remember that these types of convection are two end-member scenarios and that convection in the Earth may be somewhere between (i.e., partially layered convection). Although both models have a scientific following, as the resolution of seismic tomographic data has improved, it appears certain that lithospheric slabs descend into the lower mantle. This seems to necessitate some style of whole-mantle convection. Classical pictures of convection in the Earth show convective upcurrents coming from the deep mantle beneath ocean ridges and downcurrents returning at subduction zones. However, it now seems clear that ocean ridges are shallow passive features not related to deep convection in the Earth.

Passive Ocean Ridges

S-wave velocity distributions beneath ocean ridges are useful in distinguishing between shallow and deep sources for upwelling mantle. As shown in Figure 4.26, ocean ridges are typically underlain by broad low-velocity regions in the upper 100 km of the mantle. In all cases, the lowest velocities occur at depths less than 50 km, and regardless of spreading rate, regions with velocities 1 to 2% lower than normal are ~100 km deep (Zhang and Tanimoto, 1992). The widths of the low-velocity regions, however, correlate positively with spreading rates. This correlation may be caused by separating plates that drag the shallow asthenosphere from ridges. Supporting shallow upwelling beneath ocean ridges, ocean ridges migrate in response to stress distributions in the plates rather than to buoyancy forces in the deep mantle, as they should if they had deep roots.

The low-velocity anomalies under ocean ridges indicate they are shallow features, with shallow roots probably caused by passive upwellings produced by deformation in plates

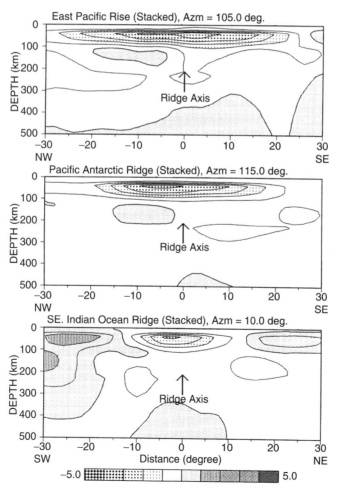

Figure 4.26 S-wave velocity distribution across three ocean ridges. Note that slow velocities are confined to shallow depths. Courtesy of Yu-Shen Zhang.

Figure 4.26, cont'd.

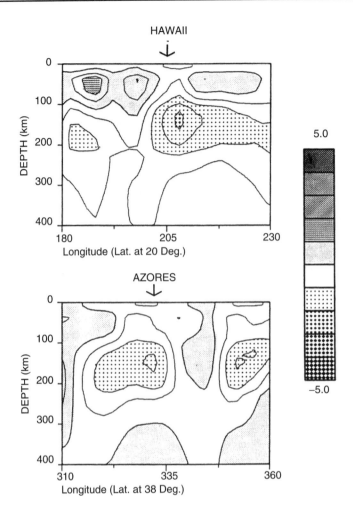

in response to sinking slabs. Because plumes are driven by density anomalies in the mantle, they should have low-velocity anomalies that extend much deeper into the mantle, as observed in Hawaii and the Azores. Both situations occur along the mid-Atlantic ridge, but only the hotspots on or near the ridge have deep low-velocity roots; the low-velocity roots of the ocean ridge are less than 100 km deep. Thus, it would appear that there is some degree of decoupling between active and passive asthenosphere at depths from 100 to 200 km. Just how this occurs and evolves with time is not well understood.

The Layered Convection Model

The layered convection model involves separately convecting upper and lower mantle reservoirs. The upper mantle, above the 660-km discontinuity, is generally equated with the geochemically depleted mantle reservoir, and the lower mantle is equated with the other mantle geochemical domains (Allegre, 1982; O'Nions, 1987). If depleted mantle is assumed to have formed chiefly by the extraction of continental crust from primitive mantle, model calculations based on Nd and Pb isotopes indicate that it must comprise between 35 and 50% of the entire mantle.

One of the major lines of evidence used to support layered mantle convection comes from the isotopic composition of rare gases in basalts (O'Nions, 1987; Allegre et al., 1995a). Nonradiogenic ^{3}He is enriched in many oceanic island basalts, and because these appear to be derived from mantle plumes, they may come from an undegassed source in the deep mantle, as described previously. Similar arguments can be made from Ar, Ne, and Xe isotopes. The simplest model to satisfy these constraints is a two-layer convective mantle with the upper layer (depleted mantle) strongly depleted in rare gases and LIL elements and the lower layer, which contains enriched components (HIMU and enriched mantle), as the site of generation of mantle plumes relatively enriched in ^{3}He.

Despite the rare gas data, layered convection has numerous difficulties. Among the most robust are the following:

1. The aspect ratio (width/depth) of convection cells should be close to unity. This is not consistent with layered convection in which convection cells bottom out at 660 km when horizontal measurements exceed 10^4 km as reflected by plate sizes.

2. Nd, Sr, and Pb isotopic data from oceanic basalts require not just two or three but several ancient mantle sources as described previously.

3. As you have seen, seismic-velocity studies indicate that descending lithosphere sinks into the lower mantle, a feature that would promote whole-mantle convection.

4. Mantle plumes appear to be derived from the lowest mantle, another feature consistent with whole-mantle convection.

As I pointed out previously, however, layering in the mantle is not an all or nothing situation, and numerous factors influence the degree of layering in the mantle (Christensen, 1995; Davies, 1995; Davies, 2002; Tackley, 2000). Among the more important are the following:

1. *Temperature.* As temperature increases, so does the Rayleigh number, and higher Rayleigh numbers tend to favor layered convection (Christensen and Yuen, 1985). This is caused because the sensitivity of an endothermic phase change, like that at the 660-km discontinuity, to retard descending slabs increases with an increasing Rayleigh number. As you shall see in Chapter 9, this may have resulted in strongly layered convection in the Archean when mantle temperatures were higher.

2. *Internal heating.* Model calculations indicate that internal heating of a substance, compared with bottom heating, increases the propensity for layered convection. Thus, the amount and distribution of radiogenic heat sources in the mantle can affect the type of convection.

3. *Exothermic phase changes.* As previously explained, all of the phase changes recognized in the upper mantle, except the 660-km perovskite phase change, are exothermic, and exothermic phase changes enhance rather than retard the movement of

descending slabs and rising plumes across the phase-change boundary. I have already mentioned the possibility that the spinel–perovskite and garnet–perovskite reactions with opposite P–T slopes may offset each other thermally such that the net effect of the 660-km transition on slabs and plumes is near zero.

4. *Plate lengths.* Models of Zhong and Gurnis (1994) show that the mass flux between the upper and lower mantle increases strongly with total plate length. For a ratio of plate length to mantle thickness of 1:1, for instance, perfectly layered convection is predicted, whereas for a ratio of 5:1, perfect whole-mantle convection should exist.

5. *Mantle viscosity.* Penetration of slabs into the lower mantle is favored by increasing viscosity with depth, a likely situation in the mantle.

6. *Slab dip angles.* In seismic tomographic profiles of subduction zones, shallow dip angles of descending slabs decrease the chances of penetrating the 660-km discontinuity without some holding time.

7. *Reaction rates.* Slow reaction rates of phase changes can temporarily retard descending slabs. This is because of the time it takes to heat the coldest parts of the slab to temperatures necessary for the reaction to proceed.

8. *Changes in chemical composition.* If, as some investigators suggest, there is an increase in the amount of Fe and perhaps Si at the 660-km discontinuity, some degree of layered convection seems necessary to preserve such a compositional boundary.

From the preceding considerations, there appears to be a real possibility that convection in the Earth involves partial layering, at least at certain times in the past. Computer modeling by Tackley et al. (1994) suggests that descending slabs may be temporarily delayed at the 660-km discontinuity, where they accumulate and spread laterally. After some critical mass is reached, however, this plate graveyard overcomes buoyancy of the phase boundary and sinks catastrophically into the lower mantle. This *avalanche of plates,* as it is commonly called, occurs only at Raleigh numbers greater than 10^6; thus, it may have been important early in the Earth's history. It should also be followed by a period of extensive plume generation after the dead slabs arrive in the D″ layer. At Raleigh numbers no greater than 10^6, a kind of "leaky" two-layer convection is predicted, which is really time-delayed whole-mantle convection. This may be the best approximation of what is going on in the mantle today.

Toward a Convection Model for the Earth

One of the long-standing controversies in earth sciences is the nature of mantle convection. Perhaps the greatest challenge in understanding mantle convection is to compile, evaluate, and reconcile the range of observations and constraints from many disciplines, especially from geochemistry and geophysics. The isotopic and geochemical data described previously, as well as heat-production considerations, suggest the existence of distinct mantle reservoirs that have retained their identity for at least 2 Gy. A major reservoir boundary is commonly placed at the 660-km discontinuity (Hofmann, 1997). The noble gas record in plume-derived basalts, which requires storage of primordial gases in the deep mantle, suggests that the deep mantle has not been extensively degassed, a feature difficult to reconcile with whole-mantle convection. Heat-flow arguments also do not favor whole-mantle convection. The sum of the heat produced by continental crust and depleted mantle extending all the way to the core–mantle boundary accounts for only a fraction of the present-day heat flow at the Earth's surface (Albarede, 1998; Turcotte and White, 2001). This seems to require a high heat-producing reservoir in the lower mantle, which again is difficult to account for if the entire mantle is convecting.

On the other hand, seismic tomography indicates that most or all descending plates penetrate the 660-km discontinuity and sink to, or close to, the core–mantle boundary (van der Hilst et al., 1997). Numerical computer models suggest that it is difficult to maintain an impermeable boundary at the 660-km discontinuity or elsewhere in the mantle (McNamara and van Keken, 2000; Tackley, 2000; Davies, 2002). Models indicate that differences between the upper and the lower mantle should disappear in a few hundred million years as long as plates continue to penetrate the boundary (Albarede and van der Hilst, 1999). Also, high-pressure mineral physics experiments with seismic observations indicate that the Clapeyron slope of the 660-km phase transition is too small (or nonexistent) to cause long-term stratification of the mantle. In addition, no convincing evidence exists for a thermal or compositional boundary at the 660-km discontinuity, of which one or both would develop if the upper and the lower mantles convected separately.

Can any model accommodate all of these apparently conflicting observations? The model of Kellogg et al. (1999) modified by Albarede and van der Hilst (1999) seems to hold promise. In this model, the mantle to a depth of approximately 2000 km is relatively uniform in major element composition and largely represents a depleted and outgassed reservoir (Fig. 4.27). A transition zone between 400 and 1000 km deep divides the mantle into two regimes with different timescales of mixing: (1) the well-mixed upper mantle (depleted mantle), which is the source of MORB, and (2) the lower mantle, which mixes at a much slower rate. The bottom layer of the mantle, which ranges from about 1700 km deep to near the core–mantle interface has a high density and contains undegassed material enriched in radiogenic heat-producing elements. Mantle plumes that give rise to oceanic-island and oceanic-plateau basalts come from the upper boundary of this deep domain rather from the D'' layer, as assumed in most plume models. It is along this boundary that descending plates accumulate (Fig. 4.27). The plates are highly depleted in such incompatible elements as K, Rb, Ba, Th, and U because of the subduction process. Numerical models of Kellogg et al. (1999) show that if the lower mantle layer is only 4% denser than the overlying mantle it will be dynamically stable and will develop substantial topographic relief (Fig. 4.27). Although such relief strongly reduces the ability to identify this interface by seismic tomographic methods, numerical modeling of these data supports the existence of an isolated lower mantle (van der Hilst and Karason, 1999; Tackley, 2000). The lower mantle layer is thin or absent beneath descending plates and thick far from these plates. Although some mantle plumes may arise from D'', most are generated at the thermal boundary zone at the top of the lower mantle. This region, which contains fragments of older crust and lithosphere mixed with lower mantle, imparts the distinctive enriched mantle and HIMU geochemical signatures to some rising plumes and their derivative basalts. If there is a depleted component in the deep mantle,

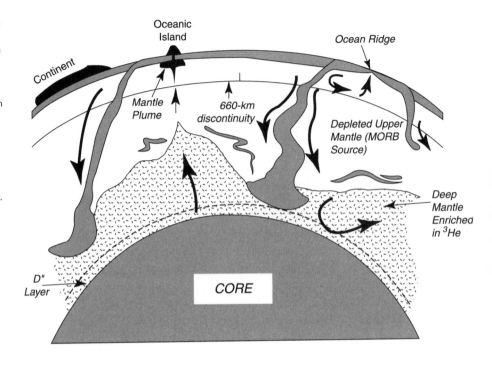

Figure 4.27 Convection model for the Earth involving a thick zone deep in the mantle that is convectively isolated from the middle and upper mantle. Mantle plumes may be produced in the D″ layer at the core–mantle interface and at the upper boundary of the deep mantle layer at local high spots. MORB, midocean-ridge basalts. Modified from Kellogg et al. (1999).

as suggested by Kempton et al. (2000), this component may also reside in this dense lower mantle layer.

Whether this model will survive the test of rapidly accumulating new data remains to be seen. However, it would appear that any model for mantle convection must be consistent with the existence of a compositionally distinct and gravitationally stabilized lower mantle.

Further Reading

Condie, K. C., 2001. Mantle Plumes and Their Record in Earth History. Cambridge University Press, Cambridge, UK, 306 pp.

Fei, Y., Bertka, C. M., and Mysen, B. O. (eds.), 1999. Mantle Petrology: Field Observations and High-Pressure Experimentation. University of Houston, Geochemical Society, Special Publication No. 6, 322 pp.

Gurnis, M., Wysession, M. E., Knittle, E., and Buffett, B. A (eds.), 1998. The Core–Mantle Boundary Region. American Geophysical Union, Geodynamic Series, Vol. 28, 338 pp.

Helffrich, G. R., and Wood, B. J., 2001. The Earth's Mantle. Nature 412, pp. 501–507.

Hemley, R. J. (ed.), 1998. Ultrahigh-Pressure Mineralogy: Physics and Chemistry of the Earth's Deep Interior. Mineralogical Society of America, Reviews in Mineralogy, Vol. 37, 671 pp.

Lillie, R. J., 1999. Whole Earth Geophysics. Prentice Hall, Upper Saddle River, NJ, 361 pp.

Schubert, G., Turcotte, D. L., and Olson, P., 2001. Mantle Convection in the Earth and Planets. Cambridge University Press, Cambridge, UK, 940 pp.

Stacey, F., 1992. Physics of the Earth, Third Edition, Brookfield Press, Brisbane, Australia, 513 pp.

The Core

5

Introduction

Seismic-velocity data indicate that the radius of the core is 3485 ± 3 km and that the outer core does not transmit secondary, or shear, waves (S-waves) (Jeanloz, 1990; Jacobs, 1992) (see Fig. 1.2 in Chapter 1). This latter observation is interpreted to mean that the outer core is in a liquid state. Supporting this interpretation are radio astronomical measurements of the Earth's normal modes of free oscillations. The inner core, with a radius of 1220 km, transmits S-waves at low velocities, suggesting that it is a solid near the melting point or partly molten. There is a sharp velocity discontinuity in primary-wave, or compressional-wave, (P-wave) velocity (0.8 km/sec; Fig. 1.2) at the inner core boundary and a low-velocity gradient at the base of the outer core. Results suggest that the top of the inner core attenuates seismic energy more than the deeper part of the inner core. Detailed analysis of travel times of seismic waves reflected from and transmitted through the core indicate that the outer liquid core is relatively homogeneous and well mixed, probably because of mixing by convection currents. Seismic data also suggest relief on the core–mantle interface is limited to about 5 km. The viscosity of the outer core is poorly known, with an estimated value of about 10^{-3} Pa sec or less than 10^4 Pa sec.

Three lines of evidence indicate that the core is composed chiefly of iron. First, the internal geomagnetic field must be produced by a dynamo mechanism, which is only possible in a liquid metal outer core. Because the fluidity and electric conductivity of the mantle are too low to produce the Earth's magnetic field, the outer core must be liquid metal for the geodynamo to operate (Jeanloz, 1990). Second, the calculated density and measured body-wave velocities in the core are close to those of iron measured at appropriate pressures and temperatures. Third, iron is by far the most abundant element in the solar system that has the seismic properties resembling those of the core.

Core Temperature

Accurate knowledge of the temperature of the core is important for constraining both the Earth's radioactivity budget and the generation of the magnetic field. Limits on the temperature profile in the core are determined through considering the solid–liquid interface at the inner core boundary from extrapolated phase equilibriums and through high-pressure experimental studies of the melting of iron and iron alloys. Depending on the approach used, significant differences exist in the estimates of core temperature. For instance, estimates of the minimum temperature at the core–mantle boundary range from 2500 to 5000° C. The core–mantle boundary and the D″ layer probably reflect large gradients in both temperature and composition. Thermal gradients are inferred from radial and lateral velocity gradients and from new experimental data on the melting point of iron at high pressures (Boehler, 1996; Ahrens et al., 2002). Melting relationships of iron and Fe-O-S compounds measured at pressures up to 2 Mbar imply a temperature discontinuity at the core–mantle boundary in excess of 1300° C.

To generate the Earth's magnetic field, the higher temperatures seem necessary, and using the high-pressure melting point data of iron, a value of 3700 ± 500° C is a reasonable estimate for the temperature at core–mantle boundary (Duba, 1992; Ahrens et al., 2002). Corresponding temperatures at the inner–outer core boundary and at the Earth's center are about 5000 and 5500° C, respectively, each with at least a 400° C uncertainty (Fig. 1.2.)

The Inner Core

Anisotropy of the Inner Core

Although seismic data indicate that the inner core is solid, on geologic timescales it may behave like a fluid, undergoing solid-state convection like the mantle. Support for this idea comes from seismic anisotropy in which the P-wave velocity is higher in the inner core along the Earth's rotational axis than in equatorial directions (Jeanloz, 1990; Tromp, 2001). It has been known for many years that body waves traveling parallel to the Earth's rotation axis arrive faster than waves traveling in the equatorial plane. Cylindrical anisotropy of a few percent with the fast axis parallel to the Earth's rotation axis and the slow axis in the equatorial plane is the preferred explanation of these observations. Seismic studies by Ishii and Dziewonski (2003) suggest that the symmetry axis in the inner core may be significantly tilted from the rotational axis. Their results also imply a seismically distinct region in the center of the Earth with a radius of about 300 km. Four mechanisms have been proposed for the inner core anisotropy: solid-state convection, solidification texturing, anisotropic growth of the inner core, and the Earth's magnetic field (Tromp, 2001). I will briefly review each of these mechanisms.

Inner core convection requires solid-state convection driven by radioactive sources in the inner core. Although still favored by some investigators, the high heat production and the high thermal conductivity of iron required make inner core convection unlikely (Yukutake, 1998). The existence of significant amounts of radioactive elements in the inner core also seems unlikely from geochemical arguments.

Solidification texturing involves the dendritic growth of iron crystals in the inner core, in which dendrites grow along an axis aligned with the direction of dominant heat flow, generally assumed to be perpendicular to the Earth's rotation axis (Bergman, 1997). The dendritic iron crystals are responsible for the inner core anisotropy, and this model is consistent with the observed depth dependence of the strength of anisotropy.

A similar model predicts that the inner core should grow faster in its equatorial regions than in its polar regions because heat transport is less effective near the poles (Yoshida et al., 1996). In this model, the viscous flow from the equator to the poles produces stresses large enough to cause alignment of iron crystals. The main uncertainty of this explanation is the viscosity of the inner core: if it is too small, the flow-induced stresses may not be large enough to align the crystals in a reasonable amount of time.

Karato (1999) has shown that the Earth's magnetic field induces a stress in the inner core called the **Maxwell stress** that causes a seismically anisotropy fabric in the inner core. In this model, the Maxwell stress squeezes iron crystals in the inner core toward the rotational axis, moving material from regions of high stress (the outer part of the inner core) to regions of low stress (near the rotational axis) (Fig. 5.1). Material that reaches the low-stress region will melt and flow outward, replacing material lost from the high-stress regions. There are two major consequences of the Maxwell stress model.

1. The flow of core material results in large strains that could be responsible for a fabric in the inner core.
2. The flow causes nonuniform release of energy at the inner–outer core boundary because of the melting and crystallization of iron.

The energy released by these reactions is likely to have an important effect on the convection pattern in the outer core; hence, the inner core could influence the geodynamo (described later in this chapter) through thermal effects.

Finally, there is at least permissive evidence for a seismic discontinuity within the inner core. Seismic waves passing along north–south paths in the inner core produce unusually broad pulse shapes at long periods and provide compelling evidence for a seismic discontinuity about 200 km below the inner core boundary (Song and Helmberger, 1998). This boundary seems to separate an isotropic upper layer from an anisotropic inner layer and may be caused by a phase change among high-pressure phases of metallic iron.

Rotation of the Inner Core

A perplexing problem that has not yet been solved is the question of whether the inner core rotates slightly faster than the mantle and crust. Small but systematic temporal variations in the travel times of seismic waves passing near the center of the Earth have been interpreted by some investigators as evidence that the solid inner core is rotating faster than the overlying mantle and crust by 0.2 to 3.0 degrees per year. However, measurements

Figure 5.1 Diagram of the inner core showing the structure and dynamics caused by the Earth's magnetic field. Shown are the magnetic field lines (thin black lines), the induced Maxwell stresses (thick black lines), and the flow lines (white lines). Courtesy of Shun-Ichiro Karato.

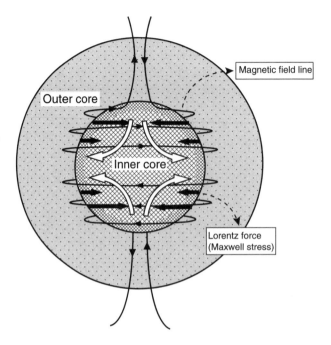

based on the Earth's large-scale free oscillations, rather than on travel times of seismic waves, seem to rule out differential rotational rates as high as 1.0 degrees per year but are marginally consistent with rates of 0.2 degrees per year (Laske and Masters, 1999). The main problem with using seismic travel times to probe the inner core is the sensitivity of the measured times to local irregularities in the core. The free oscillation method sidesteps the need to resolve these irregularities. The long-wavelength oscillations are insensitive to small-scale structures in the core. Laske and Masters (1999) conclude from their study that the inner-core mean rotation rate is 0.0 ± 0.2 degrees per year, which implies that the inner core is gravitationally "locked" to the mantle and that it does not rotate at a different rate.

Using an improved seismic method involving scattering of seismic waves by the inner core, Vidale et al. (2000) have been able to greatly improve the resolution of seismic-wave data. If the inner core rotates with respect to the mantle, waves scattered by irregularities will arrive earlier from the side of the core getting closer, whereas waves coming from the retreating side will arrive later. This is exactly what was reported by Vidale et al. (2000). Their results suggest that the inner core rotates faster than the mantle but at a considerably slower rate than predicted by earlier seismic-wave results: only 0.15 degrees per year, which is within the error of the free oscillation results described previously.

Composition of the Core

Constraints on the composition of the Earth's core come from the study of iron meteorites (which may represent samples of asteroid cores), thermodynamic model studies, and high-pressure experimental studies. Only the diamond-anvil and shock-wave experiments, however, can reach inner core pressures and temperatures (Stixrude and Brown, 1998). Three phases of iron are known at pressures that exist in the core: body-centered cubic (bcc), face-centered cubic, and hexagonal close-packed (hcp) iron. Most experimental data indicate that hcp iron is the dominant phase in the core (Tromp, 2001). Although theoretical studies agree that the hcp phase is stable at core pressures and the bcc phase is unstable, quantum mechanic studies indicate that bcc may be stable in the core (Vocadlo et al., 2003). Belonoshko et al. (2003) take this one step further based on density function theory and suggest that the hcp–bcc transition was misinterpreted as a melting transition in experimental studies (similar to the case of xenon) and the bcc iron phase is really the stable phase in the inner core. Investigators clearly need better resolution of the experimental data to solve this problem.

The presence of 5 to 10% nickel in the core is supported by the composition of iron meteorites, which may represent fragments of core material from asteroids. Seismically, however, there is no reason for nickel to be in the core because seismic velocities are essentially the same for iron and nickel. Although it is clear the core must be composed chiefly of iron, P-wave velocities and density of the outer core are about 10% lower than those of liquid iron (Jacobs, 1992). Thus, at least the outer core requires 5 to 15% of one or more low atomic number elements to reduce its density. This also means there is not a single melting temperature at a given pressure but that the core must melt over an interval given by its solidus and liquidus. Which alloying element or elements occur in the core are intimately linked with various models proposed for the origin and evolution of the core. In addition to nickel and sulfur, silicon and oxygen have received most support from geochemical modeling as core contaminants.

Another approach to estimating the composition of the core is to use the composition of the bulk Earth and the primitive mantle as estimated from meteorite compositions and determine core composition by difference. Results suggest that the total core contains about 7% silicon and traces of both sulfur (2%) and oxygen (4%) (Allegre et al., 1995). A core of this composition must have formed at low pressures.

Although most investigators now believe that sulfur, silicon, or both are the low mass elements in the outer core, considerable disagreement exists as to which of these elements dominates. Although density calculations indicate that both silicon and sulfur may be present (Sherman, 1997), metal–silicate partitioning experiments at high pressure show that both elements are mutually exclusive under the same low-oxygen fugacity in the core (Kilburn and Wood, 1997). The common presence of iron sulfides in meteorites is consistent with the presence of sulfur in the core, as is density modeling of Fe-S alloys at core pressures (Sherman, 1995). Experimental studies support the idea that the mantle is depleted in sulfur because of iron–silicate equilibriums during core formation and thus support sulfur as the dominant light element in the core (Li and Agee, 2001). If a significant amount of sulfur is in the core, it must have entered at low pressures as shown by the experimental data; hence, it supported core formation during the late stages of planetary accretion some 4.6 Ga (Newsom and Sims, 1991).

The possibility that oxygen is the dominant alloying agent in the core is based on ultra-high-pressure experimental studies and thermodynamic calculations (Kato and Ringwood, 1989; Alfe et al., 2002) because the relevant oxide phases are not stable at low pressures or found in meteorites. Experimental results document that oxygen can alloy with molten iron at pressures in excess of 10 GPa. Hence, in contrast to sulfur and nickel, alloying of oxygen would be expected well after the core began to form and perhaps after planetary accretion was complete (Ito et al., 1995). The distinction between sulfur and oxygen as the primary low-atomic number diluent in the core may be possible as investigators better constrain the timing of core formation.

It seems probable from comparing the seismically deduced density of the inner core with measured densities of iron and iron alloys at high pressures that the inner core also cannot be composed of a pure Fe-Ni alloy (Jephcoat and Olson, 1987). Like the outer core, it must contain a low-atomic number element, presumably sulfur or silicon, but in small amounts of only 3 to 7%. This also would be consistent with convection in the inner core, as mentioned previously, because these elements would lower the effective viscosity, enhancing the ability to convect. Thermodynamic calculations indicate that the bcc iron phase should be stabilized compared with the hcp phase by sulfur or silicon impurities in the inner core (Vocadlo et al., 2003). Possibly the bcc phase is responsible for the seismic complexity in the inner core. For instance, a phase change from the hcp to the bcc phase may be responsible for the seismic boundary in the inner core mentioned previously (Song and Helmberger, 1998).

Age of the Core

It is obviously not possible to date the core because investigators do not have samples that have been brought to the Earth's surface. However, there are indirect isotopic and geochemical arguments that can be used to constrain the timing of core formation. Lead isotopes provide an estimate of the age of the Earth of 4.57 to 4.45 Ga; it appears that during this time U and Pb were fractionated from each other. A significant proportion of the Earth's Pb (still at trace concentrations) may have been incorporated in the core at this time because it geochemically follows iron. If so, the Pb isotope age of the Earth suggests that separation of the core from the mantle occurred early in the Earth's history, perhaps during the late stages of planetary accretion.

More precise ages for core formation can be made using short-lived radiogenic isotopes in which parent and daughter isotopes were fractionated during accretion of the Earth. One of the most productive isotopic systems is the ^{182}Hf-^{182}W system. Although W follows iron into the core, Hf remains in the silicate fraction in the mantle. The parent isotope ^{182}Hf has a half-life of about 9 My; thus, if the core forms rapidly soon after planetary accretion, the daughter isotope ^{182}W will remain in the mantle and will show up today as tungsten anomalies in mantle xenoliths. On the other hand, if the core forms slowly through geologic time, tungsten will enter the core and there will be no anomalies in mantle xenoliths. Tungsten isotopes can also be measured in iron meteorites, which constrain the timing of core formation in the asteroid parent bodies of these meteorites. ^{182}Hf-^{182}W isotope data from iron meteorites support early segregation of parent body cores, probably in the first 50 My of planetary accretion (Harper and Jacobsen, 1996). The correct interpretation of Hf-W ages, however, depends on the initial abundance of ^{82}Hf in the solar system and the tungsten isotopic composition of chondritic meteorites and the material from which the Earth was accreted. Using Hf and W data from chondrites as a starting material for the Earth, the Earth's core must have formed within about 30 My of the beginning of accretion of bodies in the solar system (Kleine et al., 2002). Corresponding ages for the lunar core are between 26 and 33 My. Hf-W data from meteorites indicate the cores in their asteroid parent bodies formed in less than 10 My after the beginning of asteroid accretion. These ages indicate that the cores of the terrestrial planets and asteroids formed rapidly, probably during the late stages of planetary accretion.

Also supporting early core formation is the lack of change in ratios of siderophile to lithophile elements in the mantle with time (Sims et al., 1990). If the core grew gradually by blobs of liquid iron sinking through the mantle to the Earth's center, the mantle should show progressive depletion of siderophile elements (such as W, Mo, and Pb) relative to lithophile elements (such as Ce, Rb, and Ba) with time. However, it does not show such a trend. Even the oldest known basalts (~4 Ga), which carry a geochemical signature of their mantle sources, have element ratios comparable with ratios in modern basalts.

It is possible to constrain the age of the inner core by considering energy relationships associated with crystallization of the inner core. The age of the inner core depends on the heat flux from the core–mantle boundary and the concentrations of radiogenic elements in the core. Regrettably, neither of these parameters is well known. Calculations indicate that if radiogenic elements are absent in the core, the inner core cannot be older than about 2.5 Ga and it is most likely about 1 Ga (Labrosse et al., 2001). If this model were correct, it would appear that the Earth's inner core did not begin to crystallize until 2 to 3 Gy after planetary accretion terminated.

Generation of the Earth's Magnetic Field

The Geodynamo

It is well known that the Earth's magnetic field is similar to that of a giant bar magnet aligned with the rotational axis of the planet (Fig. 5.2). The magnetic lines of force trace curved paths, exiting near the South Pole and entering near the North Pole. Maps of the magnetic field at the surface show a secular variation that occurs over periods from decades to tens of thousands of years. Among the most striking changes in the last few thousand years are (1) a decrease in the dipole component of the field and (2) a westward drift of part of the field. The Earth's field can also reverse its polarity and has done so many times in the geologic past. Most evidence strongly suggests the Earth's magnetic field is generated in the fluid outer core by a dynamo-like action, although the details of how this occurs are poorly understood (Jacobs, 1992; Glatzmaier, 2002).

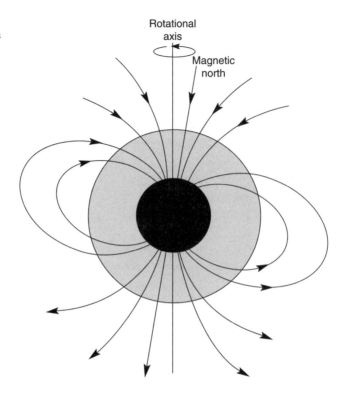

Figure 5.2 The Earth's magnetic field showing lines of equal intensity and orientation.

Fluid Motions in the Outer Core

It is possible to map fluid motions in the outer core using the flow lines of the magnetic field (Bloxham, 1992). The maps show current movements at the core surface (as vectors) and regions of upwelling and downwelling known as core spots (Fig. 5.3). The four spots that make the strongest contributions to the dipole component have been interpreted by Bloxham (1992) as the tops and bottoms of two columns of liquid that appear to touch the inner core and run parallel to the Earth's rotational axis. Liquid iron may spiral down through the two columns, creating a dynamo that concentrates magnetic flux within the columns. Magnetic field lines are "frozen" in the iron liquid, and as it moves they are carried with the liquid, thus mapping current patterns. Results show two prominent cells of circulating fluid beneath the Atlantic basin, one south and one north of the equator (Fig. 5.3). Also, there is a region of intense upwelling near the equator elongated in a north–south direction at 90° E longitude beneath the Indian Ocean, with a strong equatorial jet extending westward between the two circulating cells. This westward current may explain the slow westward drift of the magnetic field in this area. The core spots appear to be produced by intense upward and downward flow of liquid. Evidence seems to be mounting that the fall in dipole component of the magnetic field over the last few hundred years is caused by the growth and propagation of the downward flux core spots beneath Africa and the Atlantic basin. This is supported by a correlation between the intensity of the dipole component and the amount of lateral motion of these two spots in the last 300 years.

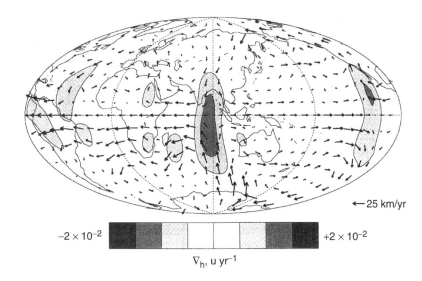

Figure 5.3 Current patterns in the outer core as shown by vectors. The gray scale shows the intensity of upcurrents and downcurrents. Upcurrents are beneath the Indian Ocean, South Africa, and the Atlantic Ocean. Downcurrents are beneath the east Pacific Ocean, northwestern Australia, southeast of Madagascar, northern Africa, and northern South America. Modified from Bloxham (1992).

Fueling the Geodynamo

The energy for the geodynamo could be gravitational, chemical, or thermal; in all cases, it is converted to heat that flows outward into the mantle (Jacobs, 1992; Glatzmaier, 2002; Stevenson, 2003). Remnent magnetism in rocks more than 3.5 Ga indicates that the Earth's geodynamo was in action by that time. Also, paleomagnetic studies show the field intensity has never varied by more than a factor of two since the Archean, so the energy needed to drive the dynamo must have been available around the same rate for at least 4 Gy.

There are two serious candidates for the **geodynamo energy source** (Kutzner and Christensen, 2000; Stevenson, 2003; Labrosse, 2003): thermal convection of the outer core and compositional convection caused by growth of the inner core. In the first case, if the liquid outer core is stirred by thermal convection, most of the heat will be carried away by convection or conduction and will not contribute to the production of the magnetic field. Growth of the inner core can supply gravitational energy in two ways: (1) latent heat of crystallization as iron crystallizes on the surface of the inner core and (2) a lighter fraction concentrated in the outer core as metal accretes to the inner core. Because of the lighter fraction buoyancy, the second option leads to compositionally driven convection.

The near exclusion of light elements from the inner core as it grows provides an important source of buoyancy for convection (Buffett, 2000; Stevenson, 2003). The light elements rise into the fluid outer core, and the dense elements (Fe, Ni, and other metals) crystallize into the inner core. Although thermal convection may also be important, it is difficult to evaluate the role of the overlying mantle on heat loss. Estimating the rate of cooling of the outer core is important because of the high thermal conductivity of liquid iron. The heat conducted through the core may be comparable with the total heat flow into the base of the mantle. Uncertainties in this heat loss make it difficult to estimate the role of thermal convection in generating the magnetic field. Convection in the core operates like a heat engine: heat is drawn from the core by the mantle and work is done to generate the magnetic field. As the inner core grows and light elements are concentrated more into the outer core, gravitational energy is released to power the geodynamo. Estimates suggest that compositional convection contributes about 80% to the power of the geodynamo and that thermal convection contributes about 20%. At earlier times in the history of the core, compositional convection should have been weaker because the inner core was smaller or nonexistent. Before formation of the inner core, the magnetic field would have been generated entirely by thermal convection.

How the Geodynamo Works

Many of the planets in the solar system, including the Earth, have magnetic fields that originate from self-sustaining dynamos (Stevenson, 2003). Some planets, such as Mars and Venus, do not have dynamos now but may have had dynamos in the distant past. Three basic ingredients are needed for an active dynamo: (1) a large volume of convecting fluid (Fe in the case of the terrestrial planets) in or near the center of the planet, (2) an energy source such as excess heat resulting in convection of the fluid, and (3) a planetary rotation to control fluid motions. It is possible with high-speed computers to construct and test numerical models that may simulate the Earth's geodynamo. Two general models have been widely discussed: polar vortex and differential rotation (Kuang and Bloxham, 1997; Olson and Aurnou, 1999; Glatzmaier, 2002; Dumberry and Bloxham, 2003). Both models involve a hypothetical tangential cylinder inside the geodynamo (Fig. 5.4). The two models call upon similar convective energy, rotation rate, and fluid properties, and they produce comparable magnetic fields. Both models delegate a special role to the solid inner core: first as an energy source as described previously, and second as a mechanical barrier to flow in the outer core.

The chief difference between the models is the assumed boundary conditions between the inner and the outer cores, which have an important effect on the tangential cylinder (Fig. 5.4). In the polar vortex model, the inner core is spun into a fast prograde rotation relative to the mantle, generating the magnetic field within the tangential cylinder. In the differential rotation model, the effect of viscosity is greatly reduced and the polar vortices are largely suppressed, causing the magnetic field to be generated outside the cylinder. This situation produces a westward drift of the magnetic field, as observed in the Earth.

Perhaps the most remarkable outcome of these two models is in showing that computer simulations can be powerful and accurate tools for relating magnetic fields to the dynamics of planetary cores.

What Causes Magnetic Reversals?

Reversals in the Earth's magnetic field occur on different timescales, and there are many aborted attempts to reverse when the field either immediately switched back or did not even reach the opposite polarity. At the other ends of the timescales are **superchrons,** which record times when the field maintained the same polarity for periods of 20 to 50 My. It is unlikely that these changes of polarity on different timescales are the result

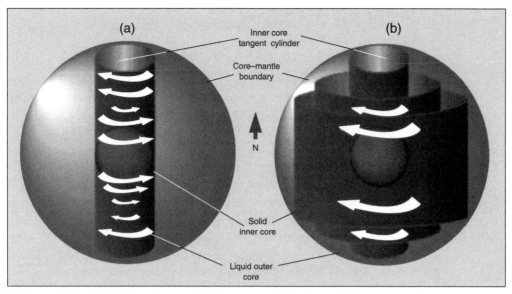

Figure 5.4   Two models for the Earth's geodynamo. (a) Polar vortex model. In this model, flow occurs inside the inner core tangent cylinder. (b) Differential rotation model. Flow occurs outside the tangent cylinder. From Olson (1997), reprinted with permission of *Nature,* copyright Macmillan Magazines Ltd., 1997.

of the same processes in the core (Jacobs, 1995). An incomplete reversal (or immediate switchback), known as an *excursion,* likely results from some local instability at the core–mantle interface and may not be worldwide. Stable reversals, on the other hand, are always global. The oldest well-documented reversed polarity interval is recorded in 2.7-Ga flood basalts from Western Australia and clearly indicates that the geodynamo was operational by this time (Strik et al., 2003).

Reversals may be initiated by changes either at the core–mantle boundary or at the inner core boundary; in both cases, it appears that some physical or chemical process arising from an energy source independent of the source that powers the geodynamo initiates the reversal (Gubbins, 1994). For instance, instabilities may be generated at the core–mantle boundary by heat loss, producing cooler, denser blobs of molten iron that sink and destabilize the main convection in the outer core. Alternatively, hot plumes rising from the inner core boundary may have the same effect. Hollerbach and Jones (1993) suggest that reversals may be caused by oscillations of the magnetic field in the outer core. Their modeling concludes that although the field oscillates strongly in the outer core, only weak oscillations should occur near the core surface, and reversals may be caused when these weak oscillations exceed some threshold value. This would occur only when the size ratio of the inner to outer core is greater than 0.25; thus, before the inner core crystallized, there should have been no reversals. Changes in the frequency of reversals can arise either from changes in the total heat flux from the core to the mantle or from instabilities associated with lateral variations at the core–mantle boundary (Gubbins, 1994).

Origin of the Core

Segregation of Iron in the Mantle

Many models have been proposed to explain how molten iron sinks to the Earth's center as the melting point of metallic iron or iron alloys is reached in the mantle. Although molten iron could collect into layers and sink as small diapirs, the high surface tension of metallic liquids relative to solid silicates should cause melt droplets to collect at grain boundaries rather than drain into layers (Stevenson, 1990; Newsom and Sims, 1991). During the late stages of planetary accretion, large amounts of molten iron should have been retained in the silicate mantle. At that time, one or both of the following must have occurred: a giant impact or increased heat retention in the Earth that led to rapid melting. At that time, most or all silicate mantle melted, producing a magma ocean as described in Chapter 10. In this deep magma ocean, droplets of liquid iron should separate from the silicate liquid and sink rapidly to the core (Stevenson, 1990; Bruhn et al., 2000) (Fig. 5.5). If the magma ocean was relatively shallow, metal droplets could accumulate at the bottom, eventually coalescing into larger blobs that could either sink as diapirs or percolate downward (Fig. 5.5).

Siderophile Element Distribution in the Mantle

It had long been generally assumed that the concentrations of siderophile elements in the mantle are higher than predicted for models of core segregation (Newsom and Sims, 1991). Because of their strong tendency to follow iron, most of these elements should have been scavenged by liquid iron as it sank to the center of the Earth during core formation. Yet the concentrations of many of siderophile elements are similar to those of primitive mantle. Murthy (1991), however, has shown that this apparent depletion may not be real. In a molten planet the size of the Earth, increasing pressure increases the melting point of metal and silicates significantly, so a large part of the Earth would be at very high temperatures (2000–4000° C). Thus, the distribution of trace elements between liquid and solid would be governed by high-temperature distribution coefficients and not by the low-temperature distribution coefficients that have been used in all previous modeling. The siderophile element distributions in the mantle calculated with high-temperature distribution coefficients is similar to the element distributions in mantle xenoliths. Thus, the problem of excess siderophile elements in the mantle goes away when appropriate high-temperature modeling is done.

Growth and Evolution of the Core

Although high-pressure phase equilibriums in the Fe-O and Fe-S systems are still not precisely known, it is informative to use the available data to track the growth of the core during the first 50 My of the Earth's history. During the late stages of planetary accretion, steepening geotherms in the Earth should intersect the melting curve of iron and silicates at relatively shallow depths in the mantle. Continued heating raises the geotherms well

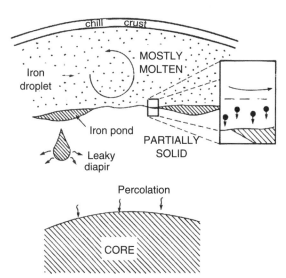

Figure 5.5 Possible core segregation processes. Molten metal droplets rapidly sink in a silicate magma ocean, large blobs of molten iron sink as diapirs, and alloying of oxygen with iron in the lower mantle triggers downward percolation of liquid alloy into the core. Modified from Stevenson (1990) and Newsom and Sims (1991).

above the melting point of both iron and silicates, producing a widespread magma ocean containing droplets of molten iron, which because of their gravitational instability begin to sink to the center of the Earth and form the core (Fig. 5.5). Sometime within the first 100 My of the Earth's history (about 4.53 Ga; Fig. 10.18), a Mars-sized planet may have collided with the Earth, forming the Moon, as explained in Chapter 10. During this collision, most of the mantle of the impactor should have accreted to the Earth and its core probably penetrated the Earth's core (Benz and Cameron, 1990), possibly transferring iron to the Earth's core. Thermal convection in the core started the geodynamo early in the Earth's history. Just when the core cooled enough for the inner core to begin to crystallize is unknown, but as previously explained, it was probably sometime between 1 and 2 Ga.

As the core continues to cool, the inner core should continue to grow at the expense of the outer core. Because the low-atomic numbered diluting element or elements (S, Si or O) are preferentially partitioned into the liquid phase, the outer core will become progressively enriched in these elements with time; thus, the melting point of the outer core should drop until the eutectic in the system is reached. The final liquid that crystallizes as the outermost layer of the core, sometime in the distant geologic future, should be a eutectic mixture.

Where Do We Go From Here?

Although the last 10 years have been extremely productive in learning more about the Earth's core, investigators have a long way to go. For instance, scientists know little about the cause of the coincidence of geoid anomalies at the surface of the Earth and at the core–mantle boundary (Chapter 4). Is this where heat is transferred to the mantle? How much interaction is there between the core and the mantle at the core–mantle boundary, and can investigators better resolve this interaction with seismic-wave studies? Although it is clear that the inner core is anisotropic, the cause of this anisotropy remains problematic. Another area about which scientists know little is the rate at which the inner core is crystallizing and how it crystallizes. Is crystallization episodic, resulting in sudden bursts of heat loss, or is it uniform and gradual? This could be important in understanding mantle-plume events, which may be triggered by sudden losses of core heat. Although investigators are beginning to understand the geodynamo in more detail, to make significant progress on this question, three-dimensional simulations are needed, which will require significant time on high-speed computers at great expense.

So unlike our understanding of the crust and mantle, which have been significantly enhanced in the last decade, the information highway for the core is just beginning to open.

Further Reading

Buffett, B. A., 2000. Earth's core and the geodynamo. Science, 288: 2007–2012.

Dehant, V., Creager, K. C., Karato, S., and Zatman, S., 2003. Earth's Core: Dynamics, Structure, Rotation. American Geophysical Union, Washington D. C., Geodynamic Series Vol. 31.

Jacobs, J. A., 1992. Deep Interior of the Earth. Chapman & Hall, London, 167 pp.

Merrill, R. T., McElhinney, M. W., and McFadden, P. L., 1996. The Magnetic Field of the Earth. Academic Press, New York, 531 pp.

Newsome, H. E., and Jones, J. H. (eds.), 1990. Origin of the Earth. Oxford University Press, Oxford, UK, 378 pp.

Tromp, J., 2001. Inner core anisotropy and rotation. Ann. Rev. Earth Planet. Sci. 29: 47–69.

The Atmosphere and Oceans

6

Introduction

Not only in terms of plate tectonics is the Earth a unique planet in the solar system; it also is the only planet with oceans and with an oxygen-bearing atmosphere capable of sustaining higher forms of life. How did such an atmosphere–ocean system arise, and why only on the Earth? Related questions are: once formed, how did the atmosphere and oceans evolve with time, and in particular when and how did free oxygen enter the system? How have climates changed with time, what are the controlling factors, and when and how was life created? What are the roles of plate tectonics, mantle plumes, and extraterrestrial impact in the evolution of atmosphere and oceans? These and related questions are addressed in this chapter.

General Features of the Atmosphere

Atmospheres are the gaseous carapaces that surround some planets and satellites, and because of gravitational forces, they increase in density toward planetary surfaces. The Earth's atmosphere is divided into six regions as a function of height (Fig. 6.1). The magnetosphere, the outermost region, is composed of high-energy nuclear particles trapped in the Earth's magnetic field. This overlays the exosphere in which lightweight molecules (such as H_2) occur in extremely low concentrations and escape from the Earth's gravitational field. Temperature decreases rapidly in the ionosphere (to about $-90°$ C) and then increases to near $0°$ C at the base of the mesosphere. It drops again in the stratosphere and then rises gradually in the troposphere toward the Earth's surface. Because warm air overlies cool air in the stratosphere, this layer is relatively stable and undergoes little vertical mixing. The temperature maximum at the top of the stratosphere is caused by absorption of ultraviolet radiation in the ozone layer. The troposphere is a turbulent region that contains about 80% of the mass of the atmosphere and most of its water vapor.

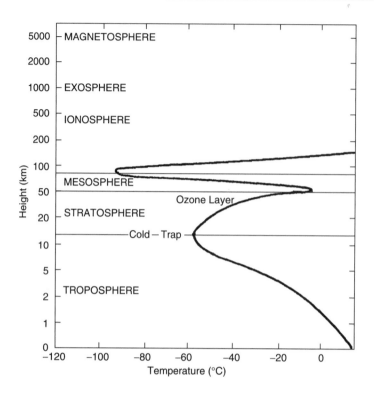

Figure 6.1 Major divisions of the Earth's atmosphere showing average temperature distribution.

Tropospheric temperature decreases toward the poles, which with vertical temperature change causes continual convective overturn in the troposphere.

The Earth's atmosphere is composed chiefly of nitrogen (78%) and oxygen (21%) with small amounts of other gases such as argon and CO_2. In this respect, the atmosphere is unique among planetary atmospheres (Table 6.1). Venus and Mars have atmospheres composed largely of CO_2; the surface pressure on Venus is up to 100 times that on the Earth, and the surface pressure of Mars is less than 10^{-2} of that of the Earth. The surface

Table 6.1	Composition of Planetary Atmospheres		
	Surface Temperature (°C)	Surface Pressure (bars)	Principal Gases
Earth (early Archean)	~85	~11	CO_2 (N_2, CO, CH_4)
Earth	−20 to 40	0.1–1	N_2, O_2
Venus	400 to 550	10–100	CO_2 (N_2)
Mars	−130 to 25	−0.01	CO_2 (N_2)
Jupiter	−160 to −90	−2	H_2, He
Saturn	−180 to −120	−2	H_2, He
Uranus	−220 to −120	−5	H_2, CH_4
Neptune	−220 to −120	−10	H_2
Pluto	−235 to −210	−0.005	CH_4

temperatures of the Earth, Venus, and Mars are also different (Table 6.1). The outer planets are composed largely of hydrogen and helium, and their atmospheres consist chiefly of hydrogen and, in some cases, helium and methane.

The concentrations of minor gases such as CO_2, H_2, and **ozone** (O_3) in the Earth's atmosphere are controlled primarily by reactions in the stratosphere caused by solar radiation. Solar photons fragment gaseous molecules (such as oxygen, H_2, and CO_2) in the upper atmosphere, producing free radicals (C, H, and O) in a process called **photolysis.** One important reaction produces free oxygen atoms that are unstable and recombine to form ozone. This reaction occurs at heights of 30 to 60 km, with most ozone collecting in a relatively narrow band from about 25 to 30 km (Fig. 6.1). Ozone, however, is unstable and continually breaks down to form molecular oxygen. The production rate of ozone is approximately equal to the rate of loss; thus, the ozone layer maintains a relatively constant thickness in the stratosphere. Ozone is an important constituent in the atmosphere because it absorbs ultraviolet radiation from the Sun, which is lethal to most forms of life. Hence, the ozone layer provides an effective shield that permits a large diversity of living organisms to survive on the Earth. It is for this reason we must be concerned about the release of synthetic chemicals into the atmosphere that destroy the ozone layer. The distributions of N_2, O_2, and CO_2 in the atmosphere are controlled by volcanic eruptions and by interactions among these gases and the solid Earth, oceans, and living organisms.

The Primitive Atmosphere

Three possible sources have been considered for the Earth's atmosphere: residual gases remaining after Earth accretion, extraterrestrial sources, and degassing of the Earth by volcanism. Of these, only degassing accommodates a variety of geochemical and isotopic constraints. One line of evidence supporting a degassing origin for the atmosphere is the large amount of ^{40}Ar in the atmosphere (99.6%) compared with the amount in the Sun or a group of primitive meteorites known as carbonaceous chondrites (both of which contain <0.1% ^{40}Ar). ^{40}Ar is produced by the radioactive decay of ^{40}K in the solid Earth and escapes into the atmosphere chiefly by volcanism. The relatively large amount of this isotope in the terrestrial atmosphere indicates that the Earth is extensively degassed of argon and, because of a similar behavior, of other rare gases.

Although most investigators agree that the present atmosphere, except for oxygen, is chiefly the product of degassing, whether a primitive atmosphere existed and was lost before extensive degassing began is a subject of controversy. One line of evidence supporting the existence of an early atmosphere is that volatile elements should collect around planets during their late stages of accretion. This follows from the low temperatures at which volatile elements condense from the solar nebula (Chapter 10). A significant depletion in rare gases in the Earth compared with carbonaceous chondrites and the Sun indicates that if a primitive atmosphere collected during accretion, it must have been lost (Pepin, 1997). The reason for this is that gases with low atomic weights (CO_2, CH_4, NH_3, H_2, etc.) that probably composed this early atmosphere should be lost even more readily than rare gases with high atomic weights (Ar, Ne, Kr, and Xe) and greater gravitational attraction. Just how such a primitive atmosphere may have been lost is not clear. One possibility is by a **T-Tauri solar wind** (Chapter 10). If the Sun evolved through a T-Tauri stage during or soon after (<100 My) planetary accretion, this wind of high-energy particles could readily blow volatile elements out of the inner solar system. Another way an early atmosphere could have been lost is by impact with a Mars-size body during the late stages of planetary accretion, a model also popular for the origin of the Moon (Chapter 10). Calculations indicate, however, that less than 30% of a primordial atmosphere could be lost during the collision of the two planets (Genda and Abe, 2003).

Two models have been proposed for the composition of a primitive atmosphere. The Oparin-Urey model (Oparin, 1953) suggests that the atmosphere was reduced and composed dominantly of CH_4 with smaller amounts of NH_3, H_2, He, and water; the Abelson model (Abelson, 1966) is based on an early atmosphere composed of CO_2, CO, water, and N_2. Neither atmosphere allows significant amounts of free oxygen, and experimental studies indicate that reactions may occur in either atmosphere that could produce the first life.

By analogy with the composition of the Sun and the compositions of the atmospheres of the outer planets and of volatile-rich meteorites, an early terrestrial atmosphere may have been rich in such gases as CH_4, NH_3, and H_2 and would have been a reducing atmosphere. One of the major problems with an atmosphere in which NH_3 is important is that this species is destroyed directly or indirectly by photolysis in as little as 10 years (Cogley and Henderson-Sellers, 1984). In addition, NH_3 is highly soluble in water and should be removed rapidly from the atmosphere by rain and solution at the ocean surface. Although CH_4 is more stable against photolysis, OH, which forms as an intermediary in the methane oxidation chain, is destroyed by photolysis at the Earth's surface in less than 50 years. H_2 rapidly escapes from the top of the atmosphere; therefore, it also is an unlikely major constituent in an early atmosphere. Models suggest that the earliest atmosphere may have been composed dominantly of CO_2 and CH_4, both important greenhouse gases (Pavlov et al., 2000; Catling et al., 2001).

The Secondary Atmosphere

Excess Volatiles

The Earth's present atmosphere appears to have formed largely by degassing of the mantle and crust and is commonly referred to as a **secondary atmosphere** (Kershaw, 1990). **Degassing** is the liberation of gases from within a planet, and it may occur directly during volcanism or indirectly by the weathering of igneous rocks on a planetary surface. For the Earth, volcanism appears to be most important both in terms of current degassing rates and calculated past rates. The volatiles in the atmosphere, hydrosphere, biosphere, and sediments that cannot be explained by weathering of the crust are known as **excess volatiles** (Rubey, 1951). These include most of the water, CO_2, and N_2 in these near-surface reservoirs. The similarity in the distribution of excess volatiles in volcanic gases to those in near-surface reservoirs (Table 6.2) strongly supports a volcanogenic origin for these gases and thus supports a degassing origin for the atmosphere.

Composition of the Early Atmosphere

Two models have been proposed for the composition of the early degassed atmosphere depending on whether metallic iron existed in the mantle in the early Archean. If metallic iron was present, equilibrium chemical reactions would liberate large amounts of H_2, CO, and CH_4 and small amounts of CO_2, water, H_2S, and N_2 (Holland, 1984; Kasting et al., 1993a). If iron was not present, reactions would liberate mostly CO_2, water, and N_2 with minor amounts of H_2, HCl, and SO_2. Because most evidence suggests that the core began to form during the late stages of planetary accretion (Chapter 5), it is possible that little if any metallic iron remained in the mantle when degassing occurred. However, if degassing began before the completion of accretion, metallic iron would have been present in the mantle and the first atmosphere would have been a hot, steamy one composed chiefly of H_2, CO_2, water, CO, and CH_4. Because the relative timing of early degassing and core formation are not well constrained, the composition of the earliest degassed atmosphere is not well known. Both core formation and most degassing were probably complete in less than 50 My after accretion, and the composition of the early atmosphere may have changed rapidly during this interval in response to decreasing amounts of metallic iron in the mantle. It is likely, however, that soon after accretion was complete around 4530 Ma (see Fig. 10.18 in Chapter 10) H_2 rapidly escaped from the top of the atmosphere and water vapor rained to form the oceans. This leaves an early atmosphere rich in CO_2, CO, N_2, and CH_4 (Holland et al., 1986; Kasting, 1993). As much as 15% of the carbon now found in the continental crust may have resided in this early atmosphere, which is equivalent to a partial pressure of CO_2, CH_4, and N_2 of about 11 bars (Table 6.1). The mean surface temperature of such an atmosphere would have been about 85° C.

Even after the main accretionary phase of the Earth had ended, major asteroid and cometary impacts continued until about 3.9 Ga, as inferred from the lunar impact record. Cometary impactors could have added more carbon as CO to the atmosphere and produced NO by shock heating of atmospheric CO_2 and N_2. A major contribution of cometary gases to the early atmosphere also solves the "missing xenon" problem. Being heavier, xenon should be less depleted and less fractionated than krypton in the Earth's atmosphere, whereas the opposite is observed (Dauphas, 2003). Any fractionation event on

Table 6.2	Excess Volatiles in Volcanic Gases and Near-Surface Terrestrial Reservoirs	
	Volcanic Gases (%)	*Near-Surface Reservoirs (%)
H_2O	83	87
CO_2	12	12
Cl, N_2, S	5	1

* Includes atmosphere, hydrosphere, biosphere, and sediments.

the early Earth would have resulted in high Xe/Kr ratios, not low ratios as observed. Noble gases such as Xe and Kr trapped in comets, however, show depletion in Xe relative to Kr. Hence, a significant contribution of cometary gases to the early atmosphere could account for the missing xenon in the Earth's atmosphere.

Growth Rate of the Atmosphere

Two extreme scenarios are considered for the growth of the atmosphere with time: the **big burp model,** in which the atmosphere grows by rapid degassing during or soon after planetary accretion (Fanale, 1971), and the **steady state model,** in which the atmosphere grows slowly over geologic time (Rubey, 1951). One way of distinguishing between these models is to monitor the buildup of ^{40}Ar and ^4He in sedimentary rocks that equilibrated with the atmosphere–ocean system through time. ^{40}Ar is produced by the radioactive decay of ^{40}K in the Earth, and as it escapes from the mantle it collects in the atmosphere. Because ^{36}Ar is nonradiogenic, the $^{40}Ar/^{36}Ar$ ratio should record distinct evolutionary paths for Earth degassing. The steady state model is characterized by a gradual increase in the $^{40}Ar/^{36}Ar$ ratio with time, and the big burp model should show initial small changes in this ratio followed by rapid increases (Fig. 6.2). This is because ^{40}Ar is virtually absent at the time of accretion; hence, the $^{40}Ar/^{36}Ar$ ratio is not sensitive to early atmospheric growth. Later in the big burp model, however, as ^{40}Ar begins to be liberated, the $^{40}Ar/^{36}Ar$ ratio grows rapidly, leveling off about 2 Ga (Sarda et al., 1985). To test these two models, it is necessary to determine $^{40}Ar/^{36}Ar$ ratios in rocks that equilibrated with the atmosphere–ocean system in the geologic past. Unfortunately, because of the mobility of argon, reliable samples to study are difficult to find. However, 2-Ga-old cherts, which may effectively trap primitive argon, are reported to have $^{40}Ar/^{36}Ar$ ratios similar to the present atmosphere (295) tending to favor the big burp model. Some argon degassing models suggest that the atmosphere grew rapidly in the first 100 My of planetary accretion followed by continuous growth to the present (Sarda et al., 1985). These models indicate a mean age for the atmosphere of 4.5 Ga, suggesting rapid early degassing of the Earth, probably beginning during the late stages of accretion. On the

Figure 6.2 Idealized evolution of atmospheric $^{40}Ar/^{36}Ar$ in the steady state and big burp models for terrestrial atmospheric growth.

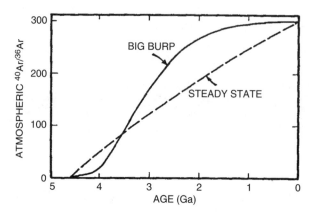

other hand, relatively young K-Ar ages of midocean-ridge basalt mantle sources (<1 Ga) show that the depleted upper mantle was not completely degassed and that it decoupled from the atmosphere early in the Earth's history (Fisher, 1985). These data, with the relatively young U-He and U-Xe ages of depleted mantle, suggest that some degassing has continued to the present. The fraction of the atmosphere released during the early degassing event is unknown but may have been substantial (Marty and Dauphas, 2002). This idea is consistent with the giant impactor model for the origin of the Moon (Chapter 10), because such an impact should have catastrophically degassed the Earth.

The Faint Young Sun Paradox

Models for the evolution of the Sun indicate that it was less luminous when it entered the main sequence 5 Ga. This is because with time the Sun's core becomes denser and therefore hotter as hydrogen is converted to helium. Calculations indicate that the early Sun was 25 to 30% less luminous than it is today and that its luminosity has increased with time in an approximately linear manner (Kasting, 1987). The paradox associated with this luminosity change is that the Earth's average surface temperature would have remained below freezing until about 2 Ga for an atmosphere composed mostly of nitrogen (Fig. 6.3). Yet the presence of sedimentary rocks as old as 3.8 Gy indicates the existence of oceans and running water. A probable solution to the faint young Sun paradox is that the early atmosphere contained a much larger quantity of greenhouse gases than it does today. For instance, CO_2 or CH_4 levels of even a few tenths of a bar could prevent freezing temperatures at the Earth's surface because of an enhanced greenhouse effect. The **greenhouse effect** is caused by gases that allow sunlight to reach a planetary surface but absorb infrared radiation reflected from the surface, which heats both the atmosphere and

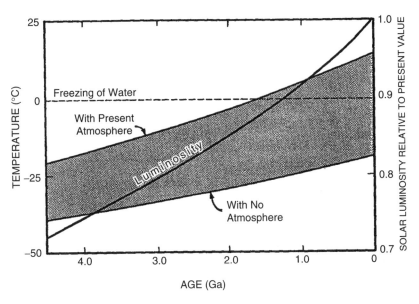

Figure 6.3 The estimated increase in solar luminosity with geologic time and its effect on the Earth's surface temperature.

the planetary surface. An upper bound on the amount of CO_2 in the early Archean atmosphere is provided by the carbon cycle and appears to be about 1 bar. Although CO_2 was undoubtedly an important greenhouse gas during the Archean, studies of a 3.5-Ga paleosol (ancient soil horizon) suggest that atmospheric CO_2 levels in the Archean were at least five times lower than required by the faint young Sun paradox (Rye et al., 1995). This constrains the Archean CO_2 levels to about 0.20 bar. The mineralogy of banded iron formation (BIF) also suggests that CO_2 levels were less than 0.15 bar 3.5 Ga. Hence, another greenhouse gas, probably CH_4, must have been the most important greenhouse gas in the Archean atmosphere (Catling et al., 2001).

Another factor that may have aided in warming the surface of the early Earth is decreased **albedo**—that is, a decrease in the amount of solar energy reflected by cloud cover. To conserve angular momentum in the Earth–Moon system, the Earth must have rotated faster in the Archean (about 14 hr/day), which decreases the fraction of global cloud cover by 20% with a corresponding decrease in albedo (Jenkins et al., 1993). However, this effect could be offset by increased cloud cover caused by the near absence of land areas—at least in the early Archean, when it is likely that the continents were submerged beneath seawater (Galer, 1991; Jenkins, 1995).

The Carbon Cycle

The most important chemical system controlling the CO_2 content of the terrestrial atmosphere is the **carbon cycle.** CO_2 enters the Earth's atmosphere by volcanic eruptions, burning of fossil fuels, uplift, erosion, and respiration of living organisms (Fig. 6.4). Of these, only volcanism appears to have been important in the geologic past, but the burning of fossil fuels is becoming more important today. For instance, records indicate that during the last 100 years, the rate of release of CO_2 from the burning of fossil fuels has risen 2.5% per year and could rise to 300% its present rate in the next 100 years. CO_2 returns to the oceans by the chemical weathering of silicates, the dissolution of atmospheric CO_2 in the oceans, and the alteration on the seafloor; of these, only the first two are significant now (Fig. 6.4). The ultimate sink for CO_2 in the oceans is the deposition of marine carbonates. Although CO_2 is also removed from the atmosphere by photosynthesis, it is not as important as carbonate deposition. Weathering and deposition reactions can be summarized as follows (Walker, 1990; Brady, 1991):

1. Weathering

$$CaSiO_3 + 2CO_2 + 3H_2O \rightarrow Ca^{+2} + 2HCO_3^{-1} + H_4SiO_4$$

2. Deposition

$$Ca^{+2} + 2HCO_3^{-1} \rightarrow CaCO_3 + CO_2 + H_2O$$

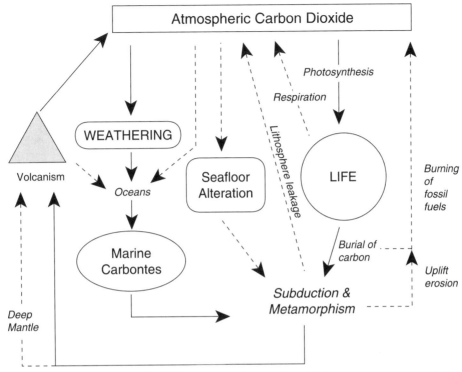

Figure 6.4   Simplified version of the carbon cycle. Solid arrows are major controls and dashed arrows are minor controls on atmospheric CO_2 levels.

The cycle is completed when pelagic carbonates are subducted and metamorphosed and CO_2 is released and reenters the atmosphere either by volcanism or by leaking through the lithosphere (Fig. 6.4). The metamorphic reactions that liberate CO_2 can be summarized by the carbonate-silica reaction as follows:

$$CaCO_3 + SiO_2 \rightarrow CaSiO_3 + CO_2$$

To maintain equilibrium in the carbon cycle, increased input of CO_2 into the atmosphere causes more weathering and carbonate deposition, thus avoiding the buildup of CO_2 in the atmosphere. As mentioned in Chapter 1, this is known as negative feedback. Various negative feedback mechanisms in the carbon cycle may have stabilized the Earth's surface temperature in the geologic past (Walker, 1990; Berner and Canfield, 1989). As an example, if the solar luminosity were to suddenly drop, the surface temperature would fall, causing a decrease in the rate of silicate weathering because of a decrease in evaporation from the oceans (and hence a decrease in precipitation). This results in CO_2 accumulation in the atmosphere, which increases the greenhouse effect and restores higher surface temperatures. The converse of this feedback would occur if the surface temperature were to suddenly increase. Although increased CO_2 in the Archean atmosphere would result in greenhouse warming of the Earth's surface, it should also cause increased weathering rates, resulting in a decrease in CO_2. However, this would only occur in the late Archean after a large volume of continental crust had stabilized above sea level to be weathered.

The Precambrian Atmosphere

By 3.8 Ga, it would appear that the Earth's atmosphere was composed chiefly of CO_2, CH_4, and perhaps N_2 with small amounts of CO, H_2, water, and reduced sulfur gases (Kasting, 1993; Des Marais, 1994; Pavlov et al., 2000). Based on the paleosol data described previously, an average CO_2 level of about 10^3 of the present atmospheric level (PAL) seems reasonable by 3.0 Ga. Photochemical models, however, indicate that methane levels approximately 1000 times the present level in the atmosphere can also explain the paleosol data (Pavlov et al., 2000; Catling et al., 2001). Such levels could readily be maintained in the Archean by methanogenic bacteria (methane-producing bacteria), which appear to have been an important part of the biota at that time (Chapter 7). With little if any land area in the early Archean, removal of CO_2 by seafloor alteration and carbonate deposition should have been more important than today. Beginning the late Archean, however, when continental cratons emerged above sea level and were widely preserved, the volcanic inputs of CO_2 were balanced by weathering and perhaps to a lesser degree by carbonate deposition.

During the Proterozoic, the increasing biomass of algae may have contributed to a CO_2 drawdown caused by photosynthesis, and rapid chemical weathering was promoted by increased greenhouse warming. However, small amounts of methane in the atmosphere (100–300 ppm) also may have contributed to maintaining a warm climate during the Proterozoic (Pavlov et al., 2003). An overall decrease in atmospheric CO_2 level with time may be related to changes in the carbon cycle, in which CO_2 is removed from the atmosphere by carbonate deposition and photosynthesis followed by burial of organic matter faster than it is resupplied by volcanism and subduction. Methane levels in the atmosphere may also drop during this time by continued photolysis in the upper atmosphere followed by hydrogen escape (Catling et al., 2001). Shallow marine carbonates deposited in cratonic basins are effectively removed from the carbon cycle and are a major sink for CO_2 from the Mesoproterozoic onward. Decreasing solar luminosity and the growth of ozone in the upper atmosphere beginning in the Paleoproterozoic also reduced the need for CH_4 and CO_2 as greenhouse gases. Rapid extraction of CO_2 by the deposition of marine carbonates and decreases of CH_4 caused by photolysis and hydrogen escape may have resulted in sufficient atmospheric cooling to cause widespread glaciation recognized in the geologic record from 2.4 to 2.3 Ga.

The Origin of Oxygen

Oxygen Controls in the Atmosphere

In the modern atmosphere, oxygen is produced almost entirely by **photosynthesis** (Fig. 6.5) through the following well-known reaction:

$$CO_2 + H_2O \rightarrow CH_2O + O_2$$

A small amount of oxygen is also produced in the upper atmosphere by photolysis of water molecules. For instance, the photolysis of water produces H_2 and oxygen ($H_2O \rightarrow H_2 + 0.5\ O_2$). Oxygen is removed from the atmosphere by respiration and decay, which can be considered the reverse of the photosynthesis reaction, and by chemical weathering. Virtually all oxygen produced by photosynthesis in a given year is lost in less than 50 years by the oxidation of organic matter. Oxygen is also liberated by the reduction of sulfates and carbonates in marine sediments. Without oxidation of organic matter and sulfide minerals during weathering, the oxygen content of the atmosphere would double in about 10^4 years (Holland et al., 1986). It has also been suggested that wildfires may have contributed in the past to controlling the upper limit of oxygen in the atmosphere. It appears that the oxygen content of the modern atmosphere is maintained at a near-constant value by various negative, short-term feedback mechanisms involving primarily photosynthesis and decay. If, however, photosynthesis were to stop, respiration and decay would continue until all organic matter on the Earth was transformed to CO_2 and water. This would occur in about 20 years and would involve only a minor decrease in the amount of oxygen in the atmosphere. Weathering would continue to consume oxygen and would take about 4 My to use up the current atmospheric supply.

Before the appearance of photosynthetic microorganisms, and probably for a considerable time thereafter, photosynthesis was not an important process in controlling atmospheric oxygen levels. In the primitive atmosphere, oxygen content was controlled by the rate of

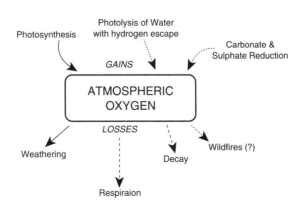

Figure 6.5 Gains and losses of atmospheric oxygen. Solid lines are major controls, long-dash lines are intermediate controls, and short-dash lines are minor controls.

photolysis of water and methane, the hydrogen loss from the top of the atmosphere, and the weathering rates at the surface. The rate at which water is supplied by volcanic eruptions is also important because volcanism is the main source of water available for photolysis. If metallic iron were in the mantle during the earliest stages of degassing of the Earth, H_2 would have been an important component of volcanic gases. Because H_2 rapidly reacts with oxygen to form water, significant amounts of volcanic H_2 would prevent oxygen from accumulating in the early atmosphere. As water instead of H_2 became more important in volcanic gases, in response to the removal of iron from the mantle as the core grew, oxygen could begin to accumulate. Early photosynthesizing organisms also may have contributed to the first oxygen in the atmosphere. Methane photolysis may contribute indirectly to the input of oxygen in the early atmosphere. The free carbon remaining after hydrogen escapes from the upper atmosphere reacts with oxygen to produce CO_2, which then can be used in photosynthesis to produce oxygen (Catling et al., 2001). As photosynthesis became more widespread, recombination of H_2 and oxygen to form water could not keep pace with oxygen input, and the oxygen level in the atmosphere must have increased. This assumes that the rate of weathering did not increase with time, an assumption supported by geologic data.

Geologic Indicators of Ancient Atmospheric Oxygen Levels

Banded Iron Formation

The distribution of BIF with geologic time provides a constraint on oxygen levels in the oceans and atmosphere. **BIF** is a chemical sediment, typically thin-bedded or laminated with less than 15% iron of sedimentary origin (Fig. 6.6). BIF also commonly contains layers of chert and generally has $Fe^{+3}/Fe^{+2} + Fe^{+3}$ ratios in the range of 0.3 to 0.6, reflecting an abundance of magnetite (Fe_3O_4). BIF is metamorphosed, and major minerals include quartz, magnetite, hematite, siderite, and various Fe-rich carbonates, amphiboles, and sulfides. Although most abundant in the late Archean and Paleoproterozoic, BIF occurs in rocks as old as 3.8 Ga (e.g., in Isua, southwest Greenland) and as young as 0.8 Ga (e.g., in the Rapitan Group, northwest Canada). The Hamersley basin (2.5 Ga) in Western Australia is the largest known, single BIF depository (Klein and Beukes, 1992).

Most investigators believe that the large basins of BIFs formed on cratons or passive margins in shallow marine environments. Many BIFs are characterized by thin, wave-like laminations that can be correlated over hundreds of kilometers (Trendall, 1983). During the late Archean and Paleoproterozoic, enormous amounts of ferrous iron appear to have entered BIF basins with calculated Fe precipitation rates of 10^{12} to 10^{15} gm/year (Holland, 1984). Although some of the Fe^{+2} undoubtedly came from weathering of the continents, most appears to have entered the oceans by submarine volcanic activity (Isley, 1995). Deposition occurred as Fe reacted with dissolved oxygen, probably at shallow depths, forming flocculent, insoluble ferric, and ferro-ferric compounds. Archean BIFs may have been deposited in deep water in a stratified ocean (Klein and Beukes, 1994).

Beginning about 2.3 Ga, the stratified ocean began to break down with the deposition of shallow-water oolite-type BIFs, such as those in the Lake Superior area.

Although Precambrian BIF clearly was a large oxygen sink, the oxygen content of the coexisting atmosphere may have been quite low. For instance, hematite (Fe_2O_3) or $Fe(OH)_3$ can be precipitated over a range in oxygen level, and early Archean BIF may have been deposited in reducing marine waters. The large amount of oxygen in late Archean and especially in Paleoproterozoic BIF, however, appears to require the input of photosynthetic oxygen. This agrees with paleontological data that indicate a rapid increase in the number of photosynthetic algae during the same period. Many BIFs contain well-preserved fossil algae remains. Thus, it appears that the increased abundance of BIFs in the late Archean and Paleoproterozoic reflects an increase in oceanic oxygen content in response to increasing numbers of photosynthetic organisms. Only after most of the BIFs were deposited and the Fe^{+2} in solution was exhausted did oxygen begin to escape from the oceans and accumulate in the atmosphere in appreciable quantities (Cloud, 1973). The drop in abundance of BIFs about 1.7 Ga reflects the exhaustion of reduced iron.

Figure 6.6   Early Archean banded iron formation from the Warrawoona Supergroup, Western Australia. Courtesy of Andrew Glikson.

Redbeds and Sulfates

Redbeds are detrital sedimentary rocks with red ferric oxide cements. They generally form in fluvial or alluvial environments. The red cements are the result of subaerial oxidation (Folk, 1976) and thus require the presence of oxygen in the atmosphere. That redbeds do not appear in the geologic record until about 2.3 Ga (Eriksson and Cheney, 1992) suggests that oxygen levels were low in the Archean atmosphere.

Sulfates, primarily gypsum and anhydrite, occur as evaporites. Although evidence of minor gypsum deposition is found in some of the oldest supracrustal rocks (~3.5 Ga), evaporitic sulfates do not become important in the geologic record until less than 2.0 Ga. Because their deposition requires free oxygen in the ocean and atmosphere, their distribution supports rapid growth of oxygen in the atmosphere beginning in the Paleoproterozoic.

Detrital Uraninite Deposits

Several occurrences of late Archean to Paleoproterozoic detrital uraninite and pyrite are well documented, the best known of which are those in the Witwatersrand basin in South Africa (~2.9 Ga), the Pilbara craton in Western Australia (3.3–2.8 Ga), and those in the Blind River–Elliot Lake area in Canada (~2.3 Ga) (Rasmussen and Buick, 1999). Detrital siderite is also documented in the Pilbara sediments. No significant occurrences are known to be younger than Mesoproterozoic, although a few minor occurrences of detrital uraninite are found in young sediments associated with rapidly rising mountain chains such as the Himalayas (Walker et al., 1983). Both uraninite and pyrite are unstable under oxidizing conditions and are rapidly dissolved. The preservation of major late Archean and Paleoproterozoic deposits of detrital uraninite and pyrite in conglomerate and quartzite indicates that weathering did not lead to total oxidation and dissolution of uranium and iron. To preserve uraninite, the partial pressure of oxygen must have been 10^{-2} PAL. The few occurrences of young detrital uraninite are metastable; the uraninite is preserved only because of extremely rapid sedimentation rates and will not survive for long.

The restriction of major detrital uraninite–pyrite deposits to more than 2.3 Ga again favors low oxygen levels in the atmosphere before this time.

Paleosols

Paleosols are preserved, ancient weathering profiles or soils that contain information about atmospheric composition (Holland, 1992). Highly oxidized paleosols retain most, if not all, iron in Fe^{+3} and Fe^{+2} compounds, whereas paleosols that formed in nonoxidizing or only slightly oxidizing environments show significant losses of iron in the upper horizons, especially in paleosols developed on mafic parent rocks. This is illustrated for a late Archean profile from Western Australia in Figure 6.7 (Yang et al., 2002). A lower atmospheric content of oxygen is necessary for the Fe to be leached from the upper paleosol horizon (MacFarlane et al., 1994). Elements such as Al, which are relatively

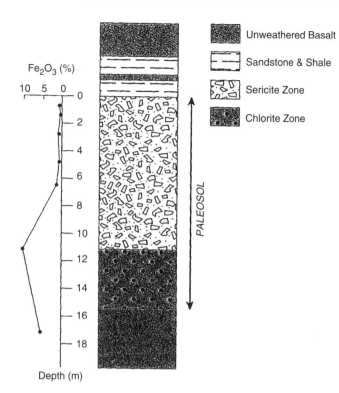

Fe$_2$O$_3$ (%)

Depth (m)

PALEOSOL

Unweathered Basalt

Sandstone & Shale

Sericite Zone

Chlorite Zone

Figure 6.7 Stratigraphic section of the 2.7-Ga Mt. Roe paleosol in Western Australia, showing the concentration of Fe$_2$O$_3$ with depth. Modified from MacFarlane et al. (1994).

immobile during weathering, are enriched in the upper horizons because of the loss of mobile elements such as Fe^{+2}. This indicates that by 2.7 Ga, the atmosphere contained little if any oxygen. Also, the presence of rhabdophane (a hydrous rare earth phosphate) with Ce^{+3} in a 2.6- to 2.5-Ga paleosol from Canada provides compelling evidence for an anoxic atmosphere by 2.5 Ga (Murakami et al., 2001).

Beginning in the Paleoproterozoic, paleosols do not show the leaching of iron. Results suggest that the oxygen level of the atmosphere rose dramatically from about 1% PAL to more than 15% of this level about 2 Ga (Holland and Beukes, 1990).

Biologic Indicators

The Precambrian fossil record also provides clues to the growth of atmospheric oxygen. Archean and Paleoproterozoic life forms were entirely prokaryotic organisms, the earliest examples of which evolved in anaerobic (oxygen-free) environments. Prokaryotes that produced free oxygen by photosynthesis appear to have evolved by 3.5 Ga. The timing of the transition from an anoxic to oxic atmosphere is not well constrained by microfossil remains but appears to have begun approximately 2.3 Ga (Knoll and Carroll, 1999). Certainly by 2.0 Ga, when heterocystous cyanobacteria appear, free oxygen was present in the atmosphere in significant amounts, in agreement with the paleosol data. The first appearance of eukaryotes from about 2.4 to 2.3 Ga indicates atmospheric oxygen had reached 1% PAL, which is necessary for mitosis to occur. The appearance of simple metazoans about 2.0 Ga requires oxygen levels high enough for oxygen to diffuse across membranes (~7% PAL).

The Carbon Isotope Record

General Features

Carbon isotopes in carbonates and organic matter offer the most effective way to trace the growth of the crustal reservoir of reduced carbon (Des Marais et al., 1992). The fractionation of ^{13}C and ^{12}C is measured by the $^{13}C/^{12}C$ ratio in samples relative to a standard such that the following is true:

$$\delta^{13}C(\text{‰}) = \{[(^{13}C/^{12}C)_{sample}/(^{13}C/^{12}C)_{std}] - 1\} \times 1000$$

This expression is used to express both carbonate (δ_{carb}) and organic (δ_{org}) isotopic ratios. The relative abundance of carbon isotopes is controlled chiefly by equilibrium isotopic effects among inorganic carbon species, fractionation associated with the biochemistry of organic matter, and relative rates of burial of carbonate and organic carbon in sediments. Most carbon in the Earth's near-surface systems is stored in sedimentary rocks; only about 0.1% is in living organisms and the atmosphere–hydrosphere. Oxidized carbon occurs primarily as marine carbonates and reduced carbon as organic matter in sediments. In the carbon cycle, CO_2 from the oceans and atmosphere is transferred into sediments as carbonate carbon (C_{carb}) or organic carbon (C_{org}), the former of which monitors the composition of the oceans (Fig. 6.4). The cycle is completed by uplift and weathering of sedimentary rocks and by volcanism, both of which return CO_2 to the atmosphere.

Because organic matter preferentially incorporates ^{12}C over ^{13}C, there should be an increase in the $^{13}C/^{12}C$ ratio (measured by $\delta^{13}C$) in buried carbon with time; indeed, this is what is observed (Worsley and Nance, 1989; Des Marais et al., 1992). $\delta^{13}C_{org}$ increases from values less than –40 per mil (‰) in the Archean to modern values of –20 to –30‰. On the other hand, seawater carbon is tracked with $\delta^{13}C_{carb}$ and remains roughly constant with time, with $\delta^{13}C_{carb}$ averaging about 0‰. The carbon cycle can be monitored by an isotopic mass balance (Des Marais et al., 1992) as follows:

$$\delta_{in} = f_{carb}\delta_{carb} + f_{org}\delta_{org}$$

Here, δ_{in} represents the isotopic composition of carbon entering the global surface environment composed of the atmosphere, hydrosphere, and biosphere. The right side of the equation represents the weighted-average isotopic composition of carbonate ($\delta^{13}C_{carb}$) and organic ($\delta^{13}C_{org}$) carbon being buried in sediments, and f_{carb} and f_{org} are the fractions of carbon buried in each form ($f_{carb} = 1 - f_{org}$). For timescales longer than 100 My, $\delta_{in} = -5‰$, the average value for crustal and mantle carbon (Holser et al., 1988). Thus, where values of sedimentary δ_{carb} and δ_{org} can be measured, it is possible to determine f_{org} for ancient carbon cycles. Note, for example, that higher values of $\delta^{13}C_{carb}$, $\delta^{13}C_{org}$, or both indicate a higher value of f_{org}.

During the Phanerozoic, there are several peaks in $\delta^{13}C_{carb}$, with the largest about 530, 400, 300, 280, and 110 Ma (Fig. 6.8). In addition, there are large peaks in $\delta^{13}C_{carb}$ about 2200 and 700 Ma (Fig. 6.9). In all cases, these peaks appear to reflect an increase in the burial rate of organic carbon (Des Marais et al., 1992; Frakes et al., 1992). This is because organic matter selectively enriched in ^{12}C depletes seawater in this isotope, raising the $\delta^{13}C$ values of seawater. In the late Paleozoic (250–300 Ma), the maxima in $\delta^{13}C_{carb}$ correspond to the rise and spread of vascular land plants, which provided a new source of organic debris for burial (Berner, 1987; Berner, 2001). Also conducive to the preservation of organic remains at this time were the vast lowlands on Pangea which appear to have been sites of widespread swamps where bacterial decay of organic matter is minimized. A drop in $\delta^{13}C_{carb}$ at the end of the Permian is not understood. Perhaps large amounts of photosynthetic oxygen generated by Carboniferous forests led to extensive forest fires that destroyed large numbers of land plants in the Late Permian. These carbon isotope excursions will be described more fully in Chapter 9.

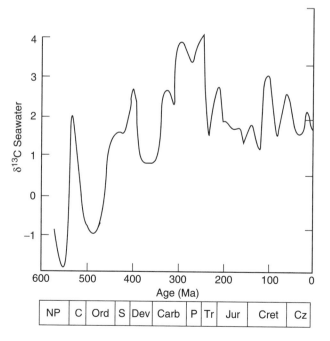

Figure 6.8 Secular changes in $\delta^{13}C$ in seawater in the last 600 My based on data from marine carbonates.

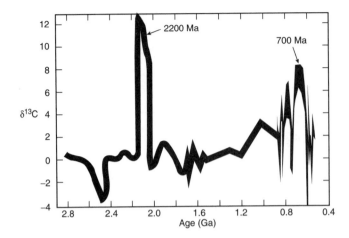

Figure 6.9 The isotopic composition of carbon in marine carbonates, between 2.8 and 0.4 Ga.

The 2200-Ma Carbon Isotope Excursion

It is now well established that a major positive excursion in $\delta^{13}C$ occurs in marine carbonates around 2200 Ma, perhaps the largest excursion in the geologic record (Karhu and Holland, 1996; Holland 2002) (Fig. 6.9). Other Paleoproterozoic carbon isotope excursions are reported in the Transvaal Group in South Africa, but it is not clear whether these are of global extent or they reflect local sedimentary conditions. The Paleoproterozoic event, which lasted about 150 My, is recorded in many sections worldwide. During this event, the fraction of carbon gases reduced to organic carbon was much larger than before or after the event (Holland, 2002). At the peak, $\delta^{13}C$ in marine carbonates was about +12, and corresponding organic carbon values, although variable, averaged about –24. If the average $\delta^{13}C$ input into the atmosphere was about –5‰, as it is today; about 50% of the carbon at that time must have been buried as organic carbon. Estimates of the total amount of oxygen released to the atmosphere during this carbon burial are 12 to 22 times present-day levels (Holland, 2002). This suggests a major period of growth in atmospheric oxygen around 2.2 Ga.

The Sulfur Isotope Record

General Features

The geochemical cycle of sulfur resembles that of carbon (Fig. 6.10). During this cycle, ^{34}S is fractionated from ^{32}S, with the largest fractionation occurring during bacterial reduction of marine sulfate to sulfide. Isotopic fractionation is expressed as $\delta^{34}S$ in a manner similar to that used for carbon isotopes. Sedimentary sulfates appear to record the isotopic composition of sulfur in seawater. Mantle ^{34}S is near 0‰, and bacteria reduction of sulfate strongly prefers ^{32}S, thus reducing $\delta^{34}S$ in organic sulfides to negative values (–180‰) and leaving oxidized sulfur species with approximately equivalent positive values (+17‰). Hence, the sulfur cycle is largely controlled by the biosphere and in particular by sulfate-reducing bacteria that inhabit shallow marine waters. Because of

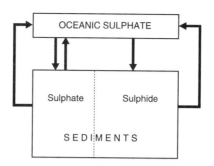

Figure 6.10 Schematic diagram of the sulfur cycle.

the mobility of dissolved sulfate ions and the rapid mixing of marine reservoirs, $\delta^{34}S$ values of residual sulfate show limited variability (+17 ± 2‰) compared with the spread of values in marine sulfides (−5 to −35‰).

How can the sulfur isotopic record help to constrain the growth of oxygen in the atmosphere? The time distribution of sulfates and sulfate-reducing bacteria should track the growth of oxygen (Ohmoto et al., 1993; Canfield et al., 2000). The overall sulfur isotope trends in marine sulfates and sulfides from 3.8 Ga to the present suggest a gradual increase in sulfate $\delta^{34}S$ and a corresponding decrease in sulfide $\delta^{34}S$ (Fig. 6.11). Changes in $\delta^{34}S$ with time reflect (1) changes in the isotopic composition of sulfur entering the oceans from weathering and erosion, (2) changes in the relative proportions of sedimentary sulfide and sulfate receiving sulfur from the ocean–atmosphere system, and (3) the temperature of seawater (Schidlowski et al., 1983; Ohmoto and Felder, 1987; Canfield et al., 2000). Because weathering tends to average sulfide and sulfate input, one or both of the latter two effects probably account for oscillations in the sulfate $\delta^{34}S$ with time. Three causes have been suggested for the tight grouping of $\delta^{34}S$ in marine sulfates and sulfides in the Archean (Canfield et al., 2000) (Fig. 6.11): (1) low fractionation of sulfur isotopes because of high sulfate reduction rates and moderate concentrations of sulfate in the oceans, (2) low fractionation of isotopes in sulfides that form a closed system of poorly mixed sediments, or (3) sulfide formation in sulfate-poor oceans. In all three scenarios, sulfate-reducing bacteria are able to reduce sulfates to sulfur efficiently, resulting in limited isotopic fractionation between sulfides and sulfates. Although with the current database a distinction among the preceding three causes is not possible, the increase in the spread of $\delta^{34}S$ values beginning from about 2.3 to 2.2 Ga is consistent with the carbon isotope data in suggesting rapid growth in oxygen in the atmosphere at this time.

Mass-Independent Fractionation

Mass-independent sulfur isotope fractionation may be extremely important in constraining the growth rate of oxygen in the Earth's atmosphere. Farquhar et al. (2000) show that the difference in abundance between ^{33}S and ^{32}S is about half that between ^{34}S and ^{32}S in rocks older than about 2.2 Ga. These differences are thought to result solely from photolysis of SO_2 and SO in the upper atmosphere (Farquhar and Wing, 2003). In a high-oxygen atmosphere

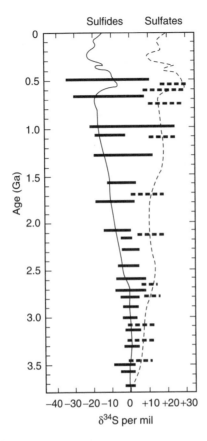

Figure 6.11   The isotopic composition of sulfur in marine sulfates and sulfides with geologic time. Modified from Schidlowski et al. (1983) and Canfield (1998).

as at present, these effects are not seen in sediments because almost all sulfur gases are oxidized to sulfuric acid and accumulate in the oceans as sulfates. In a low-oxygen atmosphere, however, where sulfur can exist in a variety of oxidation states, the probability of transferring a mass-independent fractionation signature into sediments is much greater.

You can use $\Delta^{33}S$ to track the history of oxygen in the atmosphere as follows:

$$\Delta^{33}S = \delta^{33}S - 1000((1 + \delta^{34}S/1000)^{0.515} - 1)$$

Three stages in atmospheric history are recognized when $\Delta^{33}S$ in marine sediments is plotted with age (Farquhar and Wing, 2003) (Fig. 6.12). Stage 1, from 3.80 to about 2.45 Ga, is characterized by $\Delta^{33}S$ extending to values greater than ±1.0‰; stage 2, from about 2.45 to 2.00 Ga, has a much smaller range of $\Delta^{33}S$ (–0.1 to +0.5); and stage 3, from 2.00 Ga to the present, has a very small range of $\Delta^{33}S$, from –0.1 to +0.2‰. This overall pattern supports carbon isotope data, suggesting rapid growth of oxygen in the Earth's atmosphere near 2.20 Ga. Results reported by Bekker et al. (2004) indicate that the very small range in $\Delta^{33}S$ extended to at least 2.32 Ga, suggesting rapid growth in oxygen in the atmosphere between 2.32 and 2.45 Ga. More sulfur isotope data in the age range from 2.00 to 2.50 Ga are needed to refine this estimate.

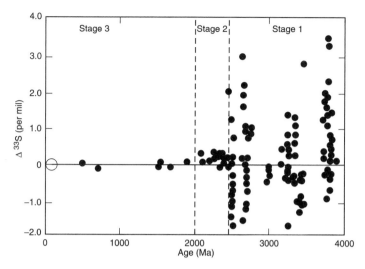

Figure 6.12 Variation in $\Delta^{33}S$ with time in marine sediments. Modified from Farquhar and Wing (2003). The large unfilled circle represents hundreds of analyses of samples less than 2000 My.

The Growth of Atmospheric Oxygen

Considering all the geologic and isotopic data, the growth of oxygen in the Earth's atmosphere can be divided into three stages (Kasting, 1987; Kasting, 1991; Farquhar and Wing, 2003): stage 1, a *reducing stage* in which free oxygen occurs in neither the oceans nor the atmosphere; stage 2, an *oxidizing stage* in which small amounts of oxygen are in the atmosphere and surface ocean water but not in deep ocean water; and stage 3, an *aerobic stage* in which free oxygen pervades the ocean–atmosphere system. These stages are diagrammatically illustrated in Figure 6.13 with corresponding geologic indicators.

During stage 1, the ozone shield required to block lethal ultraviolet radiation is absent. If such a shield existed on the primitive Earth, it must have been produced by a gaseous

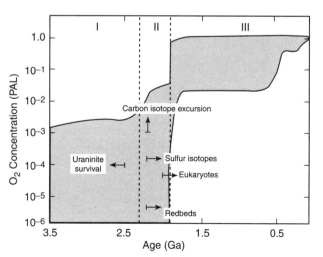

Figure 6.13 Growth in atmospheric oxygen with time. The shaded area is the range of oxygen concentrations permitted by geologic indicators. PAL, present atmospheric level. Modified from Kasting (1993).

compound other than ozone. Although marine cyanobacteria produced oxygen by photosynthesis during this stage, for large volumes of BIF, this oxygen must have combined rapidly with Fe^{+2} and been deposited. The atmosphere should have remained in stage 1 until the amount of oxygen produced by photosynthesis followed by organic carbon burial was enough to overwhelm the input of volcanic gases. The change from stage 1 to 2 occurred from about 2.3 to 2.2 Ga and is marked by the appearance of redbeds (2.3 Ga), an increase in abundance of evaporitic sulfates, the end of major detrital uraninite–pyrite deposition, a major carbon isotope excursion, and changes in sulfur isotope fractionation. At this time, photosynthetic oxygen entered the atmosphere in large amounts. What triggered this injection of oxygen from about 2.2 to 2.3 Ga is a subject of considerable controversy and will be returned to in Chapter 9. Evidence that the lower atmosphere and the surface ocean waters during this stage were weakly oxidizing, yet deep marine waters were reducing, is provided by the simultaneous deposition of oxidized surface deposits (redbeds and evaporites) and BIF in a reducing (or at least nonoxidizing) environment. Two models have been proposed for the control of atmospheric oxygen levels during stage 2. One proposes that low oxygen levels are maintained by mantle-derived Fe^{+2} in the oceans and BIF deposition, and the other relies on low photosynthetic productivity in the ocean. Which of these is most important is not yet clear. In either case, it would appear that stage 2 was short lived (Fig. 6.13).

The change from stage 2 to 3, about 2 Ga, is also marked by the near disappearance of BIF from the geologic record and the appearance of eukaryotic organisms. The onset of stage 3 is defined by the exhaustion of Fe^{+2} from the oceans such that photosynthetic oxygen levels increase and the oceans become oxidizing, and by the decreased spread in $\Delta^{33}S$ in marine sediments. Further increases liberate oxygen into the atmosphere. During this stage, an effective ozone screen develops, which is probably responsible for the rapid diversification and increase in the number of microorganisms in the Mesoproterozoic. By 540 Ma, atmospheric oxygen must have risen to at least 10% PAL to permit the appearance of carbonate shell-forming organisms.

Before leaving oxygen, a competing model of oxygen growth should be mentioned. Ohmoto (1997) proposes a model in which the oxygen level in the atmosphere remains essentially constant for 4 Gy of the Earth's history. This model questions many of the oxygen indicators and especially whether they are global in character. It also suggests that the carbon isotope peak 2.2 Ga can be explained by a drop in total carbon flux or a decrease in the deposition rate of marine carbonates. However, it offers no explanation for the change in sulfur isotopes 2.2 Ga. Although this model cannot yet be dismissed, the accumulating evidence supporting the rapid rise in atmospheric oxygen about 2.2 Ga is becoming more convincing.

Phanerozoic Atmospheric History

It is possible to track the levels of CO_2 and oxygen in the Earth's atmosphere during the Phanerozoic using the burial and weathering rates of organic carbon, Ca and Mg silicates, and carbonates as deduced from the preserved stratigraphic record. Other factors such as the effect of changing solar radiation on surface temperature and weathering rates and the use of the $^{87}Sr/^{86}Sr$ seawater curve to estimate the effect of tectonics on weathering and erosion rates can be used to fine-tune the results (Berner and Canfield, 1989; Berner, 1994; Berner, 2001). Although CO_2 shows a gradual drop throughout the Phanerozoic, it is high in the early to middle Paleozoic and shows a pronounced minimum around 300 Ma (Fig. 6.14). Variations in atmospheric CO_2 are controlled by a combination of tectonic and biologic processes, often with one or the other dominating, and by the increasing luminosity of the Sun. An increase in solar luminosity since the beginning of the Paleozoic is probably responsible for the gradual drop in atmospheric CO_2. This drawdown is caused by enhanced weathering rates brought about by the greenhouse warming of the atmosphere. The dramatic minimum in CO_2 about 300 Ma appears to be related to (1) enhanced silicate weathering and (2) the appearance and rapid development of land plants, which, through photosynthesis followed by burial, rapidly removed carbon from the atmosphere–ocean system. Rapid weathering rates in the early and middle Paleozoic probably reflect a combination of increasing solar luminosity and the fragmentation of Gondwana (early Paleozoic only), during which enhanced ocean-ridge and mantle-plume activity pumped large amounts of CO_2 into the atmosphere, increasing weathering rates. The increase in CO_2 in the Mesozoic, which resulted in worldwide warm climates, may be caused by enhanced ocean-ridge and mantle-plume activity associated with the onset of fragmentation of Pangea. A gradual drop in CO_2 in the late Cretaceous and Tertiary appears to reflect a combination of decreasing ocean-ridge activity, thus decreasing the CO_2 supply, and increased weathering and erosion rates in response to terrane collisions around the Pacific (exposing more land area for weathering).

Unlike CO_2, the variation in oxygen during the Phanerozoic appears to be controlled chiefly by the burial rate of carbon and sulfur, which in turn is controlled by sedimentation rates (Berner, 2001). Oxygen levels were relatively low and constant during the early and middle Paleozoic followed by a high peak about 300 Ma (Fig. 6.14). During the Mesozoic, oxygen gradually increased; this was followed by a steep drop in the Tertiary. The peak in oxygen in the late Paleozoic correlates with the minimum in CO_2 and appears to have the same origin: enhanced burial rate of organic carbon accompanying the rise of vascular land plants. This is evidenced today by the widespread coal deposits of this age. The drop in oxygen at the end of the Permian (250 Ma) probably reflects a change in climate to more arid conditions. Depositional basins changed from coal swamps to oxidized terrestrial basins, largely eolian and fluvial, thus no longer burying

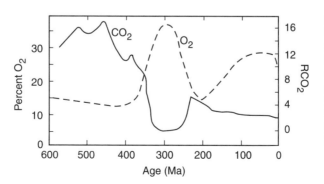

Figure 6.14 Oxygen and CO_2 atmospheric concentrations during the Phanerozoic. RCO_2 is the mass ratio of CO_2 at a given time divided by the modern value. Modified from Berner and Canfield (1989) and Berner (1994; 2001).

large quantities of carbon. Also, a dramatic drop in sea level at the end of the Permian resulted in oxidation and erosion of previously buried organic material, thus reducing atmospheric oxygen levels. Burial of land plants in coal swamps during the Cretaceous may have contributed to the increased oxygen levels at this time, and again a drop in sea level during the Tertiary exposed these organics to oxidation and erosion, probably causing the fall in atmospheric oxygen during the late Cenozoic.

The increase in atmospheric oxygen beginning in the Carboniferous may have been responsible for several important biologic changes at this time. The increase should have enhanced diffusion-dependent processes such as respiration; hence, some organisms could attain large body sizes (Graham et al., 1995; Berner, 2001). Perhaps the most spectacular examples are the insects: some forms attained gigantic sizes probably in response to increased diffusive permeation of oxygen into the organisms. Also, the increased oxygen levels would result in a denser atmosphere (21% greater than at present), which may have led to the evolution of insect flight by offering greater lift. Increased aquatic oxygen levels during the Carboniferous–Permian would permit greater biomass densities and increased metabolic rates, both of which could lead to increases in radiation of taxa and in organism sizes. Impressive examples of both of these changes occur in the brachiopods, foraminifera, and corals.

The Oceans

Oceans cover 71% of the Earth's surface and contain most of the hydrosphere. The composition of seawater varies geographically and with depth and is related principally to the composition of river water and hydrothermal water along ocean ridges. Chemical precipitation, melting of ice, and evaporation rate also affect the salinity of seawater. Dissolved nutrients such as bicarbonate, nitrate, phosphate, and silica vary considerably in concentration as a result of removal by planktonic microorganisms such as foraminifera, diatoms, and algae. Seawater temperature is rather constant in the upper 300 m (~18° C) and then drops dramatically at a depth of about 1 km, where it remains relatively constant (3–5° C) to the ocean bottom. The cold bottom water in the oceans forms in polar regions, where it sinks because of its high density, and travels toward the equator, displacing warm surface waters that travel poleward. The geologic history of seawater is closely tied to the history of the atmosphere, and oceanic growth and the concentration of volatile elements in seawater is partly controlled by atmospheric growth (Holland et al., 1986). Most paleoclimatologists agree that the oceans formed early in the Earth's history by condensation of atmospheric water. Some water, however, may have been added to the Earth by early cometary impacts. For instance, with the current rate of water accretion from small comets, it is possible to increase mean global sea level by 2 to 10 mm in about 3600 years (Deming, 1999). However, because the deuterium/hydrogen ratio in cometary water (310×10^{-6}) is considerably higher than in terrestrial water (149×10^{-6}), the contribution of cometary water to the oceans in the past probably has been small (<10%) (Robert, 2001).

Because of the relatively high CO_2 content of the early atmosphere, the first seawater to condense must have had a low pH (<7) caused by dissolved CO_2 and other acidic components (H_2S, HCl, etc.). However, this situation was probably short lived in that submarine volcanic eruptions and seawater recycling through ocean ridges would introduce large amounts of Na, Ca, and Fe. Thus, a pH between 8 and 9 and a neutral or negative Eh were probably reached rapidly in the early oceans and maintained thereafter by silicate-seawater buffering reactions and recycling at ocean ridges.

Sea Level

Factors Controlling the Sea Level

Changes in sea level leave an imprint in the geologic record by the distribution of sediments and biofacies, the aerial extent of shallow seas as shown by preserved cratonic sediments, and the calculated rates of sedimentation (Eriksson, 1999). A widely used approach to monitor sea level is to map successions of transgressive and regressive facies in marine sediments. During **transgression,** shallow seas advance on the continents, and during **regression,** they retreat. Transgressions and regressions, however, are not always accompanied by rises or drops in sea level, and to estimate sea level changes, it is necessary to correlate facies and unconformities over large geographic regions on different continents and different tectonic settings using sequence stratigraphy (Steckler, 1984; Vail and Mitchum, 1979). Using this technique, onlap and offlap patterns are identified in seismic-reflection profiles and used to construct maps of the landward extent of coastal onlap from which changes in sea level are estimated.

Although a long-term component (200–400 My) in sea level change during the Phanerozoic seems to be agreed upon by most investigators, short-term components and amplitudes of variation remain uncertain. Major transgressions are recorded in the early Paleozoic and in the late Cretaceous (Fig. 6.15). Estimates of the amplitude of the Cretaceous sea-level rise range from 100 to 350 m above the present sea level, and estimates of the early Paleozoic rise generally are tens of meters (Algeo and Seslavinsky, 1995). Cyclicity in sea level change has been proposed by many investigators with periods in the range of 10 to 80 My. The Vail sea level curve (Fig. 6.15) has been widely accepted, although the amplitudes are subject to uncertainty.

Several factors can cause short-term changes in sea level (Eriksson, 1999). Among the most important are tectonic controls, chiefly continental collisions that operate on timescales of 10^6 to 10^8 years, and glacial controls that operate on timescales of 10^4 to 10^8 years. Also, waves and tides, storm winds and hurricanes, tsunamis, melting methane hydrates (clathrates), and catastrophic sediment slumps can affect tidal range. Tectonic controls seem to produce sea level changes at rates of less than 1 cm/10^3 years, and glacial controls can produce rate changes up to 10 m/10^3 years. With the present ice cover on the Earth, the potential amplitude for sea level rise if all the ice were to melt is 200 m. The short-term drops in sea level 500, 400, and 350 Ma (Fig. 6.15) probably reflect glaciations at these times. The rise of the Himalayas in the Tertiary produced a drop in sea level of about 50 m.

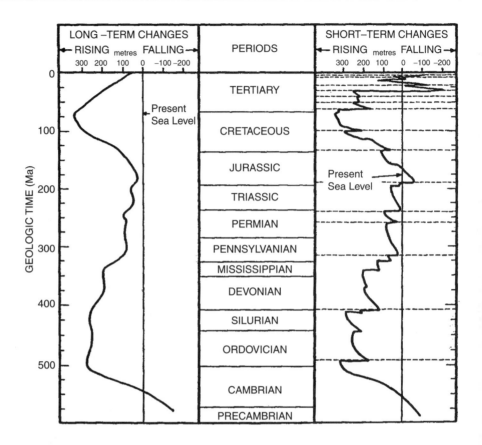

Figure 6.15 Long- and short-term changes in sea level during the Phanerozoic. Modified from Vail and Mitchum (1979).

Long-term changes in sea level are related to (1) the rates of seafloor spreading, (2) the characteristics of subduction, (3) the motion of continents with respect to geoid highs and lows, and (4) the supercontinent insulation of the mantle (Gurnis, 1993; Eriksson, 1999). One of the most important causes of long-term sea level change is the volume of ocean-ridge systems. The cross-sectional area of an ocean ridge depends on the spreading rate because its depth is a function of its age. For instance, a ridge that spreads 6 cm/year for 70 My has three times the cross-sectional area of a ridge that spreads 2 cm/year for the same amount of time. The total ridge volume at any time depends on the spreading rate (which determines the cross-sectional area) and the total ridge length. Times of maximum seafloor spreading correspond with the major transgression in the late Cretaceous, thus supporting the ridge volume model. Calculations indicate that the total ridge volume in the late Cretaceous could displace enough seawater to raise sea level about 350 m.

Subduction can also affect sea level. For instance, if additional convergent and divergent plate margins form in an oceanic plate while the spreading rate is maintained at a constant value, the average age of oceanic lithosphere decreases and leads to uplift above subduction zones, causing a drop in sea level (Gurnis, 1993). Also, when continental plates move over geoid highs, sea level falls; conversely, sea level rises when the plates move over geoid lows. As the mantle warms beneath a supercontinent, the continental plate rises isostatically with a corresponding fall in sea level.

Sea Level and the Supercontinent Cycle

Worsley et al. (1984; 1986) have proposed an intriguing model in which Phanerozoic sea level is directly tied to the supercontinent cycle, with a repeat time of about 400 My. The rapid increases in sea level observed from 600 to 500 Ma and from 200 to 100 Ma correspond to the onset of the breakup of Laurasia (Laurentia–Baltica) and Pangea, respectively, and the early Paleozoic long-term maximum coincides with the assembly of Pangea (Fig. 6.16). The low sea level about 600 Ma correlates with the assembly of Gondwana, and the low from 300 to 200 Ma correlates with the final assembly of Pangea. The low sea level accompanying the growth of Gondwana may reflect a geoid high beneath the supercontinent, whereas the high sea levels accompanying supercontinent dispersal reflect enhanced ocean-ridge activity. Supporting this idea, the number of continents with time correlates with sea level (Fig. 6.16). Also, as expected, the percentage of flooding of the continents increases during supercontinent dispersal and decreases during supercontinent stability. Because major changes in sea level appear to be related to changes in the volume of the ocean-ridge system, times of high sea level should also be times of rapid production of oceanic crust, when ridges are relatively high. Calculated rates

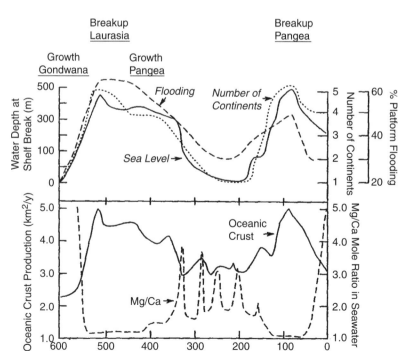

Figure 6.16 Comparison of the variation in sea level, the number of continents, the amount of flooding of shallow shelves by seawater, the production rate of oceanic crust, and the Mg/Ca mole ratio in seawater as a function of time. Also indicated are the breakup and growth stages of supercontinents. Modified from Worsley et al. (1984), Gaffin (1987), and Hardie (1996).

of production of oceanic crust, assuming they are tied to sea level changes, are shown in Figure 6.16. Note that the oceanic crust production rate increases up to a factor of two during supercontinent breakups. It should be pointed out, however, that a correlation of oceanic crust production rate to sea level change is questioned by some investigators (Rowley, 2002).

The Worsley sea level model is exciting in that it presents a potential method by which older supercontinents can be tracked from sea level data inferred from sequence stratigraphy of marine successions before 1 Ga.

Changes in the Composition of Seawater with Time

Marine Carbonates and Evaporites

Although it is commonly assumed that the composition of seawater has not changed with time, some observational and theoretical data challenge this assumption. For instance, some Archean marine carbonates contain giant botryoids of aragonite and Mg-calcite beds up to several meters thick that extend hundreds of kilometers (Grotzinger and Kasting, 1993) (Fig. 6.17). By comparison, Paleoproterozoic carbonates have less spectacular occurrences of these minerals, although cement crusts in tidal flat deposits are common. In contrast, all younger marine carbonates lack these features or, in the case of the cements, show a progressive decline in abundance with age. Although the modern ocean is oversaturated in carbonate, cement crusts are not deposited today. These secular changes strongly suggest that Precambrian seawater was greatly oversaturated in carbonate and that saturation decreased with time (Grotzinger and Kasting, 1993; Hardie, 2003). In addition, major ion concentrations and secular variations in Precambrian seawater appear to be similar to those in Phanerozoic seawater, and the Mg/Ca ratio of Precambrian seawater controlled which carbonate polymorph (calcite or aragonite) was deposited.

Another difference between Precambrian and Phanerozoic sediments is the sequence of evaporite deposition. In Paleoproterozoic sediments, halite (NaCl) is often deposited immediately on top of carbonate without intervening sulfates. Although minor occurrences of evaporitic gypsum are reported in the Archean (Buick and Dunlop, 1990), they appear to reflect the local composition of seawater controlled by local source rocks. The first extensive sulfate deposits appear in the MacArthur basin in Australia from 1.7 to 1.6 Ga, which coincides with the last examples of deposition of seafloor carbonate cements. The paucity of gypsum evaporites before 1.7 Ga may have been caused by low concentrations of sulfate in seawater before this time, or the bicarbonate/Ca ratio may have been sufficiently high so that during progressive seawater evaporation Ca was exhausted by carbonate precipitation before the gypsum stability field was reached (Grotzinger and Kasting, 1993).

Evaporite deposition also may have influenced the composition of the oceans during the Phanerozoic. The large mass of evaporites deposited during some time intervals may have been important in lowering the NaCl and $CaSO_4$ contents of world oceans (Holser, 1984). For instance, the vast volumes of salt (chiefly NaCl) deposited in the late Permian

Figure 6.17   Aragonite botryoid pseudomorphs from the late Archean (2.52-Ga) Transvaal Supergroup in South Africa. Two aragonite fans nucleated on the sediment surface and grew upward. The width of photo is about 30 cm. Courtesy of John Grotzinger.

may have caused average ocean salinity to drop significantly during the Triassic and Jurassic. Evaporite deposition has been sharply episodic, and erosion of older evaporites has been relatively slow in returning salts to the oceans. The distribution of evaporites of all ages coupled with a slow recycling rate into the oceans suggests that the mean salinity of seawater may have decreased since the Mesoproterozoic because of evaporite deposition.

Secular changes in the mineralogy of marine limestones and evaporites during the Phanerozoic may also be related to the supercontinent cycle. Periods of marine aragonite deposition are also times when $MgSO_4$ evaporites are widespread, whereas periods of dominantly calcite deposition correspond to an abundance of KCl evaporite deposition (Hardie, 1996; Horita et al., 2002). This distribution may be controlled by chemical reactions associated with hydrothermal alteration at ocean ridges in which there is a net transfer of Na + Mg + SO_4 from seawater to rock and of Ca + K from rock to seawater. The intensity of these reactions is a function of the volume of seawater circulated through the ocean-ridge system, which in turn is a function of the heat flux associated with oceanic crust production. Thus, during times of high production of oceanic crust, the oceans should be depleted in Na-Mg sulfates and enriched in Ca and K ions, as in the early

Paleozoic and again in the late Cretaceous. The Mg/Ca mole ratio in seawater can be used to track these reactions, in that at mole ratios of less than 2 only calcite is deposited, whereas at higher ratios both high-Mg calcite and aragonite are deposited. Also, during episodes of high seawater Mg/Ca ratios when aragonite is deposited, seawater has low Sr/Ca ratios because of the preferential incorporation of Sr into aragonite (Steuber and Veizer, 2002). Because the production rate of oceanic crust is tied closely to the supercontinent cycle (Fig. 6.16), the mineralogy of marine carbonates and evaporites should reflect this cycle. The Mg/Ca mole ratio of seawater calculated from the oceanic crust production rate (Fig. 6.16) agrees with the observed secular changes in the composition of carbonates and evaporites, supporting this model (Hardie, 1996; Hardie, 2003).

An alternative explanation for changes in the Mg/Ca ratio of seawater with time is that they reflect changes in the global rate of dolomite deposition (Holland and Zimmermann, 2000). For instance, a decrease in Mg/Ca would be caused by a gradual transfer of $CaCO_3$ deposition from shallow- to deepwater environments because of the proliferation of planktonic calcareous organisms.

The Biochemical Record of Sulfur

Sulfate-reducing bacteria appear to have been in place by at least 3.5 Ga as suggested by the sulfur isotope fractionation between coexisting sulfates and sulfides of this age (Fig. 6.11). The biochemical record of sulfur is consistent with sulfate-poor Archean oceans changing to modest sulfate concentrations, but not until 2.2 to 2.3 Ga was there a significant separation between $\delta^{34}S$ in sulfides and sulfates. As mentioned earlier, this probably reflects the rapid growth of oxygen in the atmosphere at this time. These results suggest that the oceans were oxygenated only to shallow depths during the Paleoproterozoic and that more extensive oxygenation did not occur until the Neoproterozoic (Anbar and Knoll, 2002). If sulfide conditions were common through much of the Proterozoic, geochemical redox indicators should provide evidence of widespread anoxia, and seawater sulfate concentrations should have been lower than at present. Both predictions are observed in black shales deposited between 1730 and 1640 Ma in Australia and Laurentia. In addition, the relatively large isotopic variability in Mesoproterozoic sulfate evaporites, not seen in Phanerozoic evaporites, suggests a smaller Mesoproterozoic global reservoir of oceanic sulfate ion.

In summary, available data point to globally extensive sulfate-reducing bacteria in low-sulfate oceans, oxygenation in the ocean insufficient to support widespread deposition of sulfate evaporites, and sulfide-rich bottom waters in at least some marine basins during most of the Proterozoic (Anbar and Knoll, 2002; Knoll, 2003).

Oxygen Isotopes in Seawater

Oxygen-18 is fractionated from oxygen-16 in the ocean–atmosphere system during evaporation, with the vapor preferentially enriched in ^{16}O. The fractionation of these isotopes decreases with increasing temperature and salinity. The fractionation of the two isotopes is measured by the deviation of $^{18}O/^{16}O$ from standard mean ocean water (SMOW) in terms of $\delta^{18}O$:

$$\delta^{18}O\ (\%o) = [\{(^{18}O/^{16}O)_{sample}/(^{18}O/^{16}O)_{SMOW}\} - 1] \times 1000$$

Because marine chert and BIF are resistant to secondary changes in $\delta^{18}O$, they are useful in monitoring the composition of seawater with geologic time (Knauth and Lowe, 1978; Gregory, 1991). The $\delta^{18}O$ of marine chert is a function of the temperature of deposition, the isotopic composition of seawater, and secondary changes in the chert. A summary of published $\delta^{18}O$ from Precambrian continental cherts and BIF clearly shows that Archean cherts have low $\delta^{18}O$ but that they overlap with Proterozoic cherts (Fig. 6.18). Although relatively few samples are available from Proterozoic cherts, there is a clear increase in Proterozoic $\delta^{18}O$ values toward the Phanerozoic values from 25 to 35‰ (Knauth and Lowe, 1978; Knauth and Lowe, 2003).

Two models have been proposed to explain increasing $\delta^{18}O$ in chert, and hence in seawater, with geologic time.

The Mantle–Crust Interaction Model (Perry et al., 1978).

If Archean oceans were about the same temperature as modern oceans, they would have a $\delta^{18}O$ of 34‰ less than chert precipitated from them, or about 14‰. For seawater to evolve to a $\delta^{18}O$ of 0‰ with time, it is necessary for it to react with substances with $\delta^{18}O$ values less than 0‰. Recycling seawater at ocean ridges and interactions of continent-derived sediments and

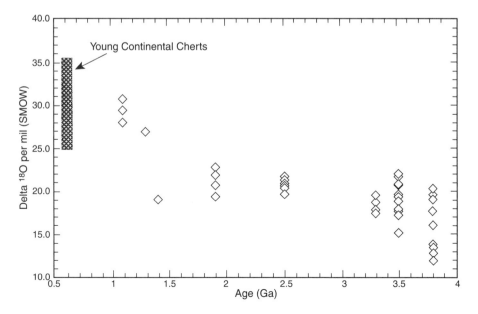

Figure 6.18 Distribution of oxygen isotopes in marine cherts with time.

river waters are processes that can raise seawater $\delta^{18}O$ values. Large volumes of seawater are recycled through the crust at ocean ridges and react with basalt that has a $\delta^{18}O$ of 5.5 to 6.0‰; continental sediments have $\delta^{18}O$ values even higher (7.0–20.0‰). This model is consistent with increasing volumes of continental crust preserved with time.

The Seawater-Cooling Model (Knauth and Lowe, 1978; Knauth and Lowe, 2003). In this model, the oxygen isotope trend is explained by a gradual cooling of the Earth's surface temperature from 60 to 85° C 3.5 Ga to present temperatures. The $\delta^{18}O$ of seawater may or may not be constant over this time; however, if not constant, it must have changed in so that it did not affected the temperature control. Chert with $\delta^{18}O$ less than 15‰ reflects formation during the deepest burial or in local areas of enhanced geothermal activity. Chert with higher $\delta^{18}O$ values formed during early diagenesis and indicates an extremely hot Archean ocean. Some investigators suggest that the evidence for gypsum deposition in the Archean is an obstacle for this model because gypsum cannot be deposited at temperatures less than 58° C. However, the only well-documented examples of Archean gypsum deposition in the Pilbara sediments appear to reflect local and not worldwide ocean compositions.

Muehlenbachs and Clayton (1976) argued persuasively that recycling seawater at ocean ridges caused the oxygen isotopic composition of seawater to remain constant with time. However, it was not until studies of ophiolite alteration (Holmden and Muehlenbachs, 1993; Lecuyer et al., 1996) that it became clear that the $\delta^{18}O$ of seawater has probably not changed in the last 2 Gy because the distribution of $\delta^{18}O$ in low- and high-temperature alteration zones in 2 Ga ophiolites is the same as that found in young ophiolites. Hoffman et al. (1986) furthermore suggest that the stratigraphic distribution of $\delta^{18}O$ in alteration zones of the 3.5 Ga Barberton greenstone belt may indicate that the oxygen isotopic composition of seawater has not changed appreciably since the early Archean. Thus, if the secular change in $\delta^{18}O$ in chert and BIF is real, it would appear that the temperature of Archean seawater must have been greater than at later times, as originally suggested by Knauth and Lowe (1978). As pointed out by some investigators, however, such an interpretation does not easily accommodate the evidence for widespread glaciation from 2.3 to 2.4 Ga, which necessitates cool oceans.

The Dolomite–Limestone Problem

It has long been recognized that the ratio of dolomite to calcite in sedimentary carbonates decreased with decreasing age for the last 3 Gy (Holland and Zimmermann, 2000). This is also reflected by the MgO/CaO ratio of marine carbonates. Short-term variations in the dolomite/calcite ratio are also resolved in the Phanerozoic carbonate record. Most investigators agree that these trends are primary and are not acquired during later diagenesis, metasomatism, or alteration. The molar MgO/MgO+CaO ratio in igneous rocks (~0.2) is close to that observed in pre-Mesozoic marine carbonates (0.2–0.3), which suggests that Mg and Ca were released during weathering in the same proportions in which they occurred in source rocks more than 100 Ma (Holland, 1984). The residence times of CO_2,

Mg^{+2}, and Ca^{+2} in seawater are short and are delicately controlled by input and removal rates of diagenetic silicates and carbonates.

The short-term variations in the dolomite/calcite ratio have been explained by changes in the distribution of sedimentary environments with time. For instance, dolomite was widespread when large evaporite basins and lagoons were relatively abundant, such as in the Permian (Sun, 1994). Although the origin of the long-term secular change in the dolomite/calcite ratio is not fully understood, one of two explanations has been most widely advocated (Holland, 1984; Holland and Zimmermann, 2000):

Change in Mg, Ca, or Both Types of Depository

Removal of Mg^{+2} from seawater may have changed from a dominantly carbonate depository in the Proterozoic to a dominantly silicate depository in the late Phanerozoic. There is a tendency for the MgO/Al_2O_3 in shales to increase over the same period, providing some support for this mechanism in that Mg appears to be transferred from carbonate to silicate reservoirs with time. Alternatively, the decrease in dolomite deposition with time may be caused by the gradual transfer of $CaCO_3$ deposition from shallow- to deepwater environments because of a growing importance of planktonic calcareous organisms.

Decrease in Atmospheric CO_2

Silicate-carbonate equilibriums depend on the availability of CO_2 in the ocean–atmosphere system. When only small amounts of CO_2 are available in the ocean, silicate-carbonate reactions shift such that Ca^{+2} is precipitated chiefly as calcite rather than in silicates. If the rate of CO_2 input into the ocean exceeds the release rate of Ca^{+2} during weathering, both Mg^{+2} and Ca^{+2} are precipitated as dolomite. Hence, a greater amount of dolomite in the Precambrian may reflect an atmosphere–ocean system with greater amounts of CO_2, which is consistent with the atmosphere model previously described.

Archean Carbonates

Another poorly understood attribute of carbonate deposition is the scarcity of Archean carbonates. Among the causes proposed for this observation are the following:

1. The pH of Archean seawater was too low for carbonate deposition.
2. Carbonates were deposited on stable cratonic shelves during the Archean and were later eroded away.
3. Carbonates were deposited in deep ocean basins during the Archean and were later destroyed by subduction.

With the first explanation, the probable high CO_2 content of the Archean atmosphere has been cited by some to explain the scarcity of Archean marine carbonates (Cloud, 1968a). The reasoning is that high CO_2 content in the atmosphere results in more CO_2 dissolved in seawater, thus lowering seawater pH and allowing Ca^{+2} to remain in solution because of its increased solubility. However, the CO_2 content of seawater is not the only factor controlling pH. Mineral stability considerations suggest that although the CO_2 to CO_3^{-2} equilibriums may have short-term control of pH, silicate equilibriums have long-term control. In seawater held at a constant pH by silicate buffering reactions, the solubility of Ca^{+2} decreases with increasing CO_2 levels. Hence, the CO_2 mechanism does not seem capable of explaining the scarcity of Archean carbonates.

As for the second explanation, carbonate rocks in Archean continental successions are rare; thus, the removal of such successions by erosion will not solve the missing Archean carbonate problem.

The third possibility tentatively holds the most promise. Because of the small number of stable cratons in the Archean, carbonate sedimentation was likely confined to deep ocean basins. If so, these pelagic carbonates would be largely destroyed by subduction, accounting for their near absence in the geologic record.

Sedimentary Phosphates

Phosphorus is a major building block of all forms of life; hence, knowledge of the spatial and temporal distribution of phosphates in sediments should provide insight into patterns of organic productivity (Follmi, 1996). It is well known that sedimentary phosphates in the Phanerozoic have an episodic distribution with well-defined periods of major phosphate deposition in the Neoproterozoic–early Cambrian, late Cretaceous–early Tertiary, Miocene, late Permian, and late Jurassic (Cook and McElhinny, 1979) (Fig. 6.19). Although less well constrained, there is also a major period of phosphate deposition from around 1.9 to 2.0 Ga. Most sedimentary phosphates occur at low paleolatitudes (<40 degrees) and appear to have been deposited in arid or semiarid climates either in narrow east–west seaways in response to dynamic upwelling of seawater or in broad north–south seaways in response to upwelling along the east side of the seaway. In both cases, shallow seas are necessary for the upwelling.

The secular distribution of phosphate deposition may be partly related to the supercontinent cycle or mantle-plume events (Chapter 9). For instance, the major peaks of phosphate deposition from 0.80 to 0.60 Ga and before 0.25 Ga correlate with the breakup of supercontinents, during which the area of shallow seas increased as continental fragments dispersed. This raised sea level, providing extensive shallow-shelf environments into which there was a transfer of deep phosphorus-rich ocean water to the shallow photic zone. The peak in phosphate about 2.0 Ga follows the rapid growth of oxygen in the atmosphere 2.2 Ga and may correlate with a mantle-plume event (Chapter 9). The rapid increase in the abundance of sedimentary phosphates at the base of the Cambrian is particularly intriguing and appears to be the largest of several phosphate "events" near the Cambrian–Precambrian boundary (Cook, 1992). The extensive phosphate deposition at this time may have triggered changes in biota, including both an increase in biomass and the appearance of metazoans with hard parts. Some of the Mesozoic–Cenozoic phosphate peaks, such as the late Cretaceous and Eocene peaks, are related to the formation of narrow east–west seaways accompanying the continuing fragmentation of Pangea.

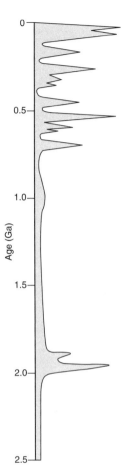

Figure 6.19 Secular variations in sedimentary phosphate deposition. Modified from Cook and McElhinny (1979).

Age (Ga)

The most dramatic example of this is the Tethyan Ocean, an east–west seaway that closed during the last 200 My, providing numerous rifted continental fragments with shallow shelves for phosphate deposition.

The Early Oceans

The earliest history of the oceans comes from an unusual source: oxygen isotopes in ancient igneous zircons (Mojzsis et al., 2001). Most or all of these zircons appear to come from granitic sources. The $\delta^{18}O$ values of detrital zircons ranging from about 2.7 to 4.4 Ga are similar to the primitive value of the Earth's mantle ($5.3 \pm 0.3\%_o$) (Valley et al., 2002) (Fig. 6.20). The zircon average $\delta^{18}O = 5.6\%_o$. Thus, the processes affecting $\delta^{18}O$ in magmatic zircons show no evidence of change during this period. Furthermore, this similarity in isotopic composition is not expected if the parent igneous rocks (granitoids) formed in the absence of water. Liquid oceans were required by 3.8 Ga

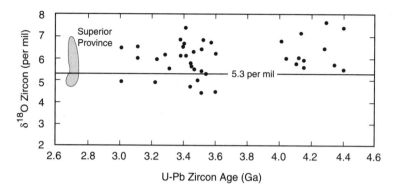

Figure 6.20 Oxygen isotopic composition of Archean magmatic zircons. The value of 5.3‰ for average mantle is also shown. Modified from Valley et al. (2002).

to precipitate chemical sediments found in southwest Greenland and by 3.5 Ga to explain pillow basalts and stromatolites. Hence, the relative constancy of $\delta^{18}O$ in all of the early Archean zircons suggests, but does not prove, that there were oceans by at least 4.4 Ga.

It is likely that the growth rate of the early oceans paralleled that of the atmosphere, perhaps with a slight delay reflecting the time it took for a steam atmosphere to cool and condense. Although continental freeboard has been used to monitor the volume of seawater with time (Wise, 1973), as explained in Chapter 8, uncertainties in the magnitude of the various factors that control freeboard render any conclusions suspect. In any case, a large fraction of the oceans probably formed soon after planetary accretion, such that by 4 Ga most of the current volume of seawater (>90%) was in place on the Earth's surface. The presence of positive Eu anomalies in marine carbonates and the occurrence of ankerite in the Warrawoona Group in Western Australia by 3.45 Ga strongly suggest that seawater was anoxic at this time (Van Kranendonk et al., 2003). Also, the presence of siderite in BIF in the Warrawoona Group suggests that hydrothermal input into the oceans was more significant at this time than later.

Paleoclimates

To better understand the history of the atmosphere and oceans, it is necessary to turn to the Earth's paleoclimatic record (Barnes, 1999). Major worldwide changes in climate are related to plate tectonics, volcanism, or astronomical phenomena. Plate tectonic processes govern the size, shape, and distribution of continents, which in turn influence ocean-current patterns. The distribution of mountain ranges, volcanic activity, and changes in sea level are also controlled by plate tectonics, and each of these factors may affect climate. Volcanism influences climate in two ways:

1. Volcanic peaks may produce a rain shadow effect similar to that of other mountain ranges.

2. Volcanism can introduce large amounts of volcanic dust and CO_2 into the atmosphere, of which the former reduces solar radiation and the latter increases the greenhouse effect.

Clearly, the distribution of continents and ocean basins is an important aspect of paleoclimatology, and reconstructions of past continental positions are critical to understanding ancient climatic regimes. Astronomical factors, such as changes in the Sun's luminosity, changes in the shape of the Earth's orbit about the Sun, and changes in the angle between the Earth's equator and the ecliptic, also influence climate.

Changes in the surface temperature of the Earth with time are related chiefly to three factors: (1) an increase of solar luminosity, (2) a variation in albedo, and (3) a variation in the greenhouse effect. Albedo and the greenhouse effect are partly controlled by other factors such as plate tectonics and global volcanism. All three factors play an important role in long-term (10^8–10^9 years) secular climate changes, and plate tectonics and volcanism appear to be important in climate changes occurring on timescales of 10^6 to 10^8 years (Crowley, 1983).

Paleoclimatic Indicators

The distribution of ancient climates can be constrained with **paleoclimatic indicators,** which are sediments, fossils, or other information sensitive to paleoclimate. A summary of major paleoclimatic indicators and their interpretation is given in Table 6.3. Some sediments reflect primarily precipitation regimes rather than temperature regimes. Coal, for instance, requires abundant vegetation and an adequate water supply but can form at various temperature regimes except arid hot or arid cold extremes (Robinson, 1973). Sand dunes may form in cold or hot arid (or semiarid) environments (as exemplified by modern dunes in Mongolia and the Sahara, respectively). Evaporites form in both hot and cold arid environments, although they are far more extensive in the former. Laterites and bauxites seem to unambiguously reflect hot, humid climates. The distribution of various floral and faunal groups may also be useful in reconstructing ancient climatic belts. Modern hermatypic hexacorals, for instance, are limited to warm surface waters (18–25° C) between the latitudes 38° N and 30° S, whereas

Table 6.3 Examples of Paleoclimatic Indicators	
Indicator	**Interpretation**
Widespread glacial deposits	Continental glaciation, widespread cold climates
Coal	Abundant vegetation, moist climate, generally midlatitudes
Eolian sandstones	Arid or semiarid, generally hot or warm temperatures
Evaporites	Arid or semiarid, generally hot or warm temperatures
Redbeds	Arid to subtropical, commonly semiarid or arid
Laterites and bauxites	Tropical, hot and humid climates
Corals and related invertebrate fauna	Subtropical to tropical, near-shore marine

Late Jurassic hermatypic corals are displaced northward about 35 degrees. It is also possible to estimate paleoclimatic temperatures with oxygen isotopes in marine carbonates and cherts, provided that the rocks have not undergone isotopic exchange since their deposition.

Glacial **diamictites** (poorly sorted sediments containing of coarse- and fine-grained clasts) and related glacial-fluvial and glacial-marine deposits are important in identifying cold climates. Extreme care must be used in identifying glacial sediments in that sediments formed in other environments can possess many of the features characteristic of glaciation. For instance, subaqueous slump, mudflow, and landslide deposits can be mistaken for tillites; indeed, some ancient glaciations have been proposed based on inadequate or ambiguous data (Rampino, 1994). It also has been proposed that fallout from ejecta from large asteroid or comet impacts can produce large debris flows with characteristics similar to glacial tillites, including striated clasts and striated pavements (Oberbeck et al., 1993). The bottom line is that no single criterion should be accepted in the identification of continental glacial deposits. Only a convergence of evidence from *widespread* locations such as tillites, glacial pavement, faceted and striated boulders, and glacial dropstones should be considered satisfactory.

Long-Term Paleoclimate-Driving Forces

Paleoclimate data are generally interpreted to mean that long-term variations in climate are driven by natural variations in the CO_2 content of the atmosphere (Crowley and Berner, 2001). Veizer et al. (2000), however, point out some serious discrepancies when CO_2 levels calculated for the Phanerozoic are compared with the paleoclimatic record. They suggest that for at least part, if not all, of the Phanerozoic an alternative driving force for climate is necessary. These investigators further proposed a cyclic component of paleoclimate with a period of about 135 My. There are no terrestrial phenomena known that recur at this frequency. Shaviv and Veizer (2003) propose that at least 66% of the variance in the paleotemperature of the Earth's surface during the Phanerozoic may be caused by variations in the galactic cosmic ray flux as the solar system passes through the spiral arms of the Milky Way with a period of about 143 My.

Reevaluation of the paleoclimate CO_2 data by Crowley and Berner (2001), however, show first-order agreement between the CO_2 record and the continental glaciation, which supports the conventional CO_2 driving force for long-term climate change. The solution to this problem will require more sophisticated climatic models that consider more climatic variables and interactions of these variables.

Glaciation

Direct evidence for major terrestrial glaciation exists for perhaps 5 to 10% of the Earth's history. At least eight major periods of continental glaciation are recognized in the geologic record (Young, 1991) (Table 6.4). The earliest glaciation, about 3 Ga, is represented only in South Africa in the Pongola Supergroup (von Brunn and Gold, 1993) and is probably

Table 6.4   Summary of Major Terrestrial Glaciations

Age (Ma)	Geographic Areas
3000	Pongola Supergroup, southern Africa
2400–2300	Huronian: Laurentia, Baltica, Siberia, South Africa
1000–900	Congo Basin, Africa; Yenisey Uplift, Siberia
760–700	Sturtian: Australia, Laurentia, South Africa
620–550	Vendian: Eurasia, South Africa, China, Australia, Antarctica, Laurentia, South America
450–400	Gondwana
350–250	Gondwana
15–0	Antarctica, North America, Eurasia

only local. The first evidence of widespread glaciation is the Huronian glaciation from 2.3 to 2.4 Ga, evidence for which has been found in Laurentia, Baltica, Siberia, and South Africa. At least two major ice caps existed in Laurentia at this time, as recorded by marine glacial deposits around their perimeters (Fig. 6.21). Although the evidence for Neoproterozoic glaciation (750–600 Ma) is overwhelming, the number, geographic distribution, and specific ages of individual glaciations are not well known (Eyles, 1993).

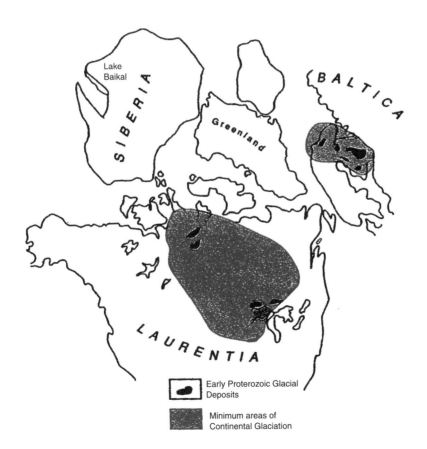

Figure 6.21 Minimum distribution of Paleoproterozoic (2.4- to 2.3-Ga) glaciation in Laurentia–Baltica.

Major Neoproterozoic ice sheets were located in central Australia, west–central Africa, the Baltic shield, Greenland and Spitzbergen, western North America, and possibly eastern Asia and South America. At least three glaciations are recognized as given in Table 6.4. Puzzling features of the Neoproterozoic glacial deposits include their association with shallow marine carbonates suggestive of warm climatic regimes, giving rise to the Snowball Earth hypothesis described later in this chapter.

During the Paleozoic, Gondwana migrated large latitudinal distances, crossing the South Pole several times. Major glaciations in Gondwana occurred in the late Ordovician, late Carboniferous, and early Permian (Fig. 6.22) and are probably responsible for drops in sea level at these times. Minor Gondwana glaciations are also recorded in the Cambrian and Devonian. Although minor glaciations may have occurred in the Mesozoic, the next large glaciations were in the late Cenozoic culminating with the multiple glaciations in the Quaternary.

Precambrian Climatic Regimes

Using paleoclimatic indicators with models of atmospheric evolution, it is possible to characterize average surface temperatures and precipitation regimes with time. The abundance of CO_2 and CH_4 in the Archean atmosphere resulted in greenhouse warming to average

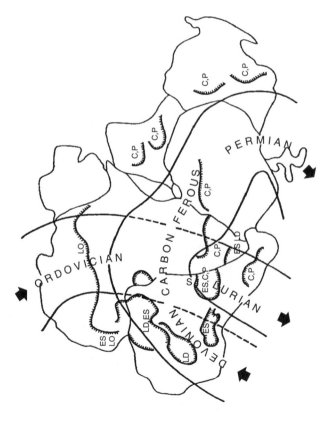

Figure 6.22 Apparent polar wander paths and major glacial centers across western Gondwana during the Paleozoic. C, Carboniferous; D, Devonian; E, early; L, late; O, Ordovician; P, Permian; S, Silurian. Modified from Caputo and Crowell (1985).

surface temperatures greater than at present (Sleep and Zahnle, 2001) (Fig. 6.23). Few climatic indicators are preserved in Archean supracrustal successions. Archean BIF is commonly thought to have been deposited in shallow marine waters and hence implies low wind velocities to preserve the remarkable planar stratification. This is consistent with a CO_2- and CH_4-rich atmosphere in which differences in temperature gradient with latitude are small as are corresponding wind velocities (Frakes et al., 1992). The earliest evaporites and stromatolitic carbonates appear about 3.5 Ga and favor generally warm climates. The relative abundance of shallow marine carbonates and the preservation of lateritic paleosols in the Proterozoic suggest a continuation of warm, moist climates after the end of the Archean. The increasing proportion of evaporites and redbeds less than 2 Ga appears to record the increasing importance of semiarid to arid climatic regimes. Colder climates as recorded by widespread glaciations alternated with warm, moist climates, as reflected by marine carbonates in the Neoproterozoic.

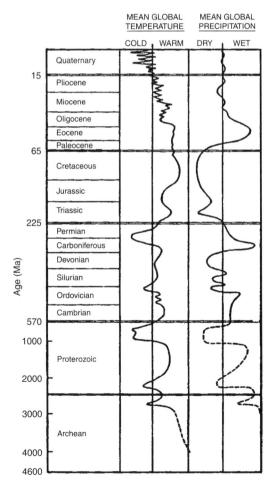

Figure 6.23 Generalized average surface temperature and precipitation history of the Earth. Modified from Frakes (1979).

Also supporting an unusually warm Archean climate are geochemical data related to weathering rates. Studies of modern weathering show that Ca, Na, and Sr are rapidly lost during chemical weathering and that the lost amount of these elements is proportional to the degree of weathering (Condie, 1993). The Chemical Index of Alteration (CIA), commonly referred to as the paleoweathering index (Nesbitt and Young, 1982), has been used to estimate the degree of chemical weathering in the source areas of shales. CIA is calculated from the molecular proportions of oxides, where CaO is the amount in only the silicate fraction and the molecular ratio is as follows:

$$CIA = [Al_2O_3/(Al_2O_3 + CaO + Na_2O + K_2O)] \times 100$$

The higher the CIA, the greater the degree of chemical weathering in sediment sources. For example, CIA values greater than 85 are characteristic of residual clays in tropical climates. Most post-Archean shales show moderate losses of Ca, Na, and Sr from source weathering, with CIA values of 80 to 95 and Sr contents of 75 to 200 parts per million (ppm). These values reflect the "average" intensity of chemical weathering of the shale sources. Most Archean shales, however, show greater losses of all three elements, with typical CIA values of 90 to 98 and Sr contents of less than 100 ppm. This suggests that the average intensity of chemical weathering may have been greater during the Archean than afterward, a conclusion consistent with a CH_4- and CO_2-rich atmosphere in the Archean.

Although data clearly indicate that Archean and Paleoproterozoic climates were warmer than present climates, at least two glaciations must somehow be accommodated. Calculations indicate that a drop of only 2° C is enough to precipitate an ice age (Kasting, 1987). Causes of the glaciations at 3.0 Ga and from 2.3 to 2.4 Ga remain illusive. The absence of significant amounts of volcanic rock beneath tillites of both ages does not favor volcanic dust injected into the atmosphere (reducing sunlight) as a cause of either glaciation. The 3.0 Ga glaciation is recorded only in eastern part of the Kaapvaal craton in South Africa and may have been caused by local conditions. However, for the Paleoproterozoic glaciation, it would appear that a widespread drawdown mechanism is needed for atmospheric CO_2. One possibility is that increasing numbers of photosynthetic microorganisms may have extracted large amounts of CO_2 from the atmosphere faster than it could be returned by chemical weathering, thus reducing the greenhouse effect. However, such a model does not explain how glaciation should end or why large amounts of marine carbonate produced during the CO_2 withdrawal do not occur beneath 2.3 to 2.4 Ga tillites. Another possibility that has not been fully explored is the possible effect of rapid continental growth in the late Archean on the CO_2 content of the atmosphere. Perhaps rapid chemical weathering of newly formed cratons reduced CO_2 levels in the atmosphere and led to cooling and glaciation. It is also possible that an early Snowball Earth (described in the next section) was responsible for the widespread Paleoproterozoic glaciation.

The Snowball Earth

There is now strong evidence for at least two global glaciations in the Neoproterozoic, one from 760 to 700 Ma (the Sturtian) and another from about 620 to 550 Ma (the Varangian or Vendian) (Hoffman et al., 1998). Although paleomagnetic data indicate that both glaciations occurred at tropical or subtropical latitudes, some ice cover extended to high latitudes (Meert and van der Voo, 1994). Support for tropical glaciation also comes from carbonate debris in glacial deposits and shallow marine carbonates overlying glacial deposits known as **cap carbonates.** Another feature of both Neoproterozoic glaciations is large, negative carbon-isotope excursions that immediately follow the glaciations as recorded in the cap carbonates (Fig. 6.24). There are two competing ideas to explain these excursions: (1) a delayed effect caused by the decrease in the burial rate of organic carbon during the glaciations or (2) a postglacial warming that caused methane hydrate destabilization during cap carbonate deposition (Jiang et al., 2003). Currently, the most popular, yet controversial, interpretation of these results is the **Snowball Earth model,** during which the entire Earth freezes over and then thaws.

Williams (1975) suggested that the obliquity of the Earth's orbit may have changed with time and that during the Neoproterozoic the equator was tipped at a high angle to the ecliptic,

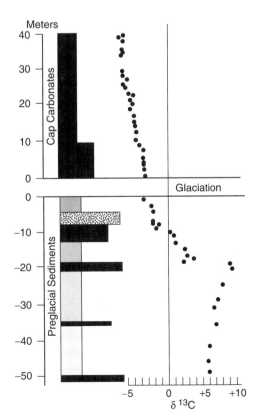

Figure 6.24 Generalized stratigraphic section of the Neoproterozoic glacial section of the Otavi Group, Namibia. Modified from Hoffman et al. (1998)

resulting in a decrease in mean annual temperature. Such a condition should favor widespread glaciation even at low latitudes. However, this model does not seem capable of explaining the angular momentum distribution in the Earth–Moon system, the relationship between obliquity and spin rate, the glaciation at middle to high latitudes, or the intimate association of glacial deposits and carbonates (Pais et al., 1999; Donnadieu et al., 2002). Global glaciation has also been explained by true polar wander during which plates migrate rapidly between high and low latitudes (Kirschvink et al., 1997). This model, however, does not account for the temporal association of glacial deposits and carbon isotope excursions, nor are midlatitudinal facies found between the diamictites and their respective cap carbonates as predicted by this model. The ocean-overturn hypothesis (Kaufman and Knoll, 1995) predicts that the carbon isotope excursions should be short-lived, whereas isotopic ages indicate durations up to 9.2 My (Hoffman et al., 1998).

Although controversial, the Snowball Earth model seems to explain most of the data, including incomplete freezing of the oceans in equatorial regions permitting the survival of multicellular life in tropical regions (Hyde et al., 2000). To produce a Snowball Earth, atmospheric CO_2 levels must drop rapidly. Contributing to this change may have been the fragmentation of the supercontinent Rodinia (Chapter 9), producing new marginal basins that were repositories for organic carbon. Falling CO_2 led to widespread cooling and glaciation extending into equatorial regions, at least in some locations. The high albedo of ice would give positive feedback to the glaciations. Decreases in bioproductivity and carbon burial during the glaciations led to negative carbon-isotope excursions in seawater, which was recorded in the cap carbonates after glaciation. But how do Snowball glaciations end? Volcanism continued to pump CO_2 into the atmosphere, enhanced by new ocean-ridge volcanism accompanying the breakup of Rodinia, yet the continents were covered with ice, so weathering was at an all-time low. This built up CO_2 in the atmosphere caused greenhouse warming and led to warm climates globally, during which the cap carbonates were deposited. Although some organisms would survive in equatorial areas and around volcanic vents, many organisms may have become extinct during the glaciations. It is difficult to evaluate such extinctions because of the lack of skeletal organisms at this time, but there are substantial changes in acritarch populations following the Vendian glaciation.

Phanerozoic Climatic Regimes

Paleoclimate Overview

Most of the early and middle Paleozoic is characterized by climates warmer than present climates, and late Paleozoic climates show wide variation (Frakes et al., 1992). Europe and North America underwent relatively small changes in climate during the Paleozoic, and Gondwana underwent several major changes. This is because the continental fragments comprising Laurasia remained at low latitudes during the Paleozoic, and Gondwana migrated large latitudinal distances. Major glaciations occurred in Gondwana in the late Ordovician, late Carboniferous, and early Permian.

The Mesozoic is characterized by warm, dry climates that extend far north and south of equatorial regions (Barron et al., 1995). Paleontological and paleomagnetic data support this interpretation. Initial Triassic climates were cool and humid like the late Paleozoic, and they were followed by a warm, drying trend until late Jurassic. The largest volumes of preserved evaporites are in the late Triassic and early Jurassic and reflect widespread arid climates on Pangea. Most of the Cretaceous is characterized by warm, commonly humid climates regardless of latitude, with tropical climates extending from 45° N to perhaps 75° S. Glacial climates are of local extent and only in polar regions. Mean annual surface temperatures during the mid-Cretaceous, as determined from oxygen isotope data, are 10 to 15° C higher than today. The late Cretaceous is a time of maximum transgression of the continents, and the warm Cretaceous climates appear to reflect a combination of increased atmospheric CO_2 levels and low albedo of the oceans (Wilson and Norris, 2001). Early Tertiary was a time of declining temperatures, significant increased precipitation, and falling sea level. Middle and late Tertiary are characterized by lower and variable temperature and precipitation regimes. These variations continued and became more pronounced in the Quaternary, leading to the alternating glacial and interglacial epochs. Four major Quaternary glaciations are recorded, with the intensity of each glaciation decreasing with time: the Nebraskan (1.46–1.3 Ma), Kansan (0.9–0.7 Ma), Illinoian (550,000–400,000 years ago), and Wisconsin (80,000–10,000 years ago). Each glaciation lasted 100,000 to 200,000 years, with interglacial periods of 200,000 to 400,000 years. Smaller glacial cycles with periods of 20,000 to 40,000 years are superimposed on the larger cycles.

Causes of Glaciation

Both early and late Paleozoic glaciations appear to be related to plate tectonics. During both of these times, Gondwana was glaciated as it moved over the South Pole (Caputo and Crowell, 1985; Frakes et al., 1992). Major glaciations occurred in central and northern Africa in the late Ordovician when Africa was centered over the South Pole (Fig. 6.22). As Gondwana continued to move over the pole, major ice centers shifted into southern Africa and adjacent parts of South America in the vv Silurian. Gondwana again drifted over the South Pole, beginning the widespread glaciation in the late Devonian and early Carboniferous in South America and central Africa, then shifting to India, Antarctica, and Australia in the late Carboniferous and Permian. Paleozoic glaciations may also be related to prevailing wind and ocean currents associated with the movement of Gondwana over the pole.

The next major glaciations began in the mid-Tertiary and were episodic with interglacial periods. Between 40 and 30 Ma, the Drake Passage between Antarctica and South America opened and the Antarctic circumpolar current was established, thus thermally isolating Antarctica (Crowley, 1983). The final separation of Australia and Antarctica about this time may also have contributed to cooling of Antarctica by inhibiting poleward transport of warm ocean currents. Glaciation appears to have begun from 12 to 15 Ma in Antarctica and by 3 Ma in the Northern Hemisphere. Initiation of Northern Hemisphere glaciation may have been response to the formation of Panama, relocating the Gulf Stream, which had previously flowed westward about the equator, to flow north along the coasts of North America and Greenland. This shift in current brought warm, moist air into the Arctic, which precipitated and began to form glaciers.

Two factors may have been important in controlling the multiple glaciations in the Pleistocene (Eyles, 1993): (1) the cooling and warming of the oceans and (2) cyclic changes in the Earth's rotation and orbit. Cyclical changes in the temperature of the oceans caused by changes in the balance between evaporation and precipitation may be partly responsible for multiple glaciations. A typical cycle might start as moist, warm air moves into the Arctic region, precipitating large amounts of snow that result in glacial growth. As oceanic temperatures drop in response to glacial cooling, evaporation rates from the ocean decrease and the supply of moisture to glaciers also decreases. Hence, glaciers begin to retreat, climates warm, and an interglacial stage begins. The warming trend has a negative feedback and eventually increases evaporation rates, and moisture is again supplied to glaciers as the cycle starts over.

Most investigators agree that cyclical changes in the shape of the Earth's orbit about the Sun and in the inclination and wobble of the Earth's rotational axis can bring about cyclical cooling and warming of the Earth. Today, the Earth's equatorial plane is tilted 23.4 degrees to the ecliptic, but this angle varies from 21.5 to 24.5 degrees with a period of about 41,000 years. Decreasing the inclination results in cooler summers in the Northern Hemisphere and favors ice accumulation. The Earth's orbit also changes in shape from a perfect circle to a slightly ellipsoidal shape with a period of 92,000 years, and the ellipsoidal orbit results in warmer-than-average winters in the Northern Hemisphere. The Earth also wobbles as it rotates with a period of about 26,000 years, and the intensity of seasons varies with the wobble. When variations in oxygen isotopes or fauna are examined in deep-sea cores, all three of these periods, known as the **Milankovitch periods,** can be detected, with the 92,000-year period prominent. These results suggest that on timescales of 10,000 to 400,000 years, variations in the Earth's orbit are the fundamental cause of Quaternary glaciations (Imbrie, 1985). Spectral analyses of $\delta^{18}O$ and orbital time series strongly support the model.

Conclusions

Although investigators understand the overall processes in the Earth's atmosphere and oceans and how they operate on short timescales, investigators are just beginning to learn about long-term changes and how the atmosphere and oceans interact with the solid Earth. To better understand this long-term history, investigators must better-understand paleoclimates. This means more precise ages from key stratigraphic sections with paleoclimatic indicators, stable isotope ratios from numerous well-dated sections of wide geographic extent, and increasingly more sophisticated numerical modeling with high-speed computers. One thing, however, is clear: the atmosphere–ocean system cannot be considered in isolation from the solid Earth. The supercontinent cycle and mantle-plume events, described in Chapter 9, may have profound effects not only on long-term variations in the atmosphere–ocean system but also on life on this planet.

Further Reading

Frakes, L. A., Francis, J. E., and Syktus, J. L., 1992. Climate Modes of the Phanerozoic. Cambridge University Press, New York, 274 pp.

Holland, H. D., 1984. The Chemical Evolution of the Atmosphere and Oceans. Princeton University Press, Princeton, NJ, 581 pp.

Huber, B. T., Wing, S. L., and MacLeod, K. G. (eds.), 1999. Warm Climates in Earth History. Cambridge University Press, Cambridge, UK, 480 pp.

Kasting, J. F., 1993. Earth's early atmosphere. Science, 259: 920–926.

Wigley, T. M. L., and Schimel, D. S., 2000. The Carbon Cycle. Cambridge University Press, Cambridge, UK, 310 pp.

Living Systems

7

General Features

Although the distinction between living and nonliving matter is obvious for most objects, it is not easy to draw this line between some unicellular organisms and large nonliving molecules such as amino acids. It is generally agreed that living matter must be able to reproduce new individuals, it must be capable of growing by using nutrients and energy from its surroundings, and it must respond in some manner to outside stimuli. Another feature of life is its chemical uniformity. Despite the great diversity of living organisms, all life is composed of a few elements (chiefly C, O, H, N, and P) grouped into nucleic acids, proteins, carbohydrates, fats, and a few other minor compounds. This suggests that living organisms are related and that they probably had a common origin. Reproduction is accomplished in living matter at the cellular level by two complex nucleic acids, ribonucleic acid (RNA) and deoxyribonucleic acid (DNA). Genes are portions of DNA molecules that carry specific hereditary information. Three components are necessary for a living system to self-replicate: RNA and DNA molecules, which provide a list of instructions for replication; proteins that promote replication; and a host organ for the RNA–DNA molecules and proteins. The smallest entities capable of replication are amino acids.

Origin of Life

Perhaps no other subject in geology has been investigated more than the origin of life (Kvenvolden, 1974; Oro, 1994). It has been approached from many points of view. Geologists have searched painstakingly for fossil evidence of the earliest life, and biologists and biochemists have provided a variety of evidence from experiments and models that must be incorporated into any model for the origin of life.

Although numerous models have been proposed for the origin of life, two environmental conditions are prerequisites to all models: (1) the elements and catalysts necessary for the production of organic molecules must be present, and (2) free oxygen, which would oxidize and destroy organic molecules, must not be present. In the past, the most popular models for the origin of life a involved primordial "soup" rich in carbonaceous compounds produced by inorganic processes. Reactions in this soup promoted by catalysts such as lightning or ultraviolet radiation produced organic molecules. Primordial soup models, however, seem unnecessary in rapid degassing of the Earth more than 4 Ga. Rapid recycling of the early oceans through ocean ridges would not allow concentrated "soups" to survive except perhaps locally in evaporite basins less than 4 Ga. Because chances are remote that organic molecules were present in sufficient amounts, in correct proportions, and in the proper arrangement, it would seem that the environment in which life formed would have been widespread in the early Archean. Possibilities include volcanic environments and hydrothermal vents along ocean ridges.

Simple amino acids have been formed in the laboratory under a variety of conditions. The earliest experiments were those of Miller (1953), who sparked a hydrous mixture of H_2, CH_4, and NH_3 to form a variety of organic molecules including 4 of the 20 amino acids composing proteins. Similar experiments, using both sparks and ultraviolet radiation in gaseous mixtures of water, CO_2, N_2, and CO (a composition more in line with that of the Earth's early degassed atmosphere) also produced amino acids, hydrocyanic acid, and formaldehydes, the latter of which can combine to form sugars. Heat also may promote similar reactions.

Role of Impacts

As indicated by microfossils, life was in existence by 3.5 Ga; carbon isotope data, although less definitive, suggests that life was present by 3.8 Ga. This being the case, life must have originated during or before the last stage of heavy bombardment of planets in the inner solar system as indicated by the impact craters on the Moon and other terrestrial planets with ancient surfaces. As an example, the impact record on the Moon shows that crater size, and hence impact energy, falls exponentially from 4.5 to about 3.0 Ga, decreasing more gradually thereafter (Sleep et al., 1989; Chyba, 1993). Similarity of crater frequency versus diameter relations for Mercury and Mars implies that planets in the inner solar system underwent a similar early bombardment history, although the Earth's history has been destroyed by plate tectonics. A decrease in impact energy with time on Earth is likely to be similar to that on the Moon except that less than 3.0 Ga energies were perhaps an order of magnitude higher on the Earth. Because the Earth's gravitational attraction is greater than that of the Moon, it should have been hit with more large objects than the Moon before 3.5 Ga. The impact record on the Moon implies that the Earth was subjected to cataclysmic impacts from about 4.0 to 3.8 Ga, preceded by a comparatively quiet period from about 4.4 to 4.0 Ga (Ryder, 2002; Valley et al., 2002) (Fig. 7.1). During the intense impact period, hundreds of impacts large enough to form mare basins (as found on the Moon) must have hit the Earth. Single, large impacts had

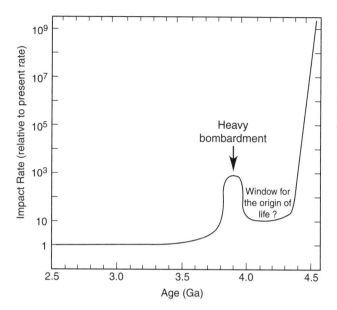

Figure 7.1 Estimates of the asteroid impact rate for the first 2 Gy of the Earth's history. Evidence of water comes from oxygen isotopes in zircons (4.4–4.0 Ga) and sedimentary rocks (Isua, 3.8–3.6 Ga). Modified from Valley et al. (2002).

only a small fraction of the energy necessary to evaporate the Earth's oceans. Large impactors, sufficient to evaporate the entire ocean, are considered rare or nonexistent less than 4.4 Ga (Zahnle and Sleep, 1997).

Although such large impacts mean that life could not form and survive in shallow aqueous environments, it may have survived in the deep ocean around hydrothermal vents. Because it appears that oceans existed on the Earth from at least 4.4 Ga (Chapter 6), life could have formed during the comparative quiet period from 4.4 to 4.0 Ga just before the cataclysmal impacts from 3.9 to 3.8 Ga (Fig. 7.1). As indicated by the oldest fossils, life was advanced by 3.5 Ga.

Another intriguing aspect of early impact is the possibility that relatively small impactors introduced volatile elements and small amounts of organic molecules to the Earth's surface that were used in the origin of life. The idea that organic substances were brought to the Earth by asteroids or comets is not new; it was first suggested in the early part of the 20th century. Lending support to the idea is the recent discovery of *in situ* organic-rich grains in Halley's comet, and data suggest that up to 25% organic matter may occur in other comets. Also, many organic compounds found in living organisms are found in carbonaceous meteorites. Some investigators propose that amino acids and other organic compounds important in the formation of life were carried to the Earth by asteroids or comets rather than formed *in situ* on the Earth (Cooper et al., 2001). Complex compounds, such as sugars, sugar alcohols, and sugar acids have recently been reported in the Murchison and Murray carbonaceous chondrites in amounts comparable with those found in the amino acids of living organisms. These compounds may have been produced by processes such as photolysis on the surfaces of asteroids or comets.

One problem with an extraterrestrial origin for organic compounds on the Earth is how to get these substances to survive impact. Even for small objects (~100 m in radius), impact should destroy organic inclusions unless the early atmosphere was dense (~10 bar of CH_4 and CO_2) and could sufficiently slow the objects before impact. However, interplanetary dust from colliding comets or asteroids could survive impact and may have introduced significant amounts of organic molecules into the atmosphere or oceans. Whether this possible source of organics was important depends critically on the composition of the early atmosphere. If the atmosphere was rich in CO_2 and CH_4 as suggested in Chapter 6, the rate of production of organic molecules was probably quite small; hence, the input of organics by interplanetary dust may have been significant.

Ribonucleic Acid World

Although it seems relatively easy to form amino acids and other simple organic molecules, how these molecules combined to form the first complex molecules, such as RNA, and then evolved into living cells remains largely unknown. Studies of RNA suggest that it may have played a major role in the origin of life. RNA molecules have the capability of splitting and producing an enzyme that can act as a catalyst for replication (Zaug and Cech, 1986) (Fig. 7.2). Necessary conditions for the production of RNA molecules in the early Archean include a supply of organic molecules, a mechanism for molecules to react

Figure 7.2
Diagrammatic representation of the RNA world. (a) RNA is produced from ribose and other organic compounds. (b) RNA molecules learn to copy themselves. (c) RNA molecules begin to synthesize proteins. (d) The proteins serve as catalysts for RNA replication and the synthesis of more proteins. They also enable RNA to make double-strand molecules that evolve into DNA. (e) DNA takes over and uses RNA to synthesize proteins, which in turn enables DNA to replicate and transfer its genetic code to RNA.

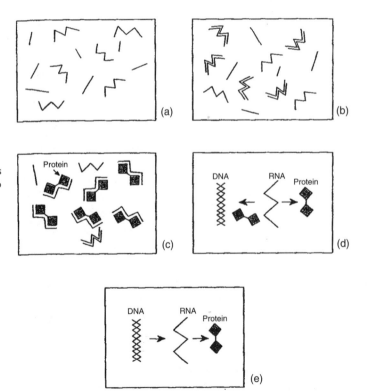

to form RNA, a container mineral to retain detached portions of RNA so that they can aid further replication, a mechanism by which some RNA can escape to colonize other populations, and some means of forming a membrane to surround a protocell wall (Nisbet, 1986; de Duve, 1995). During the Archean, hydrothermal systems on the seafloor may have provided these conditions (Corliss et al., 1981; Gilbert, 1986). In laboratory experiments, RNA splitting occurs at temperatures around $40°$ C with a pH varying from 7.5 to 9.0 and Mg in solution. The early Archean "RNA world" may have existed in clay minerals, zeolites, and pore spaces of altered volcanic rocks. The next stage in replication may have been the development of proteins from amino acids synthesized from CH_4 and NH_3. Later still, DNA must form and take over as the primary genetic library (Gilbert, 1986) (Fig. 7.2).

The next stage of development, although poorly understood, seems to involve the production of membranes, which manage energy supply and metabolism, both essential for the development of a living cell. The evolution from protometabolism to metabolism probably involved five major steps (de Duve, 1995) (Fig. 7.3). In the first stage, simple organic compounds reacted to form mononucleotides, which later were converted into polynucleotides. During the second stage, RNA molecules formed and the RNA world came into existence as illustrated in Figure 7.2. This was followed by the third stage, in which RNA molecules interacted with amino acids to form peptides. During or before this stage, the prebiotic systems must have become encapsulated by primitive fatty membranes, producing the first primitive cells. At this third stage, Darwinian competition probably began among these cells. During the fourth stage, translation and genetic code emerged through a complex set of molecular interactions involving competition and natural selection. During the final stage, the mutation of RNA genes and competition among protocells occurred. It is by this process that enzymes probably arose (Fig. 7.3). As peptides emerged and assumed their functions, metabolism gradually replaced protometabolism.

Hydrothermal Vents

Possible Site for the Origin of Life

Hydrothermal vents on the seafloor have been proposed by several investigators as a site for the origin of life (Corliss et al., 1981; Chang, 1994; Nisbet, 1995). Modern hydrothermal vents have many organisms that live in their own vent ecosystems, including a variety of unicellular types (Tunnicliffe and Fowler, 1996). Vents are attractive in that they supply the gaseous components such as CO_2, CH_4, and nitrogen species from which organic molecules can form and that they supply nutrients for the metabolism of organisms, such as P, Mn, Fe, Ni, Se, Zn, and Mo (Fig. 7.4). Although these elements are in seawater, it is difficult to imagine how they could have been readily available to primitive life at such low concentrations. Early life would not have had sophisticated mechanisms capable of extracting these trace metals, thus requiring relatively high concentrations that may exist near hydrothermal vents. One objection that has been raised to a vent origin for life is the potential problem of both synthesizing and preserving organic molecules necessary for

Figure 7.3
Diagrammatic representa-
tion of the evolution of the
earliest cells and the emergence of metabolism.
Modified from de Duve
(1995).

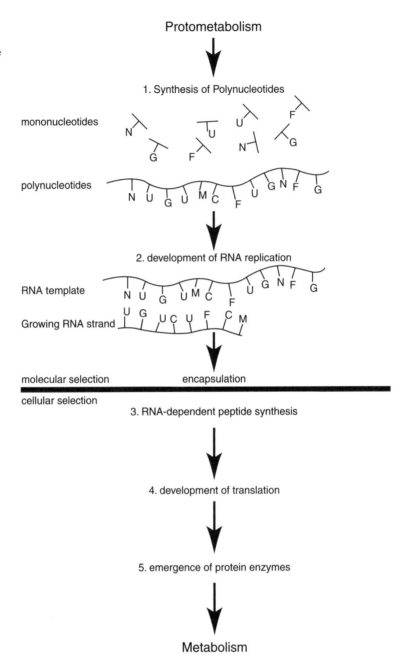

the evolution of cells. The problem is that the temperatures at many or all vents may be too high and they would destroy, not synthesize, organic molecules (Miller and Bada, 1988). However, many of the requirements for the origin of life seem to be available at submarine hydrothermal vents, and synthesis of organic molecules may occur along vent

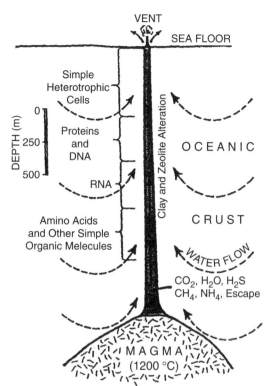

Figure 7.4 Idealized cross-section of Archean ocean-ridge hydrothermal vent showing possible conditions for the formation of life.

margins where the temperature is lower. Models by Shock and Schulte (1998) suggest that the oxidation state of a hydrothermal fluid, controlled partly by the composition of host rocks, may be the most important factor influencing the potential for organic synthesis. The probability of organic synthesis in the early Archean may have been much greater than at present because of the hotter and metal-rich komatiite-hosted hydrothermal systems.

One possible scenario for the origin of life at hydrothermal vents begins with CO_2 and N_2 in vent waters at high temperatures deep in the vent (Shock, 1992). As the vent waters containing these components circulate to shallower levels and lower temperatures, they cool and thermodynamic conditions change such that CH_4 and NH_3 are the dominant gaseous species present. Provided that suitable catalysts are available, these components can then react to produce a variety of organic compounds. The next step is more difficult to understand, but somehow simple organic molecules must react with each other to form large molecules such as peptides, as illustrated in Figure 7.3.

Experimental and Observational Evidence

Experimental results can help constrain an origin for life at hydrothermal vents. Compounds synthesized to date at conditions found at modern vents include lipids, oligonucleotides, and oligopeptides (McCollom et al., 1999). Clay minerals have been used as catalysts for the reactions. Experimental results also indicate that amino acids and mononucleotides can polymerize in hydrothermal systems, especially along the hot–cold interface of the hydrothermal fluids and cold seawater. Polymerization of amino acids to form peptides has also been reported for hydrothermal vent conditions (Ogasawara et al., 2000).

Long-chain hydrocarbons have been collected from modern hydrothermal vents along the mid-Atlantic Ridge, indicating that organic compounds can be synthesized at these vents (Charlou et al., 1998). These compounds, which have chain lengths of 16 to 29 carbon atoms, may have formed by reactions between H_2 released during serpentinization of olivine and vent-derived CO_2 at high temperatures.

First Life

One of the essential features of life is its ability to reproduce. It is probable that this ability was acquired long before the first cell appeared. Cairns-Smith (1982) has suggested that clays may have played an important role in the evolution of organic replication. Organic compounds absorbed in clays may have reacted to form RNA, and through natural selection, RNA molecules eventually disposed of their clay hosts. Because hydrothermal systems appear to have lifetimes of 10^4 to 10^5 years at any location, RNA populations must have evolved rapidly into cells or, more likely, were able to colonize new vent systems. Another possible catalyst is zeolite, which possesses pores of different shapes and sizes that permit small organic molecules to pass but that exclude or trap larger molecules (Nisbet, 1986). Zeolites are also characteristic secondary minerals around hydrothermal plumbing systems (Fig. 7.4). The significance of variable-sized cavities in zeolites is that a split-off RNA molecule may be trapped in such a cavity, where it can aid the replication of the parent molecule. Although the probability is small, it is possible that the first polynucleotide chain formed in the plumbing system of an early hydrothermal vent on the seafloor.

The first cells were primitive in that they had poorly developed metabolic systems and survived by absorbing a variety of nutrients from their surroundings (Kandler, 1994; Pierson, 1994). They must have obtained nutrients and energy from other organic substances through fermentation, which occurs only in anaerobic (oxygen-free) environments. **Fermentation** involves the breakdown of complex organic compounds into simpler compounds that contain less energy, and the energy liberated is used by organisms to grow and reproduce. Cells that obtain their energy and nutrients from their surroundings by fermentation or chemical reactions are known as **heterotrophs,** in contrast with **autotrophs** capable of manufacturing their own food. Two types of anaerobic cells evolved from DNA replication. The most primitive group, the archaebacteria, uses RNA in the synthesis of proteins, whereas the more advanced group, the eubacteria, has advanced replication processes and may have been the first set of photosynthesizing organisms.

Rapid increases in the number of early heterotrophs may have led to severe competition for food supplies. Selection pressures would tend to favor mutations that enabled heterotrophs to manufacture their own food and thus become autotrophs. The first autotrophs appeared by 3.5 Ga as cyanobacteria. These organisms produced their own food by photosynthesis, perhaps using H_2S rather than water because free oxygen, liberated during normal photosynthesis, is lethal to anaerobic cells. How photosynthesis evolved is unknown, but perhaps the supply of organic substances and chemical reactions became less plentiful as heterotrophs increased in number and selective pressures increased to develop alternative energy sources. Sunlight would be an obvious source to exploit. H_2S may have been plentiful from hydrothermal vents or decaying organic matter on the seafloor, and some cells may have developed the ability to use this gas in manufacturing food. As these cells increased in number, the amount of H_2S would not be sufficient to meet their demands and selective pressures would be directed toward alternate substances, of which water is the obvious candidate. Thus, mutant cells able to use water may have outcompeted forms only able to use H_2S, leading to the appearance of modern photosynthesis.

Possibility of Extraterrestrial Life

There is considerable interest today in the possibility of life on other bodies in the solar system and elsewhere in the galaxy. Most astrobiologists agree that any body on which life may exist must have a fluid medium, a source of energy, and conditions and components compatible with polymeric chemistry (Irwin and Schulze-Makuch, 2001). For carbon-based life, possible energy sources include sunlight (from central stars in other planetary systems), thermal gradients, kinetic motion, and magnetic fields. Although water is generally assumed to be the fluid medium, mixtures of water and various hydrocarbons or other organic compounds may be suitable. With the appropriate conditions, the emergence of self-organizing systems is regarded as inevitable by many scientists (Kauffman, 1995).

Irwin and Schulze-Makuch (2001) propose a five-scale rating for the plausibility of life on other bodies, as summarized for bodies in the solar system in Table 7.1. Life is most likely to exist where water, organic chemistry, and one or more energy sources occur. This is termed category I, and only the Earth falls in this category. Category II

Table 7.1   Plausibility of Life Ratings*

Rating	Characteristics	Examples
I	Water, available energy, organic compounds	Earth
II	Evidence for past or present existence of water, available energy, inference of organic compounds	Mars, Europa, Ganymede
III	Extreme conditions, evidence of energy sources and complex chemistry possibly suitable for life forms unknown on the Earth	Titan, Triton, Enceladus
IV	Past conditions possibly suitable for life	Mercury, Venus, Io
V	Conditions unfavorable for life	Sun, Moon, outer planets

* Modified from Irwin and Schulze-Makuch, 2001.

includes bodies in which these conditions existed and may still exist above or beneath the surface. In category 2 are Mars and Europa, a satellite of Jupiter. Category III includes bodies with nonwater liquid and energy sources, such as the Jovian satellite Triton. Category IV includes bodies on which conditions for life may have existed but no longer exist, such as Venus. These bodies have such extreme conditions that water and organic molecules are unlikely to survive. Finally, in category V, conditions are so extreme that it is unlikely life ever existed on such bodies.

The most compelling factor for life on the Martian surface is the abundance of frozen water and the likelihood that Mars had running water early in its history (Carr, 1996). Although microbial life may have existed when water was on the planetary surface, it is unlikely that life persisted to the present on the Martian surface. However, as water began to freeze and form permafrost, life may have retreated to the deep subsurface and continued to thrive as microbes do today in deep fractures and mines on the Earth (Stevens and McKinley, 1995). Subsurface geothermal areas may provide a suitable habitat for chemoautotrophic microbes that survive by oxidizing free hydrogen to water in the presence of CO_2 from volcanic eruptions. Alternatively, or in addition, microbes might reduce CO_2 to CH_4. Solar energy has been available continuously on Mars and may still sustain life along the fringe areas of the polar icecaps. Although microstructures found in Martian meteorites (meteorites from the Martian surface) originally described as possible microfossils (McKay et al., 1996) are likely of inorganic origin, Mars is clearly a body that meets the requirements for life in the past with the possible survival of some forms to the present.

Europa, Jupiter's smallest major satellite, is covered by an icy shell up to 200 km thick overlying a salt-bearing ocean (Carr et al., 1998). If water was ever liquid at the surface of Europa, theories of the origin of life on the Earth could apply to this unusual satellite. Experimental evidence allows a methanogen-driven biosphere to exist on Europa, despite the low temperatures. Oxygen may be produced in the European atmosphere by the Jovian magnetic field, thus permitting the operation of oxidation–reduction chemical cycles (Chyba, 2000). The relatively high density and the presence of a surficial electromagnetic field are consistent with a liquid core that could generate internal energy, thus providing another source of energy for microbes in the overlying ocean. Clearly, Europa needs to be explored for the possibility of microbial life forms that may have persisted to the present.

Isotopic Evidence of Early Life

Metabolic activity in organisms produces distinct patterns of isotopic fractionation in such elements as carbon, nitrogen, and sulfur. The geologic record of these fractionations can be obtained by analyzing organic matter of biologic origin preserved in sedimentary rocks. Enzymatic processes discriminate against ^{13}C in the fixation of atmospheric CO_2, causing a difference of up to 5% in the isotopic composition biologic and nonbiologic carbon. Microbial **methanogens,** which produce methane gas during their metabolism, are responsible for isotopic fractionation up –40 per mil (‰) from inorganic carbon, reaching $\delta^{13}C$ values in some forms as low as –60‰. When found in buried carbon in the geologic record, these distinct, light-carbon isotope signatures (–30 to –60‰) are considered diagnostic of methanogen activity at the time of burial in extreme environments, such as high saline, high temperature, and variable pH environments (Mojzsis and Harrison, 2000). **Methanotrophs,** which use methane in their metabolism, may also be in these environments.

Carbon isotopic data from early Archean carbon is well known to be isotopically light, consistent with a biologic origin (Schidlowski et al., 1983; Des Marais, 1997). Carbonaceous inclusions trapped in minerals such as apatite are particularly informative because they may have remained unchanged since burial. In the earliest known sediments from Isua in southwest Greenland, carbonaceous inclusions have $\delta^{13}C$ values of -30 to $-40‰$ (Mojzsis et al., 1996; Rosing, 1999). The simplest interpretation of these data is that the microorganisms in these early sediments were metabolically complex, perhaps comprising phosphate-utilizing photoautotrophs and chemoautotrophs. Furthermore, the strongly negative $\delta^{13}C$ values suggest the presence of photosynthesizing methanogenic and methanotrophic bacteria on the Earth by 3.85 Ga.

First Fossils

Two lines of evidence are available for the recognition of the former existence of living organisms in early Archean rocks: microfossils and organic geochemical evidence including carbon isotopes described previously (Schopf, 1994). Many microstructures preserved in rocks can be mistaken for cell-like objects (inclusions, bubbles, microfolds, etc.), and progressive metamorphism can produce structures that look remarkably organic and can destroy real microfossils. Therefore, caution must be exercised in accepting microstructures as totally biologic.

The oldest well-described assemblage of microfossil-like structures comes from cherts in the 3.5-Ga Barberton greenstone belt in South Africa (Fig. 7.5). The oldest unambiguous structures of organic origin are 3.5-Gy-old stromatolites from the Pilbara region of Western Australia (Walter, 1994) (Fig. 7.6). Three types of microstructures, ranging from less than 1 mm to about 20 mm, have been reported from the Barberton sequence and from other Archean sediments. These are rod-shaped bodies, filamentous structures, and spheroidal bodies. The spheroidal bodies are similar to alga-like bodies from Proterozoic assemblages and are generally interpreted as such. Of the two known types of cells, prokaryotic and eukaryotic, only prokaryotic types are represented among Archean microfossils. **Prokaryotes** are primitive cells that lack a cell wall around the nucleus and are not capable of cell division; **eukaryotes** possess these features and hence are capable of transmitting genetic coding to various cells and to descendants.

Biomarkers

Biomarkers are geologically stable compounds, mostly lipids, of known biologic origin. The oldest known biomarkers come from 2.7-Ga shales from northwestern Australia (Brocks et al., 1999). These occurrences confirm the existence of photosynthetic cyanobacteria by the late Archean. Such biomarkers are widespread in the 2.5-Ga

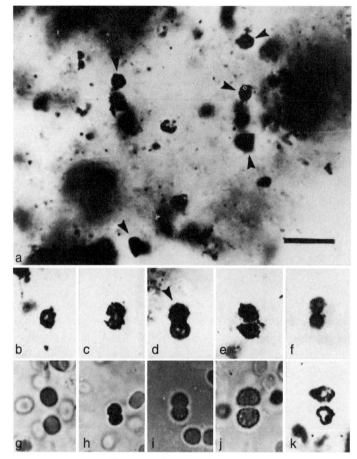

Hamersley iron formation in Western Australia and may have provided the free oxygen for precipitation of the banded iron formation. Included in the biomarkers are 2-Me bacterio-heopanepolyols, membrane lipids synthesized in large quantities only in cyanobacteria, and steranes, molecules derived from sterols. It is probable that at least some oxygen entered the atmosphere in the late Archean to support these forms of cyanobacteria.

Paleosols

Although microorganisms existed in the oceans from at least 3.8 Ga onward, it is uncertain when they first came onto dry land. Two late Archean paleosols have been described from South Africa and Western Australia, both of which appear to contain evidence of biologic activity on land. Paleosols in the Schagen area of South Africa formed from 2.7 to 2.6 Ga have negative $\delta^{13}C$ (–30 to –35‰) suggestive of a biologic origin (Watanabe et al., 2000). Supporting an important role of cyanobacteria in the Schagen soil is a similar fractionation factor between the Archean atmosphere and the soil mat (12‰), which is similar to the

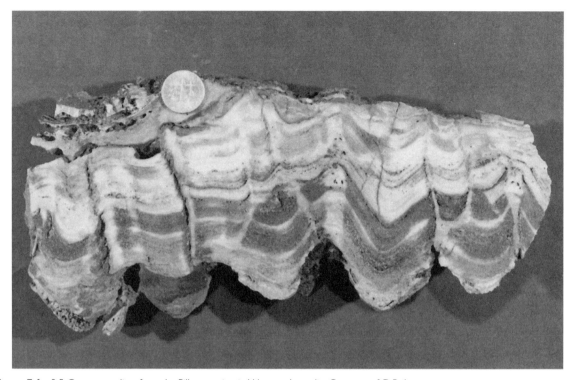

Figure 7.6   3.5-Ga stromatolites from the Pilbara region in Western Australia. Courtesy of D.R. Lowe.

fractionation value today. The existence of land-based photosynthesizing cyanobacteria by 2.7 Ga also may indicate that the ozone shield had started to develop. Organic carbon from the Mt. Roe paleosol in Western Australia records strong negative $\delta^{13}C$ values (–33 to –51‰), suggesting in this case the presence of methanotrophs in the soil (Rye and Holland, 2000). Methanotrophs furthermore imply significant levels of methane in the Archean atmosphere as described in Chapter 6. Filamentous bacteria in mats from the 2.9-Ga Mozaan Group in South Africa are the oldest known occurrences of microbial mats in siliciclastic sediments (Noffke et al., 2003). Textures in these mats resemble trichomes of modern cyanobacteria, chloroflexi, or sulfur-oxidizing proteobacteria. Mineralogical, isotopic, and geochemical analyses of these mats are consistent with a biologic origin for the filament-like textures in the mats.

Thus, evidence suggests that cyanobacteria had moved from the oceans onto the continents by the late Archean.

Origin of Photosynthesis

Oxygenic photosynthesis, in which water is the electron donor in the photosynthesis reaction, probably developed later than anoxygenic photosynthesis (Nisbet and Sleep, 2001).

Anoxygenic photosynthesis makes use of light in the longer-visible and near-infrared spectrum. Examples are anoxygenic purple bacteria, which absorb light at wavelengths of at least 900 nm, and oxygenic green bacteria, which have an absorption maximum of around 750 nm in living cells. Anoxygenic photosynthesis can use a variety of electron donors in various bacteria, including hydrogen, H_2S, sulfur, iron, and various organic compounds. Again supporting the hydrothermal vent model for the origin of life, all of these substances occur in deep-sea hydrothermal systems (Nisbet, 2002). Experimental results suggest that purple, nonsulfur bacteria can oxidize Fe^{+2} to brown Fe^{+3} and reduce CO_2 to cell material, implying that oxygen-dependent biologic iron oxidation was possible before the appearance of oxygenic photosynthesis (Widdel et al., 1993). This being the case, it is possible, if not probable, that bacteria were important in the deposition of banded iron formation in the Archean.

Unlike anoxygenic photosynthesis, oxygenic photosynthesis uses visible light in the more energetic parts of the spectrum and, of course, uses water as the electron donor. Green sulfur bacteria and cyanobacteria both use iron–sulfur centers as electron acceptors, whereas in purple bacteria, pigments and quinones are used as electron acceptors. The existence of both types of electron acceptors in oxygenic photosynthesis suggests an origin by genetic transfer among cooperating or closely juxtaposed cells, each using anoxygenic photosynthesis. Perhaps the key component in oxygenic photosynthesis is the oxygen complex that is part of an Mn complex exploiting the transition from Mn_4O_4 to Mn_4O_6. The presence of Mn is again consistent with hydrothermal systems, but the environment needs to be oxygen rich.

The evolution and structure of microbial mats may have paralleled the evolution of photosynthesis, with newer bacteria progressively occupying the more productive but more dangerous uppermost level in the mats where the light is more intense. In this model, it is possible that the prephotosynthetic mats were composed of hyperthermophiles (heat-loving forms) with sulfate processors on the top and archaea beneath that recycled redox power. This scenario may have allowed the occupation of hydrothermal vents and eventually the open sea away from volcanic heat sources (Nisbet and Sleep, 2001). More than 3.5 Ga, a new cyanobacteria component appeared, possibly from genetic exchange between coexisting purple and green bacteria living on the redox boundary of a microbial mat. This component used water, CO_2, and sunlight to photosynthesize and may have spread rapidly, filling many ecological niches in the oceans.

Tree of Life

You can learn about the origin and evolution of life from two sources:

1. Historical information can be deduced by comparing sequences of nucleic acids contained in the genomes of living organisms by constructing family trees based on observational differences (Doolittle, 1999).
2. You can piece together the evolutionary history of life from the fossil record.

By combining the results of these two approaches, you can construct a robust framework to infer the timing of the major evolutionary events in the history of life (Bengtson, 1994; Farmer, 1998). The **universal phylogenetic tree** constructed from comparisons of ribosomal RNA indicates that life can be divided into three general categories (Fig. 7.7): bacteria, archaea, and eukarya. Branches within the bacteria and archaea domains are short, suggesting relatively rapid evolution of the subgroups. Cyanobacteria appeared by 2.7 Ga (Fig. 7.7). In contrast, the branches separating the three major domains are long, indicating both greater evolutionary distances and rapid divergence. Although the placement of the root of the tree is uncertain, it likely lies somewhere near the midpoint of the bacteria and archaea domains. Ribosomal RNA is a slowly evolving molecule and is considered important in studying early events in biosphere evolution. Although the fossil record is poor in microbial life forms, it supports the genetic relationships deduced from sequencing RNA.

The earliest branches of the bacteria and archaea domains include **hyperthermophiles,** which are forms that grow at temperatures greater than 80° C. In addition to exhibiting the highest temperature tolerances, the deepest branching organisms are **chemoautotrophs** (i.e., microbes that synthesize organic molecules from inorganic materials). These combined properties of the earliest microorganisms are widely assumed to be those of the last common ancestor of living organisms. Placement of the "root of life" within the hyperthermophile bacteria is consistent with the model for the origin of life in hydrothermal vents on the sea floor. Oxygenic photosynthesis first appeared in cyanobacteria and was later transferred to plants (in eukarya) through lateral gene transfer, symbiotic association with primitive plants, or both.

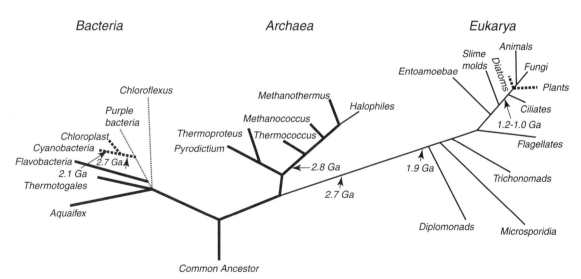

Figure 7.7   Phylogenetic "tree of life" based on a comparison of 16S and 18S ribosomal RNA. Also shown are isotopically dated events from the fossil record. The thick black line indicates hyperthermophiles, the dashed line is for oxygenic photosynthetic forms, and the dotted line marks anoxygenic photosynthetic forms.

Stromatolites

Stromatolites are finely laminated sediments composed chiefly of carbonate minerals that have formed by the accretion of both detrital and biochemical precipitates on successive layers of microorganisms (commonly cyanobacteria) (Fig. 7.6). They exhibit a variety of domical forms and range in age from about 3.5 Ga to modern (Grotzinger and Knoll, 1999). Two parameters are especially important in stromatolite growth: water currents and sunlight. There are serious limitations to interpreting ancient stromatolites in terms of modern ones, however. First, modern stromatolites are not well understood and occur in a great variety of aqueous environments (Walter, 1994; Grotzinger and Knoll, 1999). The distribution in the past is also controlled by the availability of shallow, stable shelf environments; the types of organisms producing the stromatolites; the composition of the atmosphere; and perhaps the importance of burrowing animals. It is possible to use stromatolite reefs to distinguish deepwater from shallow-water deposition because reef morphologies are different in these environments.

Although some early Archean laminated carbonate mats appear to be of inorganic origin, by 3.2 Ga, well-preserved organism-built stromatolites were widespread. The oldest relatively unambiguous stromatolite 3.5 Ga occurred near the town of North Pole in Western Australia, and its age was constrained by U-Pb zircon dates from associated volcanics (Buick et al., 1995; Van Kranendonk et al., 2003) (Fig. 7.6). Early Archean stromatolites were probably built by anaerobic photoautotrophs with mucus sheaths (Walter, 1994; Hofmann et al., 1999). These microbes were able to cope with high salinities, desiccation, and high sunlight intensities as indicated by their occurrence in evaporitic cherts. Late Archean stromatolites are known from both lagoon and near-shore marine environments, and some are similar to modern stromatolites, suggesting that they were constructed by cyanobacteria. Paleoproterozoic stromatolites appear to have formed in peritidal and relatively deep subtidal environments and mostly appear to be built by cyanobacteria.

Stromatolites increased in number and complexity from 2.2 to about 1.2 Ga, after which they decreased rapidly (Grotzinger, 1990; Walter, 1994) (Fig. 7.8). Whatever the cause or causes of the decline, it is most apparent initially in quiet subtidal environments and spreads later to the peritidal realm. Numerous causes have been suggested for the decline, of which the two most widely cited are (1) grazing and burrowing of algal mats by the earliest metazoans and (2) decreasing saturation of carbonate in the oceans, resulting in decreasing stabilization of algal mats by precipitated carbonate. Not favoring the grazing idea, the rapid increase in number and diversity of metazoan life forms begins after much of the decline in stromatolites (i.e., about 600 Ma). The extent to which stromatolites can be used to establish a worldwide Proterozoic biostratigraphy is a subject of controversy, which revolves around the roles of the environment and diagenesis in determining stromatolite shape and the development of an acceptable taxonomy. Because the growth of stromatolites is at least partly controlled by organisms, it should be possible to construct a worldwide biostratigraphic column. Another controversial subject is that of how stromatolite height is related to tidal range. Cloud (1968b) suggests that the height

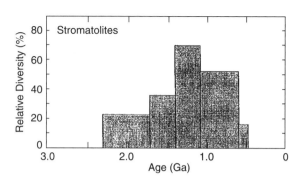

of intertidal stromatolites at maturity reflects the tidal range, whereas Walter (1994) suggests that the situation is more complex. The distribution of laminations in stromatolites also has been suggested as a means of studying secular variation of the Earth–Moon system, as described in Chapter 10.

Appearance of Eukaryotes

It was not until about 2 Ga, however, that microbes entirely dependent on the use of molecular oxygen appeared in the geologic record (Runnegar, 1994). This correlates with a rapid growth of oxygen in the atmosphere. These are eukaryotes, advanced cells with a cell nucleus enclosing DNA and with specialized organs in the cell. Eukaryotes also are able to sexually reproduce. RNA studies of living unicellular eukaryotes suggest that they are derived from archaebacterial prokaryotes about 2.7 Ga (Brocks et al., 1999). Although the earliest fossil eukaryotes appear about 1.9 Ga, they did not become widespread in the geologic record until 1.7 to 1.5 Ga. The oldest fossil thought to represent a eukaryote is *Grypania* from 1.9-Gy-old sediments in Michigan. *Grypania* is a coiled, cylindrical organism that grew to about 50 cm in length and 2 mm in diameter (Fig. 7.9). Although it has no certain living relatives, it is regarded as a probable eukaryotic alga because of its complexity, structural rigidity, and large size.

RNA studies of modern eukaryotes suggest that the earliest forms to evolve were microsporidians, amoebae, and slime molds. Between 1.3 and 1.0 Ga, eukaryotes began to accelerate, leading to the appearance of red algae. A minimum date for this radiation is given by well-preserved, multicellular red alga fossils from 1.2 to 1.0 Ga sediments in Arctic Canada. According to RNA data, this was followed by major radiation, leading to ciliates, brown and green algae, plants, fungi, and animals.

RNA studies suggest that modern algal and plant chloroplasts have a single origin from a free-living cyanobacterium. The symbiotic theory of cell evolution suggests that the remarkable complexity of eukaryotes requires time and probably developed from symbioses between prokaryotes and amitochondriate protists (Cavalier-Smith, 1987). Purple bacteria were probably acquired first to provide mitochondria, and photosynthetic prokaryotes were acquired later to form chloroplasts. It is likely that host–photosymbiont

Figure 7.9   *Grypania,* the oldest known fossil of eukaryotic algae from the 1.9-Ga Negaunee Iron Formation, Marquette, Michigan. Scale × 2. Courtesy of Bruce Runnegar.

relationships take long periods to develop, perhaps hundreds of millions of years. Such evolution may have occurred during a period of remarkable stability in the carbon cycle from about 1.9 to 0.8 Ga (Brasier and Lindsay, 1998).

Origin of Metazoans

Metazoans (multicellular animals) appear to have evolved from single-celled ancestors that developed a colonial habit. The adaptive value of a multicellular way of life relates chiefly to increases in size and the specialization of cells for different functions. For instance, more suspended food settles on a large organism than on a smaller one. Because all cells do not receive the same food input, food must be shared among cells and a "division of labor" develops among cells. Some concentrate on food gathering, others on reproduction, and others on protection. At some point, when intercellular communication was well developed, cells no longer functioned as a colony of individuals but as an integrated organism.

The trace fossil record suggests that metazoans were well established by 1000 Ma (Fig. 7.10), and the great diversity of metazoans of this age suggests that more than one evolutionary line led to multicellular development. Findings of leaf-shaped fossils in north China suggest that some form of multicellular life had evolved by 1.7 Ga (Shixing and Huineng, 1995). On the basis of their size (5–30 mm long), probable development of organs, and possible multicellular structures, these forms are likely benthic multicellular algae (Fig. 7.11). Although metazoans appeared by 1.7 Ga, they did not become widespread until less than 1 Ga. Because of an inadequate fossil record, investigators cannot trace these groups of multicellular organisms back to their unicellular ancestors.

Neoproterozoic Multicellular Organisms

Although most paleontologists regard Ediacaran fossils as metazoans, some have suggested that some or all may represent an extinct line of primitive plant-like organisms similar to algae or fungi (Seilacher, 1994; Narbonne, 1998). However, there are more similarities of the Ediacarans to primitive invertebrates than to algae or fungi (Runnegar, 1994; Weiguo, 1994). From the widespread fossil record, some 31 Ediacaran species have been described, including forms that may be ancestors to flatworms, coelenterates, annelids, soft-bodied arthropods, and soft-bodied echinoderms. The most convincing

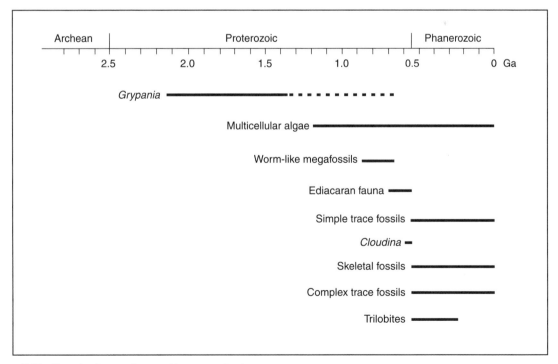

Figure 7.10 Time distribution of various Proterozoic fossil groups.

Figure 7.11 A carbonaceous multicellular fossil *(Antiqufolium clavatum)* from the 1.7-Ga Tuanshanzi Formation, north China. Scale bar = 2 mm. Courtesy of Zhu Shixing.

Figure 7.11 A carbonaceous multicellular fossil *(Antiqufolium clavatum)* from the 1.7-Ga Tuanshanzi Formation, north China. Scale bar = 2 mm. Courtesy of Zhu Shixing.

evidence for Neoproterozoic animals comes from trace fossils associated with the Ediacaran fauna. Looping and spiraling trails up to several millimeters in width and strings of fecal pellets point to the presence of soft-bodied animals with a well-developed nervous system, asymmetry, and a one-way gut.

Reported U-Pb zircon ages from ash beds associated with Ediacaran fossils in Namibia in southwest Africa indicate that this fauna is no older than 550 Ma and that some forms are as young as 543 Ma (Grotzinger et al., 1995). This places their age just before the Cambrian–Precambrian boundary 540 Ma (Fig. 7.10). Prior to these isotopic ages, a large gap was thought to have existed between the Ediacaran fossils and the diverse invertebrate forms that suddenly appear in the Cambrian. It now appears that some of the shelly Cambrian forms overlap with the Ediacaran forms (Kimura and Watanabe, 2001). These new findings support the idea, but do yet prove, that some Ediacaran forms were ancestors to some Cambrian invertebrates.

Cambrian Explosion and Appearance of Skeletons

All of the major invertebrate phyla (except the Protozoa) made their appearances in the Cambrian, a feature sometimes referred to as the **Cambrian explosion** (Weiguo, 1994; Bengtson, 1994). This increase in the number and the diversity of organisms is matched by a sharp increase in the diversity of trace fossils and the intensity of bioturbation (the churning of subaqueous sediments by burrowing organisms). One of the immediate results of bioturbation is the return of buried organic matter to the carbon cycle and hence a decrease in the net release of oxygen because of decay. The Cambrian explosion may have been triggered by environmental change near the Proterozoic–Cambrian boundary and later amplified by interactions within reorganized ecosystems (Knoll and Carroll, 1999). Two biologic inventions permitted organisms to invade sediments (Fischer, 1984): First, the development of exoskeletons allowed organisms (such as trilobites) to dig using appendages; second, the appearance of coeloms permitted worm-like organisms to penetrate sediments. Although calcareous and siliceous skeletons did not become widespread until the Cambrian, the oldest known metazoan with a mineralized exoskeleton is *Cloudina,* a tubular fossil of worldwide distribution predating the base of the Cambrian at least 10 My (appearing about 550 Ma) (Fig. 7.10).

The reason that hard parts were developed in so many groups about the same time is a puzzling problem in the Earth's history. Possibly it was for armor that would protect against predators. However, one of the earliest groups to develop a hard exoskeleton was the trilobites, the major predators of the Cambrian seas (Fig. 7.10). Although armor has a role in the development of hard parts in some forms, it is probably not the only or the original reason for hard parts. The hard parts in different phyla developed independently and are made of different materials. More plausible ideas are that hard parts are related to an improvement in feeding behavior, locomotion, or support. As an example of improved feeding behavior, in brachiopods, the development of a shell enclosed the filter-feeding "arms" and permitted the filtration of larger volumes of water, similar in principle to how a vacuum cleaner works. Possibly, the appearance of a hard exoskeleton in trilobites permitted a faster rate of locomotion by extending the effective length of limbs. Additional structural support may have been the reason that an internal skeleton developed in some echinoderms and in corals.

Still another factor that may have contributed to the Cambrian explosion is the growth of oxygen in the atmosphere. A continuing increase in the amount of fractionation of sulfur isotopes between sulfides and sulfates beginning in the early Cambrian may record increased levels of oxygen in the atmosphere caused by repeated cycles of oxidation of H_2S (Canfield and Teske, 1996). An increase in oxygen levels at this time may have contributed to the rapid growth and diversification of organisms in the Cambrian.

Precise U-Pb zircon ages that constrain the base of the Cambrian to about 545 Ma have profound implications for the rate of the Cambrian explosion (Bowring et al., 1993). These results show that the onset of rapid diversification of phyla probably began within 10 My of the extinction of the Ediacaran fauna. All of the major groups of marine invertebrate organisms reached or approached their Cambrian peaks from 530 to 525 Ma, and some taxonomic groupings suggest that the number of Cambrian phyla exceeded the number of phyla known today. Assuming the Cambrian ended between 510 and 505 Ma, the evolutionary turnover among the trilobites is among the fastest observed in the Phanerozoic record. Using the new age for the base of the Cambrian, the average longevity for Cambrian trilobite genera is only about 1 My, much shorter than previously thought.

Evolution of Phanerozoic Life Forms

In recent years, progress in both systematic and developmental genetics has revolutionized scientists' perspective on animal relationships and provided new ideas about early animal evolution. With most of the 35 modern metazoan phyla represented or inferred by the end of the Cambrian from the fossil record, the age, rate of divergence, and number of branches in the phylogenetic tree are beginning to emerge (Knoll and Carroll, 1999). New molecular phylogenies based on 18S ribosomal RNA sequences suggest that most animals fall into three major groups: the deuterostomes, the lophotrochozoans, and the ecdysozoans. One implication of this new phylogeny is that early-diverging groups within each group may display ancestral and derived characters. At present, however, the three-branched tree provides no evidence for the kind of animal that could represent the last common ancestor.

From the fossil record, it is possible to track the major evolutionary development of life. Cambrian faunas are dominated by trilobites (~60%) and brachiopods (~30%), and by late Ordovician, most of the common invertebrate classes that occur in modern oceans were well established. Trilobites reached the peak of their development during the Ordovician, with a great variety of shapes, sizes, and shell ornamentation. Bryozoans, which represent the first attached communal organisms, appeared in the Ordovician. Graptolites, cephalopods, crinoids, echinoderms, mollusks, and corals also began to increase in number at this time. Vertebrates first appeared during the Ordovician as primitive fish-like forms without jaws. Marine algae and bacteria continued to be the important plant forms during the early Paleozoic.

The late Paleozoic is a time of increasing diversification of plants and vertebrates and of decline in many invertebrate groups. Brachiopods, coelenterates, and crinoids all increase in abundance in the late Paleozoic, followed by a rapid decrease in number at the end of the Permian. The end of the Paleozoic was also a time of widespread extinction, with trilobites, eurypterids, fusulines, and many corals and bryozoa becoming extinct. Insects appeared in the late Devonian. Fish greatly increased in abundance during the Devonian and Mississippian, amphibians appeared in the Mississippian, and reptiles emerged in the Pennsylvanian. Plants increased in number during the late Paleozoic as they moved into terrestrial environments. Psilopsids are most important during the Devonian, with lycopsids, ferns, and conifers becoming important thereafter. Perhaps the most important evolutionary event in the Paleozoic was the development of vascular tissue in plants, which made it possible for land plants to survive under extreme climatic conditions. Seed plants also began to become more important relative to spore-bearing plants in the late Paleozoic and early Mesozoic. The appearance and rapid evolution of amphibians in the late Paleozoic was closely related to the development of forests, which provided protection for these animals. The appearance for the first time of shell-covered eggs and of scales (in the reptiles) allowed vertebrates to adapt to a greater variety of climatic regimes.

During the Mesozoic, gymnosperms rapidly increased in number, with cycads, ginkgoes, and conifers becoming the most important. During the early Cretaceous, angiosperms (flowering plants) made their appearance and rapidly grew in number thereafter. The evolutionary success of flowering plants is because of the development of a flower and enclosed seeds. Flowers attract birds and insects that provide pollination, and seeds may develop fleshy fruits that, when eaten by animals, disperse the seeds. Marine invertebrates, which decreased in number at the end of the Permian, made a comeback in the Mesozoic (such as bryozoans, mollusks, echinoderms, and cephalopods). Gastropods, pelecypods, foraminifera, and coiled cephalopods were particularly important Mesozoic invertebrate groups. Arthropods in the form of insects, shrimp, crayfish, and crabs also rapidly expanded during the Mesozoic. Mesozoic reptiles are represented by a great variety of groups, including dinosaurs. Dinosaurs had herbivorous and carnivorous types and marine, terrestrial, and flying forms. Birds and mammals evolved from reptilian ancestors in the early Jurassic. The development of mammals was a major evolutionary breakthrough in that their warm-blooded nature allowed them to adapt to a great variety of natural environments (including marine) and that their increased brain size allowed them to learn more rapidly than other vertebrates. During the Cenozoic, mammals evolved into a large number of groups, filling numerous ecological niches. Man evolved in the anthropoid group about 4 Ma. The vertebrate groups characteristic of the Mesozoic continued to increase in number, and angiosperms expanded exponentially.

Biologic Benchmarks

Many benchmarks have been recognized in the appearance and evolution of new life forms on the Earth, as summarized in Figure 7.12. The first is that life on the Earth originated probably between 4.4 and 4.0 Ga after the cessation of large impact events at the surface. The first autotrophs may have appeared by 3.8 Ga; by 3.5 Ga, anaerobic prokaryotes were widespread and stromatolites, probably constructed by cyanobacteria, had appeared. Sulfate-reducing bacteria were also present by 3.5 Ga, although probably not widespread. About 2.4 Ga, oxygen levels in the atmosphere had increased enough for eukaryotes to appear, although the oldest eukaryotic fossils appeared 1.9 Ga. The oldest metazoan fossils 1.7 Ga suggest that metazoans may have evolved soon after the appearance of eukaryotic cells. Unicellular eukaryotes became the dominant life forms from 1.7 to 1.5 Ga and stromatolites peaked in abundance and diversity about 1.2 Ga. Soft-bodied metazoans increased rapidly in number between 550 and 1 Ga, culminating with the Ediacaran fauna from 600 to 543 Ma. *Cloudina,* the first metazoan with an exoskeleton, appeared about 550 Ma.

Important Phanerozoic benchmarks include the explosion of marine invertebrates between 540 and 530 Ma and the appearance of vertebrates (hemichordates) about 530 Ma, land plants 470 Ma, vascular plants 410 Ma, amphibians 370 Ma, reptiles and the amniote egg about 330 Ma, insects 310 Ma, mammals 215 Ma, flowering plants 210 Ma, and man about 4 Ma (*Australopithecus* 4 Ma and *Homo* 2 Ma).

Mass Extinctions

Another important and controversial topic dealing with living systems is that of **mass extinctions,** when many diverse groups of organisms become extinct over short periods.

Figure 7.12 Biologic benchmarks in the geologic past.

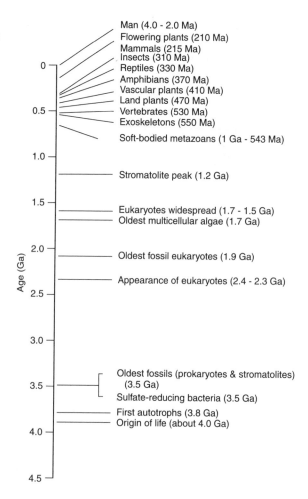

Man (4.0 - 2.0 Ma)
Flowering plants (210 Ma)
Mammals (215 Ma)
Insects (310 Ma)
Reptiles (330 Ma)
Amphibians (370 Ma)
Vascular plants (410 Ma)
Land plants (470 Ma)
Vertebrates (530 Ma)
Exoskeletons (550 Ma)
Soft-bodied metazoans (1 Ga - 543 Ma)

Stromatolite peak (1.2 Ga)

Eukaryotes widespread (1.7 - 1.5 Ga)
Oldest multicellular algae (1.7 Ga)

Oldest fossil eukaryotes (1.9 Ga)

Appearance of eukaryotes (2.4 - 2.3 Ga)

Oldest fossils (prokaryotes & stromatolites) (3.5 Ga)
Sulfate-reducing bacteria (3.5 Ga)
First autotrophs (3.8 Ga)
Origin of life (about 4.0 Ga)

Age (Ga)

Important mass extinctions occurred eight different times during the Phanerozoic (McLaren and Goodfellow, 1990). Mass extinction episodes affect a variety of organisms, marine and terrestrial, stationary and swimming, carnivores and herbivores, protozoans and metazoans. Hence, the causal processes do not appear to be related to specific ecological, morphological, or taxonomic groups.

Causes of mass extinction fall into three groups: extraterrestrial, physical, and biologic. Extraterrestrial causes that have been suggested include increased production of cosmic and X-radiation from nearby stars, increased radiation during reversals in the Earth's magnetic field, and climatic changes caused by supernova events or by impact on the Earth's surface. Among the physical environmental changes proposed to explain extinctions are rapid climatic changes, reduction in oceanic salinity caused by widespread evaporite deposition, fluctuations in atmospheric oxygen level, and changes in sea level. Collision of continents may also lead to extinction of specialized groups of organisms over longer periods, as described in Chapter 9. Rapid changes in environmental factors lead to widespread extinctions, and gradual changes permit organisms to adapt and may lead to diversification. Correlation analysis between times of microfossil extinctions in deep-sea sediments and polarity reversals of the Earth's magnetic field does not support a relationship between the two (Plotnick, 1980). As described later in this chapter, a great deal of evidence seems to support impact on the Earth's surface as a cause for the extinctions at the Cretaceous–Tertiary (K–T) boundary. One of the most significant problems in constraining the causes of mass extinction is precise isotopic dating of time boundaries with the U-Pb system in zircons. Precise dates require zircon-bearing ash beds that bound mass extinction events, a situation uncommonly found in key stratigraphic sections. Even when these conditions exist, mass extinction events cannot as yet be dated to better than about 500,000 years.

Although it is clear that no single cause is responsible for all major extinctions, many mass extinctions share common characteristics suggestive of a catastrophic event. Perhaps no other subject in geology has received more attention or has been more controversial than the question of what catastrophic process or processes are responsible for some major mass extinctions. In this section, I review the characteristics of mass extinctions, concentrating on the K–T boundary extinctions, which have attracted the most interest.

Mass Extinctions and Originations

Episodic Distributions

Eight mass extinctions are recognized in the Phanerozoic, and the same peaks are found for terrestrial and marine organisms, indicating that the major extinctions affected organisms on land and in the sea at the same time (Sepkoski, 1989; Benton, 1995; Foote, 2003) (Fig. 7.13). The late Carboniferous, late Jurassic, and early Cretaceous extinctions are more prominent for terrestrial than for marine organisms. An apparently high extinction rate in the early Cambrian may not be real but may reflect the low diversity of organisms at that time. Five major mass extinctions are recognized in the data (Fig. 7.13): the late Ordovician, late Devonian, late Permian, late Triassic, and late Cretaceous. Of these, the Permian extinction rate is highest, with a mean family extinction rate of 61% for all life, 63% for terrestrial organisms, and 49% for marine organisms (Benton, 1995). Peaks in the origin of species generally follow extinction peaks by less than 30 My, which appears to reflect the time necessary to fill abandoned ecological niches with new species. Extinction events tend to occur during times of high preservation potential, and origination events following these extinctions are more likely to occur at times more poorly represented by the fossil record (Foote, 2003).

Role of Supercontinents

Plate tectonics has played a major role in the origination and extinction of some groups of organisms. Modern ocean basins are effective barriers not only to terrestrial organisms but also to most marine organisms. Larval forms of marine invertebrates, for instance,

Figure 7.13 Rates of the origination and the extinction of marine animals during the Phanerozoic. Modified from Foote (2003).

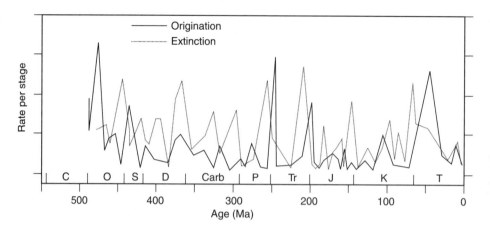

can only survive several weeks, which means they can travel only 2000 to 3000 km with modern ocean-current velocities. Hence, the geographic distribution of fossil organisms provides an important constraint on the sizes of oceans between continents in the geologic past. As illustrated in Figure 7.14, when a supercontinent fragments (Fig. 7.14a), organisms that cannot cross the growing ocean basin become isolated. This can lead to the evolution of diverse groups on each of the separating continents (Hallam, 1974). This affects both marine and terrestrial organisms. Examples are the diversification and specialization of mammals in South America and in Africa following the opening of the South Atlantic in the late Cretaceous. Similar changes occurred in bivalves of East Africa and India after the separation of these continents in the Cretaceous and Tertiary. Continental collision (Fig. 7.14b), on the other hand, removes an oceanic barrier, and formerly separate faunal and floral groups mix and compete for the same ecological niches. This leads to the extinction of groups that do not successfully compete. Collision also creates a new land barrier between coastal marine populations. If these marine organisms cannot migrate around the continent (perhaps because of climatic barriers), diversification occurs and new populations evolve along opposite coastlines. As an example, during the Ordovician collisions in the North Atlantic, many groups of trilobites, graptolites, corals, and brachiopods became extinct. The formation of new arc systems linking continents has the same effect as a collision. For instance, when the Panama arc was completed in the late Tertiary, mammals migrated between North and South America, which led to the extinction of large endemic mammals in South America. Panama also separated marine populations, which led to the divergence of Pacific and Caribbean marine organisms.

The biogeographic distribution of Cambrian trilobites indicates the existence of several continents separated by major ocean basins during the early Paleozoic. Major faunal province boundaries commonly correlate with suture zones between continental blocks brought together by later collisions. Wide oceans are implied in the Cambrian between Laurentia (mostly North America) and Baltica (northwest Europe), Siberia and Baltica, and China and Siberia; paleomagnetic data support this interpretation. Minor faunal·

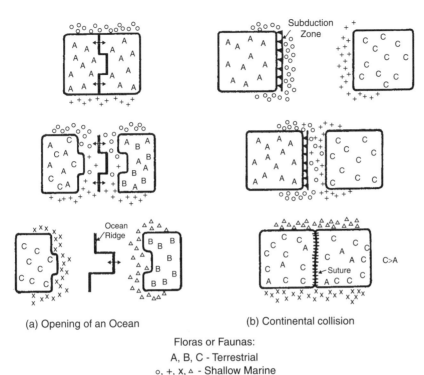

(a) Opening of an Ocean   (b) Continental collision

Floras or Faunas:
A, B, C - Terrestrial
o, +, x, ᐃ - Shallow Marine

provincialism within individual blocks probably reflects climatic differences. Studies of early Ordovician brachiopods indicate they belonged to at least five distinct geographic provinces, which was reduced to three provinces during the Baltica–Laurentia collision in the late Ordovician. The opening of the North Atlantic in the Cretaceous resulted in the development of American and Eurasian invertebrate groups from an originally homogeneous Tethyan group. Similar ammonite populations from East Africa, Madagascar, and India indicate that only shallow seas existed between these areas during the Jurassic.

The similarity of mammals and reptiles in the Northern and Southern Hemispheres more than 200 Ma demands land connections between the two hemispheres (Hallam, 1973). The breakup of Pangea during the early to middle Mesozoic led to diversification of birds and mammals and the evolution of unique groups of mammals (e.g., marsupials) in the Southern Hemisphere (Hedges et al., 1996). On the other hand, North America and Eurasia were not completely separated until early Tertiary, accounting for the overall similarity of Northern Hemisphere mammals today. When Africa, India, and Australia collided with Eurasia in the mid-Tertiary, mammalian and reptilian orders spread both ways, and competition for the same ecological niches was keen. This competition led to the extinction of 13 orders of mammals.

Plant distributions are also sensitive to plate tectonics. The most famous are the *Glossoptera* and *Gangemoptera* flora in the Southern Hemisphere. These groups range in age from Carboniferous to Triassic and occur on all continents in the Southern

Hemisphere and in northeast China, confirming that these continents were connected at this time. The complex speciation of these groups could not have evolved independently on separate landmasses. The general coincidence of late Carboniferous and early Permian ice sheets and the *Glossoptera* flora appears to reflect adaptation of *Glossoptera* to relatively temperate climates and its rapid spread over high latitudes during the Permian. The breakup of Africa and South America is reflected by the present distributions of the rain forest tree *Symphonia globulifera* and the semiarid leguminous herb *Teramnus uncinatus,* both of which occur at similar latitudes on both sides of the South Atlantic (Melville, 1973). Conifer distribution in the Southern Hemisphere reflects continental breakup, as evidenced by the evolution of specialized groups on dispersing continents after supercontinent breakup in the early Mesozoic.

Although a correlation between rate of taxonomic change of organisms and plate tectonics seems to be well established, the causes of such changes are not always agreed upon. At least three factors have been suggested to explain rapid increases in the diversity of organisms during the Phanerozoic (Hallam, 1973; Hedges et al., 1996).

1. An increase in the aerial distribution of particular environments increases the number of ecological niches in which organisms can become established. For instance, an increasing diversification of marine invertebrates in the late Cretaceous may reflect increasing transgression of continents, resulting in an aerial increase of the shallow-sea environment in which most marine invertebrates live.
2. Continental fragmentation leads to morphological divergence because of genetic isolation.
3. Some environments are more stable than others in terms of such factors as temperature, rainfall, and salinity. Studies of modern organisms indicate that stable environments lead to intensive partitioning of organisms into well-established niches and to corresponding high degrees of diversity. Environmental instability, decreases in environmental area, and competition of various groups of organisms for the same ecological niche can result in extinctions.

Glaciation and Mass Extinctions

Perhaps the best example of a mass extinction that may be related to glaciation was at the end of the Ordovician (Sheehan, 2001). At this time, about 26% of families and 49% of genera became extinct (Sepkoski, 1996). Two environmental changes associated with the Ordovician glaciation may be responsible for the late Ordovician extinction (Sheehan, 2001): (1) cooling global climates and (2) sea level decline, which drained the large shallow seas and eliminated habitats of shallow marine life. In some regions, such as Scandinavia, sea level drawdown was more than 100 m. The Ordovician extinction was severe in terms of number of taxa lost but was less severe in terms of ecological consequences. Unlike many mass extinctions, there was relatively little evolutionary innovation during the recovery that followed the Ordovician extinction. Newly developing communities were drawn largely from surviving taxa that had previously lived in similar ecological settings.

During the Ordovician extinction, strong thermal gradients appear to have been established between the poles and the equator, resulting in the movement of cold polar seawater into deep tropical seawater. At the end of the glaciation, sea level rebounded rapidly and temperature gradients returned to normal, with the entire glaciation lasting, perhaps, only 500,000 years. At the onset of glaciation, initial extinctions occurred chiefly in the receding shallow seas. Cool temperatures in the open ocean led to the development of the cool-water *Hirnantia* fauna. At the end of the glaciation, sea level rise and atmosphere warming resulted in a second pulse of extinctions, leading to the rapid extinction of open marine faunas including the *Hirnantia* fauna.

Impact-Related Extinctions

It is important to distinguish between the ultimate cause of mass extinctions, such as impact, and the immediate cause or causes, such as rapid changes in environment that kill plants and animals in large numbers worldwide. Impact of an asteroid or comet on the Earth's surface clearly will cause changes in the environment, some of which could result in extinctions of various groups of organisms (Toon et al., 1997). The timescale of associated environmental changes ranges from minutes or hours to perhaps a million years, as summarized in Table 7.2.

As an example, the major consequences of a collision of 10-km-diameter asteroid with the Earth's surface that may contribute to mass extinctions are summarized as follows.

1. *Darkness.* Fine dust and soot particles would spread worldwide in the upper atmosphere, cutting out sunlight for a few months after the impact. This would suppress photosynthesis and initiate a collapse to food chains, causing death by starvation to many groups of organisms. For instance, major groups that became extinct at the end of the Cretaceous, such as most or all dinosaurs, marine planktonic and nektonic organisms, and benthic filter feeders, were in food chains tied directly with living plants. Organisms less affected by extinction, including marine benthic scavengers, deposit feeders, and small insectivorous mammals, are in food chains that depend on dead plant material.
2. *Cold.* The dust would produce darkness and be accompanied by extreme cold, especially in continental interiors far from the moderating influence of oceans. Within two or three months, continental surface temperatures would fall to $-20°$ C (Fig. 7.15).
3. *Increased greenhouse effect.* If an asteroid collided in the ocean, both dust and water vapor would be spread into the atmosphere. Fallout calculations indicate that after the dust settles, water vapor would remain in the upper atmosphere for decades, producing an enhanced greenhouse effect that could raise surface temperatures well in excess of the tolerance limits of many terrestrial organisms. A CO_2 greenhouse effect could last for hundreds of years.

Table 7.2	Environmental Changes from the Impact of a Large Comet or Asteroid (10–15 km in Diameter)	
Environmental Agent	**Timescale**
Fireball irradiance	Minutes
Impact-related wildfires	Hours
Winds and tsunamis	Hours
Dust veil (cold, darkness)	Months
Acid rain (nitric, sulfuric)	Many months
Stratospheric aerosols (cold)	Decades
Ozone depletion (Ultraviolet exposure)	Decades
H_2O greenhouse effect	Decades
Poisons and mutagens	Years to millennia
CO_2 greenhouse effect	Millennia
Impact-triggered volcanism	Millennia (?)
Major climatic changes	Millions of years

4. *Acid rain.* Energy liberated during the impact may cause atmospheric gases to react, producing nitric and sulfuric acids and various nitrogen oxides. Hence, on extinctions, another side effect of an impact is the possibility of acid rain. These rains could last as long as a year and would lower the pH of surface water in the upper 100 m sufficiently to kill a large number of planktonic organisms.

5. *Wildfires.* The soot particles reported in clays at the K–T boundary may be the result of widespread wildfires ignited by infrared radiation from the initial impact and lasting many hours. Studies of the soot particles indicate that they come chiefly from the burning of coniferous forests. Charcoal is even more definitive and can be used to locate the region in which impact-ignited fires occurred (Belcher et al., 2003).

6. *Toxic seawater.* Asteroid collision should also result in the introduction of a variety of trace elements, many of which are toxic, into the oceans (such as Hg, Se, Pb,

Figure 7.15 Change in terrestrial surface temperature of the oceans and continents following impact with an asteroid about 10 km in diameter. Atmospheric dust density following impact is 1 gm/cm². Modified from Toon (1984).

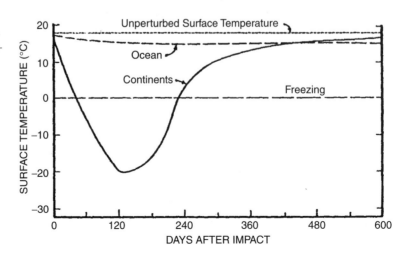

and Cd). Organisms living in surface marine waters would be exposed to these toxic elements for hundreds of years, contributing perhaps to their extinction.

Extinctions at the Cretaceous–Tertiary Boundary

Organisms Affected

Although most or all dinosaurs did not survive the K–T boundary, numerous species, such as lizards, frogs, salamanders, fish, crocodiles, alligators, and turtles, show no significant affects across this boundary. At the generic level, the terrestrial K–T extinction was only about 15%. One of the exciting controversies in geology today is that of whether all dinosaurs disappeared suddenly at the end of the Cretaceous. Teeth of 12 dinosaur genera have been described above the K–T boundary in Paleocene sediments in eastern Montana (Sloan et al., 1986). If these teeth are in place and have not been reworked from underlying Cretaceous sediments, they clearly indicate that at least some dinosaurs survived the K–T extinction. If, on the other hand, they are reworked, then the oldest *in-place* dinosaur remains are late Cretaceous in age. Suggesting these dinosaur teeth are not reworked, reworked remains of a widespread species of late Cretaceous mammals common in the underlying Cretaceous sediments are not found in these earliest Paleocene rocks of Montana. Also, the dinosaur teeth do not have eroded edges as they would if they were redeposited. The other possible earliest Tertiary dinosaur remains are known from India, Argentina, and New Mexico. It would appear that debate will continue about the precise age of the final dinosaur extinctions until articulated dinosaur skeletons are found above the K–T boundary layer.

The marine extinctions at the end of the Cretaceous are far more spectacular than the terrestrial extinctions, involving many more species and groups of animals (McLaren and Goodfellow, 1990; Erwin, 1993). At the family level, the marine extinction rate is about 15%, and at the generic and specific levels, it is about 70%. Major groups to disappear at the close of the Cretaceous are the ammonites, belemnites, inoceramid clams, rudistid pelecypods, mosasaurs, and plesiosaurs. The extinction of benthic forms preferentially takes place in groups that had free-swimming larvae. Among benthic forms such as most mollusks, bryozoans, and echinoids, extinction rates were low, whereas calcareous phytoplankton and planktonic foraminifera show sharp extinctions. The abrupt disappearance of the plankton appears to have stressed marine communities by removing much of the base of the marine food chain. Particularly affected were reef-living bivalves, most oysters, clams, ammonites, many corals, most marine reptiles, and many fishes.

Also at the K–T boundary, selective killing agents resulted in widespread losses of swamp forests and in the extinction of most of the insect-pollinated angiosperms (Sweet, 2001). The preservation of fern spores across the boundary, however, suggests that understory plants (ferns, bryophytes, etc.) survived the mass extinction event. Gymnosperm pollens in North America decrease in abundance in a northerly direction in postextinction sediments, indicating widespread losses of gymnosperms.

Seeking a Cause

Evidence for Impact. *Iridium anomalies.* Some of the strongest evidence for impact is the enrichment of Ir in a clay layer at the K–T boundary at many locations worldwide (Alvarez et al., 1990). The age of this clay layer has been measured at several places on the continents, and the average age is 65 Ma, precisely the age of the K–T boundary dated from deep-sea sediments. This enrichment, known as an **iridium anomaly** (Fig. 7.16), cannot be produced from crustal sources because of the exceedingly low Ir content of crustal rocks, but it could come from the collision of an asteroid. Following impact, Ir, which is volatile, would have been injected into the stratosphere and spread over the globe, gradually settling in dust particles over a few months.

Glass spherules. Spherules are glassy droplets (a few tenths of a millimeter in diameter) of felsic composition commonly found in K–T boundary clays (Maurasse and Sen, 1991). By analogy with tektites, which are impact glasses with diameters up to a few centimeters found on the Earth, the small spherules at the K–T boundary appear to have formed by the melting of crustal rocks followed by rapid chilling as they were thrown into the atmosphere.

Soot. Soot or small carbonaceous particles are also widespread in K–T boundary clays and may be the remains of wildfires that spread through forests following impact (Wolbach et al., 1985). The absence of charcoal in boundary-layer clays, however, means there were no global wildfires associated with the K–T impact (Belcher et al., 2003).

Shocked quartz. Some of the strongest evidence for impact is the widespread occurrence of shocked quartz in K–T boundary clays (Bohor et al., 1987; Hildebrand et al., 1991).

Figure 7.16 Ir anomaly at the Cretaceous–Tertiary boundary in the Raton basin in northeastern New Mexico. Data are from a drill core. Also shown is the ratio of angiosperm pollen to fern spores in the section. ppt, parts per trillion. Modified from Orth et al. (1987).

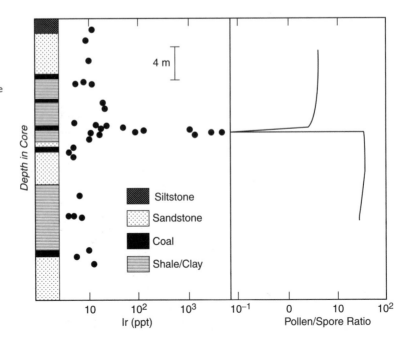

Shock lamellae in quartz are easily identified (Fig. 7.17) and are produced by a high-pressure shock wave passing through the rock. Such shocked quartz is common around nuclear weapon test sites and around well-documented impact sites such as Meteor Crater in Arizona.

Stishovite. Stishovite is a high-pressure polymorph of silica formed during impact and has been found in the K–T boundary-layer clay (McHone et al., 1989). Like shocked quartz, it has only been reported at known impact and nuclear explosion sites.

Chromium isotopes. The isotopic composition of Cr from K–T boundary clays from Denmark and Spain is different from the isotopic composition of terrestrial Cr but similar to that of carbonaceous chondrites (Shukolyukov and Lugmair, 1998). This similarity strongly suggests an asteroid impact source for the Cr in the K–T boundary-layer clay. Meteorite fragments have been described from K–T boundary-layer clays in the North Pacific Ocean that may represent samples of a K–T asteroid impactor (Kyte, 1998).

Earth-Crossing Asteroids. Is it possible from our understanding of asteroid orbits and how they change because of collisions in the asteroid belt that asteroids collided with the Earth? The answer is yes. Today, there are approximately 50 Earth-crossing asteroids with diameters greater than 1 km and a total population of more than 1000. **Earth-crossing asteroids** are capable of colliding with the Earth with only small perturbations of their orbits. To be effective in K–T mass extinctions, the "killer asteroid" would have to be at

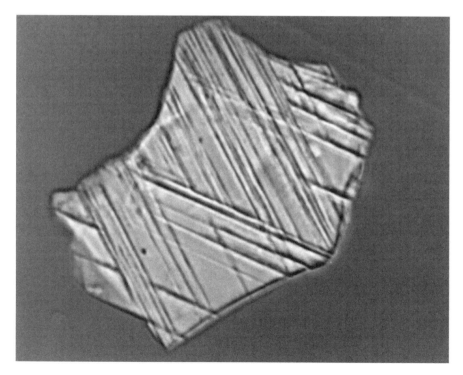

Figure 7.17 Scanning electron microscopy photo of a hydrogen fluoride-etched shocked quartz grain from the Cretaceous–Tertiary boundary claystone at Teapot Dome, Wyoming. Two major sets of planar deformation features (shock lamellae) are displayed in this view. Open planar features of the two major sets originally contained a silica glass phase, which has been removed by the acid etching. Grain is 72 μm in maximum dimension. Courtesy of Bruce Bohor.

least 10 km in diameter. Today, there are about 8 Earth-crossing asteroids with diameters of at least 10 km, and probability calculations suggest that about 10 asteroids of this size have collided with the Earth since the end of the Precambrian (McLaren and Goodfellow, 1990).

Comets. About 3 long-period and 10 short-period (less than 20 comets per year) comets pass inside the Earth's orbit each year. Because long-period comets have velocities greater than asteroids, collisions with the Earth should liberate approximately an order of magnitude more energy. Statistical calculations indicate that Earth-crossing comets make about nine returns to the inner solar system before being ejected into different orbits, and this corresponds to a mean lifetime of about 500,000 years. Estimates based on the frequency of cometary showers suggest that about 50% of the preserved craters on the Earth are the products of cometary rather than asteroid or meteorite impact. A major shower involving approximately 10^9 comets more than 3 km in diameter would result in about 20 impacts on the Earth's surface and should occur every 300 to 500 My, and smaller showers (~10^8 comets) involving about two impacts should occur every 30 to 50 My. The probable ages of known impact structures (and impact glasses) on the Earth's surface suggest episodes of cometary impacting with major peaks 99, 65, and 35 Ma. Recognized mass extinctions that annihilate 50 to 95% of ecologically diverse lower taxa are recognized 93, 66, and 36 Ma. The similarity in timing is consistent with the possibility that cometary showers may be responsible for these extinctions. A major problem with the periodic extinctions model, however, is that periodicity has not been recognized in the Paleozoic and early Mesozoic. It would appear that any explanation of extinctions involving cometary collision must also explain why these extinctions only started in the last 100 My.

Flood Basalts. Evidence has been proposed in recent years to support a volcanic cause for K–T extinctions. The discovery of enriched Ir in atmospheric aerosols erupted from Kilauea in Hawaii indicates that Ir may be concentrated in oceanic plume-fed volcanic eruptions. Although glass spherules can be formed during eruptions of basalt, their distribution is localized around eruptive centers. Large eruptions of flood basalts have been suggested as causes of mass extinctions; consistent with this possibility, major flood-basalt eruptions show a periodicity of about 30 My with some peaks coinciding with some major extinction peaks (Courtillot and Cisowski, 1987; Courtillot et al., 1999b). Major flood-basalt eruptions produce 1-2×10^6 km^3 of magma and erupt over short periods of less than 1 My. Isotopic dating of the Deccan Traps in India indicates they erupted 66 to 65 Ma, near the K–T boundary and that volcanism occurred in three periods, each lasting 50,000 to 100,000 years. The first and most significant of these eruptions from 65.6 to 66.1 Ma occurred before the extinction of dinosaurs, and fossil remains of sauropods, carnosaurs, and mammals are found between the first two lava eruptions. A rapid decline in the ^{187}Os/^{188}Os ratio in marine sediments during this time correlates with a short period of global warming of 3 to 5° C (Ravizza and Peucker-Ehrenbrink, 2003). This confirms that the major pulse of Deccan volcanism occurred *before* most of the K–T extinctions and thus could not be the cause of these extinctions.

Environmental damage produced by the rapid eruption of large volumes of flood basalt is chiefly because of toxic gases (principally SO_2 and halogens) and sulfate aerosols. This requires subaerial eruption. In addition, variable amounts of ash may be introduced into the troposphere. Large volcanic eruptions are capable of introducing large quantities of sulfate aerosols into the atmosphere, which could cause immense amounts of acid rain, reduce the pH of surface seawater, add both volcanic ash and CO_2 to the atmosphere, and perhaps deplete the ozone layer. Erupted ash could further reinforce a global cooling trend, and the combined effect of these events could result in widespread extinctions, extending 1 million years or so.

Model calculations indicate that tremendous volumes of SO_2 and halogens may be introduced into the atmosphere during single, large flood-basalt eruptions. Also, data from the eruption of the Laki fissure in Iceland in 1783 document severe air pollution and plant damage in Europe related to this small flood-basalt eruption (Grattan, 2003). Hence, large flood-basalt eruptions should have severe consequences on global climate and would probably produce acid rain, ozone damage, and increased reflectance of solar radiation, rapidly cooling in the hemisphere affected. All of these may contribute to mass extinctions. Although flood-basalt eruptions may inject large volumes of toxic aerosols into the stratosphere, affecting global climate, the effect of CO_2 emitted during such eruptions in warming the atmosphere may be rather minor. Calculations show that even for a relatively large eruption such as the Deccan Traps, the surface temperature of the planet would be raised less than $2°$ C over 4×10^5 years (Caldeira and Rampino, 1991). This indicates that the K–T extinctions were not caused by global warming resulting from volatiles released by the eruption of the Deccan Traps.

Conclusions. So where do we stand in understanding the K–T extinctions today? Clearly, some extinctions occurred in the 10 My before the K–T boundary, and these appear to be related to terrestrial causes such as a fall in sea level or a temperature drop. However, the numerous extinctions that occurred from 65 to 66 Ma seem to require a catastrophic cause. In Table 7.3, various evidences for impact are compared with flood-basalt eruption to explain the K–T extinctions. Although it would appear that both impact and volcanic causes may explain the Ir anomalies, only impact can readily account for the wide distribution of glass spherules and soot and the presence of shocked quartz and stishovite.

Chicxulub and the Cretaceous–Tertiary Impact Site

There is another question related to the asteroid impact model: where is the impact crater? If the impact was in the ocean, any crater formed on the seafloor was probably subducted in the last 65 My. Although the search for a crater of the right age and size (≥ 100 km in diameter) on the continents is still continuing, the current best candidate is the 180-km Chicxulub crater in Yucatan (Hildebrand et al., 1991) (Fig. 7.18). Consistent with a location in the Caribbean for a K–T impact site is the common presence of shocked quartz and spherules in Caribbean K–T boundary clays. Both of these features require at least some continental crust, and the Caribbean basin contains both oceanic and

Table 7.3 Comparison of Impact and Volcanic Models for Cretaceous–Tertiary Boundary Extinctions		
Observational Evidence	**Impact**	**Flood-Basalt Eruption**
Ir anomaly	Yes: Asteroid Maybe: Comet	Possible but not likely
Glass spherules	Yes: Impact melts	No: Only local
Shocked quartz	Yes: Common at impact sites	No
Stishovite	Yes: Found at some impact sites	No
Soot	Yes: From widespread fires	No: Fires only local
Worldwide distribution of evidence	Yes	No
Summary	Acceptable: Accounts for all observational evidence	Rejected: Cannot explain shocked quartz, stishovite, or soot

Partly from Alvarez (1986).

continental crust. Impact breccia deposits also are widespread in Cuba and Haiti at the K–T boundary, supporting a Caribbean impact site (Fig. 7.19). The Chicxulub crater is the only known, large example of a Caribbean crater. It occurs in late Cretaceous marine carbonates deposited on the Yucatan platform on continental crust (Fig. 7.18). Some of the crater is now filled with impact breccias and volcanic rocks (Fig. 7.20), the latter of which have chemical compositions similar to the glass spherules found in K–T boundary clays. Melted crustal rocks from the crater also have high Ir contents. Supporting Chicxulub as the K–T impact site are precise $^{40}Ar/^{39}Ar$ ages from 65 to 66 Ma from glassy melt rock in the crater, which within the limits of error, is the age the K–T boundary

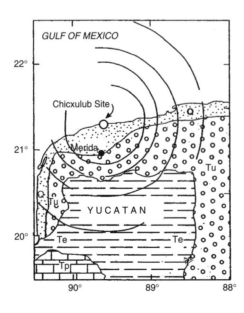

Figure 7.18 Simplified geologic map of the region around Merida, Mexico, showing the location of the Chicxulub crater buried with Tertiary sediments. Rings are gravity anomalies related to the impact. Q, Quaternary alluvium; Te, Eocene sediments; Tp, Paleocene sediments; Tu, Upper Tertiary sediments.

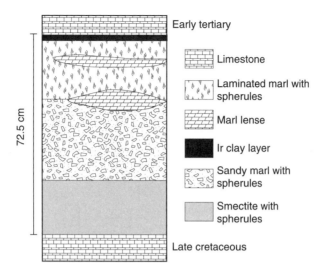

Figure 7.19 Early tertiary

Limestone

Laminated marl with spherules

Marl lense

Ir clay layer

Sandy marl with spherules

Smectite with spherules

Late cretaceous

72.5 cm

Figure 7.19 Section of the Cretaceous–Tertiary boundary layer at Belloc, Haiti. The layer is 72.5 cm thick with Ir concentrated in the thin clay layer at the top. Modified from Maurrasse and Sen (1991).

(Swisher et al., 1992). U-Pb isotopic ages from shocked zircons in K–T boundary-layer clays from widely spaced locations in North America yield ages identical to those of shocked zircons from the Chicxulub crater of 65 Ma, again supporting a Chicxulub source for the fallout (Kamo and Krogh, 1995).

A major environmental consequence of asteroid impact if it occurred in the oceans is a tsunami. **Tsunamis** are rapidly traveling sea waves caused by catastrophic disturbances such as earthquakes and volcanic eruptions. Calculations indicate that the asteroid that formed Chicxulub crater should have produced successive tsunamis up to 100 m high with periods of less than one hour, and these would have flooded the coastlines around the Gulf of Mexico and Atlantic within a day of the impact (Matsui et al., 1999). Tsunamis should cause widespread erosion and deposit poorly sorted sediments in tidal and beach

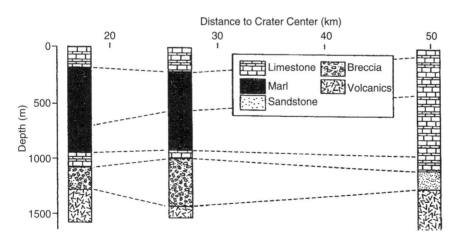

Distance to Crater Center (km)

Depth (m)

Limestone Breccia

Marl Volcanics

Sandstone

Figure 7.20 Correlation of stratigraphic sections across the Chicxulub crater from borehole data.

environments; such deposits are known at the K–T boundary from the Caribbean, the Gulf of Mexico, and the Atlantic Coast from points as far away as New Jersey and Denmark.

Possibility of Multiple Cretaceous–Tertiary Impacts

Studies of the K–T boundary in Mexico report more than one ejecta layer, introducing the possibility of more than one impact at or near the K–T boundary (Adatte et al., 2003). In northeastern Mexico, two to four ejecta layers occur at the K–T boundary that span at least 300,000 years. The oldest layer is dated 65.27 ± 0.03 Ma based on sediment accumulation rates and magnetostratigraphy. The Ir anomaly occurs in only one horizon, tentatively identified as the third and final impact event 64.9 Ma (Adatte et al., 2003; Keller et al., 2003). A multiple impact scenario is consistent with the ejecta evidence, with the first impact between 65.4 and 65.2 Ma correlating with a decrease in primary productivity and the onset of terminal decline in planktonic foraminifera. This impact may be the Chicxulub impact. The final K–T impact 64.9 Ma is associated with the last pulses of Deccan volcanism, and both impact and volcanism may have contributed to the major mass extinctions at the end of the Cretaceous. If this scenario is confirmed by rigorous isotopic dating, the Chicxulub impact would appear to predate the K–T boundary at least 300,000 years and may have occurred during intense Deccan volcanism, both contributing to high biotic stress levels and to greenhouse warming but not to major mass extinctions.

Impact and a 580-Ma Mass Extinction

Biostratigraphic and chemostratigraphic results from Neoproterozoic successions in Australia that contain Ediacaran fossils and acritarchs show a striking correlation among acritarch changes, a short-lived excursion in the carbon isotopic record of kerogens, and an ejecta layer from the 580-Ma Acraman impact in southern Australia (Grey et al., 2003). Acritarchs changed from assemblages dominated by long-range simple spheroids to a diverse assemblage characterized by short-range, large, complex shapes. At this time, 57 species of acritarchs make their first appearance in the geologic record. In addition, a negative carbon isotopic excursion occurred just before the acritarch radiation. Although the Snowball Earth hypothesis predicted postglacial biotic changes, the acritarch radiations did not occur until long after the Vendian glaciation ended. It is possible, if not likely, that the global extinction and recovery of acritarchs were associated with the Acraman impact 580 Ma in southern Australia.

Permian–Triassic Extinction

General Features

Killing about 60% of all organisms on the Earth is remarkably difficult, yet this happened at the end of the Permian (McLaren and Goodfellow, 1990; Erwin, 1994; Erwin, 2003). This is the closest metazoans have come to being exterminated during the past 500 My.

Also at the end of the Permian, Pangea was nearly complete, the sea level dropped significantly, evaporite deposition was widespread, and global warming occurred. The Siberian flood basalts, which have been dated 251.4 ± 0.3 Ma, erupted in less than 1 My at the Permian–Triassic (P–Tr) boundary (Bowring et al., 1998; Kamo et al., 2003).

The pattern of P–Tr extinctions is complex, with some groups disappearing well below the boundary, others at the boundary, and still others after the boundary. The marine fossil record provides the most complete record of the P–Tr extinctions. In south China, where a complete marine succession across the P–Tr boundary is exposed, 91% of all invertebrate species disappeared, including 98% of ammonoids, 85% of bivalves, and 75% of shallow-water fusulinids. Precise isotopic dating at the P–Tr boundary suggests that many of these extinctions occurred in less than 2 My and perhaps less than 1 My. Many groups of sessile, filter feeders disappeared at this time, as did marine invertebrates living in near-shore tropical seas. In terms of terrestrial animals, 80% of the reptile families and 6 of the 9 amphibian families disappeared at the end of the Permian. Among the insects, 8 of the known 27 orders disappeared in late Permian, 4 suffered serious declines in diversity, and 3 became extinct during the Triassic. Land plants showed little evidence of extinction at the P–Tr boundary, although pollens underwent significant change.

A major period of global warming is the most obvious climatic signal associated mass extinctions at the end of the Permian (Wignall, 2001). Evidence includes the migration of calcareous algae to Boreal latitudes, the loss of high latitude floras, and the development of intermediate latitude paleosols at high latitudes. Oxygen isotope data indicate that equatorial temperatures rose as much as $6°$ C at the P–Tr boundary, and Sr isotopes in marine carbonates suggest a corresponding rise in atmospheric CO_2. The inferred global warming may also have led to the widespread marine anoxia characteristic of late Permian sediments (Isozaki, 1997). This superanoxic event first appeared in deepwater, pelagic cherts, expanding at the end of the Permian to oxygen-poor shallow marine seas (Knoll et al., 1996). Two effects of the warming may have been responsible for the mass extinctions at the end of the Permian. First, the decrease in temperature gradient in the oceans from equator to pole led to a decrease in oceanic circulation. Second, the lower solubility of oxygen in the warmer seawater may have led to the widespread anoxia. The ultimate source of elevated CO_2 in the atmosphere, however, is unknown. Eruption of the Siberian flood basalts is one possibility, and destabilization of gas hydrates is another. Global warming may have led to the catastrophic release of large volumes of methane from gas hydrates, which could have been directly responsible for mass extinctions (Ryskin, 2003).

One problem still not resolved is the precise age of the P–Tr boundary. Bowring et al. (1998) propose an age of 251.0 ± 0.3 Ma based on zircon ages from ash beds at Meishan in China that bracket the P–Tr boundary. Dating of zircons from the same region by Mundil et al. (2001), however, report problems with Pb loss and inheritance in the zircons, resulting in a considerable scattering of the ages. When results from only single zircons are used, the age of the P–Tr boundary increases to about 253 Ma, 2 My older than the previous results. Results of Mundil et al. (2001) also indicate that it is not possible to obtain robust, high-accuracy ages from the Meishan zircons; thus, ages from other sites will be necessary to precisely determine the age of the P–Tr boundary.

Evidence for Impact

Unlike the K–T boundary, there is no evidence at the P–Tr boundary for impact fallout. No convincing Ir anomalies or shocked quartz has been found in boundary clays. New evidence for impact was proposed by Becker et al. (2001) and Poreda and Becker (2003) based on the presence of fullerenes (C_{60} and C_{70}) from boundary sections in Japan and south China. The fullerenes occur with trapped argon and helium, similar to occurrences the K–T sections. For these results to be accepted, however, they first must be reproduced in an independent laboratory. There are several problems with the impact model besides the lack of critical evidence. For instance, deepwater anoxic sediments spanned large areas of the Permian ocean, and there is no evidence that these sediments were disturbed during the mass extinctions as would be expected with a large impact. Also, there appear to be multiple negative peaks in $\delta^{13}C_{org}$ at the P–Tr boundary, which are difficult to reconcile with a single impact.

Evidence for Volcanism

Another cause considered for the Permian extinctions is the eruption of the Siberian flood basalts 251 Ma (or 253 Ma), releasing large amounts of sulfate aerosols and CO_2, the latter causing enhanced greenhouse warming (Wignall, 2001). Extinctions could be related both to changing climatic regimes and to the injection of sulfate aerosols into the atmosphere, resulting in acid rains (Renne et al., 1995). However, it is difficult to explain all of the extinctions by this mechanism, especially those that occurred long before the eruption of the Siberian basalts.

Summary

Many investigators lean toward multiple causes for the P–Tr extinctions (Erwin, 1994; Erwin, 2003). Important contributing factors may have been the following: (1) the loss of shallow marine habitats in response to falling sea level; (2) the completion of Pangea, which increased weathering rates and drew down atmospheric CO_2; (3) the eruption of the Siberian flood basalts, causing global warming and acid rains; and (4) the rapid transgression in the early Triassic, which destroyed near-shore terrestrial habitats and perhaps liberated large volumes of methane from gas hydrates. High-resolution dating of critical boundary sections shows that the mass extinctions occurred in a series of stages, each with a duration of hundreds of thousands if not tens of thousands of years (Wignall, 2003). This observation tends to favor terrestrial causes rather than a single impact for the P–Tr mass extinctions.

Further Reading

Bengtson, S. (ed.), 1994. Early Life on Earth. Columbia University Press, New York, 630 pp.

Courtillot, V., 1999. Evolutionary Catastrophes: The Science of Mass Extinction. Cambridge University Press, Cambridge, UK, 188 pp.

Dressler, B. O., Grieve, A. F., and Sharpton, V. L, 1992. Large Meteorite Impacts and Planetary Evolution. Geological Society of America, Spec. Paper 358.

Dyson, F., 1999. Origins of Life, Second Edition. Cambridge University Press, Cambridge, UK, 110 pp.

Jakosky, B., 1998. The Search for Life on Other Planets. Cambridge University Press, Cambridge, UK, 336 pp.

Schopf, J.S. (ed.), 1983. Earth's Earliest Biosphere, Its Origin and Evolution. Princeton University Press, Princeton, NJ, 543 pp.

Crustal and Mantle Evolution

8

Introduction

One of the unique features of the Earth is its crust. None of the other terrestrial planets seem to have a crust similar that of the Earth, and the reason for this is related to plate tectonics, which in turn is related to the way the mantle cools and how it has evolved. Most investigators agree that the history of the Earth's crust and mantle are closely related and that many of the features found in the crust are controlled by processes in the mantle. In this chapter, I will show you just how closely the evolution of the crust is tied to that of the mantle and will review how this dynamic system has evolved as the Earth has cooled over the last 4.6 Ga. As a starting point, consider the thermal history of the Earth.

Earth's Thermal History

There are two lines of evidence that indicate the mantle was hotter during the Archean. The most compelling evidence is that the present heat loss from the Earth is approximately twice the amount of heat generated by radioactive decay, which requires the excess heat to come from the cooling of the Earth. U, Th, and K isotopes provide most of the Earth's radiogenic heat (Chapter 2); from the decay rates of these isotopes, it would appear that heat production in the Archean was three to four times higher than the rate today (Richter, 1988). A second argument for higher Archean mantle temperatures is the presence of high-Mg komatiites in Archean greenstones, which, as you shall see later, require higher mantle temperatures for their generation. Estimates of the average Archean mantle temperature at 3 Ga vary from about 100 to 300° C higher than modern temperatures (Fig. 8.1).

It is commonly thought that the most realistic approach to calculating the Earth's thermal history is by using parameterized convection models in which expressions are solved that relate convective heat transport both to the temperature difference between the surface

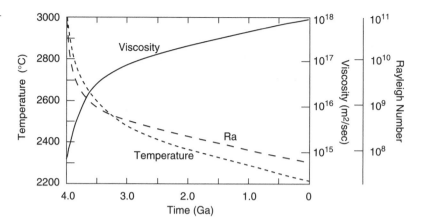

Figure 8.1 Average temperature, viscosity, and Rayleigh number of the mantle with time. Modified from McGovern and Schubert (1989).

and the interior of a convecting cell in the mantle and to the viscosity distribution in that cell. Results for whole-mantle convection show that mantle temperature, heat production, and Rayleigh number (Chapter 4) decrease with time and that mantle viscosity increases with time, all in accordance with a cooling Earth (McGovern and Schubert, 1989) (Fig. 8.1). Early in the Earth's history when the mantle was very hot, viscosity was low and convection rapid, perhaps chaotic, as dictated by the high Rayleigh numbers. During this period, around 500 My, the Earth cooled rapidly, followed by a gradual cooling of about 100° C/Gy to the present.

Earth's Primitive Crust

Origin of the First Crust

With our ever-increasing database, it is possible to address several outstanding questions related to the origin of the Earth's early crust:

1. Was the first crust of local or global extent?
2. When and by what process did the first crust form?
3. What was the composition of the early crust?
4. When and how did oceanic and continental crustal types develop?

As you shall see later in this chapter, the oldest preserved fragments of continental crust from 4.0 to 3.8 Ga are chiefly of tonalitic (tonalite-trondhjemite-granodiorite, or TTG) gneisses containing fragments of komatiite and basalt (amphibolite), some of which may be remnants of the early oceanic crust. Model lead ages of the Earth and isotopic ages from meteorites suggest that the earliest terrestrial crust may have formed just after or during the late stages of planetary accretion, about 4.5 Ga. Although the original extent of early Archean continental fragments is not known, they comprise less than 10% of preserved Archean crust (Fig. 8.2). The sparsity of rocks older than 3.0 Ga may be related

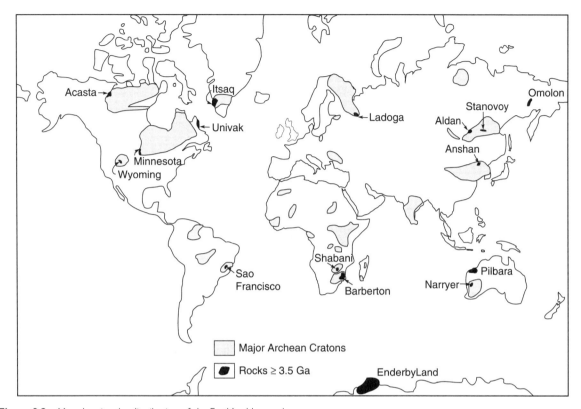

Figure 8.2 Map showing the distribution of the Earth's oldest rocks.

to losses resulting from the recycling of early crust into the mantle, as described later in the chapter. It may be possible to learn more about the first terrestrial crust by comparing the lunar crust and the crusts of other terrestrial planets. The lunar highlands, for instance, are remnants of an early lunar crust (~4.4 Ga), which covered most or all of the Moon's surface (Taylor, 1982). Studies of topographic features and cratering histories of Mercury and Mars also suggest the preservation of widespread primitive crusts. If the early history of the Earth was similar to these planets, it may have had an early crust that covered most its surface. Because remnants of this crust do not appear to be preserved in the continents, if such a crust existed on the Earth soon after planetary accretion, it must have been destroyed by continuing impacts on the surface, by recycling into a magma ocean, or both.

Theories for the origin of the Earth's crust fall into three broad categories: (1) heterogeneous accretion of the Earth, (2) impact models, and (3) terrestrial models. In the heterogeneous accretion model, the last compounds to condense from the solar nebula produce a thin veneer on planetary surfaces rich in alkali and other volatile elements, which may form or evolve into the first crust. A major problem with this model is that many nonvolatile elements such as U, Th, and rare earth elements (REEs), which are concentrated in the core and lower mantle in a heterogeneously accreted Earth, are today concentrated in the crust. This necessitates magmatic transfer from within the Earth, thus producing a crust of magmatic origin. Also, as described in Chapter 10, heterogeneous accretion of the Earth faces other geochemical problems.

Several models have been proposed for crustal origin either directly or indirectly involving the impact of accreting objects. All call upon surface impacting that leads to melting in the mantle, producing either mafic or felsic magmas that rise to form a crust. Large impacts may have produced mare-like craters on the terrestrial surface that were filled with impact-produced magmas (Grieve, 1980). If the magmas or their differentiation products were felsic, continental nuclei may have formed and continued to grow by magmatic additions from within the Earth. Alternatively, if the impact craters were flooded with basalt, they may have become oceanic crust. Although initially attractive, impact models face many difficulties in explaining crustal origin. For instance, most or all of the basalts that flood lunar mares formed were later than the impacts and were not related directly to impacting. Also, only relatively small amounts of magma were erupted into lunar mare craters. Perhaps the most significant problem with the lunar mare analogy is that mare basins formed in still older anorthositic crust.

Models that call upon processes operating within the Earth have been the most popular in explaining the origin of the Earth's early crust. Textures and geochemical relationships indicate that the early anorthositic crust on the Moon is a product of magmatic processes, favoring a similar origin for the Earth's earliest crust. It is likely that enough heat was retained in the Earth, after or during the late stages of planetary accretion, that the upper mantle was partially or entirely melted. Complete melting of the upper mantle would result in a magma ocean, which upon cooling should produce a widespread crust. Even without a magma ocean, extensive melting in the early upper mantle should produce large quantities of magma, some of which rise to the surface to form an early basaltic crust. Whether or not plate tectonics was operative at this time is not known. However, some mechanism of plate creation and recycling must have been operative to accommodate the large amounts of heat loss and vigorous convection in the early mantle.

Composition of the Primitive Crust

Numerous compositions have been suggested for the Earth's earliest crust. Partly responsible for diverging opinions are the different approaches to estimating composition. The most direct approach is to find and describe a relict of the primitive crust (≥ 4.4 Ga). Although some investigators have not given up on this approach, the chances that a remnant of this crust is preserved seem small. Another approach is to deduce the composition from studies of the preserved Archean crust. However, compositions and field relations of rock types in the oldest preserved Archean terranes may not be representative of the earliest terrestrial crust. Another approach has been to assume that the Earth and the Moon have undergone similar early histories and hence to go to the Moon, where the early record is well preserved, to determine the composition of the Earth's primitive crust. Geochemical models based on crystal-melt equilibriums and a falling geothermal gradient with time have also been used to constrain the composition of the early terrestrial crust.

Felsic Models

Some models for the production of a primitive felsic or andesitic crust rely on the assumption that low degrees of partial melting in the mantle will be reached before high degrees; hence, felsic magmas should be produced before mafic ones. Other models call upon fractional crystallization of basalt to form andesitic or felsic crust. Shaw (1976) proposed that the mantle cooled and crystallized from the center outward, concentrating incompatible elements into a near-surface basaltic magma layer. This layer underwent fractional crystallization, resulting in the accumulation of an anorthositic scum in irregular patches and in residual felsic magmas that crystallize to form the first stable crust by about 4 Ga.

Two main obstacles face the felsic crustal models. First, the high heat generation in the early Archean probably produced large degrees of melting of the upper mantle; hence, it is unlikely that felsic melts could form directly. Although felsic or andesitic crust could be produced by fractional crystallization of basaltic magmas, this requires a large volume of basalt, which itself probably would have formed the first crust.

Anorthosite Models

Studies of lunar samples indicate that the oldest rocks on the lunar surface are gabbroic anorthosites and anorthosites of the lunar highlands, remnants of a widespread crust formed about 4.4 Ga (Taylor, 1982). This primitive crust appears to have formed in response to catastrophic heating that led to the widespread melting of the lunar interior and the production of a voluminous magma ocean. As the magma ocean rapidly cooled and underwent fractional crystallization, pyroxenes and olivine sank and plagioclase (and some pyroxenes) floated, forming a crust of anorthosite and gabbroic anorthosite. Impact disrupted this crust and produced mare craters; these craters were later filled with basaltic magmas (3.9–2.5 Ga).

Most early Archean anorthosites are similar in composition (i.e., high An content, associated chromite) to lunar anorthosites and not to younger terrestrial anorthosites. It is clear from field relationships, however, that these Archean anorthosites are not remnants of an early terrestrial crust because they commonly intrude tonalitic gneisses. If, however, the Earth had an early melting history similar to that of the Moon, the first crust may have been composed dominantly of gabbroic anorthosites. In this scenario, preserved early Archean anorthosites may represent the last stages of anorthosite production, which continued after both mafic and felsic magmas were being produced.

The increased pressure gradient in the Earth limits the stability range of plagioclase to depths considerably shallower than those on the Moon. Experimental data suggest that plagioclase is not a stable phase at depths greater than 35 km in the Earth. Hence, if such a model is applicable to the Earth, the anorthosite fraction, either as floating crystals or as magmas, must find its way to shallow depths to be preserved. The most serious problem with the anorthosite model, however, is related to the hydrous nature of the Earth. Plagioclase will readily float in an anhydrous lunar magmatic ocean, but even small amounts of water in the system causes it to sink (Taylor, 1987; Taylor, 1992). Hence in the terrestrial system, where water was probably abundant in the early mantle, an anorthosite scum on a magma ocean would not form.

Basalt and Komatiite Models

In terms of understanding the Earth's early thermal history and the geochemical and experimental database related to magma production, it seems likely that the Earth's primitive crust was mafic to ultramafic in composition. If a magma ocean existed, cooling would produce a widespread basaltic crust, perhaps with komatiite components. Without a magma ocean (or after its solidification), basalts again may have composed an important part of the early crust. The importance of basalt and komatiite in early Archean greenstone successions attests to their probable importance on the surface of the Earth before 4 Ga.

Earth's Oldest Rocks and Minerals

The oldest preserved rocks occur as small, highly deformed terranes tectonically incorporated within Archean crustal provinces (Fig. 8.2). These terranes are generally less than 500 km across and are separated from surrounding crust by shear zones. Although the oldest known rocks on the Earth are about 4.0 Ga, the oldest minerals are detrital zircons from the 3-Ga Mount Narryer quartzites in Western Australia. Detrital zircons from these sediments have U-Pb ion probe ages ranging from about 3.5 Ga to 4.4 Ga, although only a small fraction of the zircons are older than 4.0 Ga (Froude et al., 1983; Nutman, 2001). Nevertheless, these old zircons are important in that they indicate the presence of felsic sources, some of which contained domains up to 4.4 Ga. These domains may have been remnants of continental crust, although the lateral extent of any given domain could have been much smaller than microcontinents such as Madagascar and the Lord Howe Rise.

The oldest isotopically dated rocks on the Earth are the Acasta gneisses in northwest Canada (Fig. 8.3). These gneisses are a heterogeneous assemblage of highly deformed TTG tectonically interleaved on a centimeter scale with amphibolites, ultramafic rocks, granites, and—at a few locations—metasediments (Bowring et al., 1989; Bowring, 1990). Acasta amphibolites appear to represent basalts and gabbros, many of which are deformed dykes and sills. The metasediments include calc-silicates, quartzites, and biotite–sillimanite schists. The rare occurrence of the tremolite-serpentine-talc-forsterite assemblage in ultramafic rocks indicates that the metamorphic temperature was in the range from 400 to 650° C. U-Pb zircon ages from the tonalitic and amphibolite fractions of the gneiss range from 4.03 to 3.96 Ga, and some components, especially the pink

Figure 8.3   The 4.0 Ga Acasta gneisses from the Archean Slave province, northwest of Yellowknife in the Northwest Territory, Canada. This outcrop, with the founder Sam Bowring, shows interlayered tonalite-trondhjemite-granodiorite and granite (light bands).

granites, have ages as low as 3.6 Ga. Thus, it would appear that this early crustal segment evolved over about 400 My and developed a full range in composition of igneous rocks from mafic to K-rich felsic types. Because of the severe deformation of the Acasta gneisses, the original field relations among the various lithologies are not well known. However, the chemical compositions of the Acasta rocks are much like those of less deformed Archean greenstone-tonalite-trondhjemite-granodiorite assemblages, suggesting a similar origin and tectonic setting.

The largest and best-preserved fragment of early Archean continental crust is the Itsaq Gneiss Complex in Southwest Greenland (Nutman et al., 1996; Nutman et al., 2002). In this area, three terranes have been identified, each with its own tectonic and magmatic history, until their collision about 2.7 Ga (Friend et al., 1988) (Fig. 8.4). The Akulleq terrane is dominated by the Amitsoq TTG complex, most of which formed from 3.9 to 3.8 Ga and underwent high-grade metamorphism at 3.6 Ga. The Akia terrane in the north comprises 3.2 to 3.0 Ga tonalitic gneisses deformed and metamorphosed at 3.0 Ga; the Tasiusarsuaq terrane, dominated by 2.9 to 2.8 Ga rocks, was deformed and metamorphosed when the terranes collided in the late Archean. Although any single terrane records less than 500 My of precollisional history, collectively, the terranes record more

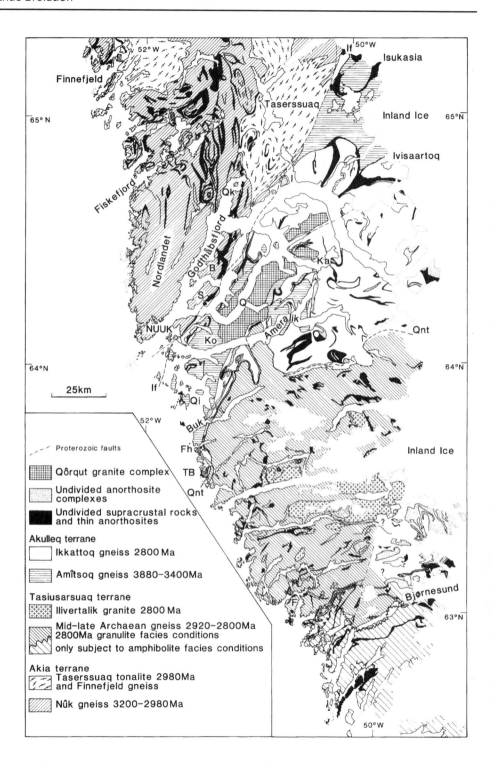

Finnefjeld

65° N

52° W

50° W

If

Isukasia

Taserssuaq

Inland Ice

65°N

Ivisaartoq

Fiskefjord

Qk

Nordlandet

Godthåbsfjord

B

Ka

S

Q

NUUK

Ko

Ameralik

Qnt

64°N

64°N

If

Qi

52° W

Buk

Fh

TB

Qnt

Inland Ice

Bjørnesund

63°N

F

50° W

25km

Proterozoic faults

Qôrqut granite complex

Undivided anorthosite complexes

Undivided supracrustal rocks and thin anorthosites

Akulleq terrane

Ikkattoq gneiss 2800 Ma

Amîtsoq gneiss 3880–3400Ma

Tasiusarsuaq terrane

Ilivertalik granite 2800 Ma

Mid-late Archaean gneiss 2920–2800Ma
2800Ma granulite facies conditions
only subject to amphibolite facies conditions

Akia terrane

Taserssuaq tonalite 2980Ma and Finnefjeld gneiss

Nûk gneiss 3200–2980Ma

than 1 Gy of history before their amalgamation in the late Archean. Each of the terranes also contains remnants of highly deformed supracrustal rocks. The most extensively studied is the Isua sequence in the Isukasia area in the northern part of the Akulleq terrane (Fig. 8.4). Although highly altered by submarine metasomatism, this succession comprises (from bottom to top) basalts and komatiites with intrusive ultramafics interbedded with banded iron formation, intrusive sheets of tonalite and granite, basalts and ultramafic rocks, mafic volcanogenic turbidites, and basalts with interbedded banded iron formation (Rosing et al., 1996). Remapping of the Isua succession suggests that at least some of the schists are highly deformed tonalitic gneisses or pillow basalts (Fedo et al., 2001). Carbonates in the Isua succession are now considered to be mostly, or entirely, metasomatic in origin; some may represent the products of seafloor alteration. The Isua succession is similar to island arc greenstone successions or perhaps to ocean-ridge successions, thus supporting the existence of plate tectonics in the early Archean.

The Pilbara craton in Western Australia also comprises a group of accreted terranes, the most widespread of which is the Warrawoona terrane, which formed between 3.7 and 3.2 Ga. Although extensively altered by submarine processes, the Warrawoona sequence is the best preserved early Archean greenstone (Barley, 1993; Krapez, 1993). It rests unconformably on an older greenstone–TTG complex with a U-Pb zircon age of about 3.5 Ga (Buick et al., 1995). This is important because it indicates not only that was the Warrawoona deposited on still older continental crust but also that land emerged above sea level by 3.46 Ga in this region.

Work in the Pilbara indicates the existence of three separate terranes with unique stratigraphy and deformational histories (Van Kranendonk et al., 2002): an eastern terrane (3.72–2.85 Ga), a western terrane (3.27–2.92 Ga), and the Kuranna terrane (≤3.29 Ga). The oldest supracrustal rocks in the eastern Pilbara terrane (the Coonterunah and Warrawoona Groups, 3.5–3.3 Ga) were deposited unconformably on older felsic crust about 3.72 Gy in age. The Warrawoona Group is divided into three volcanic cycles: (from base to top) the Talga-Talga (3.49–3.46 Ga), Salgash (3.46–3.43 Ga), and Kelly (3.43–3.31 Ga) Subgroups (Fig. 8.5). These dominantly mafic rocks include chert beds, containing the Earth's oldest stromatolites, and are interbedded with felsic volcanics erupted intermittently between 3.49 and 3.43 Ga. The Pilbara successions appear to be remnants of one or more oceanic plateaus erupted on thin continental crust.

The Barberton greenstone in southern Africa is one of the most studied early Archean greenstones. With coeval TTG plutons, the Barberton succession formed from 3.55 to 3.2 Ga (Kamo and Davis, 1994; Kroner et al., 1996). It includes four tectonically juxtaposed terranes with similar stratigraphic successions in each terrane (Lowe, 1994b). Each succession (known as the Onverwacht Group) begins with submarine basalts and komatiites of the Komati Formation (Fig. 8.5), an Archean mafic plain succession that could represent remnants of an oceanic plateau. Overlying the mafic plain succession are the Hooggenoeg and Kromberg Formations, a suite of felsic to basaltic submarine volcanics, fine-grained volcaniclastic sediments, and cherts, possibly representing an oceanic arc. The terminal Moodies Group (not shown in Fig. 8.5), which includes orogenic sediments, may have been deposited during amalgamation of the four terranes just after 3.2 Ga.

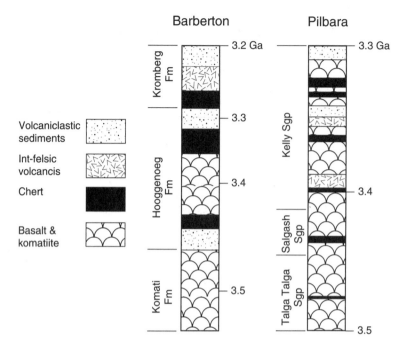

Unlike most late Archean greenstones, many of which evolved in less than 50 My, early Archean greenstones had long histories of more than 500 My before colliding and stabilizing as part of a continent (Condie, 1994). In the Barberton greenstone, individual cycles lasted from 50 to 80 My and included rifting and eruption of thick successions of mafic flows, magmatic quiescence with deposition of chemical sediments, and crustal thickening caused by intrusion of TTG plutons. Unlike late Archean terranes, which accreted into cratons almost as they formed, early Archean terranes appear to have bounced around like bumper cars for hundreds of millions of years. Why they did not accrete into continents is an important question that remains unresolved. Perhaps there were too few of these terranes and collisions were infrequent. Alternatively, most of these terranes may have been recycled into the mantle before having a chance to collide and make a continent.

No crust is known to have survived that is older than the 4.0 Ga Acasta gneisses in Canada. However, evidence of even older crust is provided by detrital zircons in metasediments from the Mount Narryer and Jack Hills areas in Western Australia (Amelin et al., 1999; Wilde et al., 2001; Nutman, 2001). Detrital zircons with ages up to 4.4 Ga have been reported from these sediments. One deep purple zircon measuring 220 by 160 microns, with internal complexities or inclusions, has a concordant $^{207}Pb/^{206}Pb$ age of 4404 ± 8 Ma, interpreted as the age of crystallization of this zircon (Wilde et al., 2001) (Fig. 8.6a). This is the oldest reported mineral age from the Earth. Although this represents only one scanning high resolution ion microprobe (SHRIMP) analysis, it is not affected by cracks and has a relatively small error. The other $^{207}Pb/^{206}Pb$ ages from

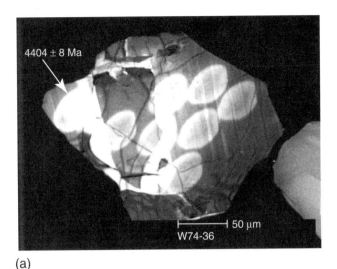

(a)

Figure 8.6 (a) Cathodoluminescence image of early Archean zircon W74-36 from Western Australia. $^{207}Pb/^{206}Pb$ ages shown for each SHRIMP analysis. (b) Combined Concordia plot for grain W74/2-36. Courtesy of Simon Wilde.

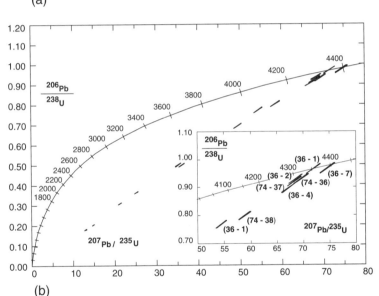

(b)

this zircon are at least 4.3 Ga and may represent actual geologic events (Fig. 8.6b), possibly triggered by asteroid impact, whereas the more discordant ages may represent Pb loss during impact events. REE distributions in this zircon show enrichment in heavy REEs, a positive Ce anomaly and a negative Eu anomaly. These REE distributions indicate that the zircon crystallized from an evolved granite melt. This observation is important because it means that evolved granites were produced on the Earth by 4.4 Ga. Furthermore, coupled with the oxygen isotopic results (Chapter 6) from this sample (Mojzsis et al., 2001), it appears that the granitic melt was produced by partial melting of older crust, either continental crust or hydrothermally altered oceanic crust (Wilde et al., 2001; Mojzsis et al., 2001).

Crustal Origin

The probable characteristics of the early oceanic and continental crust are summarized in Table 8.1. Oceanic crust is generated today at ocean ridges by partial melting of the upper mantle, and there is no reason to believe that early oceanic crust did not form the same way. When the first oceanic crust formed is unknown because it was undoubtedly recycled into the mantle, but it is likely that it crystallized from a magma ocean soon after planetary accretion. Because of the greater amount of heat in the Archean upper mantle, oceanic crust may have been produced four to six times faster than at present and thus would have been considerably thicker than modern oceanic crust. Like modern oceanic crust, however, it was probably widely distributed on the Earth's surface.

The Earth may be the only terrestrial planet with continental crust. If so, what is unique about the Earth that gives rise to continents? Two factors immediately stand out:

1. The Earth is the only planet with significant amounts of water.
2. It may be the only planet on which plate tectonics has been operative.

An important constraint on the origin of Archean continents is the composition of Archean TTG. Experimental data favor an origin for Archean TTG by partial melting of amphibolite or eclogite in the presence of significant amounts of water (Rapp and Watson, 1995). Without water, magmas of TTG compositions cannot form. The production of large amounts of Archean continental crust requires the subduction of large quantities of hydrated basalt and large quantities of water. Hence, with the possible exception of Venus, the absence of continental crust on other terrestrial planets may reflect the small amounts of water and the absence of plate tectonics on these planets.

It is possible that the earliest felsic crust developed from mafic oceanic plateaus, either by partial melting of the thickened mafic roots of the plateaus or by melting of slabs subducted around their margins. In either case, the resulting TTG magmas rise and underplate mafic and komatiitic rocks, some of which are preserved today in greenstone belts.

Table 8.1 Characteristics of the Earth's Early Crust		
	Oceanic Crust	**Continental Crust**
First appearance	~4.5 Ga	~4.3 Ga
Where formed	Ocean ridges	Submarine plateaus
Composition	Basalt	TTG
Lateral extent	Widespread, rapidly recycled	Local, rapidly recycled
How generated	Partial melting of ultramafic rocks in upper mantle	Partial melting of wet mafic rocks with garnet left in residue

TTG, tonalite-trondhjemite-granodiorite.

Large granite plutons do not appear in the geologic record until about 3.2 Ga and do not become important until after 2.6 Ga. Geochemical and experimental data suggest that these granites are produced by partial melting or fractional crystallization of TTG (Condie, 1986). It was not until TTG was relatively widespread that granites appeared in the geologic record. Thus, the story of early continental crust is the story of three rock types, basalt, tonalite, and granite, listed in the general order of appearance in the Archean geologic record. Field relations in most Archean granite–greenstone terranes also indicate this order of relative ages. It would appear that early Archean basalts were hydrated by seafloor alteration and that they partially melted later, either in descending slabs or in thickened root zones of oceanic plateaus, producing TTG magmas. TTG, in turn, was partially melted or fractionally crystallized to produce granites.

Thus, unlike the first oceanic crust, which probably covered much or all of the Earth's surface, the first continental crust probably had a more local extent associated with subduction zones and oceanic plateaus. Now that I have covered continental crust, the next question is as follows: How and at what rate did continents grow?

How Continents Grow

General Features

Although most investigators agree that the production of post-Archean continental crust is related to subduction, how continents are produced in arc systems is not well understood. Oceanic terranes such as island arcs and oceanic plateaus may be important building blocks for continents as they collide and accrete to continental margins. However, these terranes are largely mafic (Kay and Kay, 1985; DeBari and Sleep, 1991) yet upper continental crust is felsic, indicating that oceanic terranes must have undergone dramatic changes in composition to become part of the continents. Although details of the mechanisms by which mafic crust evolves into continental crust are poorly known, delamination of the lower crust during or soon after collision may play a role. Perhaps colliding oceanic terranes partially melt and felsic magmas rise to the upper continental crust, leaving a depleted mafic restite in the lower crust. Because investigators do not see seismic evidence for thick, depleted continental roots beneath recently accreted crust, if this mechanism is important, the depleted root must delaminate and sink into the mantle, perhaps during plate collisions.

Various mechanisms have been suggested for the growth of the continents, the most important of which are magma additions by crustal underplating and by terrane collisions with continental margins (Rudnick, 1995). Magma from the mantle may be added to the crust by underplating, involving the intrusion of sills and plutons (Fig. 8.7). Magma additions can occur in a variety of tectonic environments, the most important of which are arcs, continental rifts, and over-mantle plumes. Up to 20% of the crust in the Basin and Range Province in Nevada was added during the Tertiary by juvenile volcanism and plutonism (Johnson, 1993). Large volumes of juvenile magma from the mantle are added to both oceanic and continental margin arcs. Major continental growth by this mechanism

A. Magmatic Underplating

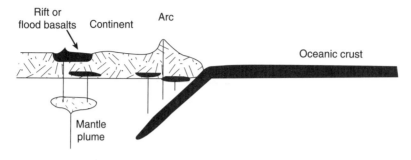

B. Terrane Collisions

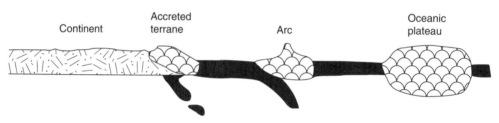

Figure 8.7 Mechanisms of continental growth.

can occur during seaward migration of subduction zones, when arc magmatism must keep up with slab migration. However, two important observations indicate that island arcs are not important components in the lower continental crust:

1. They are generally less than 25 km thick and as such should be subducted (Cloos, 1993).
2. Lower crustal xenoliths contain too much Ni, Cr, and Co for an arc source (Condie, 1994; Abbott, 1996).

Growth by Mafic Underplating

Field relationships in exposed lower crustal sections, such as the Ivrea Zone in Italy, suggest that many mafic granulites are intrusive gabbros and that additions of mafic magmas to the lower continental crust may be important. Also, a high-velocity layer at the base of Proterozoic shields has been interpreted as a mafic underplate (Durrheim and Mooney, 1991) (Fig. 8.8). This accounts for a difference in the average thickness of Archean shields (~35 km) and Proterozoic shields (~45 km). Why Archean continents were not underplated with mafic magma is not understood but may be related to the thick, depleted lithospheric roots beneath Archean shields that somehow protect the crust from underplating.

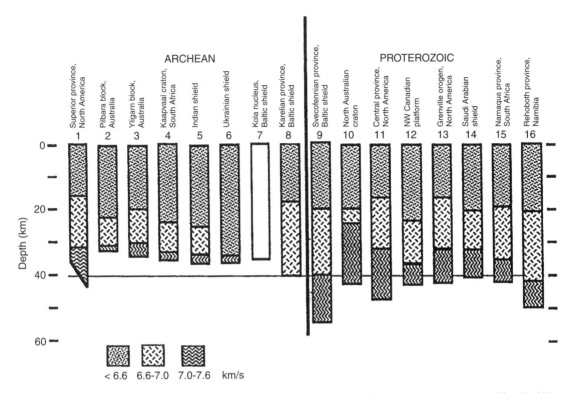

Figure 8.8 Seismic primary-wave, or compressional-wave, velocity layers of Archean and Proterozoic crustal provinces. Note the high-velocity roots of the Proterozoic provinces. Courtesy of Walter Mooney.

One of the first lines of evidence that the lower continental crust has grown, at least partly, through the underplating of basaltic magma comes from seismology (Nelson, 1991). Strong reflectors, widespread in the lower continental crust, have generally been interpreted as flows, sills, and layered mafic intrusions. Supporting this conclusion, some shallow reflectors have been traced to the surface and identified (Percival et al., 1992). Single basalt flows can be up to 100 m thick and cover several thousand square kilometers. Crustal xenolith studies and studies of uplifted segments of the lower crust clearly show mafic intrusions in the lower crust, some considerably younger than their host rocks (Rudnick, 1995; Rudnick and Fountain, 1995).

Another potentially important source of juvenile crust is crustal underplating associated with the eruption of continental flood basalts. The largest data source for mafic underplating of continental crust comes from the North Atlantic Igneous province, where both seismic studies and deep-sea drilling help to constrain the volumes of underplated igneous rock (Fitton et al., 1998). Results of these studies show that enormous volumes of mafic magma (both as basalt flows and intrusives) were emplaced along the transition between continental and oceanic crust during the rifting of Laurentia from Baltica. Both submarine and subaerial volcanism occurred along more than 2600 km of the propagating coastlines in the North Atlantic (the coasts of Greenland and North Europe) during the early Tertiary, mainly concentrated in a 1- to 2-My period about 57 Ma. Similar widespread volcanism, but more prolonged, occurred along the eastern coast of North America during the opening of this part of the Atlantic in the Mesozoic (White et al., 1987).

Oceanic Plateaus and Continental Growth

Some or most oceanic plateaus have arc systems erupted along their margins, such as the Solomon arc along the southern margin of the Ontong-Java Plateau and the Lesser Antilles arc along the eastern edge of the Caribbean Plateau. During collision and obduction, these arcs are obducted at shallow levels onto continental crust. Although in some instances thick slices of oceanic plateau can also be obducted (20 km for the Caribbean Plateau) or underplated (10 km for the Siletz terrane in Oregon), arcs should greatly dominate in obducted fragments. Perhaps this is the reason for the relative abundance of arc-related greenstones in the geologic record.

So, what is the fate of oceanic plateaus that collide with continents and lose their surficial arcs by obduction? Perhaps they are accreted to continental margins and with time they evolve into lower continental crust. Although crustal thickening during collision could produce eclogites in the root zones of accreted plateaus, leading to minor recycling of plateaus into the mantle as suggested by Saunders et al. (1996), a significant volume of these plateaus may be accreted to the continents. This idea has important implications for continental development in that the lower continental crust would comprise chiefly accreted and underplated oceanic plateaus, whereas the upper continental crust would form by subduction-related processes, perhaps beginning before accretion of oceanic plateaus to continental margins. Most of the upper continental crust, however, must develop after collision by subduction-related magmatism, described later in this chapter.

If oceanic plateaus are important in the lower crust, mafic xenoliths derived from the lower crust should carry the geochemical signatures of oceanic-plateau basalts. Unfortunately, results are complicated by arc-derived basalts and midocean-ridge basalts (MORB) in oceanic plateaus, upper crustal contamination of basaltic magmas, remobilization of elements during high-grade metamorphism and metasomatism, and possibly by later plume-derived and asthenosphere-derived magmas injected into the lower crust (Rudnick, 1992; Downes, 1993; Rudnick and Fountain, 1995). Also, at given xenolith localities, the range of compositions can be large, reflecting a mixture of plume- and arc-derived components in the lower crust. Clearly, it is going to be a challenging problem to use Nd, Sr, and Pb isotopes to sort accreted plateau sources from later underplated plume magmas in lower crustal xenolith suites.

The most convincing test for oceanic-plateau accretion as a process of continental growth is to identify young, accreted oceanic plateaus in the process of evolving into continental crust. It is well established that the Mesozoic Cordilleran crust in northwestern North America is relatively immature and, on the whole, more mafic in composition than Precambrian cratonic crust (Condie and Chomiak, 1996; Patchett and Gehrels, 1998).

Wrangellia, the dismembered oceanic plateau accreted to northwestern North America in the late Cretaceous, may provide a test for this continental evolution model (Condie, 1997). Using results from several seismic reflection sections from the Lithoprobe project in Canada (Varsek et al., 1993), Wrangellia can be traced laterally at depth into central British Columbia (Fig. 8.9). The eastern margin of Wrangellia is a major transcurrent fault system (the Fraser–Pasayten fault system) separating the accreted oceanic plateau from the Precambrian craton. Although greenstone basalts with both arc and plume geochemical characters occur in Wrangellia, the average composition of these basalts clearly is plume-like in character. Since accretion to the North American craton, the upper crust in Wrangellia has been intruded by the Coast Range batholith, a complex of felsic plutons carrying a subduction-zone geochemical signature (Nb and Ta depletion) and juvenile isotopic characters (Samson and Patchett, 1991). Also, the Cascade–Garibaldi arc system and its predecessors have been erupted on the surface of Wrangellia. With the emplacement of the Coast Range batholith and the eruption of arc volcanics, the upper crust of Wrangellia has become more felsic and appears to be evolving into more typical upper-continental crust. Perhaps Wrangellia provides us with a young and still-evolving example of how continental crust is formed from two sources: the lower crust from an accreted oceanic plateau and the upper crust from subduction-related processes.

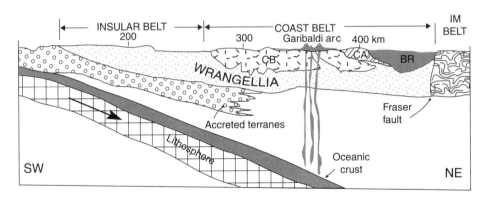

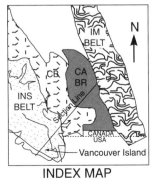

INDEX MAP

Figure 8.9 Interpretive geologic cross-section of southeastern British Columbia, Canada, showing the possible extent of Wrangellia. BR, Bridge River terrane; CA, Cadwallader terrane; CB, Coast Range batholith; IM, Intermontane; INS, insular (Alexander + Wrangellia). Modified from Clowes et al., 1992, and Condie, 1997.

Growth by Plate Collisions

Most of the Cordilleran and Appalachian orogens in North America represent collages of oceanic terranes added by collision either at convergent plate boundaries or along transform faults. Lithologic associations, chemical compositions, terrane life spans, and tectonic histories of Cordilleran terranes in northwest North America are consistent with collisional growth of the continent in this area (Condie and Chomiak, 1996). In addition to the Wrangellia collision described previously, remnants of other oceanic plateaus and juvenile oceanic arcs are important in the Cordillera. Although some of these terranes began to evolve into continental crust before accretion to North America, most terranes probably began this evolution at or not long before the time they accreted to the continent. This was accomplished largely by incompatible element enrichment resulting from subduction-related processes associated with collisionally thickened crust. Similar mechanisms have been proposed for Archean continents (Percival and Williams, 1989).

Continental Growth Rates

Continental growth is the net gain in mass of continental crust per unit of time. Because continental crust is both extracted from and returned to the mantle, continental growth rate can be positive, zero, or even negative. Many models of continental growth rate have been proposed (Fig. 8.10). These are based on one or a combination of (1) Pb, Sr, and Nd isotopic data from igneous rocks; (2) Sr isotopic ratios of marine carbonates; (3) the constancy of continental freeboard through time; (4) Phanerozoic crustal addition and subtraction rates; (5) the aerial distribution of isotopic ages; and (6) estimates of crustal recycling rates into the mantle. Most investigators have proposed growth models in which the cumulative volume of continental crust has increased with time at the expense of primitive mantle, leaving a complementary depleted mantle behind. Many models

Figure 8.10 Examples of published continental growth rate models. See text for explanation.

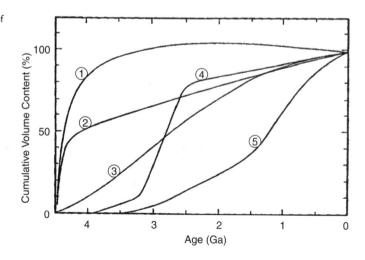

tacitly assume that the aerial extent of continental crust of different ages preserved today directly reflects crustal growth through time, an assumption investigators now know is unrealistic. The earliest models for continental growth were based chiefly on the geographic distribution of isotopic ages on the continents (Hurley and Rand, 1969). These models suggested that continents grew slowly in the Archean and rapidly after 2 Ga (curve 5, Fig. 8.10). Researchers now realize that this is not a valid approach to estimating crustal growth rates because many of the Rb-Sr and K-Ar dates used in such studies have been reset during later orogenic events, and the true crustal formation age is older than the reset dates. On the opposite extreme are models that suggest rapid growth early in the Earth's history, followed by extensive recycling of continental sediment back into the mantle (Reymer and Schubert, 1984) (curves 1 and 2, Fig. 8.10). In the model proposed by Fyfe (1978) (curve 1, Fig. 8.10), the volume of continental crust in the Proterozoic exceeds that present today. Recycling into the mantle is necessary in the latter two models because there are no large volumes of old continental crust on the Earth today. Other growth models fall between these extremes and include approximately linear growth with time (curve 3, Fig. 8.10) and episodic growth in which continents grow rapidly during certain periods, such as in the late Archean (Taylor and McLennan, 1985) (curve 4, Fig. 8.10).

Although an episodic distribution of isotopic ages has been recognized since the classic paper of Gastil (1960), it has been only since the early 1990s that the episodic growth of juvenile continental crust has been recognized (Condie, 1998). The distribution of U-Pb zircon ages coupled with Nd isotopic data suggest two major peaks in juvenile crust production rate, one at 2.7 Ga and another at 1.9 Ga (Fig. 8.11). In addition,

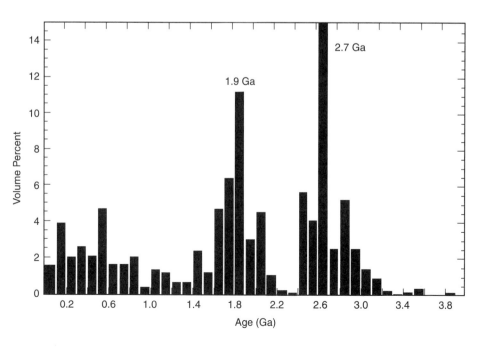

Figure 8.11 Volume distribution of juvenile continental crust based on a total continental crust volume of 7.177×10^9 km^3. Juvenile crust ages are U-Pb zircon ages used with Nd isotopic data and lithologic associations. Modified from Condie (1998; 2000).

smaller peaks may be present about 2.8, 2.5, 2.0, 1.7, and 1.45 Ga, and two peaks occur in the Phanerozoic.

Role of Recycling

The distribution of juvenile continental crust shown in Figure 8.11 is a function of two factors: (1) the rate of continental extraction from the mantle, and (2) the rate of the recycling of continent back into the mantle. The difference in these two rates is the **net continental growth rate,** which can also be considered as the preservation rate of continental crust. The degree and rate at which continental crust is returned to the mantle is a subject of considerable disagreement, and radiogenic isotopic data do not clearly resolve the problem (Armstrong, 1991). One of the major lines of evidence used to argue against a large volume of early Archean continental crust is the small volume of preserved crust at least 3.5 Ga, either as crustal blocks or as sediments. Although small in volume, continental crust at least 3.5 Ga is widespread with one or more examples known on all continents (Fig. 8.2), a feature that could indicate that the early Archean continental crust was more extensive than the present crust. Recycling of this early crust into the mantle is not surprising, as proposed by Armstrong (1981), in that a hotter mantle in the early Archean should lead to faster and more vigorous convection, which could result in faster crustal recycling than seen today.

The Nd isotopic composition of both detrital and chemical sediments also supports extensive sediment recycling. For instance, on a plot of stratigraphic age versus Nd model age, progressively younger sediments deviate farther from the equal age line (Fig. 8.12). This reflects recycling of older continental crust and a progressively greater proportion of older crust entering the sediment record with time. Another process that may have

Figure 8.12 Comparison of stratigraphic and Nd model ages of fine-grained terrigenous sediments. Data compiled from many sources.

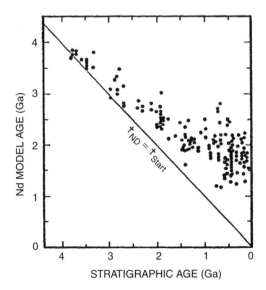

contributed to early crustal recycling is the bombardment of the Earth's surface with asteroid-size bodies (McLennan, 1988). It is well established that planets in the inner solar system underwent intense bombardment with large impactors until about 3.9 Ga, and such impact on the terrestrial surface could have aided in the destruction and recycling of early continental crust.

There are two processes by which modern continental crust may be destroyed and returned to the mantle (Reymer and Schubert, 1984; von Huene and Scholl, 1991; Armstrong, 1991). These are (1) sediment subduction and subduction erosion, and (2) delamination and sinking of the lower crust into the mantle during collisional orogeny. **Subduction erosion** is the mechanical plucking and abrasion along the top of a descending slab which causes a trench's landward slope to retreat shoreward. Subduction erosion and sediment subduction are both potentially important processes in subduction zones. Studies of modern arcs indicate that about half of the ocean-floor sediment arriving at trenches is subducted and does not contribute to the growth of accretionary prisms (von Huene and Scholl, 1991). At arcs with significant accretionary prisms, 70 to 80% of incoming sediment is subducted, and at arcs without accretionary prisms, all of the sediment is subducted. The combined average rates of subduction erosion (0.9 km^3/year) and sediment subduction (0.7 km^3/year) suggest that, on average, 1.6 km^3 of sediment are subducted each year.

Delamination is the decoupling and sinking of the lower crust and lithosphere from the overlying crust. The most important driving force for delamination is the negative buoyancy of the lower continental lithosphere primarily because of the eclogite phase transition (Meissner and Mooney, 1998; Schott and Schmeling, 1998). Delamination leads to crustal uplift as the dense lithosphere is replaced by lower-density asthenosphere. Where negative buoyancy exists in continental lithosphere, whether delamination occurs or not depends on the existence of a suitable zone of decoupling between the less dense portions of the lithosphere and the mantle lithosphere. Viscosity-depth calculations show low-viscosity zones at three depths (Meissner and Mooney, 1998): (1) the base of the felsic upper crust, (2) just above the Mohorovicic discontinuity (Moho) in the lower crust, and (3) tens of kilometers below the Moho. These low-viscosity zones, in addition to being zones of potential decoupling, are avenues for lateral flow during both compressive and extensional deformation. Supporting the possibility of modern collisional delamination is a major vertical seismic gap in the western Mediterranean basin. Tomographic images in this region indicate the presence of a high-velocity slab beneath low-velocity mantle, interpreted to be a piece of delaminated continental mantle lithosphere (Seber et al., 1996).

Veizer and Jansen (1985) have shown that various tectonic settings on the Earth have finite lifetimes in terms of recycling. Contributing to their life spans are rates of uplift and erosion, as well as subduction and delamination. Active plate settings, such as oceanic crust, arcs, and back-arc basins, are recycled much faster than continental cratons. Most active plate settings have recycling half-lives of less than 50 My, whereas collisional orogens and cratons have half-lives of more than 350 My, with finite lifetimes (or oblivion ages) of less than 100 My and more than 1000 My, respectively (Fig. 8.13). Because in the geologic record small remnants of rocks are formed at active plate

Figure 8.13 Mass–age distributions of tectonic settings. Key to tectonic settings with corresponding half-lives: 1, oceanic crust (50 My); 2, arc (80 My); 3, passive margin (100 My); 4, collisional orogen (250 My); 5, craton (350 My). Data from Veizer and Jansen (1985).

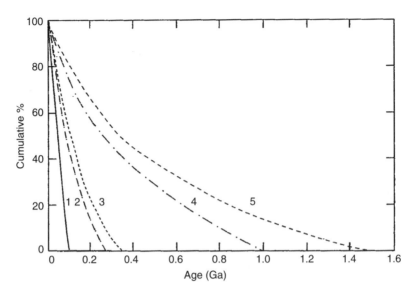

Nd Isotopes and Detrital Zircons

Perhaps the best way to recognize juvenile continental crust is with Nd isotopes. Positive E_{Nd} values are interpreted to reflect derivation from juvenile sources, whereas negative E_{Nd} values reflect derivation from enriched sources (Arndt and Goldstein, 1987). Ages of crust must be determined by some other isotopic system, the most accurate of which is the U-Pb system in zircons. It is now widely recognized that both positive and negative E_{Nd} values are recorded in Precambrian crustal rocks, even in the Earth's oldest known rocks, the Acasta gneisses (Fig. 8.14). The vertical arrays of E_{Nd} on a E_{Nd}-time plot may be explained by the mixing of crustal (negative E_{Nd}) and juvenile mantle melts (positive E_{Nd}). If this is the case, most individual data points cannot be used to estimate the rates and the amounts of continental growth and corresponding mantle depletion (Bowring and Housh, 1995). Calculated E_{Nd} values as high as +3.5 in 4.0 to 3.80 Ga rocks indicate the presence of a strongly depleted mantle reservoir at that time. The isotopic composition of this reservoir changed little during the Precambrian, and some of it may be the source of modern ocean-ridge basalts. This implies that the isotopic composition of the depleted reservoir was buffered by the addition of an enriched mantle component or continental crust. Otherwise, the Nd isotopic composition would have evolved along a steep slope, similar to the lunar growth curve in Figure 8.14. Investigators now recognize that continental crust was available for recycling by 4 Ga and probably by 4.4 Ga, as evidenced by the detrital zircon age from Western Australia.

The wide range of E_{Nd} in the Acasta gneisses and other early Archean rocks requires extreme early fractionation accompanied by efficient recycling to generate these differences

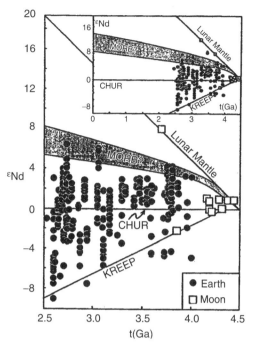

Figure 8.14 Distribution of E_{Nd} values in Archean rocks. Inset shows the evolution of the midocean-ridge basalt source (depleted mantle), chondritic uniform reservoir (CHUR, similar to primitive mantle), lunar mantle, and KREEP (potassium, rare earth elements, and phosphorous), a high component found on the lunar surface. Courtesy of S. A. Bowring.

by 4 Ga and to prevent unobserved isotopic evolution of the depleted mantle reservoir (Bowring and Housh, 1995). As a result of the relative enrichment of Nd in the continents, it is easier to buffer mantle evolution with subducted continental sediments than with an enriched mantle reservoir because of the much smaller volumes of crust required. Armstrong (1981; 1991) proposed a model for continental growth whereby the amount of recycling decreased with age parallel to the cooling of the Earth. It is interesting that the calculated growth curve for E_{Nd} in his model falls near the upper limit of E_{Nd} data in Figure 8.14 (bottom of the MORB field). Thus, the paucity of early Archean rocks in Precambrian shields may reflect an efficiency of recycling rather than a lack of production of early continental crust. The gradual increase in E_{Nd} values in the last 4 Gy is consistent with a gradual increase in the volume of preserved continental crust with time as a result of cooling of the mantle and consequent decreases in recycling rates. As shown in Figure 8.11, however, there are episodic spikes superimposed on the increasing preservation curve for continental crust.

Although detrital zircons with ages below 3.9 Ga have been reported from metasediments with depositional ages from 3.0 to 3.6 Ga in Australia, China, Greenland, and Labrador, these zircons are exceedingly rare. The most likely interpretation of these ages is that the zircon provenance areas with ages of at least 3.9 Ga were small and that they grew rapidly only after 3.9 Ga (Nutman, 2001). If the sparsity of zircons in these early sediments with ages of at least 3.9 Ga is representative, they support growth for continental crust in the early Archean rather than extensive recycling of crust older than 3.9 Ga.

Freeboard

The **freeboard** of continents, the mean elevation above sea level, is commonly assumed to be constant with time (Reymer and Schubert, 1984; Armstrong, 1991). Supporting this assumption for the Phanerozoic, collisional stages of supercontinent growth coincide with globally low sea levels and periods of active ocean-ridge activity correlate with high sea levels and transgressions (Eriksson, 1999). If freeboard has been constant with time and ocean ridges have diminished in volume because of cooling of the mantle, then the continents must have grown at a steady rate to accommodate the decreasing volume of seawater (curve 2, Fig. 8.10).

Although the constant freeboard model is supported by observations from the Phanerozoic, investigators now realize that this assumption may be inaccurate for the Precambrian (Galer, 1991). One problem is the thick Archean lithosphere, because a thicker lithosphere tends to offset the effects of shrinking ocean-ridge volume. Another is thicker Archean oceanic crust, which would make the Archean oceanic lithosphere more buoyant and thus add to the volume discrepancy required by a larger volume of Archean ocean ridges. Also, freeboard is especially sensitive to asthenosphere temperature, which was greater in the Archean. For instance, if the upper mantle temperature in the early Archean was 1600° C (about 200° C hotter than at present), continental crust would have been considerably below sea level (Galer, 1991). In terms of the geologic record, evidence for subaerial weathering and erosion first appears in rocks about 3.8 Ga; thus before this time, continents may have been deeply submerged.

The freeboard concept and a constant or variable freeboard over post-Archean time are closely linked to the growth of continents. However, because of the uncertainties in the freeboard of continents with time, and especially because researchers do not yet fully understand the magnitudes of all the controlling factors, extreme caution should be used when using freeboard to constrain continental growth models.

Nb/U and Nb/Th Ratios and Continental Growth

As originally suggested by Hofmann et al. (1986), it may be possible to use the Nb/U ratio of oceanic basalts to monitor the growth rate of continental crust and thus to test the two end-member models of crustal growth. Because U is relatively mobile, especially after about 2 Ga when the Earth's atmosphere became oxygenated, Th is often used as a proxy for U in tracking continental growth rates (Sylvester et al., 1997). Because Nb, U, and Th are incompatible in the mantle, with time they have been transferred into the continental crust (Hofmann et al., 1986; Campbell, 2002). If the continental crust formed as one large burst before 3.5 Ga, the Nb/U and Nb/Th ratios in the mantle from which this crust was extracted should be high and then decrease with time as the depleted mantle mixes with primitive mantle and recycled continent. Alternatively, if continental crust grew continually over time with little recycling into the mantle, the oldest oceanic basalts should have Nb/U and Nb/Th ratios similar to primitive mantle and these ratios should increase with time.

Greenstone basalts with probable mantle-plume sources show a relatively sudden increase in the Nb/Th and Nb/U ratios about 2 Ga (Collerson and Kamber, 1999; Condie, 2003a) (Fig. 8.15). This indicates (Condie, 2003a) that one or both of the following may explain the rather abrupt increase in Nb/Th ratio about 2 Ga: (1) rapid growth of continental crust in the late Archean, increasing the extent of depletion in the mantle reservoir from which continental crust is extracted; or (2) a rapid change in the mantle reservoir from which continental crust is extracted, from a relatively enriched reservoir to a more depleted reservoir. Rapid growth of continental crust at 2.7 Ga has the advantage in that it is consistent with Nd isotopic data supporting rapid continental growth at this time. A plot of E_{Nd} or E_{Hf} of greenstone oceanic basalts with time shows a trend similar to that of Nb/Th, with positive epsilon values decreasing with age and a possible rapid increase in epsilon values at 2 Ga (McCulloch and Bennett, 1994; Bowring and Housh, 1995). Thus, Nb/Th, E_{Nd} and E_{Hf} all seem to be consistent with episodic growth of the continental crust, with a major peak in growth in the late Archean.

As pointed out by other investigators (Bowring and Housh, 1995), continental growth is not an all or nothing process but involves both continuous continental growth and recycling into the mantle. A continental growth curve can be constructed by assuming that 0% crust was present when the mantle had chondritic Nb/Th ratios (near 8) and 100% was present when the mantle volume from which continental crust is extracted had ratios equal to normal MORB today (19) (Collerson and Kamber, 1999). This assumes that the upper mantle is the reservoir from which continental crust has been extracted. Using these assumptions, as much as 25% of the current volume of continental crust may have formed in the early Archean, and subsequently or concurrently with formation, most of it must have been recycled back into the mantle. Preserving continental crust requires the development of cratons with thick lithospheric roots, and the first, probably small, cratons appeared about 3 Ga (Condie, 1997). Earlier, continental crust may have formed

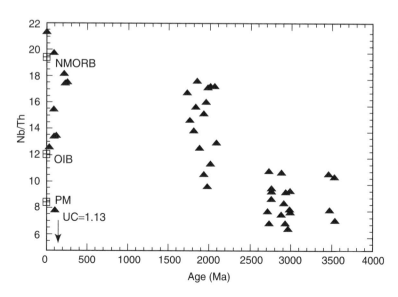

Figure 8.15 Mean values of Nb/Th versus age from greenstone basalts and komatiites from nonarc oceanic tectonic settings. NMORB, normal mid-ocean-ridge basalts; OIB, oceanic-island basalt; PM, primitive mantle; UC, average upper-continental crust. Data from Condie (2003a).

by partial melting of amphibolites in oceanic arcs or oceanic plateaus. Rapid uplift, erosion, and collision of these terranes could have rapidly recycled felsic components into the upper mantle. It was not until 2.7 Ga that the first cratons grew in size and thickness sufficiently to resist recycling into the mantle. The first supercontinent or supercontinents may have formed at this time (Aspler and Chiarenzelli, 1998; Condie, 1998). Hence, although significant volumes of continental crust may have formed in the early Archean, rapid convection in the hotter mantle led to rapid recycling of this crust into the upper mantle.

Continental Growth in the Last 200 Million Years

Reymer and Schubert (1984) were the first to estimate the growth rates of young continental crust by estimating the various crustal addition and subtraction rates. In Table 8.2, I present an updated and revised information for continental crust produced in the last 200 My. The results for arc and oceanic-plateau crustal addition rates include both arcs and plateaus in ocean basins today and an estimate of the volume of arcs and plateaus accreted to continental margins in the last 200 My. Hence, the arc accretion rate (1.6 km³/year) is considerably greater than the 1.1 km³/year value of Reymer and Schubert based on "nonaccreted" arcs only. The estimate of 0.5 km³/year for oceanic plateaus is a minimum value because the volume of plateaus accreted to continents is not well known. The volume of magma underplating continents is assumed to be equivalent to twice the volume of flood basalts erupted in the last 200 My, and the oceanic hotspot volcanic rate is from Reymer and Schubert (1984). The volume of subducted sediments and the material recycled into the mantle by subduction erosion are from von Huene and

Table 8.2 **Growth Rate of Continents in the Last 200 Million Years**	
	Rate (km³/year)
Gains	
Island arcs	1.6
Oceanic plateaus	0.8
Oceanic hotspot volcanism	0.2
Continental underplating	0.2
Total	2.8
Losses	
Sediment subduction	0.7
Subduction erosion	0.9
Delamination	0.1 (0–0.2)
Total	1.7
Net growth rate (2.8 km³/year minus 1.7 km³/year)	1.1

Partly from Reymer and Schubert (1984) and von Huene and Scholl (1991). Arcs, oceanic plateaus, and hotspot volcanics include an estimate of those accreted to the continents in the last 200 My.

Scholl (1991). The amount of crust returned to the mantle by delamination is poorly known, with estimates ranging from none to perhaps 0.2 km^3/year (I have assumed a value of 0.1 km^3/year). It is interesting that despite the major revisions in rates of gains and losses to the continental crust in the last 200 My, the net continental growth rate (gains minus losses) of 1.1 km^3/year is similar to the value originally proposed by Reymer and Schubert (1984) (1.06 km^3/year).

It is of interest to see whether this rate of crustal growth can account for the volume of juvenile continental crust formed in the last 200 My, estimated from precise geochronology and Nd isotopic studies. It would appear that about 6% of the continental crust was extracted from the mantle in the last 200 My (Fig. 8.11). Hence, for a total volume of continental crust of 7.18×10^9 km^3 (Cogley, 1984), 430×10^6 km^3 formed in the last 200 My. This agrees within a factor of two of the volume calculated using a rate 1.1 km^3/year (Table 8.2) of 220×10^6 km^3.

Toward a Continental Growth Rate Model

Approach

The net continental growth rate in the geologic past, which can also be considered the preservation rate of continental crust, critically depends on two factors (Condie, 1990a): (1) the proportion of reworked crust within a given crustal province and (2) the growth intervals assumed. The volume of juvenile crust, *g,* extracted from the mantle during a specified time interval, Δt, is given by the following:

$$g = a - r + m$$

Here, *a* is the volume of crust formed during Δt, *r* is the volume of reworked crust that must be subtracted, and *m* is the volume of crust formed in Δt but now tectonically trapped as blocks in younger crust. The value of *a* is determined from the scaled area of crustal provinces–terranes of a given age times the average crustal thickness. The final crustal thickness includes a 10-km-thick restoration of crust lost by erosion for a total average thickness of 45 km. Values for *r* and *m* are estimated from published Nd isotopic data, U-Pb zircon ages, and detailed geologic maps. Growth is considered in 100 My increments. The amount of reworked crust in a given crustal province is estimated from Nd isotopic data, assuming the mixing of juvenile and evolved end members (Arndt and Goldstein, 1987) and the distribution of detrital and xenocrystic zircon ages. Volumes of reworked crust are redistributed into appropriate earlier growth intervals.

Model

Because of the extensive reworking of older crust in some areas, only the most intense and widespread orogenic events are used in delineating the distribution of juvenile crust. Most crustal provinces less than 2.5 Ga contain variable amounts of reworked older crust. Early Archean (>3.5 Ga) crustal provinces, although widely distributed, are small (chiefly <500 km across) and may represent remnants of the Earth's early continents.

Late Archean provinces (3.0–2.5 Ga) are widespread on all continents and probably underlie much of the platform sediment (or ice) in Canada, Africa, Antarctica, and Siberia. Paleoproterozoic provinces (2.0–1.7 Ga) are widespread in North America and in the Baltic shield in Europe, and they form less extensive but important orogens in South America, Africa, and Australia. Mesoproterozoic provinces (chiefly 1.3–1.0 Ga) occur on almost all continents, where they form narrow belts along which the Mesoproterozoic supercontinent Rodinia was sutured. Unlike older crustal provinces, little juvenile crust is known of this age. Neoproterozoic provinces (0.8–0.55 Ga) are of importance only in South America, Africa, southern Asia, and perhaps Antarctica.

Isotopic age data have been compiled, and major juvenile crustal provinces are shown on an equal-area map projection of the continents in Plate 5. Aerial distributions indicate that approximately 39% of the continental crust formed in the Archean, 31% in the Paleoproterozoic, 12% in the Meso- and Neoproterozoic, and 18% in the Phanerozoic. Clearly portrayed on the map is the strikingly irregular distribution of juvenile crust with time. Whereas most of Africa, Antarctica, Australia, and the Americas formed in the Archean and Paleoproterozoic, much of Eurasia formed in the Meso- and Neoproterozoic and the Phanerozoic.

Secular Changes in the Crust

Upper Continental Crust

As a constraint on the evolution of the crust–mantle system, it is important to know whether the composition of continental crust has changed with time. Some investigators have used fine-grained terrigenous sediments to monitor changes in upper crustal composition. Justification of this approach relies on the mixing of sediments during erosion and sedimentation such that shale, for instance, reflects the composition of a large geographic region on a continent. Only elements relatively insoluble in natural waters, and thus transferred in bulk to sediments, can be used to estimate crustal composition (such elements as Th, Sc, and REEs). Some studies have compared sediments of different ages from different tectonic settings, thus erroneously identifying secular changes in crustal composition that reflect different tectonic settings (Gibbs et al., 1986). The prime example of this is the comparison of Archean greenstone sediments (dominantly volcanogenic graywackes) with post-Archean cratonic sediments (dominantly cratonic shales) and the interpretation of differences in terms of changes at the Archean–Proterozoic (A–P) boundary (Veizer, 1979; Taylor and McLennan, 1985). Another limitation of sediments in monitoring crustal evolution results from the recycling of older sediments, which leads to a buffering effect through which changes in juvenile crustal composition may not be recognizable (Veizer and Jansen, 1985; McLennan, 1988). For sediment geochemical results to be meaningful, sediments should be grouped by average grain size and by lithologic association, which in turn reflect tectonic setting.

Major Elements

There have been numerous studies of secular changes in major element distributions, most concentrated on detrital sediments. Some investigators have suggested that the proportions of sedimentary rocks have changed with time and are responsible for possible secular changes in major element concentrations (Engel et al., 1974; Schwab, 1978; Ronov et al., 1992). However, all of these studies suffer from the same problem: they indiscriminately lump sediments from different tectonic settings. Thus, although significant changes in lithologic proportions and element distributions have been proposed, their existence is questionable.

Only three major element trends are well documented from the sampling of bedrock in Precambrian shields, described in Chapter 2. Results suggest that Ti increases and Mg decreases and that the K/Na ratio increases near the end of the Archean (Condie, 1993) (Fig. 8.16). The changes in Ti and Mg appear to reflect a decrease in the amount of komatiite and high-Mg basalt in continental sources after the Archean. The increase in the K/Na ratio is caused by an increase in K and a decrease in Na in average post-Archean granitoids, which dominate in the upper crust.

Rare Earth and Related Elements

Again based on bedrock sampling from Precambrian shields, at the end of the Archean, average upper-continental crust increases in P, Nb, and Ta and to some degree in Zr and Hf (Condie, 1993). Also, Rb/Sr and Ba/Sr ratios increase at this time. The Th/Sc ratio shows only a moderate increase at the A–P boundary (Fig. 8.16). Although the sediment

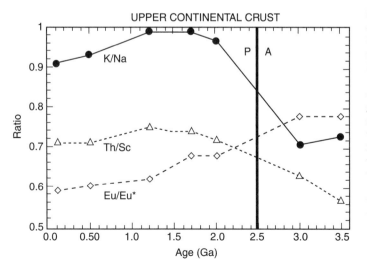

Figure 8.16 Distribution of K/Na, Th/Sc, and Eu/Eu* in the upper continental crust with time. Eu/Eu* is a measure of the Eu anomaly; positive Eu anomalies have values greater than 1, and negative anomalies have values less than 1. $Eu/Eu^* = Eu_n/(Sm_n \times Gd_n)^{0.5}$, where n stands for chondrite-normalized values. Estimated errors for ratios range from 1 to 2 times the size of the plotting symbol. A, Archean; P, Proterozoic. Data from Condie (1993).

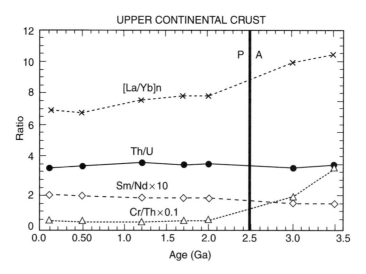

data suggest that Th/U ratio increases with time, there is no evidence for secular variation of this ratio in the shield data (Fig. 8.17).

At the end of the Archean, there is a notable decrease in the fractionation of REEs, as shown by a decrease in the (La/Yb)n ratio and a small increase in the Sm/Nd ratio in the upper crust (Fig. 8.17). Both of these ratios change in opposite directions to those proposed by Taylor and McLennan (1985) based on sediment data. The Sm/Nd ratio of continental shales has remained about constant with time at 0.18 and agrees with the average ratio of upper crust today.

Taylor and McLennan (1985) have long maintained that post-Archean sediments differ from Archean sediments in the presence of a negative Eu anomaly. However, it is difficult to test their conclusion because it is based on comparing largely greenstone sediments from the Archean with cratonic sediments in the post-Archean. Precambrian shield results indicate that both Archean shales and Archean upper crust have sizable negative Eu anomalies (small Eu/Eu* ratios) and that there is only a modest increase at the end of the Archean (Condie, 1993; Gao and Wedepohl, 1995) (Fig. 8.16). This increase reflects the importance of Eu anomalies in felsic igneous rocks and sediments derived there from after the end of the Archean. One contributing factor is that unlike Archean TTG, post-Archean TTG generally has significant negative Eu anomalies. Negative Eu anomalies, however, are not limited to post-Archean rocks, and both Archean shales and granites typically show sizable Eu anomalies.

Thus, although it seems certain that post-Archean upper-continental crust has a larger Eu anomaly than its Archean counterpart, it is clear that (1) both Archean upper crust and Archean sediments have negative Eu anomalies and (2) cratonic shales give only a weak suggestion of the increasing Eu anomaly in upper crust formed after the end of the Archean.

Ni, Co, and Cr

Decreases in Ni, Co, and Cr and in Cr/Th, Co/Th, and Ni/Co ratios are observed at the A–P boundary in both upper continental crust and fine-grained cratonic sediments (Taylor and McLennan, 1985; Condie, 1993) (Figs. 8.17 and 8.18). These changes appear to reflect a decrease in the amounts of both high-Mg basalt and komatiite in continental sources after the end of the Archean. A striking example of this decrease occurs in Precambrian cratonic sediments of the Kaapvaal craton in southern Africa (Condie and Wronkiewicz, 1990). A decrease in the Cr/Th ratio in Kaapvaal sediments near the A–P boundary appears to reflect a decrease of komatiite and high-Mg basalt sources.

Sr Isotopes in Marine Carbonates

The isotopic composition of Sr in seawater is controlled chiefly by the isotopic composition of rivers entering the oceans and the hydrothermal input at ocean ridges (Veizer, 1989). Today, the river influx greatly dominates the control of the seawater $^{87}Sr/^{86}Sr$ ratio. Because of the long residence of Sr in the oceans (~4 My) compared with ocean mixing times (~1000 years), the isotopic composition of modern seawater is uniform (about 0.7099) even in partially landlocked seas such as the Black Sea. Because marine carbonates record the seawater Sr isotopic composition, they have been used to monitor the composition of seawater in the geologic past. When land areas are extensive and elevations are relatively high, weathering and erosion transport large amounts of continental Sr into the oceans, which has relatively high $^{87}Sr/^{86}Sr$ ratios. In contrast, when sea level is high and ocean ridges, mantle plumes, or both are widespread and active, the input of mantle Sr into seawater is enhanced. Many of the long-term changes in the Sr isotopic composition of seawater may be related to the supercontinent cycle and perhaps

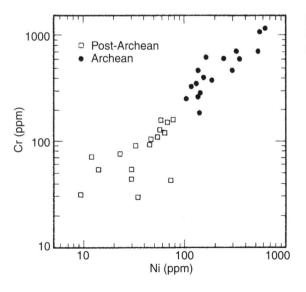

Figure 8.18 Cr-Ni distribution in cratonic fine-grained detrital sediments.

to mantle-plume events (Chapter 9). In principle, investigators should be able to track the formation and destruction of a supercontinent with the Sr isotopic record in seawater: during supercontinent formation, when land area is increasing, the $^{87}Sr/^{86}Sr$ ratio of seawater should increase, and during fragmentation, when ocean-ridge and mantle-plume activity increase, the ratio should decrease.

Although the Sr isotopic record of seawater for the Phanerozoic is relatively well known (Veizer et al., 1999), except for the Neoproterozoic, the record for Precambrian seawater is poorly known (Asmerom et al., 1991; Jacobsen and Kaufman, 1999). Although several orders of variation in the Sr isotopic composition of ancient carbonates is recognized, only the first-order changes that occur over 500 My or more can be resolved in the Precambrian record. Veizer and Compston (1976) were the first to suggest that the growth rate of continental crust can be tracked with Sr isotopes using the $^{87}Sr/^{86}Sr$ ratio in marine carbonates. Because the $^{87}Sr/^{86}Sr$ ratio in marine carbonates reflects chiefly the balance of Sr from continental and deep-sea hydrothermal (mantle) sources entering the oceans, relatively high $^{87}Sr/^{86}Sr$ ratios in marine carbonates indicate sources with high Rb/Sr ratios (continental crust), whereas low ratios reflect dominant mantle input. Before 2.5 Ga, $^{87}Sr/^{86}Sr$ ratios fell near the mantle growth curve (~0.702), indicating that the oceans were largely buffered by mantle input and that little continental crust existed, at least above sea level (Fig. 8.19). After the major period of continental crustal growth at 2.7 Ga, however, the $^{87}Sr/^{86}Sr$ ratio in seawater increased rapidly as increased amounts of continental Sr entered the oceans (dashed line, Fig. 8.19).

Figure 8.19 The record of the $^{87}Sr/^{86}Sr$ ratio in seawater since the late Archean using marine carbonates as a proxy. Also shown is the mantle growth curve. Gray patterns show the distribution of carbonate isotopic ratios. Carbonate Sr isotopic data from Mirota and Veizer (1994).

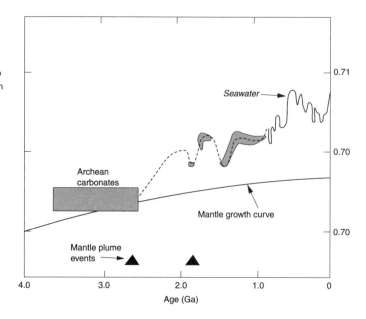

Alkaline Igneous Rocks

Alkaline igneous rocks, such as trachytes, phonolites, basanites, kimberlites, and carbonatites occur on cratons and in some continental rifts and oceanic islands. They do not, however, become important in the geologic record until after 200 Ma. Although alkaline igneous rocks are reported in late Archean greenstones, they are rare, and only a small number of occurrences are known in the Proterozoic and Paleozoic. Several factors probably contribute to a decrease in the proportion of alkaline igneous rocks in the Precambrian. Partly contributing to this distribution, continental alkaline igneous centers are small and they are readily removed during uplift and erosion. Veizer et al. (1992) show, for instance, that the half-life for the loss of carbonatites by erosion is only about 450 My. Hence, they have a low probability of survival from the Precambrian. Alkaline igneous rocks in oceanic islands in accreted terranes of any age are rarely reported, perhaps because they do not survive erosion. Another factor contributing to the rarity of Archean alkaline igneous rocks is the degree of melting in the mantle source. Because small degrees of melting (<5%) of the upper mantle are required to produce these magmas, it is not unexpected that they should be rare in the Archean when mantle temperatures were higher, leading to larger degrees of melting.

Anorthosites

Archean anorthosites often occur in or are associated with greenstone belts, and they typically have megacrystic textures with equidimensional calcic ($>An_{80}$) plagioclase crystals 5 to 30 cm in diameter in a mafic groundmass (Ashwal and Myers, 1994). Some are associated with cumulate chromite. Field, textural, and geochemical relationships show that most are genetically related to greenstone basalts; they appear to represent subvolcanic magma chambers that fed submarine eruptions. Parental magmas are Fe- and Ca-rich tholeiites that may have fractionated from komatiitic magmas.

In contrast, most post-Archean anorthosites are associated with anorogenic granites and syenites and contain much less calcic plagioclase (chiefly An_{40-60}) (Wiebe, 1992). They are generally interlayered with gabbros and norites and exhibit cumulus textures and rhythmic layering. Many bodies, which range from 10^2 to 10^4 km^2 in surface area, are intruded into older granulite-facies terranes, and some are highly fractured. Gravity studies indicate that most bodies are from 2 to 4 km thick and are sheet-like in shape, suggesting that they represent portions of stratiform igneous intrusions. The close association of Proterozoic anorogenic granites and anorthosites in the Grenville province in eastern Canada suggests a genetic relationship between these rock types. Geochemical and isotopic studies, however, indicate that the anorthosites and granites are not derived from the same parent magma by fractional crystallization or from the same source by partial melting. Data are compatible with an origin for the anorthosites as cumulates from fractional crystallization of high-Al_2O_3 tholeiitic magmas produced in the upper mantle (Emslie, 1978; Wiebe, 1992). The granitic magmas appear to be produced by partial melting of lower crustal rocks, the heat coming from associated basaltic magma that gives rise to the anorthosites by fractional crystallization.

It is still not clear why Archean anorthosites appear to have been produced in oceanic, tectonic settings, whereas most post-Archean anorthosites clearly formed in older continental crust.

Ophiolites

As explained in Chapter 3, a complete ophiolite includes (from bottom to top) ultramafic tectonite, layered and nonlayered gabbros and ultramafic rocks, sheeted diabase dykes, and pillow basalts. If you adhere strictly to this definition, the oldest known ophiolites are about 2 Ga, with the possible exception of the Dongwanzi occurrence in China at 2.5 Ga. Although some Archean ophiolites have been described, they lack one or more of the ophiolite components and hence may not be fragments of oceanic crust (Bickle et al., 1994). Again, except for the Dongwanzi body (Kusky et al., 2001), harzburgite tectonites are missing in all rock packages described as Archean ophiolites and only two convincing sheeted-dyke complexes have been described from Archean terranes. One dyke swarm is in the Dongwanzi ophiolite. The other is in the Kam Group near Yellowknife in northwest Canada, in which the dykes have been intruded through older felsic crust and hence are unlike true ophiolite dyke swarms (Helmstaedt and Scott, 1992).

Why are complete ophiolites rare or missing in the Archean? As described in this and other chapters, it seems likely that plate tectonics was operating in the Archean. This leaves two options:

1. Archean ophiolites have been overlooked or not recognized as such.
2. Remnants of Archean oceanic crust do not look like complete post-Archean ophiolites.

Considering the detailed geologic mapping in Archean greenstones, it is becoming less likely that ophiolites have been overlooked. If higher heat production in the Archean mantle produced thicker oceanic crust because of the greater amount of melting beneath ocean ridges, complete ophiolites may not have been preserved by obduction or underplating mechanisms (Moores, 2002). Because ophiolites are generally less than 7 km thick, a tectonic slice off the top of Archean oceanic crust that was 20 km thick may only include the pillow basalt unit, and pillow basalts are prolific in Archean greenstones. If indeed this were the case, researchers are immediately faced with the question of how to distinguish slices of oceanic crust from slices of oceanic plateaus or arcs in Archean greenstone belts. They must be careful in using basalt geochemistry because plume and depleted mantle components may not have been as well established in the Archean mantle and depleted and enriched reservoirs may have been intimately mixed in the upper mantle (Bowring and Housh, 1995). One possible way to distinguish between oceanic crust and oceanic plateaus may be the presence of komatiites. If komatiites require plume sources, as seems likely, only those greenstone successions without komatiites could be candidates for Archean oceanic crust. Certainly renewed effort seems warranted in the hunt for remnants of Archean oceanic crust.

Blueschists

Blueschists are formed in association with subduction and continental collision and reflect burial to high pressures at relatively low temperatures. It has long been recognized that blueschists older than about 1000 Ma are apparently absent in the geologic record (Ernst, 1972). Those with aragonite and jaditic clinopyroxene, which reflect the highest pressures, are confined to arc terranes less than 200 Ma. Three general ideas have been proposed for the absence of pre-1000-Ma blueschists:

1. Steeper geotherms beneath pre-1000-Ma arcs prevented rocks from entering the blueschist pressure–temperature stability field.
2. Uplift of blueschists led to recrystallization of lower pressure mineral assemblages.
3. Erosion removed old blueschists.

It may be that all three of these factors contributed to the absence of pre-1000-Ma blueschists. Before 2 Ga, steeper subduction geotherms may have prevented blueschist formation. After that time, however, when geotherms were not much steeper than at present, the second two factors may have controlled blueschist preservation. Calculated pressure–temperature–time trajectories for blueschists suggest that they may increase in temperature before uplift (see Fig. 2.10 in Chapter 2), resulting in recrystallization of blueschist-facies assemblages to greenschist- or amphibolite-facies assemblages. After 500 My of uplift and erosion, only the latter two assemblages would be expected to survive at the surface. Uplift and erosion after continental collisions may also remove blueschists. Even in young collisional mountain chains such as the Himalayas, only a few minor occurrences of blueschist have not been removed by erosion.

Summary of Chemical Changes at the Archean–Proterozoic Boundary

The compositional changes in the upper continental crust that occurred at or near the A–P boundary can be related to one or more of four evolutionary changes in the Earth's history (Condie, 1993). Three of these are direct consequences of cooling of the mantle with time. First is the **komatiite effect,** which results from relatively large production rates of komatiites and high-Mg basalts in the Archean. Some of these rocks are trapped in the upper continental crust, either tectonically or by underplating with granitoids. Second is the **TTG effect,** resulting from the voluminous production of TTG in the Archean because of partial melting of amphibolite (or eclogite) in which amphibole, garnet, or both are left in the residue. Third is the **subduction effect,** in which, after the Archean, greenstone basalt sources change from unmetasomatized to metasomatized mantle. At the same time, TTG production shifts from the partial melting of wet mafic crust to the fractional crystallization of basaltic magmas. Finally, some changes seem to be related to a **paleoweathering effect,** in which the intensity of chemical weathering decreased after the Archean, as explained in Chapter 6.

Komatiite Effect

The high Mg, Cr, Co, and Ni in both Archean shales and in Archean upper-continental crust argues for the existence of komatiites and high-Mg basalts in Archean crustal sources but not in post-Archean crustal sources (Taylor and McLennan, 1985; Wronkiewicz and Condie, 1989). This effect also strongly contributes to the high Cr/Th, Co/Th, and Ni/Co ratios and low Th/Sc ratio in Archean upper-continental crust and Archean sediments. The sublinear variation in Cr and Ni in post-Archean shales can be explained by mixing felsic and mafic end members (Fig. 8.18). Archean shales do not define a linear array in Ni-Cr space, which may be caused by selective enrichment of Cr in Archean shales either by the selective adsorption in clays during weathering or by the presence of fine-grained detrital chromite in Archean sediments.

As previously described, greater amounts of komatiite and high-Mg basalt in the Archean reflect higher mantle temperatures at that time.

Tonalite-Trondhjemite-Granodiorite Effect

Several important geochemical differences exist between Archean and post-Archean TTG (Martin, 1993; Martin, 1994). As reflected by their large amounts of Na-plagioclase, Archean TTG is high in Na compared with K and Ca (Fig. 8.20). Almost all Archean TTG fall in the tonalite–trondhjemite field and have a limited compositional distribution. On the other hand, post-Archean granitoids are calc-alkaline, ranging in composition from diorite to granite with an abundance of granite–granodiorite. Most striking among the trace elements is the relative depletion in Y and heavy REEs in Archean TTG compared with post-Archean calc-alkaline granitoids. This distinction is particularly well defined on an La/Yb versus Yb plot (Fig. 8.21). The Archean REE patterns are strongly fractionated with low Yb contents. In contrast, post-Archean granitoids show only moderate REE fractionation and a range of Yb values.

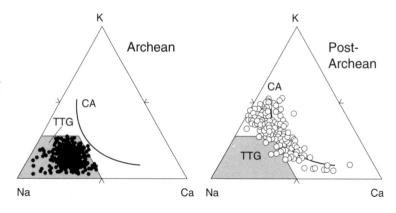

Figure 8.20   Na-Ca diagrams showing the distribution of Archean and post-Archean granitoids. CA, calc-alkaline trend; TTG, tonalite-trondhjemite-granodiorite field. Courtesy of Herve Martin.

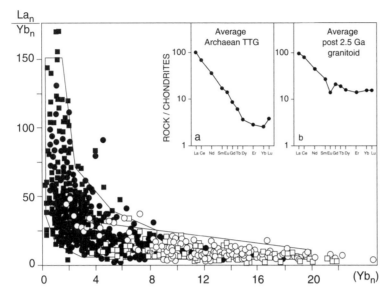

Figure 8.21 La_n/Yb_n ratio versus Yb_n content of granitoids. La_n and Yb_n are values normalized to chondritic meteorites. Solid symbols are Archean and open symbols post-Archean. TTG, tonalite-trondhjemite-granodiorite. Courtesy of Herve Martin.

The relatively low, heavy REE and Y contents of Archean upper-continental crust are well established and reflect heavy REE-depleted TTG, which dominates in the Archean upper-continental crust. Heavy REE depletion in TTG appears to require a mafic source in which amphibole, garnet, or both, which concentrate heavy REEs, remain in the residue after melt extraction (Martin, 1993; Martin, 1994). Low Nb and Ti and low Sm/Nd ratios in Archean upper-continental crust also probably reflect fractionation of REEs by amphibole- or garnet-rich residues.

These differences in REEs and related element behavior in the upper continental crust can be explained if Archean TTG is produced by partial melting of wet, mafic crust with amphibole, garnet, or both left in the restite; most post-Archean felsic magmas are produced by fractional crystallization at shallower depths (Martin, 1994). Thus, garnet and amphibole retain Y and heavy REEs in the source of the Archean magmas but do not play roles in the formation of the post-Archean magmas. In the Archean, subducted oceanic crust was relatively young (<30 My) and warm, and it reached melting conditions during slab dehydration because of steeper subduction geotherms. These conditions may also have existed in the deep root zones of oceanic plateaus or thickened continental crust. Melting occurs in the stability field of garnet and amphibole, and one or both of these phases are left in the restite. In contrast, modern subducted oceanic crust is old and cool, and it dehydrates before melting. In this case, fluids released into the mantle wedge promote melting, leading to the production of basalts that later undergo fractional crystallization to produce felsic magmas. Thus, decreasing geotherms at convergent plate margins may have led to a change in the site of subduction-related magma production from the descending slab, the deep roots, or both of continental crust or oceanic plateaus in the Archean to the mantle wedge thereafter.

Subduction Effect

Taylor and McLennan (1985; 1995) suggested that large ion lithophile (LIL) element enrichment in post-Archean sediments is inherited from granites produced by intracrustal melting following rapid continental growth at the end of the Archean. Although the widespread occurrence of cratonic sediments beginning in the Paleoproterozoic may reflect the rapid growth of cratons in the late Archean, as previously described, the increase in LIL elements in both sediments and average upper-continental crust is not necessarily tied to a changing mass of continental crust. It may reflect the formation of an enriched source for granitoids beginning in the Proterozoic irrespective of continental growth rate. Also, the TTG/granite ratio in the upper continental crust is roughly constant from the early Archean onward (Condie, 1993), but this does not favor an increasing importance of granite in upper crustal sources after the end of the Archean.

Enrichment in LIL elements in felsic igneous rocks and in upper continental crust at the end of the Archean is accompanied by an increase in the size of the Eu anomaly and small decreases in such ratios as K/Rb, Ba/Rb, and La/Th (Condie, 1993). Similar but less pronounced changes in LIL element distributions occur in greenstone-related basalts and andesites at the end of the Archean. The increase in LIL elements in arc-related igneous rocks beginning in the Paleoproterozoic may be related to an increase in depth of devolatilization of descending slabs, which increased the total volume of mantle wedge metasomatized by escaping fluids. Because of the greater volume of metasomatized wedge, post-Archean subduction-related magmas may carry a more prominent subduction-zone geochemical signature (LIL element enrichment relative to Nb and Ta) than Archean subduction-related magmas. Another possibility is that Archean TTG was produced by the partial melting of the unmetasomatized roots of thickened oceanic plateaus, and later the source shifted to metasomatized mantle wedges (Condie, 1992a).

Secular Changes in the Mantle

Although there is much written about the role of plate tectonics and the nature of mantle convection in the early part of the Earth's history, there is no consensus of opinion on these processes. From radiogenic isotopic ratios in young basalts from oceanic regimes, several mantle components have been defined as described in Chapter 4 (Hart et al., 1992; Hofmann, 1997). However, because many of the isotopes, such as Sr, Pb and U, are subject to remobilization during secondary processes, it has proved difficult to extend these definitions to Precambrian basalts. One way around this problem is to use relatively immobile incompatible elements or, better yet, immobile incompatible element ratios (Weaver et al., 1987; Condie, 2003a). Incompatible element ratios also have the advantage that they do not change with time as isotopic ratios do. I suggested that it is possible to characterize at least some isotopic mantle domains with four incompatible element ratios: Nb/Th, Zr/Nb, Zr/Y, and Nb/Y (Condie, 2003a; Condie, 2004). To minimize the effect of the degree of melting, alkali basalts should not be used to identify mantle sources. Although these ratios cannot identify mantle geochemical reservoirs with the resolution of radiogenic isotopes, in some cases they can be used to distinguish plume from nonplume mantle sources, and they may be useful in distinguishing plume-head from plume-tail components. However, they cannot, without additional data, distinguish subduction-derived basalts from plume-derived basalts contaminated with continental crust or subcontinental lithosphere. If age relationships are well known in a given greenstone succession, the Zr/Y–Nb/Y and Nb/Th–Zr/Nb relationships may be useful in tracking plume histories, supercontinent breakup, and mixed deep and shallow depleted mantle sources. They should be used, however, only when stratigraphic and lithologic assemblages are well characterized and then only to supplement and complement other data sources.

Tracking Mantle-plume Sources into the Archean

Using the high field strength element ratios, there are two general populations of late Archean nonarc-related basalts (Fig. 8.22). Most of the data plot in a broad mixing array between enriched and deep depleted mantle end members is suggestive of plume-head sources. Some of the greenstone basalts plot near the enriched component, suggesting interactions with continental crust, subcontinental lithosphere, or both. However, most samples plot in a restricted

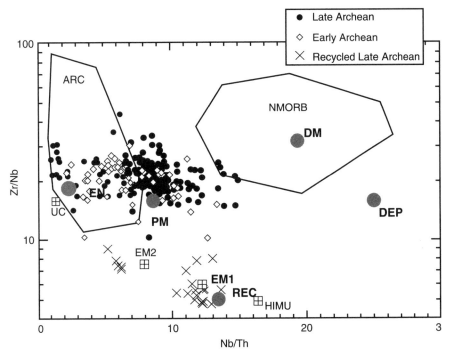

Figure 8.22 Nb/Th-Nb/Zr graph showing the distribution of nonarc-related Archean greenstone basalts. DEP deep depleted mantle end members; DM, shallow depleted mantle source; EM1 and EM2, enriched mantle components; EN, enriched mantle end members; HIMU, high U/Pb ratio; NMORB, normal midocean-ridge basalts; PM, primitive mantle; REC, recycled component; UC, average upper-continental crust. Modified from Condie (2003a).

area near primitive mantle composition, and there is a surprising lack of data that spread toward the deep depleted mantle component. A second population, represented only by the late Archean data, falls near the recycled component, suggestive of affinities to oceanic islands with basalts derived from recycled sources. Perhaps the most striking observation is that there is no clear indication for a shallow depleted mantle source in the Archean.

Three possibilities merit consideration to explain the unusual distribution of the Archean data:

1. The sampling of Archean greenstones is not representative of the range of compositions that existed in mantle sources.
2. The source region in the deep mantle from which Archean plumes were derived was more thoroughly mixed than the region from after the Archean.
3. The major source of Archean plumes contained a significant contribution of "primitive mantle."

Although investigators cannot eliminate the possibility that the Archean geochemical database of nonarc greenstone basalts is not representative, this seems more unlikely with increasing numbers of high-quality analyses of Archean basalts (Condie, 1994; Condie, 2003a; Kerrich et al., 1999; Polat et al., 1999).

The higher temperatures in the Archean mantle could have led to greater degrees of mixing than occurred later, as originally pointed by Blichert-Toft and Albarade (1994) based on radiogenic isotopes. If so, the composition of the Archean mantle may have been more uniform than at later times. However, this effect would be expected to be greatest in the early Archean when mantle heat production was higher than it was in the late Archean. Yet early and late Archean nonarc-related basalts show a similar distribution on the Nb/Th–Zr/Nb diagram.

Could the relatively large number of basalt samples falling near primitive mantle composition reflect significant input of "primitive mantle" to the sources of these basalts? Most or all of these basalts appear to come from Archean mantle plumes; could this reflect a less fractionated state of the deep mantle during the Archean, when "primitive mantle" domains existed in large volumes? Although it is not possible to resolve this question with the current database, one thing is clear from the results: if Archean plumes contain a major input from primitive mantle, this provides an important constraint on convection and fractionation in the Archean mantle.

An even more perplexing problem is the apparent absence of contributions of shallow depleted mantle to any Archean greenstone basalts, be they of arc or nonarc affinity. What does the absence of Archean basalts with shallow depleted mantle sources mean in terms of oceanic crust production and preservation during the Archean? At least four possibilities need to be considered:

1. Ocean-ridge basalts have not yet been analyzed from Archean greenstones.
2. Archean oceanic crust was almost entirely recycled into the mantle, and none is preserved (or at least recognized as yet) in Archean greenstones.
3. Shallow depleted mantle was not preserved in the Archean because of recycling and mixing with enriched components (continental crust and subcontinental lithosphere).

4. Archean ocean-ridge basalts were derived from the same geochemical reservoir from which mantle plumes were derived.

As mentioned previously, explanation 1 is becoming increasingly less likely as the geochemical database of Archean basalts increases. Recycling of all or most Archean oceanic crust into the mantle, explanation 2, also seems improbable considering the probable increase in the production rate of oceanic crust in the Archean because of higher mantle temperatures and considering the thicker and more buoyant oceanic lithosphere in the Archean (Davies, 1992). Both of these factors make it more difficult to subduct Archean oceanic lithosphere, thus increasing the chances of tectonic emplacement of pieces of oceanic crust on the continents. Explanation 3 is possible in that layered convection in the Archean may have led to the recycling of continental crust into the shallow mantle (perhaps as subducted sediments). However, in terms of high field strength element ratios, if this were the case, mixing in the shallow mantle should result in a curvilinear array of data between enriched and shallow depleted mantle end members. Although there is a suggestion of mixing between enriched and deep depleted mantle components in Figure 8.22, there is little if any suggestion of mixing between enriched and shallow depleted mantle components.

Could it be that Archean ocean-ridge basalts were derived from a source similar to that from which Archean mantle plumes were derived rather than from a shallow depleted mantle source? Until 2.7 Ga, when the first large pulse of continental crustal production probably occurred (Condie, 1998), it is possible that there was little if any shallow depleted reservoir in the mantle. Perhaps some of the Archean nonarc-related greenstones were produced at ocean ridges rather than at mantle plumes. If primitive mantle was the main source component in Archean plumes, perhaps it was also the most important component in the upper mantle from which ocean-ridge basalts were derived. Clearly geochemistry, by itself, cannot identify the tectonic setting of Archean basalts. This being the case, investigators will have to use tools besides geochemistry to recognize Archean ocean-ridge basalts.

How Hot Was the Archean Mantle?

Most petrologists agree that the dramatic decrease in the abundance of komatiites at the end of the Archean reflects a decrease in mantle temperature. However, where and how komatiites are produced in the mantle is not agreed upon. Although generally assumed to form by partial melting of the mantle under dry conditions, experimental studies show that if water is present, the liquidus temperature of komatiitic magma is reduced substantially (Grove and Parman, 2004). However, experimental data also show that it is difficult to erupt wet komatiitic lavas because they should erupt explosively and show widespread textural evidence of degassing or they should solidify before eruption (Stone et al., 1997; Arndt et al., 1998). Although pyroclastic komatiites may reflect wet magmatic eruptions, they are rare in the geologic record. Also, the effect of water on liquidus temperature is reduced at the high degrees of melting (30–50%) necessary to produce komatiitic magmas.

In addition, trace element distributions and isotopic characteristics of komatiites indicate that their source was depleted in incompatible elements, including water, before magma formation, leaving a dry source to melt. Cr content of olivine in equilibrium with chromite is also sensitive to melting temperature. Olivines with Cr contents greater than 1550 parts per million (ppm) crystallize at temperatures above 1500° C, whereas those with Cr contents less than 1000 ppm crystallize at temperatures below 1250° C (Li et al., 1995). Komatiites typically have Cr contents greater than 2000 ppm, confirming they are high-temperature rather than low-temperature melts. This is in contrast to low-temperature, high-Mg island arc magmas that have olivine with Cr contents less than 800 ppm.

For these reasons, Arndt et al. (1998) concluded that most komatiites are derived from dry mantle sources; thus, komatiite liquidus temperatures, as deduced from MgO content, can be used to constrain source temperatures in the Archean mantle and to track the temperature of komatiite sources with time (McKenzie and Bickle, 1988; Abbott et al., 1994). Because of alteration, however, it is difficult to estimate the original MgO content of the komatiite magmas. Maximum percentages of MgO range from 29 to 32%, with the lower percentage from the least altered komatiites (Nisbet et al., 1993). The eruptive temperature of a komatiite with 30% MgO is about 1600° C (Fig. 8.23). In comparison, the hottest modern magmas are basaltic komatiites (MgO = 20%) from Gorgona Island, erupted at 1400° C, and typical basalts are not erupted at temperatures greater than 1300° C. If 1600° C is taken as the maximum eruption temperature of komatiites in the Archean, following a mantle adiabat suggests that their mantle source would have a temperature near 2000° C (Fig. 8.23). Because this temperature is greater than that predicted by the secular cooling curve of the mantle, it would appear that komatiites come from hot, rising plumes with temperatures 200 to 300° C hotter than surrounding mantle, similar to the temperature difference calculated between modern plumes and surrounding mantle. Campbell and Griffiths (1992a) have argued that komatiites are a minor component in Archean greenstones and that Archean tholeiites more closely reflect the temperature of the upper mantle. Because most Archean tholeiites have MgO contents of 6 to 10% MgO, similar to modern oceanic basalts, it would seem that most of the late Archean upper mantle was less than 50° C hotter than the modern upper mantle of about 1400° C.

Because garnet is stabilized in the mantle relative to pyroxenes, the solubility of garnet in silicate liquids is reduced, resulting in magmas that have low concentrations of Al_2O_3, similar to many Al-depleted komatiites (Herzberg, 1995). Because CaO varies much less with residual mineralogy, the CaO/Al_2O_3 ratio is strongly pressure dependent and a Al_2O_3-CaO/Al_2O_3 plot can be used to estimate the depth of melting of primary picrite and komatiite magma. As shown in Figure 8.24, high-Mg picrites and komatiites less than 100 Ma have large Al_2O_3 values and CaO/Al_2O_3 ratios near 1. In contrast, late Archean komatiites have intermediate values of these indices, and early Archean komatiites have low Al_2O_3 and high CaO/Al_2O_3. This suggests that young komatiites come from depths of up to 100 km, late Archean komatiites from depths of 150 to 200 km, and early Archean komatiites from depths of 300 to 450 km (Herzberg, 1995).

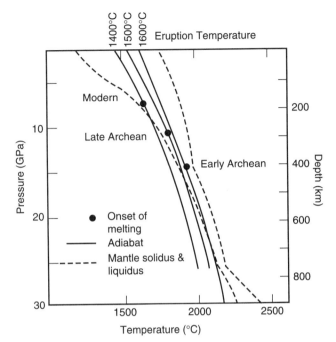

Figure 8.23
Pressure–temperature diagram showing mantle liquidus and solidus and various adiabats for basalt and komatiite.

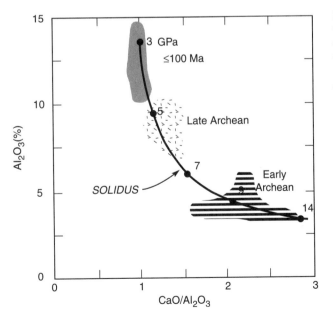

Figure 8.24 Al_2O_3 versus CaO/Al_2O_3 graph showing the mantle solidus as a function of increasing pressure. Pressure given in gigapascals. Also shown is the distribution of komatiite and picrite as a function of age (patterned areas). Modified from Herzberg (1995).

Many petrologists agree that komatiites are produced in the hot, central tail region of mantle plumes, and associated basalts probably come from the cooler plume head (Campbell and Griffiths, 1992b). Each follows a different adiabat (Fig. 8.23). With the compositional-pressure data in Figure 8.24, this suggests that both temperature and depth of magma segregation in plumes has decreased with time, presumably in response to the cooling of the mantle. If this is true, the secular compositional changes in komatiites may be the result of a hotter bulk Earth rather than just hotter plume sources in the Archean.

Mantle Lithosphere Evolution

The continental lithosphere has had a complicated history, perhaps involving more than one growth mechanism and changes in growth mechanisms at the end of the Archean. The post-Archean subcontinental lithosphere may chiefly represent the remnants of "spent" mantle plumes. That is, mantle plumes that rose to the base of the continental lithosphere—partially melted, producing flood basalts—and the restite remained behind as a lithospheric underplate. Seismic reflection results, however, suggest that at least some of the continental lithosphere represents remnants of partially subducted oceanic lithosphere. In northern Scotland, for instance, dipping reflectors in the lower lithosphere are thought to represent fragments of now eclogitic oceanic crust, a relic of pre-Caledonian oceanic subduction (Warner et al., 1996). Isotopic data from xenoliths also indicate that some asthenosphere is added directly to the subcontinental lithosphere.

The relationship of mantle lithosphere to its overlying crust is important in understanding the origin of continents and whether plate tectonics and crustal growth have changed with time. As explained in Chapter 4, average subcontinental lithospheric mantle composition is broadly correlated with the age of overlying crust, suggesting that mantle lithosphere and overlying crust formed approximately the same time (Griffin et al., 1999). In terms of Al and Ca, lithospheric mantle has become progressively less depleted from the Archean to the Phanerozoic (Fig. 8.25). High Ca and Al contents reflect rather primitive, unfractionated mantle, and low values indicate depleted mantle in which significant amounts of melt have been removed. Garnet lherzolite xenoliths from young extensional regimes are chemically similar to primitive mantle, indicating small amounts of melt extraction. In contrast, Archean xenoliths have low Ca and Al, indicating significant amounts melt loss. Although the details of this secular evolution in composition are not well understood, in a general way it probably reflects the cooling of the mantle with time (Griffin et al., 2003). Higher degrees of melting in the Archean led to greater melt production, and as the mantle cooled with time, melt production gradually decreased.

Were Mantle Plumes More Widespread in the Archean?

For greenstone belts that have geochemical or lithologic data available, it would appear that about 35% of late Archean (3.0–2.5 Ga) greenstones have mantle-plume affinities (Condie, 1994; Kerrich et al., 1999; Polat et al., 1999; Hollings and Kerrich, 1999; Tomlinson and Condie, 2001). In contrast, about 80% of early Archean (>3.0 Ga)

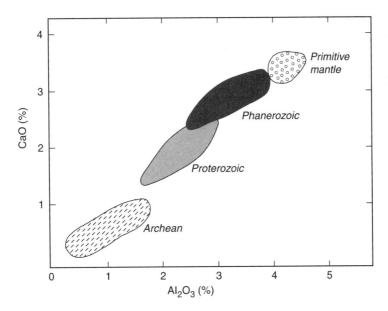

Figure 8.25 $CaO-Al_2O_3$ plot showing the range of mantle lithosphere compositions calculated from garnet xenocrysts. Modified from O'Reilly et al. (2001).

greenstones have plume affinities, although there are relatively few early Archean examples upon which to base this estimate. These proportions of Archean "plume-related" greenstones are considerably greater than those found in post-Archean greenstones (Condie, 1994; Condie, 1997; Condie, 1999).

The observation that plume-related basalts and komatiites are a major component of Archean greenstone belts and appear to be more frequent in the Archean than afterward rests partly on two major assumptions:

1. Archean basalts with oceanic-plateau affinities are truly part of oceanic plateaus; they are not oceanic crust generated at ocean ridges.
2. The relative abundance of greenstone belts with plume affinities is not a relict of preservational bias in the Archean.

As described previously, ocean-ridge and oceanic-plateau basalts may be difficult to distinguish from each other, especially in the Archean when the production rate of oceanic crust was probably greater than it is today.

So where do investigators draw the line between oceanic-plateau and ocean-ridge basalts in greenstone successions? Two observations favor plume affinities for Archean, mafic plain greenstones. First, komatiites, which reflect higher mantle temperatures than basalts and greater melting depths, occur almost exclusively in these types of greenstone belts. If komatiites require a mantle-plume source as most data suggest, then komatiite-bearing greenstone belts should reflect mantle-plume sources. Second, in terms of Nb/Th and other trace element distributions (Fig. 8.22), basalts with ocean-ridge geochemical affinities are not recognized in Archean greenstones.

The bottom line is that with the current level of understanding, mantle plumes may have been more frequent in the Archean than afterward.

Evolution of the Crust–Mantle System

Archean Plate Tectonics

If the Earth has steadily cooled with time, as is widely accepted, there is good reason to suspect that plate tectonics may have been different in the Archean. For instance, a hotter mantle in the Archean would have produced more melt at ocean ridges and hence a thicker oceanic crust. Model calculations indicate that the Archean oceanic crust should have been about 20 km thick (maybe thicker in the early Archean) compared with the present thickness of about 7 km (Sleep and Windley, 1982). Because oceanic crust is less dense than the mantle, the Archean oceanic lithosphere would have been more buoyant and thus more difficult to subduct. A hotter mantle would also convect faster, probably causing plates to move faster; thus, oceanic lithosphere would have less time to thicken and become gravitationally unstable. Paleomagnetic results from 2.7 Ga rocks in the Pilbara craton of Western Australia confirm that late Archean plate motions were considerably faster than those today (Strik et al., 2003). Thicker oceanic crust and faster plate motions would tend to increase the frequency of buoyant subduction in the Archean (de Wit, 1998).

Considering the density and thickness distribution of lithosphere and crust, modern oceanic lithosphere reaches neutral buoyancy in about 20 My (Davies, 1992). In a hotter mantle, it takes oceanic plates longer to become neutrally buoyant; thus in the late Archean, neutral buoyancy would not be reached until the plate is about 80 My (Fig. 8.26). After the lithosphere becomes negatively buoyant, it is not immediately subducted. From the current rate of plate production at ocean ridges (2.5 km²/year) and the total area of seafloor (about 3×10^{14} m²), the mean age of oceanic crust when it *begins* to subduct is about 75 My. On the other hand, hotter mantle plates in the Archean should move faster in response to faster convection, and thus they should subduct sooner. For instance, in the late Archean, they should subduct about 30 My. Hence, although the time currently needed

Figure 8.26 Effect of mantle cooling on the time needed for a plate to reach neutral buoyancy and the time needed to subduct. Modified from Davies (1992).

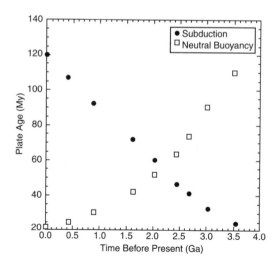

to reach neutral buoyancy is less than the subduction age, as needed for modern-style plate tectonics, in the Archean plates would be ready to subduct *before* they reached neutral buoyancy. The crossover in these two ages is between 2 and 2.5 Ga (Fig. 8.26).

Although it would have been possible for plates to buoyantly subduct in the Archean, they would not have been moving fast enough to remove the excess Archean heat (Davies, 1992; Davies, 1993). How, then, can this excess Archean heat be lost? Although, as described previously, it seems that mantle-plume activity was probably greater in the Archean than afterward, plumes cannot be a substitute for heat loss by plate tectonics because they bring heat into the mantle from the core and only marginally increase heat loss from the top of the mantle. In effect, plates cool the mantle and plumes cool the core. What is required is a mechanism in the lithosphere that promotes heat loss at the surface. At least three possibilities merit consideration:

1. A greater total length of the Archean ocean-ridge system could help to alleviate the problem (Hargraves, 1986). For instance, if Archean heat production was 3 times that of the present heat production, the total ocean-ridge length must have been 27 times the present length to accommodate the additional heat loss.
2. The inversion of basalt to eclogite in buoyantly subducted Archean oceanic crust may have increased the density of the lithosphere sufficiently for it to subduct or delaminate. This mechanism is attractive in that thick Archean oceanic crust provided a large reserve of mafic rocks at depth that could convert to eclogite. Experimental data indicate that conditions were particularly favorable for eclogite formation and delamination from depths of 75 to 100 km (Rapp and Watson, 1995).
3. The oceanic mantle lithosphere may have been negatively buoyant because of latent heat loss associated with the extraction of melt at ocean ridges (Davies, 1993). Two competing factors affect the density of the oceanic mantle lithosphere. The extraction of melt leaves a residue depleted in garnet and thus less dense. On the other hand, the latent heat of melting is carried away by the melt, leaving a colder and denser residue. If the latent heat effect dominated in the Archean, the oceanic lithosphere could have become negatively buoyant.

Toward an Archean Model

Although plate tectonics undoubtedly operated in the Archean, it must have differed in some fundamental aspects from modern plate tectonics to effectively remove the excess heat from the Earth's upper boundary layer (Durrheim and Mooney, 1994). There is an extensive database to constrain mantle and crustal evolution both in the Archean and afterward. Perhaps the single most important factor leading to evolutionary changes is the cooling of the Earth. Most of the changes that occurred near the end of the Archean probably reflected cooling of the mantle through threshold temperatures of various mantle processes, which led to changes not only in how the mantle evolves but also to changes in how the crust and atmosphere–ocean system evolves. Consider a possible model for the Archean that accommodates the constraints described in this chapter.

If fragments of oceanic plateaus formed a significant proportion of Archean greenstones, as seems likely, mantle-plume activity may have been more widespread than at present. Perhaps the first continental nuclei were oceanic plateaus produced from mafic and komatiitic melts extracted from mantle plumes. The plume model provides a means of obtaining komatiites and basalts from the same source: komatiites from the high-temperature plume tail and basalts from the cooler head (Campbell et al., 1989). Spent plume material could have underplated the oceanic plateaus, causing significant thickening of the lithosphere and thus accounting for the thick lithosphere beneath most Archean cratons (Wyman and Kerrich, 2002). This material would have been depleted from melt extraction and thus would have been relatively low in garnet, which would have made it buoyant. Partial melting of the thickened mafic roots of the oceanic plateaus could have created TTG magmas, producing the first differentiated continental nuclei (Fig. 8.27). Buoyant subduction around the perimeters of the plateaus should have resulted in lateral continental growth and thickening of the lithosphere (Abbott and Mooney, 1995). Thin slabs of oceanic crust may have been obducted, remnants of which should have been preserved in some Archean greenstones. Because these slabs were thin, they would not have included complete ophiolites and perhaps not even sheeted dykes. Collision of oceanic plateaus and island arcs also have contributed to the growing continental nuclei, and both of these should have been represented in Archean greenstones. Descending slabs would have sunk because of their negative buoyancy, because of eclogite production in buoyantly subducted oceanic crust, or both (Fig. 8.27). TTG could have been produced in the mafic descending slabs, in the thickened mafic continental roots, or both. Regardless, garnet must have been left in the restite to explain the heavy REE depletion in Archean TTG. A subduction zone geochemical component would have been imparted to the thickening mantle lithosphere by dehydration and partial melting of descending slabs. Some TTG may also have been produced in delaminated fragments of eclogite coming from oceanic crust plastered beneath the continental nuclei by buoyant subduction.

Nonplate Tectonic Models for the Archean

Although the evidence would seem to be overwhelming for a plate-tectonic Earth during the Archean, not all investigators share this interpretation. Before leaving the subject of

Figure 8.27 Model for the formation of Archean continents. TTG, tonalite-trondhjemite-granodiorite.

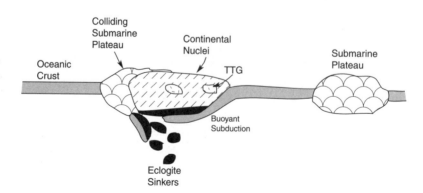

crust–mantle evolution, I need to point out some of the difficulties suggested for plate tectonics during the Archean. Some investigators have suggested that the relative abundance of komatiites and high-Mg basalts in the Archean require different tectonic settings than those found on the Earth today (Hamilton, 1998; Bleeker, 2002). Another difference commonly cited is that many Archean greenstone successions were erupted and deposited unconformably on older felsic crust, unlike young greenstone successions. The alleged absence of passive-margin sequences, ophiolites, blueschists, paired metamorphic belts, accretionary prisms, and melange has also been cited as evidence against an Archean plate-tectonic Earth. With the exception of blueschists and paired metamorphic belts, however, most investigators have recognized these components in Archean greenstones. Hamilton (1998) points out that greenstone belts are anastomosing networks of upright synforms produced by granitic diapers unlike most post-Archean greenstones.

So what kinds of models are proposed to accommodate these differences (alleged and real)? Most of the "nonplate" group has suggested some delamination model for the Archean in which the root zones of continental crust invert to eclogite and sink with associated mantle lithosphere (Zegers and van Keken, 2001). Widespread delamination of the crust and mantle heats and partially melts the lower crust and produces TTG magmas.

No one doubts that there are differences between Archean rocks and post-Archean rocks. The question is whether these are differences of kind or degree. Certainly, the Archean Earth was hotter, accounting for the relative abundance of komatiites, the absence of blueschists, and perhaps the absence of Phanerozoic-type ophiolites. But might not plate tectonics still operate? Most investigators agree that the similarities between Archean granite–greenstone terrains are greater than the differences, and most of these differences can be explained by a combination of (1) a hotter Archean Earth and (2) a lack of preservation in the geologic record.

Further Reading

Bowring, S. A., and Housh, T., 1995. The Earth's early evolution. Science, 269: 1535–1540.

Condie, K. C. (ed.), 1992. Proterozoic Crustal Evolution. Elsevier, Amsterdam, 538 pp.

Condie, K. C. (ed.), 1994. Archean Crustal Evolution. Elsevier, Amsterdam, 542 pp.

Condie, K. C., 2001. Mantle Plumes and Their Record in Earth History. Cambridge University Press, Cambridge, UK, 306 pp.

Davies, G. F., 1999. Dynamic Earth Plates, Plumes, and Mantle Convection. Cambridge University Press, Cambridge, UK, 458 pp.

Rudnick, R. L., 1995. Making continental crust. Nature, 378: 571–577.

Schubert, G., Turcotte, D. L., and Olson, P., 2001. Mantle Convection in the Earth and Planets. Cambridge University Press, Cambridge, UK, 940 pp.

Taylor, S. R., and McLennan, S. M., 1985. The Continental Crust: Its Composition and Evolution. Blackwell Scientific Publications, Oxford, 312 pp.

The Supercontinent Cycle and Mantle-Plume Events

9

Introduction

Supercontinents have aggregated and dispersed several times during geologic history, although our geologic record of supercontinent cycles is only well documented for the last two cycles: Gondwana–Pangea and Rodinia (Hoffman, 1989; Rogers, 1996). It is generally agreed that the supercontinent cycle is closely tied to mantle processes, including both convection and mantle plumes. However, the role that mantle plumes may play in fragmenting supercontinents is still debated.

Condie (1998) and Isley and Abbott (1999) have presented arguments that mantle-plume events have been important throughout the Earth's history and may account for the episodicity of continental growth as described in Chapter 8. Although the meaning of **mantle-plume event** varies in the scientific literature, I shall constrain the term to refer to a short-lived mantle event (≤100 My) during which many mantle plumes bombard the base of the lithosphere. During a mantle-plume event, plume activity may be concentrated in one or more mantle upwellings, as during the mid-Cretaceous mantle-plume event some 100 Ma, when activity was focused mainly in the Pacific mantle upwelling. However, as I have pointed out (Condie, 1998; Condie, 2000), alleged Precambrian mantle-plume events at 2.7 and 1.9 Ga correlate with maxima in worldwide production rate of juvenile crust; thus, these events may not have been confined to one or two mantle upwellings.

One of the first models presented to explain episodic continental growth was that of McCulloch and Bennett (1994). They proposed a nonrecycling model involving three reservoirs: continental crust, depleted mantle, and primitive mantle. It assumes that the volume of depleted mantle increases with time in a stepwise manner, which is linked to major episodes of continental crust formation at 3.6, 2.7, and 1.8 Ga. The isotopic and trace element composition of the upper mantle is buffered by the progressive extraction of continental crust and the increasing size of the depleted mantle reservoir.

Stein and Hofmann (1994) were among the first to advocate that episodic instability at the 660-km seismic discontinuity controls the growth of continental crust. They suggested that convection patterns changed in the mantle from layered convection (the normal case), when the growth rates of continental crust were relatively low, to whole-mantle convection when the growth rates were high. Whole-mantle convection occurs in short-lived episodes during which subducted slabs accumulated at the 660-km discontinuity catastrophically sink into the lower mantle in a manner similar to that proposed by Tackley et al. (1994). One of the important features of the Stein-Hofmann model is that during periods of whole-mantle convection, plumes rise from the D" layer above the core and replenish incompatible elements to the upper mantle, which has been depleted by oceanic crust and arc formation.

Based on the same theme of instability at the 660-km discontinuity and using parameterized mantle convection, Davies (1995) proposed catastrophic global magmatic and tectonic events at a spacing of 1 to 2 Gy. The favored models show layered convection, which becomes unstable and breaks down episodically to whole-mantle convection as in the Stein-Hofmann model. During the catastrophic mantle overturns, hot lower mantle material is transferred to the upper mantle and may be responsible for rapid episodic growth of juvenile crust, as well as for replenishing the upper mantle with incompatible elements.

Peltier et al. (1997) extended thermal constraints to more thoroughly evaluate the catastrophic mantle models. These investigators quantified the physical processes that control the Rayleigh number at the 660-km discontinuity, which in turns controls the frequency of slab avalanches at this discontinuity. They also suggested a correlation between the avalanche events and the supercontinent cycle. Their results imply that slab avalanches occur at a spacing of 400 to 600 My and that they are brought about by the growth of an instability in the thermal boundary layer at the 660-km discontinuity. During and after slab avalanches, a large mantle downwelling is produced directly above the avalanches; this downwelling attracts fragments of continental lithosphere, thus leading to the formation of a supercontinent.

Based on the episodic occurrence of juvenile crust and associated mineral deposits, Barley et al. (1998) proposed a global tectonic cycle beginning in the late Archean with the breakup of a supercontinent. Enhanced magmatism from 2.8 to 2.6 Ga results from a global mantle-plume event. I also proposed (Condie, 1998) a model to explain the episodic growth of juvenile crust based on episodic mantle-plume events, which will be described in more detail later in this chapter.

Supercontinent Cycle

The most detailed and extensive coverage of the supercontinent cycle comes for the recent supercontinent Pangea. Pangea at 200 Ma was centered approximately over the African geoid high (Fig. 4.4a), and the other continents moved from this high during the breakup of Pangea. Because the geoid high contains many of the Earth's hotspots and is characterized by low seismic-wave velocities in the deep mantle, it is probably hotter than average, as explained in Chapter 4. Except for Africa, which still sits over the geoid high, continents seem to be moving toward geoid lows that are also regions with relatively few hotspots and high lower-mantle velocities, all of which point to cooler mantle (Anderson, 1982). These relationships suggest that supercontinents may affect the thermal state of the mantle, with the mantle beneath continents becoming hotter than normal, expanding, and producing the geoid highs (Anderson, 1982; Gurnis, 1988). This is followed by increased mantle-plume activity, which may fragment supercontinents or at least contribute to the dispersal of cratons.

Supercontinent Cycle in the Last 1000 Million Years

In the last 1000 My, three supercontinents have come and gone. The Meso- and Neoproterozoic supercontinent Rodinia formed as continental blocks collided primarily along what today is the Grenville orogen, which extends from Siberia along the coasts of Baltica, Laurentia, and Amazonia into Australia and Antarctica (Hoffman, 1991; Condie, 2002a) (Fig. 2.28). Gondwana formed chiefly between 600 and 500 Ma (Fig. 2.27), and Pangea formed between 450 and 300 Ma (Fig. 2.26).

Rodinia

Although Rodinia appears to have assembled largely between 1100 and 1000 Ma (Fig. 9.1), some collisions, such as those in the northwest Grenville orogen (eastern Canada) and collisions between the South and Western Australia plates (Rivers, 1997; Condie, 2003b; Meert and Torsvik, 2003; Pesonen et al., 2003) began as early as 1300 Ma. Relatively minor collisions between 1000 and 900 Ma, collisions such as Rockall–Amazonia and Yangtze–Cathaysia, added the finishing touches on Rodinia. Paleomagnetic data suggest that with the exception of Amazonia most or all of the cratons in Africa and South America were never part of Rodinia (Kroner and Cordani, 2003). These latter cratons, however, remained relatively close to each other from the Mesoproterozoic onward. Rodinia began to fragment from 800 to 750 Ma with the separation of Australia, east Antarctica, south China, and Siberia from Laurentia. Extensive dyke swarms emplaced at 780 Ma in western Laurentia may record the initial breakup of Rodinia in this area (Harlan et al., 2003). Although most fragmentation occurred between 900 and 700 Ma, the opening of the Iapetus Ocean began about 600 Ma with the separation of Baltica–Laurentia–Amazonia. In addition, small continental blocks, such as Avalonia–Cadomia and several blocks from western Laurentia, were rifted away as recently as 600 to 500 Ma (Condie, 2003b).

As described in Chapter 8, Sr isotopes of marine carbonates, as proxies for seawater, can be useful in tracking the history supercontinents. As an example, consider Rodinia. It would appear that the increase in the Sr isotopic ratio of marine carbonates between 1030 and 900 Ma records the last stages in the formation of Rodinia (Fig. 9.2). The Sr isotopic ratio decreases in seawater from about 0.7074 at 900 Ma to a minimum of 0.706 from 850 to 775 Ma (Jacobsen and Kaufman, 1999). This dramatic decrease

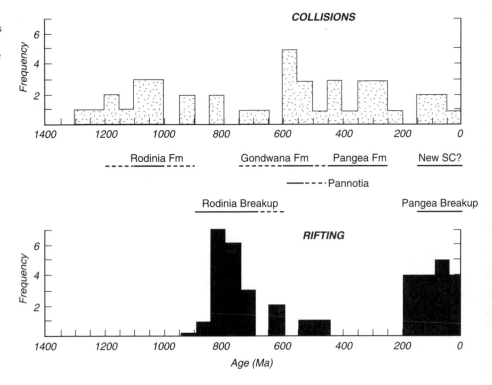

Figure 9.1 Distribution of rifting and collisional ages used in the construction of supercontinent cycles in the last 1 Gy. Fm, formation; SC, supercontinent. Data references in Condie (2002a).

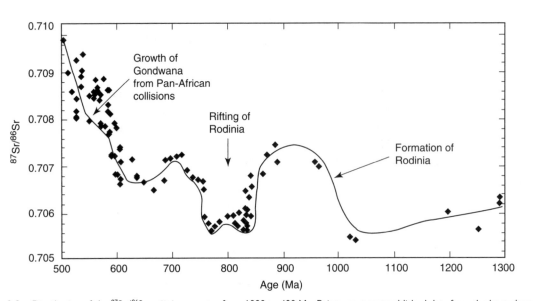

Figure 9.2 Distribution of the $^{87}Sr/^{86}Sr$ ratio in seawater from 1000 to 400 Ma. Points represent published data from the least altered marine limestones. Modified from Condie (2003b).

probably records the breakup of Rodinia with increased input of mantle Sr accompanying the breakup. The minimum is followed by a small but sharp increase in radiogenic Sr, leveling off between about 700 and 600 Ma. This small increase may reflect some of the early plate collisions in the Arabian–Nubian shield and elsewhere. The most significant change in the Sr isotopic ratio of Neoproterozoic seawater occurs between 600 and 500 Ma when the $^{87}Sr/^{86}Sr$ ratio rises to near 0.7095 in only 100 My. This rapid increase corresponds to the Pan-African collisions leading to the formation of Gondwana. As collisions occurred, land areas were elevated and a greater proportion of continental Sr was transported into the oceans.

Gondwana and Pangea

The formation of Gondwana immediately followed the breakup of Rodinia with some overlap in timing between 700 and 600 Ma (Fig. 9.1). The short-lived supercontinent Pannotia, which formed as Baltica, Laurentia, and Siberia briefly collided with Gondwana between 580 and 540 Ma (Dalziel, 1997), assembled and fragmented during the final stages of Gondwana construction.

Pangea began to form about 450 Ma with the Precordillera–Rio de la Plata, Amazonia–Laurentia, and Laurentia–Baltica collisions (Li and Powell, 2001) (Fig. 9.1). It continued to grow by collisions in Asia, of which the last major collision produced the Ural orogen between Baltica and Siberia about 280 Ma. It was not until about 180 Ma that Pangea began to fragment with rifting of the Lhasa and west Burma plates from Gondwana. Major fragmentation occurred between 150 and 100 Ma, with the youngest fragmentation—that is, the rifting of Australia from Antarctica—beginning about 100 Ma. Small plates, such as Arabia (rifted at 25 Ma) and Baja California (rifted at 4 Ma) continue to be rifted from Pangea. Although often overlooked, there are numerous examples of continental plate collisions that paralleled the breakup of Pangea. Among the more important are the China–Mongolia–Asia (150 Ma), west Burma–Southeast Asia (130 Ma), Lhasa–Asia (75 Ma), India–Asia (55 Ma), and Australia–Indonesia (25 Ma) collisions. In addition, numerous small plates collided with the Pacific margins of Asia and North and South America between 150 and 80 Ma (Schermer et al., 1984). These collisions in the last 150 My may represent the beginnings of a new supercontinent (Condie, 1998). If they do, the breakup phase of Pangea and the growth phase of this new supercontinent significantly overlap in time (Fig. 9.1).

During the last 500 My, Sr isotopes in marine carbonates have shown considerable variation (Fig. 9.3). Overall, they parallel the complex Gondwana–Pangea supercontinent history. The minima at 450 and 150 Ma may reflect the fragmentation of Laurasia (Laurentia–Baltica) and Pangea, respectively. The high isotopic ratios in the last 60 My probably reflect the collision of India with Asia and the uplift of the Himalayas (Harris, 1995). Other Sr isotopic peaks in the Paleozoic may reflect continental collisions in the assembly of Pangea, such as the Taconic orogeny in the Ordovician (400 Ma), the Acadian orogeny in the Devonian (360 Ma), the Hercynian orogeny (about 300 Ma), and the collision that produced the Ural Mountains (about 280 Ma).

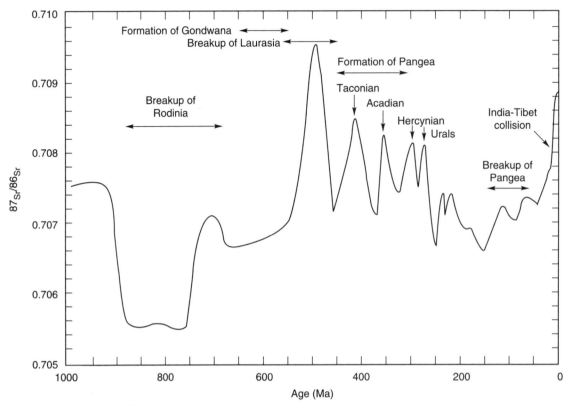

Figure 9.3   Distribution of the $^{87}Sr/^{86}Sr$ ratio in seawater from 1000 Ma to the present based on marine carbonates. The curve is a visual fit of data from Veizer et al. (1999) and references therein.

Juvenile Continental Crust and the Supercontinent Cycle

An outstanding question is whether or not there is a relationship among the episodic growth of continental crust (Fig. 8.11), the supercontinent cycle, and possible mantle-plume events. Does juvenile crust production correlate with the accumulation or breakup phase of supercontinents, or does it occur independently of the supercontinent cycle? Geologic data support the existence of at least two supercontinents before Rodinia—one (or more) at the end of the Archean and one in the early Paleoproterozoic (Hoffman, 1989; Rogers, 1996; Aspler and Chiarenzelli, 1998; Pesonen et al., 2003). In an attempt to more precisely evaluate possible relationships between the supercontinent cycle and the peaks in juvenile crust production, U-Pb zircon ages that reflect either rifting or collisional phases in continental cratons, as well as juvenile crust ages, have been compiled and are summarized in Figure 9.4. Breakup ages include only those ages that have been inter-preted by investigators to have fragmented continental blocks (Condie, 2002a).

Ages from Archean cratons suggest that the first supercontinent (or supercontinents [Aspler and Chiarenzelli, 1998]) formed during the frequent collisions and suturing of older continental blocks and juvenile oceanic terranes (principally arcs and

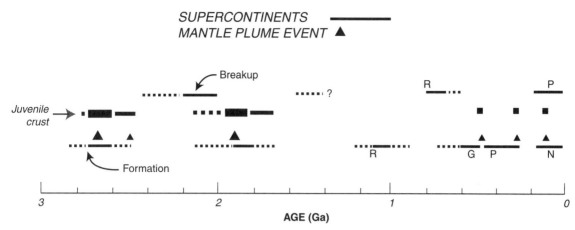

Figure 9.4   Formation and breakup of supercontinents in the last 3.0 Gy. Also shown are times of the maximum production rates of juvenile continental crust and proposed catastrophic mantle-plume events. G, Gondwana; N, new supercontinent; P, Pangea; R, Rodinia. Data from Condie (1998; 2001).

oceanic plateaus) between 2750 and 2650 Ma (Fig. 9.4). In Laurentia, Siberia, and Baltica, collisions were chiefly between 2725 and 2680 Ma, and in Western Australia and southern Africa, most collisional ages fall between 2680 and 2650 Ma. Paleomagnetic data indicate that at least three large supercratons existed at this time (Pesonen et al., 2003). The late Archean peak in juvenile crust production rate is also centered at 2700 ± 50 Ma, thus confirming a strong correlation between supercontinent formation and juvenile continental crust production.

Zircon ages suggest that although the final breakup of late Archean supercratons occurred between 2200 and 2300 Ma, rifting and accompanying dyke swarm injection and mafic magmatic underplating of the continents began at 2450 Ma (Pesonen et al., 2003). Collisional ages, furthermore, indicate the formation of a Paleoproterozoic supercontinent between 1900 and 1800 Ma, with most collisions in Laurentia, Baltica, and Siberia occurring near 1850 Ma (Condie, 2002b). Some collisions began as early as about 2100 Ma (West Africa and Amazonia) and, at least in Laurentia and Baltica, continued until about 1700 Ma. Although the Paleoproterozoic peak in crustal production preceded the collisional peak by 50 My, there is considerable overlap between supercontinent formation and juvenile crust production. In any case, peak crustal production does not correlate with supercontinent fragmentation in pre-1.0 Ga supercontinents.

Mantle Plumes and Supercontinent Breakup

One question not fully understood is the role of mantle plumes in the supercontinent cycle. Are they responsible for fragmenting supercontinents, or do they play a more passive role? Many investigators doubt that mantle convection provides sufficient forces to fragment continental lithosphere and that mantle plumes play an active role

(Storey, 1995). Because plumes have the capacity to generate large quantities of magma, it should be possible to track the role of plumes in continental breakup by the magmas they have left behind as flood basalts and giant dyke swarms.

Gurnis (1988) published a numerical model based on feedback between continental plates and mantle convection, whereby supercontinents insulate the mantle causing the temperature to rise beneath a supercontinent. This results in a mantle upwelling that fragments and disperses the supercontinent. Beginning with a supercontinent with cold downwellings along each side, a hot upwelling generated beneath the supercontinent by its insulation effect fragments the supercontinent (Fig. 9.5a and 9.5b). After the breakup, two smaller continental cratons begin to separate rapidly as the hot upwelling extends to the surface between the two plates, producing a thermal boundary layer (Fig. 9.5b). Both plates rapidly move toward the cool downwellings (vertical arrow, Fig. 9.5c). Approximately 150 My after the breakup, the two continental fragments collide over a downwelling (Fig. 9.5d). Nearly 450 My after the breakup, a new thermal upwelling develops beneath the new supercontinent, and the supercontinent cycle starts over (Fig. 9.5e).

The breakup of Gondwana provides a means of testing the timing of plume magmatism and supercontinent fragmentation (Storey, 1995; Dalziel et al., 2000). The initial rifting stage that began 180 Ma produced a seaway between West (South America and Africa)

Figure 9.5 Computer-generated model of super-continent breakup and formation of a new super-continent. Frame of reference is fixed to the left corner of the diagram, and the right continent moves with respect to the left continent, which is stationary. Modified from Gurnis (1988).

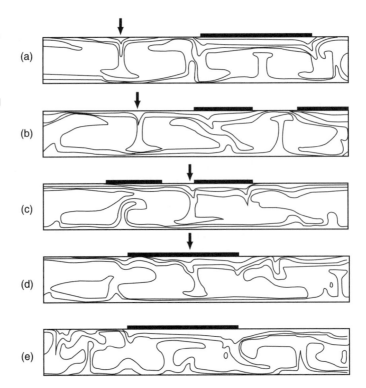

and East Gondwana (Antarctica, India, and Australia) (Fig. 9.6). Seafloor spreading began in the Somali, Mozambique, and Weddell Sea basins by 156 Ma (Fig. 9.6b). Approximately 130 Ma, South America separated from Africa–India and Africa–India separated from Antarctica–Australia (Fig. 9.6c). The breakup was complete by 100 Ma when Australia separated from Antarctica and Madagascar and the Seychelles separated from India as it migrated northward on a collision course with Asia (Fig. 9.6d). Precise isotopic dating suggests that continental separation is closely associated with plume volcanism (Fig. 9.7). In most cases, volcanism begins 3 to 15 My before a breakup; in most instances, such as the Deccan and Parana provinces, the most intense volcanism accompanies initial fragmentation of the supercontinent. The onset of major volcanism in the Deccan Traps is coeval with continental breakup, and intense volcanism continues

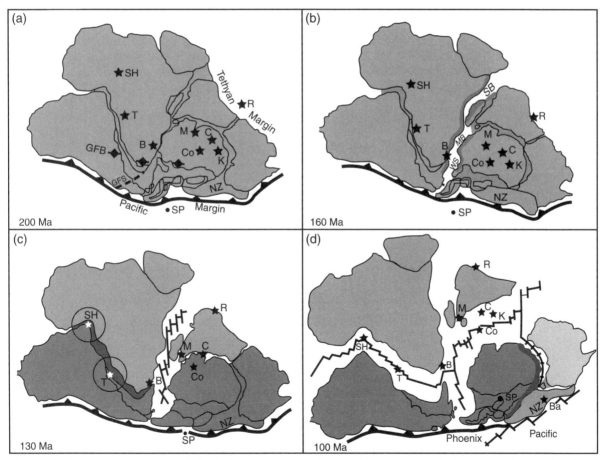

Figure 9.6 Gondwana reconstructions during the last 200 Ma. Shown also are subduction zones (barbed lines), major hotspots (stars), and inferred sizes of plume heads (circles). Ocean ridges are diagrammatic. B, Bouvet; Ba, Balleny; C, Crozet; Co, Conrad; GFB, Gondwana orogen; GFS, Gastre fault system; K, Kerguelen; M, Marion; MB, Mozambique basin; NZ, New Zealand; R, Reunion; SB, Somali basin; SH, St. Helena; SP, South Pole; T, Tristan; WS, proto-Weddell Sea. Modified from Storey (1995).

for more the 20 My. In the case of Iceland, melt production began 60 Ma, followed by extensive rifting at 55 Ma, and the first oceanic crust formed about 53 Ma as Greenland and Norway separated (Larsen et al., 1998). In Afar (Ethiopia), oceanic volcanism has not yet begun in the Afar depression. The time between the onset of flood basalt eruption and the production of oceanic crust ranges from less than 5 My in the Parana and Deccan to 13 My for the Karoo and 25 My for the Central Atlantic province (Fig. 9.7).

The opening of other basins, such as the Red Sea, the Gulf of Aden, the Arabian Sea, and the Indian Ocean, appears to be related to plume volcanism. Except for the Siberian and Emeishan Traps in eastern Asia, all major flood basalt provinces in the last 200 My are associated with the opening a new ocean basin (Coffin and Eldholm, 1994; Courtillot et al., 1999). The location of plume impacts on the lithosphere may not have been random or uniform in the mantle. In some instances, as illustrated by the breakup of Gondwana, plume impacts were centrally located under supercontinents (Fig. 9.6). In all cases cited previously, rifting did not exist before flood basalt eruption or it jumped to a new location at or before a major eruption began. If the plume-head model for flood-basalt magma generation is accepted, basalt eruption, uplift (if any), and rifting are all related to rising plume heads, yet they occur in slightly different time sequences in different areas. Most ocean basins not lined with subduction zones may have been shaped by the episodic impact of large plume heads in the interior or at the edges of continents (Courtillot et al., 1999).

Large Plates and Mantle Upwelling

The insulating properties of large plates, continental or oceanic, result from the lithosphere that inhibits mantle convection currents from reaching the surface of the Earth

Figure 9.7 Timing of supercontinent breakup and plume volcanism associated with several large igneous provinces. Modified from Courtillot et al. (1999).

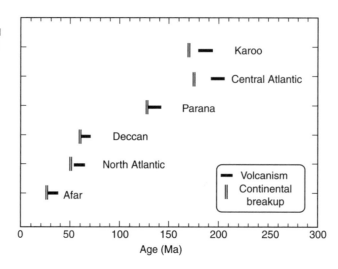

(Gurnis, 1988). An equally, if not more, important effect is that large plates prevent the mantle beneath them from being cooled by subduction. Numerical models show that a large plate becomes increasingly effective as an insulator when its width is much greater than the depth of the convecting layer (Lowman and Jarvis, 1996; Lenardic and Kaula, 1996). The net result of this effect is that large mantle upwellings develop beneath large plates. If a large plate happens to carry a supercontinent, the upwelling may weaken and eventually break up the supercontinent.

The models of Lowman and Jarvis (1999) have been useful in quantifying the relationships among mantle upwelling, supercontinent fragmentation, and whole-mantle versus layered-mantle convection. Their results indicate that supercontinent rifting varies with the mode of mantle heating (basal versus internal radioactive heat sources) and the supercontinent aggregation history. Whether tensile stresses in the interior of super-continents exceed the yield stress of the lithosphere of about 80 MPa depends on the continental aggregation history, the supercontinent size, the Raleigh number of the convecting mantle, the amount of radioactive heating in the mantle, and the viscosity distribution with depth. In the Lowman-Jarvis models, subduction-related forces are at least as important as mantle upwelling in supercontinent breakup. As observed in the breakup of Gondwana, whole-mantle model results predict that plate velocities should be rapid after supercontinent breakup, reducing speed thereafter. For layered-mantle convection, the Lowman-Jarvis models require unreasonably long periods to generate stresses necessary to rift supercontinents (>600 My). Model supercontinents survive for more than 500 My when internal heating in the mantle is 40% or less, but they survive less than 250 My when models have 80% internal heating of the mantle.

Patterns of Cyclicity

The timing of the breakup and dispersal of supercontinents in the last 1000 My does not support a simple supercontinent cycle in which a breakup phase is always followed by a growth phase, the growth phase by a stasis phase, and the stasis phase by another breakup phase (Condie, 2002a). Rather, the data suggest that two types of supercontinent cycles may be operating: (1) a sequential breakup and assembly cycle and (2) a supercontinent assembly cycle only. In the sequential cycle, a supercontinent breaks up over a geoid high (mantle upwelling) (Anderson, 1982; Lowman and Jarvis, 1999) and the pieces move to geoid lows, where they collide and form a new supercontinent partly during but chiefly after supercontinent breakup (Hoffman, 1991). The formation of Rodinia followed by its breakup and then by the assembly of Gondwana is an example of the sequential cycle (Fig. 9.1). Up to a 100-My overlap may occur between each stage of the cycle. The breakup of Pangea, still going on in East Africa, and the possible formation of a new supercontinent with collisions in Southeast Asia seem to overlap in time but nevertheless probably belong to the sequential cycle. The Rodinia–Gondwana cycle from the first breakup of Rodinia to the final aggregation of Gondwana lasted about 300 My (800–500 Ma), and the Pangea–new supercontinent cycle has been in operation for about 200 My.

The second type of supercontinent cycle, which characterizes the growth of Rodinia (1100–1000 Ma) and Pangea (450–300 Ma), appears to involve only the formation of a supercontinent without the fragmentation of another supercontinent. But how can investigators explain such a cycle? Perhaps the answer is that an earlier supercontinent did not fully fragment; thus, the later supercontinent involved relatively few collisions of large, residual continental blocks. In the case of Pangea, Gondwana did not fragment before becoming part of Pangea. Pangea is really the product of continued growth of Gondwana. Thus, Pangea formed from an already existing supercontinent that collided with three large residual fragments left from the breakup of Rodinia (Laurentia, Baltica, and Siberia). In a similar manner, Rodinia may have formed from relatively few residual continental blocks that survived the incomplete breakup of a Paleoproterozoic supercontinent. Condie (2002b) has shown from the distribution of sutures in Rodinia that the predecessor supercontinent did not fully fragment. At least two large fragments, Atlantica (Amazonia, Congo, Rio de la Plata, and West and North Africa) and Arctica (Laurentia, Siberia, Baltica, and north China) survived the breakup of the Paleoproterozoic supercontinent.

This immediately presents the problem of why some supercontinents do not fully fragment. Based on the models of Lowman and Jarvis (1999) and Lowman and Gable (1999) (described previously), supercontinent fragmentation depends on supercontinent size. Small supercontinents do not produce sufficient mantle shielding to be fragmented. Only when supercontinents reach large sizes such as Rodinia and Pangea can they completely fragment. Why should some supercontinents grow to large sizes and others remain relatively small? One possibility is that supercontinent size is related to the geographic distribution of subduction zones over which supercontinent growth is centered. If subduction zones are strung out in a linear, disconnected array rather than grouped in a few closely connected regions on the Earth's surface, a large supercontinent would not form over the subduction zones at one point. Rather, two or three relatively linear supercontinents of smaller size may form, and because these supercontinents do not provide adequate thermal shielding to the underlying mantle, they do not fragment. These survivors later collide to form a new supercontinent; thus, complete breakup of a supercontinent is not required for supercontinent formation in the second type of supercontinent cycle.

First Supercontinent

One of the intriguing yet puzzling questions of any of the episodic models for production of continental crust is that of how and why the first supercontinent formed. There are no robust data that support the existence of a supercontinent before the late Archean, and even then evidence points to several supercratons rather than one large supercontinent (Bleeker, 2003). There are about 35 Archean cratons today, and most or all appear to be rifted fragments of larger landmasses. Bleeker (2003) suggested that these cratons can be grouped into clans based on their degrees of similarity. There are at least three clans, each of which seems to come from a different supercraton. The Slave, Dharwar, Zimbabwe, and Wyoming cratons appear to be fragments of one supercraton that stabilized about 2.6 Ga and broke up between 2.2 and 2.0 Ga. Superior, Rae, Kola, Hearne, and Volga may have been part of a second supercraton, and Kaapvaal and Pilbara may have been part of a third.

For supercratons or supercontinents to form, they require a significant volume of continental crustal fragments that survive recycling into the mantle. Before the late Archean, the high mantle temperatures and inferred large mantle convection rates probably rapidly recycled continental crust, presumably before continental pieces had time to collide to make supercratons or a supercontinent (Armstrong, 1991; Bowring and Housh, 1995). So what happened in the late Archean that led to formation of the first supercratons?

One possibility is that a slab avalanche in the mantle at 2.7 Ga (described later) led to the production of large volumes of continental crust in a relatively short period (≤ 100 My). If this was the case, the first supercratons would form in response to the first major mantle-plume event. Mantle plumes can produce juvenile crust in two ways: directly by the production of oceanic plateaus or indirectly by heating the upper mantle and increasing the production rate of ocean crust because of increased convection rates, increasing the total length of the ocean-ridge system, or increasing both (Larson, 1991a). The increased production rates of oceanic crust are accompanied by increased subduction rates and hence increased rates of production of juvenile continental crust in arc systems. Also contributing to the growth of late Archean supercratons is the thick Archean subcontinental mantle lithosphere that is relatively buoyant (Griffin et al., 1998), thus resisting subduction during plate collisions. As pointed out in Chapter 8, the frequency of late Archean greenstones with oceanic-plateau geochemical affinities supports the idea that oceanic plateaus were a major contributor to the production of late Archean supercratons.

Supercontinents, Mantle Plumes, and Earth Systems

One of the most exciting aspects of the supercontinent cycle and episodic mantle-plume activity is the consequences they may have had in the Earth's history and especially the effects on paleoclimates and the biosphere. In this section, I review various feedback associated with supercontinent formation and breakup and with mantle-plume events that may affect near-surface systems in the planet.

Supercontinent Formation

Supercontinent assembly affects the carbon cycle in several ways (Kerr, 1998; Condie et al., 2000) (Fig. 9.8). Continental collisions are initially a net source of CO_2 because of the burial, thermal destruction, or both of sedimentary organic matter and carbonates within collisional zones (Bickle, 1996) (paths b and c, Fig. 9.8). Continued uplift of a supercontinent accelerates the erosion of sedimentary rocks and their carbon (paths d

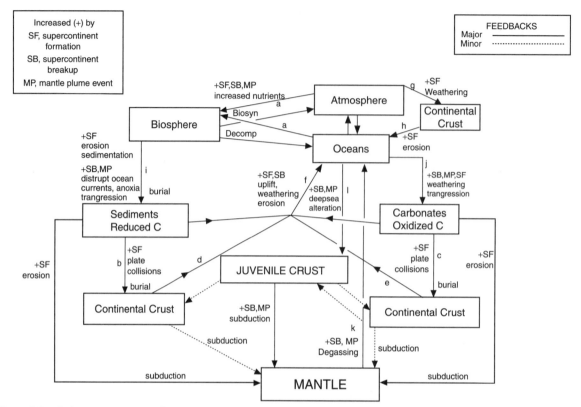

Figure 9.8 Carbon reservoirs in the Earth showing possible effects of supercontinents and mantle plumes. Each box represents a carbon reservoir. Juvenile crust = oceanic crust + oceanic plateaus + island arcs. Numbered paths refer to text descriptions. Biosyn, biosynthesis; decomp, decomposition. Modified from Condie et al. (2000).

and e, Fig. 9.8). Whether this carbon source changes the $\delta^{13}C$ of seawater depends on the ratio of reduced carbon ($\delta^{13}C = -20$ to $-40‰$) to oxidized carbon ($\delta^{13}C = 0‰$) recycled back into the oceans (path f, Fig. 9.8). For example, if both carbonate and organic carbon are recycled in approximately the same ratio as their ratio before supercontinent formation, the $\delta^{13}C$ of seawater will not change (Des Marais et al., 1992) (Chapter 6). As the surface area of a growing supercontinent increases, weathering of surface rocks withdraws more CO_2 from the atmosphere, transferring it to the continents (path g, Fig. 9.8), where it is eventually returned to the oceans by erosion (path h, Fig. 9.8). Increased erosion also releases more nutrients (e.g., phosphorus), increasing biologic productivity (paths h and a, Fig. 9.8). The nutrient source and CO_2 sinks can draw down atmospheric CO_2 levels, favoring cooler climates that intensify ocean circulation and thus increase nutrient upwelling and marine productivity. Intense drawdown of CO_2 with increasing albedo caused by the increasing land/ocean ratio can lead to widespread glaciation. The preceding factors collectively promote increased burial rates of organic carbon, relative to carbonates, and thus may raise the $\delta^{13}C$ value of seawater. However, uplift of collisional mountain belts during supercontinent formation can recycle older carbon depleted in ^{13}C. For instance, a dramatic drop of δ^{13}C in marine carbonates about 55 Ma coincides with the initial uplift of the Himalayas in response to the India–Tibet collision and may reflect recycling of carbon depleted in ^{13}C (Beck et al., 1995). In addition, the final stages in the formation of Pangea were accompanied by compressive stresses around the margin of most of the supercontinent, leading to significant uplift and erosion. Faure et al. (1995) suggested that this enhanced erosion may be responsible for a pronounced minimum in seawater δ^{13}C at 250 Ma. Hence, it appears that the control of δ^{13}C in seawater during supercontinent formation reflects a delicate balance between carbon burial and carbon recycling.

Two ideas have been proposed for a possible role of gas hydrates in global climate change (Kvenvolden, 1999). The first is the direct injection of methane—or more likely its oxidized equivalent, CO_2—into the ocean–atmosphere system as gas hydrates dissolve during warm climatic regimes. This would provide strong, positive feedback for global warming. The second is that continental-margin gas hydrates release methane during the falling sea level, which generally accompanies global cooling. Such cooling, for instance, could occur during glaciation or supercontinent formation. However, with the present reserves of gas hydrates, neither of these effects should have a significant influence on climate change or on sea level (Kvenvolden, 1999; Bratton, 1999). If gas hydrates were widespread during the Precambrian, supercontinent formation could lead to gas hydrate evaporation as the sea level drops, which would introduce biogenic carbon as CO_2 into the atmosphere, increasing both organic and carbonate burial rates and increasing greenhouse warming (Haq, 1998). Also, because gas hydrates contain carbon with negative δ^{13}C values (averaging about –60‰), they may offset any increase in the δ^{13}C because of organic carbon burial.

As the sea level falls during supercontinent formation, the ensuing regression restricts the deposition of shelf carbonates and mature clastic sediments, and the emerging shelves can accommodate the deposition of extensive evaporites. Organic carbon sedimentation occurs farther offshore or in freshwater basins within the interior of the supercontinent (Berner, 1983). Overall, supercontinent formation promotes higher rates of erosion and sedimentation (path i, Fig. 9.8), which correlate with organic carbon burial rates, and platform carbonate deposition becomes more restricted. The result is that periods of supercontinent formation favor relatively high ratios of organic versus carbonate sedimentation and burial. If this is the case, positive carbon isotopic anomalies should develop in seawater during supercontinent formation, if other processes do not obscure this effect.

Supercontinent Breakup

Supercontinent breakup creates new, narrow ocean basins with restricted circulation and hydrothermally active spreading centers (Kerr, 1998; Condie et al., 2000). These features promote anoxia in the deep ocean (path i, Fig. 9.8). The actively eroding escarpments along new rift margins contribute sediments to these basins, and marine transgressions increase the rate of burial of organic and carbonate carbon on stable continental shelves. The amount of shallow marine carbonate deposition (path j, Fig. 9.8), however, critically depends on the redox stratification of the oceans because reducing environments are not conducive to carbonate precipitation. Should anoxic deep-ocean water invade the shelves, it would facilitate organic carbon burial on the shelves, including the deposition of black shale and the accumulation of gas hydrates. This, in turn, should lead to enhanced growth of oxygen in the atmosphere. Perhaps the most striking example of this was the rapid growth of oxygen from 2.2 to 2.3 Ga accompanying the breakup of the late Archean supercratons.

The increase in the length of the ocean-ridge network that accompanies supercontinent fragmentation promotes increased degassing of the mantle, including CO_2 (path k, Fig. 9.8). Increasing atmospheric CO_2 levels and rising sea level promote warmer climates, resulting in increased weathering rates (Berner and Berner, 1997) (path g, Fig. 9.8) as well as the potential for the marine water column to become stratified and for deep water to become anoxic (path i, Fig. 9.8). Increasing carbonate in the oceans with a growing ocean-ridge system would also enhance the rates of the removal of seawater carbonate by deep-sea alteration (path l, Fig. 9.8). To the extent that these developments enhance the fraction of carbon buried as organic matter, they would also lead to an increase in the $\delta^{13}C$ of seawater because ^{12}C is preferentially incorporated into organic carbon (Des Marais et al., 1992; Melezhik et al., 1999).

Mantle-Plume Events

During a mantle-plume event, ascending plumes warm the upper mantle and lithosphere and thereby elevate the seafloor by thermal expansion and create oceanic plateaus by the eruption of large volumes of submarine basalt. Rising sea level triggers marine transgressions (Larson, 1991b) (path i, Fig. 9.8). Oceanic plateaus can locally restrict ocean currents (Kerr, 1998), thus promoting local stratification of the marine water column leading to anoxia (path i, Fig. 9.8). Plume volcanism and associated extensive hydrothermal activity exhale both CO_2 and reduced constituents into the atmosphere–ocean system (Larson, 1991b; Caldeira and Rampino, 1991; Kerr, 1998). The increased CO_2 flux warms the climate and enhances weathering rates (Berner and Berner, 1997) (path g, Fig. 9.8). During mantle-plume events when anoxia is widespread in the oceans, gas hydrates could form in large volumes if the oceans are not warm enough to dissolve the hydrates.

Biologic productivity during a mantle-plume event is enhanced by several factors, such as increased concentrations of CO_2, increased nutrient fluxes from both hydrothermal activity (such as CO_2, CH_4, phosphorus, iron, and trace metals [Sb, As, and Se]) and enhanced weathering, and elevated temperatures because of CO_2-driven greenhouse warming (path a, Fig. 9.8). Studies of modern microbial mats show that the rate of carbon fixation in these organisms is higher for greater levels of CO_2 in the atmosphere (Rothschild and Mancinelli, 1990). Hence, an increase in hydrothermal venting associated with a mantle-plume event could lead to an increase in the biomass, at least in photosynthesizing microorganisms and in organisms that live around hydrothermal vents on the seafloor.

Carbonate precipitation is enhanced by increased chemical weathering and by marine transgressions (path j, Fig. 9.8). Increased hydrothermal activity on the seafloor should also increase the rate of deep-sea alteration, which in turn should increase the removal rate of carbonate from seawater (path l, Fig. 9.8). The liberation of large amounts of SO_2 into the oceans by increased hydrothermal activity might decrease ocean pH, the effect of which would be to dissolve marine carbonates, particularly adjacent to high-temperature emanations (Kerr, 1998). However, a more acid ocean would also dissolve more Na and Ca, increasing the oceanic pH again. If the oceans are relatively reducing, hydrothermal exhalation of Fe^{2+} promotes siderite deposition, for example, adjacent to banded iron formations (BIFs) (Beukes et al., 1990). This, in turn, promotes carbonate deposition overall, because siderite is less soluble than calcite and dolomite. Organic matter burial is enhanced by increased productivity, marine transgression, and the expansion of anoxic waters, in particular onto continental shelves (Larson, 1991b; Kerr, 1998) (path i, Fig. 9.8). In summary, phenomena associated with mantle plumes promote the formation and deposition of both organic and carbonate carbon. It has been proposed that the relative deposition of carbonates and organic carbon reflect redox buffering of the crustal and surface environment by the redox state of the upper mantle (Holland, 1984). Redox buffering by the mantle should be even stronger during mantle-plume events.

The subduction of carbon may also play a role in determining the response of the carbon cycle to mantle-plume events. During most of the Earth's history, the relative rates of the subduction of carbonate and reduced carbon reflected their relative crustal abundances. If they did not, the mean $\delta^{13}C$ values of crustal versus mantle carbon reservoirs would differ substantially today, because the preferential subduction of either oxidized or reduced carbon would have made the crust–mantle exchange of carbon an isotopically selective process. However, the $\delta^{13}C$ values of the total crust and mantle carbon reservoirs are identical within the uncertainties of measurements (Holser et al., 1988). Therefore, subduction has not favored either carbonates or organic carbon.

Mantle-Plume Events Through Time

Isley and Abbott (1999; 2002) and Ernst and Buchan (2003) have used the distribution of komatiites, flood basalts, mafic dyke swarms, and layered mafic intrusions in the geologic record to identify mantle-plume events in the Precambrian. Analysis of the age distribution of giant dyke swarms indicates numerous plume-head events in the last 3.5 Ga, with no plume-free intervals greater than about 200 My (Ernst and Buchan, 2001; Ernst and Buchan, 2003; Prokoph et al., 2004). Although time series analyses of the data clearly show that mantle-plume activity is strongly episodic, the frequency of events depends on the database used. The time series analysis by Isley and Abbott (2002) shows major mantle-plume events from 2.75 to 2.70, 2.45 to 2.40, 1.8 to 1.75, 1.65, and 0.08 to 0.1 Ga and numerous minor or possible events (Fig. 9.9). The 2.75 to 2.70 Ga

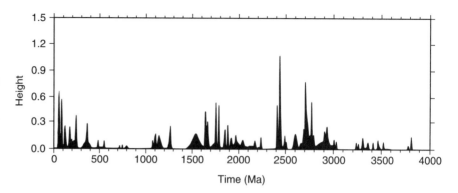

Figure 9.9 Distribution of mantle-plume events deduced from time series analysis of plume proxies from Abbott and Isley (2002). Peak height depends on the number of mantle-plume proxies and the errors of the age, the latter of which is set at 5 My.

events correspond with the peak in juvenile crust formation, and peaks from 1.80 to 1.75 Ga correlate with widespread continental growth in southern Laurentia and southern Baltica. The large 2.43 Ga peak does not correlate with a juvenile crust formation event, yet appears to record widespread mantle-plume activity. The strongest cyclicity of plume events reported by Prokoph et al. (2004) is 170, 650 (730–550), 250 to 220, and 330 My. In addition, there are several short cycles in the last 100 My.

Mid-Cretaceous Event

The effects of a mid-Cretaceous mantle-plume event (Chapter 4) are concentrated chiefly in and around the Pacific basin (Larson, 1991a). The paleotemperature curve for the last 150 My shows a broad increase from 150 to 100 Ma then a steady decrease to the present (Fig. 9.10). The broad temperature peak from 110 to 90 Ma cannot be explained by supercontinent fragmentation alone but also requires excess CO_2 in the atmosphere (Larson, 1991b; Barron et al., 1995). Approximately two to six times the present content of CO_2 is required to raise mid-Cretaceous temperatures to the observed levels (Barron et al., 1995). Mantle plumes in the Pacific basin may have contributed to an increase in CO_2. Models by Caldeira and Rampino (1991) show that a mid-Cretaceous mantle-plume event may have produced atmospheric CO_2 levels 4 to 15 times the modern preindustrial value. This would result in global warming of 3 to 8° C over today's mean temperature. The continuing fragmentation of Pangea could contribute another 5° C of warming. The computer models of Barron et al. (1995) show that a combination of increased atmospheric CO_2 and increased poleward heat flux are necessary to explain the global distribution of warm climates in the mid-Cretaceous. The greater poleward heat flux is necessary to prevent the tropical oceans from overheating because of the increased CO_2 content.

The approximately 125-m increase in eustatic sea level that reached a maximum about 90 Ma (Fig. 9.10) may be related to increased ocean-ridge activity, displacement of seawater by oceanic plateaus, and uplift of the oceanic lithosphere over mantle plumes (Larson, 1991a; Larson, 1991b; Kerr, 1998). It would appear that the effect of a mantle-plume

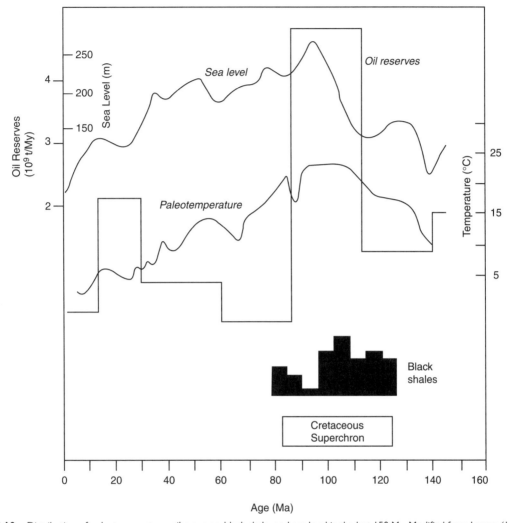

Figure 9.10   Distribution of paleotemperature, oil reserves, black shale, and sea level in the last 150 My. Modified from Larson (1991b).

event is superimposed on that of supercontinent breakup. The gradual decrease in sea level after 90 Ma is not predicted by the mantle-plume model but is probably related to continental collisions associated with the development of a new supercontinent (including India–Tibet, Australia–Southeast Asia, and the Mesozoic terrane collisions in the American Cordillera).

Extensive deposition of black shale is recorded worldwide from about 125 to 80 Ma and may reflect increased CO_2 related to a mid-Cretaceous mantle-plume event (Jenkyns, 1980). Black shales are generally interpreted to result from anoxic events caused by increased organic productivity and poor circulation in basins on continental platforms.

As explained earlier, a mantle-plume event can supply both of these requirements: directly by the hydrothermal spring input of methane into the oceans and indirectly by increasing sea level and the frequency of partially closed basins on the continental shelves (Kerr, 1998). The upwelling of trace metals and nutrients from the deep oceans may have increased the habitat of some marine organisms and may have led to the extinction of other organisms, especially those becoming extinct near the Cenomanian–Turonian boundary about 90 Ma (Wilde et al., 1990). About 60% of the world's oil reserves were generated between 110 and 88 Ma (Irving et al., 1974), consistent with an abundance of black shale. There is also a broad peak in natural gas reserves about the same time (Bois et al., 1980). A mid-Cretaceous mantle-plume event may have contributed carbon and other nutrients, such as phosphorus and iron, for the expansion of phytoplankton. Supporting this prediction are results from Cretaceous shales and marine carbonates, which typically have trace metal contents up to 100 times the background levels (Duncan and Erba, 2003). The high stand of sea level vastly increased the continental shelf area covered with shallow seas, providing appropriate depositional environments for hydrocarbon precursors. Also consistent with a mantle-plume event about 100 Ma is the peak in seawater $\delta^{13}C$, consistent with extensive burial of carbon (Fig. 9.11).

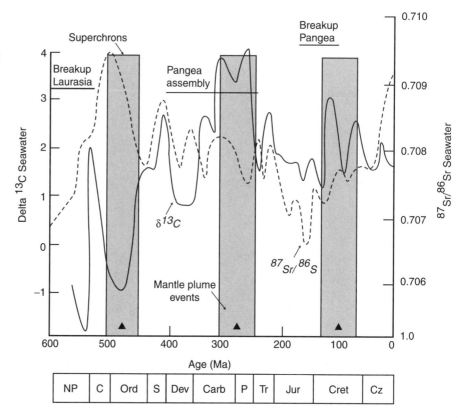

Figure 9.11 Secular changes in $\delta^{13}C$ and $^{87}Sr/^{86}Sr$ ratios of seawater in the last 600 My, based on data from marine carbonates. Also shown are Phanerozoic superchrons. Sr and carbon isotopic data from Veizer et al. (1999).

Late Paleozoic Event

The Permo-Carboniferous reversed superchron centered about 280 Ma (Fig. 9.11) also may record a mantle-plume event. Many geologic features similar to those associated with a mid-Cretaceous mantle-plume event occurred during the Permo-Carboniferous superchron (Larson, 1991b; Tatsumi et al., 2000). Paleoclimates at this time, however, were mixed. Swampy, tropical, and wet climates characterized the Northern Hemisphere, whereas in the Southern Hemisphere, Gondwana underwent widespread glaciation (Crowell, 1999). Also, about 50% of the world's coal reserves formed during the Permo-Carboniferous (Bestougeff, 1980) and large reserves of natural gas appear to have formed at this time. The volume of coal is particularly striking in that coal contains about nine times the fossil fuel energy of oil and gas combined, and it does not migrate from the source rock. There is also a dramatic minimum in the CO_2 level of the atmosphere from 300 to 280 Ma (Fig. 6.14) as calculated from buried carbon, chiefly the burial of vascular land plants (Berner, 1994). Consistent with the increased burial rate of carbon from 300 to 250 Ma is the peak doublet in $\delta^{13}C$ of seawater, which rises to a value of about 4‰ (Veizer et al., 1999) (Fig. 9.11). The rapid decrease in the $^{87}Sr/^{86}Sr$ ratio of seawater during this interval may represent a response to enhanced input of mantle Sr through plume and ocean-ridge volcanism during a mantle-plume event. The minimum $^{87}Sr/^{86}Sr$ ratio about 250 Ma could reflect a combination of factors including waning mantle-plume activity and the final collisions of continental plates forming Pangea, both of which would enhance the continental Sr isotopic signature.

There is also a sea level high centered about 280 Ma, consistent with a mantle-plume event (Fig. 9.12). This high sea level is readily apparent in the Sloss cratonic sequences in North America (Sloss, 1972). The Permo-Carboniferous reversed superchron correlates with the Absaroka transgressive sequence, which began at the same time as the superchron. After the peak about 280 Ma, sea level fell to an all-time minimum at the Permian–Triassic boundary (250 Ma) (Fig. 9.12). Although not fully understood, this fall may have occurred in response to waning plume activity and the final growth of Pangea, each providing positive feedback to a drop in sea level.

Ordovician Event

The recognition of a superchron in the Ordovician centered about 480 Ma presents the possibility of yet another mantle-plume event in the Phanerozoic (Johnson et al., 1995). Consistent with a mantle-plume event at this time is a peak in eustatic sea level (Fig. 9.12). Black shales are also widespread in the Ordovician. A steep fall in the $^{87}Sr/^{86}Sr$ ratio of seawater at this time could be a response to enhanced input of mantle Sr accompanying the plume event (Fig. 9.11). Puzzling, however, is the minimum in $\delta^{13}C$ of seawater (about −1) (Fig. 9.11). Perhaps uplift associated with Taconic collisions in Laurentia–Baltica recycled buried organic carbon, thus enriching seawater in ^{12}C. Alternatively, maybe oxidized carbon was preferentially buried on widespread continental shelves. If so, however, why did these environments favor carbonate deposition and not

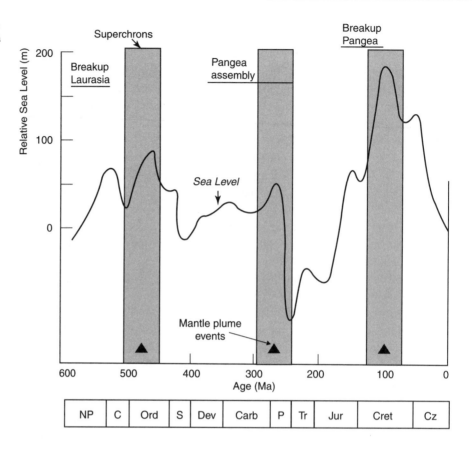

Figure 9.12 Eustatic sea level relative to present sea level over the last 600 My. Data from Haq (1991) and Algeo and Seslavinsky (1995).

organic carbon? The rapid increase in $\delta^{13}C$ toward the end of the superchron leads into late Ordovician–Silurian glaciation, and it is interpreted by Kump et al. (1999a) as a response to increased weathering of carbonate platforms caused by a glacial–eustatic sea level drop (reaching a minimum in the Silurian, Fig. 9.12) rather than as a response to increased burial of organic carbon.

1.9 Ga Event

Sea Level

Although well-preserved shallow marine sedimentary successions are widespread in the Phanerozoic and Neoproterozoic, only small remnants of older Precambrian successions remain in the geologic record. For this reason, it is not possible to use sequence stratigraphy to estimate the sea level in most sedimentary rocks older than about 800 Ma (Eriksson, 1999). Remnants of Meso- and Paleoproterozoic marine intracratonic, passive margin, and platform sediments, however, are widespread on the continents; hence, their aerial distribution as a function of time yields an important insight into the relative elevation of sea level with time. My colleagues and I estimated the abundance of preserved sediments from these tectonic regimes during the Precambrian as a guide to relative sea level (Condie et al., 2000). Using a 700 My half-life for erosion, the restored aerial distribution of these sediments has prominent peaks at 1.9 and 1.7 Ga (Fig. 9.13) and at 600 Ma and a smaller peak about 2.5 Ga (Condie et al., 2000). This suggests that shallow marine sediments were more widespread on the continents at these times than at other times during the Precambrian and, by inference, that sea level was high. It is significant that one of the highest peaks for restored intracratonic sediment occurs at 1.9 Ga, corresponding to a mantle-plume event that I suggested (Condie, 1998).

The peak in the abundance of shallow marine sediments at 1.9 Ga suggests that a 1.9 Ga mantle-plume event "overpowered" supercontinent formation at this time, significantly raising the sea level. This may reflect the relative timing of the two events. Supercontinent formation with many craton and arc collisions from 1.85 to 1.70 Ga occurred on the end of the 1.9 Ga mantle-plume event and may have contributed to the lowering of the sea level following the mantle-plume event.

Also supporting a high sea level about 1.9 Ga is the widespread occurrence of submarine flood basalts of this age erupted on continental platforms. Many examples of such basalts occur in the Ungava orogen in Quebec, the Birimian in West Africa, and the Baltic shield in Scandinavia (Arndt, 1999). This suggests that continental shelves were extensively inundated at 1.9 Ga.

Black Shales

There is a correlation between a 1.9 Ga mantle-plume event and the distribution of black shales during the Precambrian (Condie et al., 2000). A cumulative thickness histogram shows a clear maximum in black shale abundance from 1.9 to 1.7 Ga with smaller peaks about 2.1 Ga and 600 Ma and the suggestion of a peak at 2.7 Ga (Fig. 9.14). The relatively small cumulative thickness of black shales older than 2 Ga may reflect removal by erosion of older successions. A similar distribution is found for the ratio of black shale to total shale with time (Condie et al., 2000). When black shale is plotted as a time series weighted by errors in ages and by the ratio of black shale to total shale, a strong peak occurs at 1.9 Ga with smaller peaks at 1.7, 2.0, and 0.6 Ga (Condie et al., 2000) (Fig. 9.13).

Paleoclimate

The Chemical Index of Alteration (CIA), described in Chapter 6, has been used to estimate the degree of chemical weathering in the source areas of shales. Although CIA data from shales show considerable scatter in some stratigraphic sections, perhaps because of later remobilization of Ca, Na, or K, there is a peak in CIA about 1.9 Ga and another at 1.7 Ga (Condie et al., 2000) (Fig. 9.13). These peaks in CIA suggest that paleoclimates were unusually warm at these times, a feature consistent with an increased input of greenhouse gases (principally CO_2) into the atmosphere. This is to be expected during

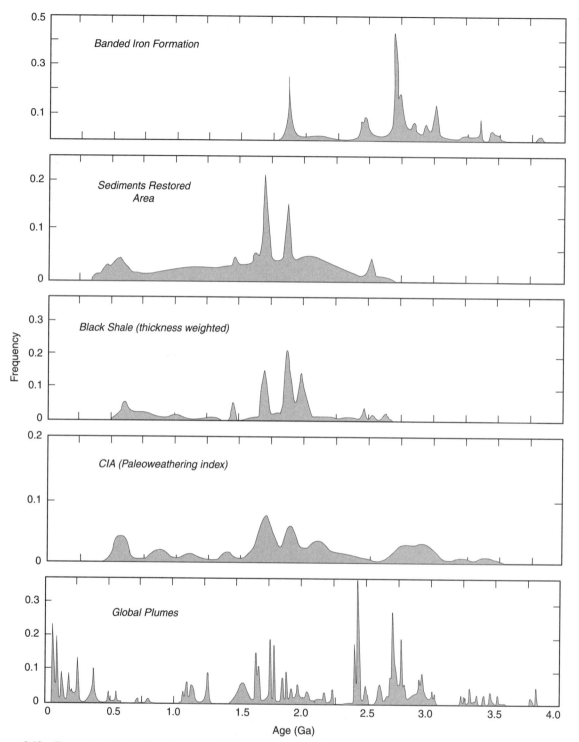

Figure 9.13 Time series of banded iron formations, black shale, Chemical Index of Alteration in intracratonic shales, and restored intracratonic sediments in the Proterozoic compared with global plumes from Isley and Abbott (1999). Time series generated by summing Gaussian distributions of unit area using mean ages and standard deviations given in Condie et al. (2000). CIA = a [$Al_2O_3/(Al_2O_3 + CaO + Na_2O + K_2O) \times 100$] molecular ratio, with CaO representing the silicate fraction only. Modified from Condie et al. (2000).

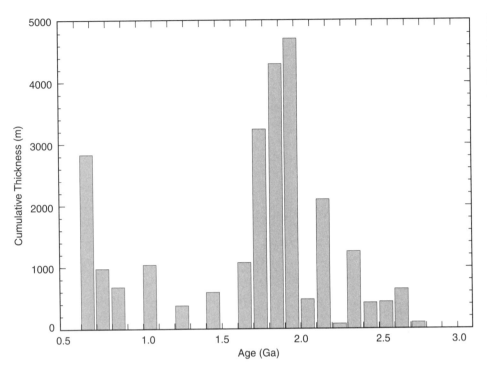

Figure 9.14 Age distribution of black shale in the Precambrian passive margin, intracratonic, and platform successions. Frequency is the cumulative thickness of black shale.

mantle-plume events. The origin and significance of the 1.7 Ga CIA peak is not yet understood. It seems to correlate with intracratonic sediment and black shale peaks about the same time, indicating a warm climatic regime and relatively high sea level at 1.7 Ga. Isley and Abbott (1999) have proposed a mantle-plume event about 1.65 Ga based on the distribution of plume-related igneous rocks in the geologic record (Fig. 9.9).

Banded Iron Formation

The last major period of BIF deposition was about 1.9 Ga when the large BIFs of the Labrador Trough in northern Quebec, the Animikie basin in Minnesota, and the Nabberu basin in Western Australia were deposited (Klein and Beukes, 1992). As shown by Isley and Abbott (1999), this peak in BIF deposition correlates with a 1.9 Ga mantle-plume event and suggests a cause-and-effect relationship (Fig. 9.13). A similar correlation with mantle-plume events has been suggested for the voluminous BIFs at 2.7 and 2.5 Ga (Barley et al., 1997; Isley and Abbott, 1999).

A mantle-plume event can account for several features of BIF deposition. First, the enhanced submarine volcanism and hydrothermal venting associated both with ocean-ridge and oceanic-plateau volcanism during a mantle-plume event may be the source of iron and silica for the BIF. Furthermore, the elevated sea level caused by a mantle-plume event provides extensive shallow marine basins along stable continental platforms necessary to preserve BIFs against later subduction. This applies to either the upwelling or the hydrothermal plume models of deposition. The end of the BIF event at 1.9 Ga may be related to either of those or, more likely, a combination of the following: (1) a decrease in the concentration of ferrous iron in the oceans, resulting from decreasing amounts of submarine hydrothermal activity as a mantle-plume event declines in intensity, or (2) increasing oxygenation of deep-ocean waters, including, perhaps, the introduction of dense plumes of sulfate-rich water.

Sedimentary Phosphates

Although sedimentary phosphates do not become widespread until after about 800 Ma, some important deposits occur in the Paleoproterozoic (Cook and McElhinny, 1979). These are in Australia in the Rum Jungle (1.9 Ga) and Broken Hill (1.9–1.8 Ga) areas, in the Animikie–Gunflint successions in Minnesota (~1.9 Ga), in the Vayrylankyla area of Finland (2.0–1.9 Ga), and in the Yenisey province of Siberia (1.85 Ga) (Cook and McElhinny, 1979; Needham et al., 1988; Rosen et al., 1994; Nutman and Ehlers, 1998) (Fig. 6.19). Most of these phosphates were deposited at or near 1.9 Ga and hence may correlate with a 1.9-Ga mantle-plume event. The source of the increased phosphorus and widespread anoxia in seawater at this time could be from submarine hydrothermal systems associated with a mantle-plume event. This idea also explains the association of some phosphates with BIFs in that dissolved phosphate is strongly absorbed on ferric oxides under aerobic conditions (Berner, 1999).

Stromatolites

Stromatolites (Chapter 7) are widespread in the Proterozoic with a prominent peak (or peaks) in distribution from about 1.9 to 1.8 Ga. Maxima at this time are found in the number of stromatolite occurrences, the diversity of stromatolites, and the number of occurrences of microdigitate stromatolites (Grotzinger and Kasting, 1993; Hofmann, 1998) (Fig. 9.15 and Fig. 9.16b). In addition to stromatolites, there are maxima in the reported occurrences of microfossils, oncoids, and chemofossils (biogenic chemical remains) about the same time (Hofmann, 1998). The peaks in abundance and diversity of stromatolites about 1.9 Ga may reflect a combination of global warming, high sea level stand, and enhanced input of CO_2 into the sedimentary cycle. All of these may be related to a possible mantle-plume event at 1.9 Ga. Grotzinger and Knoll (1999) suggested that the degree of carbonate saturation in seawater is important in controlling stromatolite diversity. During mantle-plume events, seawater carbonate saturation may increase significantly, increasing both the availability of carbonate and the proportion of shallow platforms for the deposition and preservation of carbonates. Also, high sea level stands create widespread shallow tidal flats in which both Ca^{+2} and HCO_3^{-1} ions increase in concentration in seawater because of evaporation.

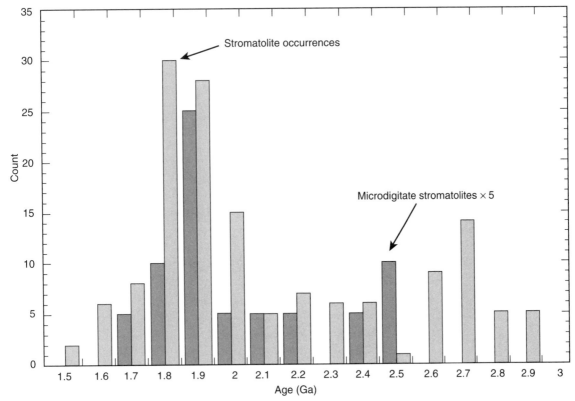

Figure 9.15   Distribution of the reported number of occurrences of total stromatolites and of microdigitate stromatolites during the Precambrian. Data from Grotzinger and Kasting (1993) and Hofmann (1998).

Massive Sulfate Evaporites

The first massive sulfate evaporites in the geologic record occurred from 1.8 to 1.6 Ga following a possible 1.9 Ga mantle-plume event (Grotzinger and Kasting, 1993). Widespread sulfate deposition follows a possible mantle-plume event, which is consistent with the following sequence of events:

1. Large amounts of sulfur were injected into the oceans as sulfide during the mantle-plume event, but only some of the sulfur was deposited as iron sulfides on the deep seafloor (Canfield, 1998).
2. The oceans became oxic as submarine volcanic input related to the plume event subsided.
3. As carbonate levels decreased in seawater because of falling CO_2 input from plume-generated volcanism, marine carbonate deposition became less important and Ca^{+2} ion became available to precipitate as sulfates (Grotzinger and Kasting, 1993).

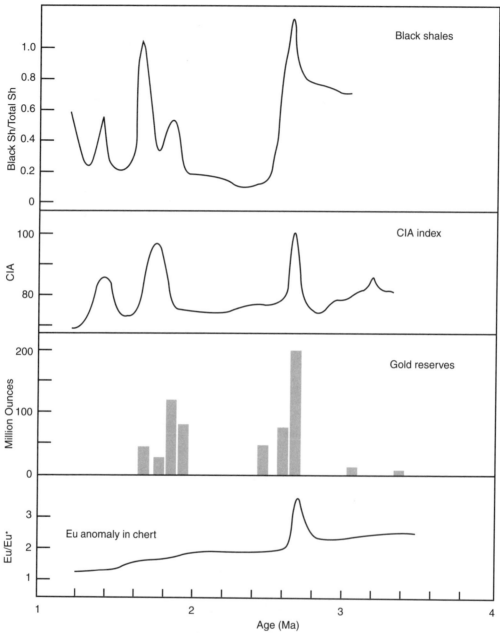

Figure 9.16(a). Geologic changes that may correlate with a possible mantle-plume event at 2.7 Ga. CIA, Chemical Index of Alteration.

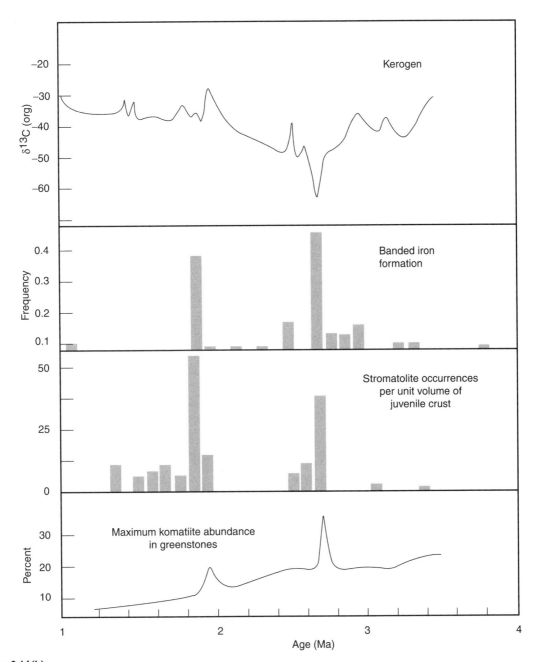

Figure 9.16(b).

Increasing levels of sulfate in the oceans at this time is also supported by sulfur isotopes in which the range of $\delta^{34}S$ increased (less than –20 to more than +20‰) after 2.2 Ga (Canfield, 1998). Also, the completion of a supercontinent may have provided numerous partially closed basins for evaporite deposition, as it did on Pangea during the Permian and Triassic.

Carbon and Sulfur Isotopes

The effect of a mantle-plume event on the biogeochemical cycles of carbon and sulfur can be explored by considering their isotopic records. As described in Chapter 6, these cycles consist of elemental reservoirs linked by processes that either transport or chemically transform the elements. Both elements are rapidly cycled through the biosphere, which converts a small amount to reduced species by using reducing power provided by weathering, thermal sources, and oxygenic photosynthesis. The amounts of organic carbon, carbonate, sulfides, and sulfate buried depend upon both the elemental fluxes through the surface environment and the burial rates in a range of sedimentary environments that favor the sedimentation of either oxidized or reduced species. The burial of both reduced carbon and sulfides is favored by reducing marine environments (Berner, 1983).

The net effect of mantle plumes and supercontinent formation at 1.9 Ga may have been to introduce significant volumes of CO_2 into the atmosphere–ocean system, increasing depositional rates of both organic and carbonate carbon (Condie et al., 2000). Although data suggest that a major, positive carbon isotopic excursion occurs in marine carbonates about 2.2 Ga (Karhu and Holland, 1996; Melezhik et al., 1999), no anomaly is observed in marine carbonates at 1.9 Ga (Fig. 6.9) Therefore, the relative burial rates of both reduced and oxidized carbon changed little at this time even as their absolute rates increased. The absence of a significant 1.9 Ga carbon isotopic anomaly also means that the burial of plume-related carbon masked the burial of supercontinent-related carbon, the latter of which should favor the burial of organic carbon. The breakup of the late Archean supercratons from 2.2 to 2.3 Ga, however, may have led to the enhanced burial of organic carbon, thus causing rapid growth of oxygen in the atmosphere at this time. After 2.2 Ga, the carbon isotopic record indicates that the fraction of carbon buried as organic matter dropped from as much as 50% of the global carbon flux to less than 20% (Karhu and Holland, 1996). After 2 Ga, the fraction of carbon buried as organic matter remained remarkably constant for hundreds of millions of years (Fig. 6.9).

The range of sulfide $\delta^{34}S$ increased beginning about 2.1 Ga and is consistent with an increase in seawater sulfate levels (Knoll and Canfield, 1998; Condie et al., 2000) (Fig. 6.11). After about 2 Ga, sulfide $\delta^{34}S$ became more negative, indicating that a substantial sulfate reservoir existed and that the burial rate of sulfate increased relative to the burial rate of sulfide during that interval. This isotopic trend is consistent with the observed peak in the distribution of sulfate evaporites between 1.8 and 1.6 Ga.

The carbon and sulfur isotopic trends in Paleoproterozoic sediments are consistent with a high sea level stand at 1.9 Ga (Condie et al., 2000). This is followed by a sea level decline and a decline in platform sedimentation in favor of shallow-to-emergent coastal environments that accumulated more oxidized, evaporitic sulfates. This view is corroborated by the decline, from 1.9 to 1.7 Ga, in stromatolite diversity, in the abundance of preserved BIFs and black shale, and in $\delta^{13}C_{org}$ sulfide $\delta^{34}S$. The trend toward greater rates of sulfate deposition between 1.8 and 1.6 Ga is indicated by increased abundance of platform sulfate-rich evaporites and generally lower sulfide $\delta^{34}S$.

Case for a 1.9 Ga Mantle-Plume Event

Consistent with a mantle-plume event at 1.9 Ga are the following: (1) a high sea level as inferred from the relative abundance of intracratonic, passive margin, and platform sediments; (2) a relatively high abundance of black shale; (3) a peak in CIA in shale, implying unusually warm paleoclimates; (4) a peak in the abundance of BIFs; (5) an abundance of shallow marine phosphate deposits; and (6) a peak in the number of reported occurrences of and in the diversity of stromatolites.

The peak in the distribution of shallow marine sediments at 1.9 Ga suggests that the mantle-plume event overpowered supercontinent formation, significantly raising sea level. The black shale and CIA peaks may reflect the introduction of massive amounts of plume-derived CO_2 into the atmosphere–ocean system, increasing depositional rates of carbon and increasing global warming. Increased black shale deposition at 1.9 Ga is primarily caused by anoxia driven onto stable continental shelves. The peak in CIA suggests that paleoclimates were warm at 1.9 Ga, a feature consistent with increased input of greenhouse gases (CH_4 and CO_2) into the atmosphere.

Increased BIFs and marine phosphate deposition at 1.9 Ga reflect increased input of iron and phosphorus into deep anoxic oceans by submarine hydrothermal activity followed by upwelling (or spreading of hydrothermal plumes) into shallow, oxidizing seas on continental shelves. A peak in reported occurrences and diversity of stromatolites about 1.9 Ga required seawater on continental shelves that was greatly oversaturated in $CaCO_3$ and warm climates, both consistent with high sea level stand and enhanced CO_2 input into the oceans from submarine volcanism and hydrothermal vents accompanying the alleged mantle-plume event. The first massive sulfate evaporites in the geologic record from 1.8 to 1.6 Ga follow the 1.9 Ga mantle-plume event, reflecting the oxidizing conditions and the greater availability of Ca^{+2} ions as carbonate deposition declined during mantle-plume waning.

2.7 Ga Event

Based on the distribution of plume-generated mafic igneous rocks, Isley and Abbott (1999) proposed two mantle-plume events in the late Archean from about 2.45 and 2.7 Ga (Fig. 9.9). There is a correlation between the mantle-plume event proposed Isley and Abbott (1999) and myself (Condie, 1998) at 2.7 Ga and the peaks in the abundance of BIFs, CIA distribution in shale, Eu anomalies in hydrothermal chert, komatiite abundance, gold production, black shale abundance, stromatolite occurrences, and negative excursion in $\delta^{13}C$ in organic matter (Fig. 9.16). The peak in BIF abundance may reflect a source of iron from hydrothermal vents associated with plume magmatism. Chemical sediments, such as chert and BIFs, may retain the rare earth element (REE) patterns (including Eu anomalies) of the seawater from which they were deposited, making it possible to track the composition of seawater. REEs deposited near hydrothermal springs on the seafloor have positive Eu anomalies, which reflect the temperature of the water. The peak in Eu in chert at 2.7 Ga (Fig. 9.16a) may reflect higher temperatures related to hotter magma systems associated with widespread mantle-plume activity. The peak in gold production at 2.7 Ga is probably related to the peak in juvenile crust production at this time. The peak in relative komatiite abundance in greenstones at 2.7 Ga would appear to reflect widespread mantle plumes at this time (Fig. 9.16b).

The large volumes of CO_2 introduced into the atmosphere–ocean system during this event may have increased the depositional rates of both organic and carbonate carbon and may have caused global warming, accounting for the CIA and black shale maxima (Fig. 9.16a). The negative $\delta^{13}C$ excursion in organic carbon at 2.7 Ga (Fig. 9.16b) may reflect either enhanced methanogen activity or extensive methane hydrate destabilization (Hinrichs, 2002), both of which could be a consequence of a mantle-plume event. As at 1.9 Ga, the number of occurrences of marine stromatolites shows a peak at 2.7 Ga (Hofmann, 1998), again perhaps recording enhanced input of CH_4 and CO_2 into seawater. Decreasing amounts of CO_2 pumped into the atmosphere by plume magmas, negative feedback of continental weathering, and increasing albedo caused by the newly formed late Archean supercratons appear to have decreased atmospheric CO_2 levels sufficiently to cool worldwide climates after the 2.7 Ga mantle-plume event. This may have led to widespread glaciation from 2.4 to 2.3 Ga (Young, 1991).

Other Possible Mantle-Plume Events

Time series analysis of mantle-plume proxies suggests a major mantle-plume event about 2.45 Ga (Fig. 9.9). Sea level appears to have been high on both the Kaapvaal and Pilbara cratons at this time, consistent with a mantle-plume event (Eriksson et al., 1999; Eriksson et al., 2002). Also supporting a plume event at this time are the large volumes of BIFs deposited on both cratons (Barley et al., 1997; Isley and Abbott, 1999). If the production of significant volumes of juvenile continental crust is a necessary consequence of a major mantle-plume event, as I have suggested (Condie, 1998), the 2.45 Ga event probably does not qualify as a major event because the volume of known juvenile crust of this age is relatively minor. Perhaps the 2.45 Ga event is a subsidiary event associated with the 2.7 Ga mantle-plume event during which large volumes of juvenile continental crust were produced.

Although peaks in black shale abundance occur about 2.2 Ga and 600 Ma (Fig. 9.14), neither of these times has been recognized as a mantle-plume event (Condie, 1998; Isley and Abbott, 1999). Both of these peaks, however, correlate with breakups of supercontinents, the late Archean supercratons from 2.2 to 2.0 Ga and Rodinia from 800 to 600 Ma (Condie, 1998). It is possible that the increased burial of organic carbon in black shales at these times is the result of supercontinent breakup. Particularly important may have been the formation of numerous partially closed basins leading to widespread anoxic environments (Kerr, 1998). Supporting the supercontinent breakup model, both of these times record positive $\delta^{13}C$ anomalies in marine carbonates (Fig. 6.9), which reflect enhanced burial rates of carbon (Karhu and Holland, 1996; Des Marais, 1997; Kaufman, 1997). The appearance of multicellular organisms in the late Neoproterozoic again suggests increasing oxygen levels in the atmosphere (Knoll and Canfield, 1998). Hence, the 600-Ma black shale peaks may correlate with increasing diversification of oxygen-dependent biota in response to oxygen liberated into the atmosphere–ocean system.

Why is there no evidence of growth in atmospheric oxygen at 1.9 Ga, when even greater amounts of organic carbon appear to have been buried? Possible factors contributing to minimal atmospheric oxygen input at this time include the following:

1. An increase in total surface area exposed to weathering as the Paleoproterozoic supercontinent grew may have enhanced the removal of oxygen in weathering products (including oxidation of recycled organic carbon).
2. Oxidation of reduced volcanic gases emitted by widespread submarine hydrothermal vents associated with a mantle-plume event could consume free oxygen in the oceans.
3. Because geologic indicators of atmospheric oxygen level are not sensitive to increases when oxygen levels that are 10% of the present atmospheric level are reached, increases in the oxygen level at 1.9 Ga may not be recognized in the geologic record.

There is a broad peak in CIA from 700 to 600 Ma (Fig. 9.13), suggesting widespread warm climates at this time. Perhaps increased ocean-ridge volcanism resulting from the growth of the worldwide ocean-ridge system as Rodinia broke up put significant amounts of CO_2 into the atmosphere–ocean system, which in turn, caused greenhouse warming and led to an overall increase in the intensity of rock weathering. Marine transgressions during breakup of this supercontinent also may have contributed to global warming.

What Causes a Mantle-Plume Event?

Superchrons and Mantle Plumes

Coincident with a mid-Cretaceous mantle-plume event from 120 to 80 Ma is the Cretaceous **superchron,** when few magnetic reversals occurred for a period of 30 to 40 My (Figs. 9.10–9.12). Larson and Olson (1991) show that magnetic reversal frequency correlated inversely with inferred mantle-plume activity for the past 150 My. To explain this correlation, these authors proposed a model whereby the removal of large amounts of heat from the core not only fuels the plume event but also stops, or greatly retards, the frequency of magnetic reversals in the magnetic field. They suggested that as mantle plumes rise from the D″ layer, this layer thins, cooling the core by allowing heat to be conducted more rapidly across the core–mantle boundary. Outer core convection then increases to restore the abnormal heat loss, causing a decrease in the frequency of magnetic reversals. In effect, this switches the Earth's geodynamo from a reversing alternating current state to a nonreversing direct current state. When the convective activity increases above some threshold value, a magnetic superchron is initiated.

Although this model has been widely accepted, other explanations for superchrons exist. Loper (1992) suggested that the thermal inertia of the D″ layer may not rapidly change the core's heat flux when mantle plumes are formed. Cox (1969) proposed a model for magnetic field reversals caused by perturbations of the outer core, which feed sufficient reverse flux into the geodynamo to overcome the main dipole. In this model, increased convective vigor generates more frequent instabilities in the fluid outer core, leading to an increase in reversal rate. Courtillot and Besse (1987) suggested a different model whereby the decrease in reversal rate just before the mid-Cretaceous superchron was caused by a decrease in core convection because of a build up of heat at the core–mantle boundary. This excess heat was released with the plumes that rose from D″ near the end of the superchron. The loss of heat increased core convection, which in turn increased reversal rates.

Using the Global Paleomagnetic Database, Johnson et al. (1995) identified two other superchrons in the last 600 My. One, the Permo-Carboniferous reversed superchron, is well established; the other, in the early Ordovician, may be a previously unrecognized superchron (Figs. 9.11 and 9.12). Each of the Phanerozoic superchrons had a 30 to 50 My duration and a spacing of 200 My. If each superchron correlated with a mantle-plume event, the spacing between events was considerably less than the spacing between the late Archean and the Paleoproterozoic mantle-plume events. Data suggest a period of enhanced production of juvenile crust in the late Paleozoic, at least in Asia (Sengor et al., 1993). In my mantle-plume model (Condie, 1998), the three Phanerozoic superchrons may have reflected mantle-plume events, although ones less intense than their counterparts in the Precambrian. As previously mentioned, the decrease in intensity results from cooling of the mantle and consequent weakening of the 660-km discontinuity, which would result in more frequent and less intense slab avalanches (described in the next section) in agreement with observations. An interesting test of this idea is to check for magnetic reversals in rock successions that formed during the 2.7 and 1.9 Ga mantle-plume events. If the model is feasible, superchrons should have occurred near each of these peaks. Although single magnetic reversals have been identified in igneous rocks formed at or near the 1.9 Ga peak (Morgan, 1985; Zhai et al., 1994; Buchan et al., 1996), multiple reversals in single stratigraphic successions at or near these peaks have not been resolved.

Another curious feature associated with the Cretaceous superchron was an abrupt tilting of about 20 degrees in the Earth's rotational axis (Prevot et al., 2000). This observation is based on high-quality paleomagnetic data. The tilting, which culminated at 110 Ma, appears to reflect major reorganization of mass in the mantle, perhaps caused by catastrophic avalanching of descending slabs into the lower mantle. The strict stability of the time-averaged position of the rotation axis before and after this tilting implies the existence of steady convection that does not affect mass distribution in the mantle (Prevot et al., 2000).

Slab Avalanches

Most models for the episodic growth of continents involve catastrophic sinking of slabs through the 660-km seismic discontinuity in the upper mantle, which upon arrival at the D″ layer above the core initiate a mantle-plume event (Stein and Hofmann, 1994; Peltier et al., 1997; Brunet and Machetel, 1998; Condie, 1998). Although seismic tomographic results clearly suggest that descending slabs sink into the lower mantle today, this may not have been the case in the geologic past when the Earth was hotter. Christensen and Yuen (1985) have shown that in a hotter mantle with a larger Rayleigh number, such as probably existed in the Archean, the amount of leakage across the 660-km discontinuity is considerably reduced, perhaps resulting in layered convection. This model, however, does not consider the effect of the garnet reaction at this depth, which, as described in Chapter 4, tends to offset the effect of the spinel–perovskite reaction. Computer models of mantle evolution suggest that increased internal heating of the mantle strongly favors layering in the mantle (Zhao et al., 1992); this would be the case during the Archean, when heat production by radiogenic isotopes was more pronounced than it is today.

Layered convection throughout most of the Archean may have prevented the occurrence of slab avalanches. It may not have been until the late Archean when the 660-km discontinuity became less "robust" that slabs fell through to the lower mantle (Peltier et al., 1997; Condie, 1998). Cooling of the Earth also may have been responsible for the shutdown or decrease in the intensity of slab avalanches after 1.9 Ga. As the mantle temperature and Rayleigh number decreased with time, slabs should more easily penetrate the 660-km discontinuity, leading eventually to whole-mantle convection.

Numerical simulations by Yuen et al. (1993) show that at the higher temperatures that existed in the Archean, the mantle would convect more chaotically, a type of convection known as *hard turbulence*. Their results show that during hard turbulence with a higher Raleigh number, which is also temperature dependent, phase transitions such as the perovskite transition at 660 km become stronger barriers and result in layered convection. This implies that subducted slabs may accumulate at the 660-km boundary, which eventually leads to gravitational instability and catastrophic collapse of the slabs through the phase boundary. It is this catastrophic collapse that may trigger mantle-plume events in the D″ layer.

Core Rotational Dynamics

Another possible mechanism by which a mantle-plume event could be triggered is by the resonance of tidal waves in the fluid outer core (Greff-Lefftz and Legros, 1999). When the core rotational frequency and solar tidal waves are in resonance, frictional power may be converted into heat, destabilizing the D″ layer above the core and generating many

Figure 9.17 Temporal evolution of the visco-magnetic dissipative power at the core–mantle boundary induced by the lunar–solar tidal potential. Modified from Greff-Lefftz and Legros (1999).

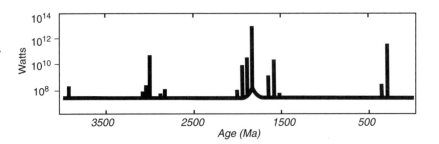

mantle plumes. Numerical models predict two major resonances in the past, one about 3 Ga and another about 1.8 Ga (Fig. 9.17).

These times correspond closely with the observed peaks in juvenile crust production at 2.7 and 1.9 Ga. During core resonance periods, the temperature near the inner core boundary should increase, an effect that could stop inner core growth and produce new momentum equilibrium for the geodynamo. This, in turn, could lead to a decrease in magnetic reversal frequency, thus accounting for the superchrons associated with the Phanerozoic mantle-plume events.

Effects of Impacts on Mantle-Plume Events

A question of considerable interest is that of whether or not impacts on the Earth's surface can trigger or strengthen mantle-plume events, where strength refers to plume temperature in excess of surrounding mantle. Abbott and Isley (2002) have shown that strong mantle-plume events correlate with the terrestrial impact record to better than a 99% confidence level. A time series analysis indicates the existence of 10 major peaks in impact activity over the last 3.5 Ga. Of these peaks, 9 of the 10 have a counterpart in mantle-plume distribution with time. Results of this study suggest that asteroid and cometary impacts may increase the amount of magma production from already active mantle plumes. Possible mechanisms for this effect include cracking of the lithosphere, providing channels for magma movement, and impact-triggered mixing at the core–mantle boundary, producing hotter mantle plumes. In addition, oblique impacts may impart high shear stresses at the core–mantle boundary, triggering avalanches along the boundary (Muller, 2002). These avalanches would cause sudden temperature changes in the D″ region, perhaps triggering a mantle-plume event.

Mantle-Plume Events and Supercontinents

A Plume–Supercontinent Connection?

A possible relationship of mantle-plume events to the supercontinent cycle is summarized in Figure 9.18. The timing of various events within this cycle is constrained chiefly

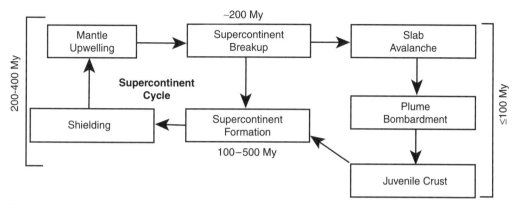

Figure 9.18   Box diagram showing a possible relationship of the supercontinent cycle to mantle-plume events. Modified from Condie (1998; 2000).

by data from two sources: (1) isotopic ages of juvenile continental crust and (2) results of computer simulations of mantle processes (Tackley et al., 1994; Condie, 1998; Condie, 2000). Beginning with a supercontinent, computer models suggest that it takes on average 200 to 400 My for shielding of a large supercontinent to cause a mantle upwelling beneath it (Lowman and Jarvis, 1996). The upwelling breaks the supercontinent over approximately a 200-My period. This breakup may be the trigger for a mantle-plume event by producing a slab avalanche through the 660-km discontinuity. The time from when a slab avalanche begins to when juvenile crust is produced is probably quite short, less than 100 My. This is because slabs can sink to the bottom of the mantle in 100 My or less (Larson and Kincaid, 1996), and in a mantle in which viscosity increases with depth, mantle plumes can rise to the base of the lithosphere in a few million years (Larsen and Yuen, 1997). This scenario suggests the correlation of slab avalanches with supercontinent formation rather than with supercontinent breakup. If supercontinents are really cyclic, it is clear that they are not periodic. Also, supercontinent breakup and aggregation overlap by 50 to 100 My, perhaps with the proportion of overlap increasing in the Phanerozoic.

As I have suggested (Condie, 1998), the duration of supercontinent formation and the total life span of supercontinents decrease with age. The duration of supercontinent formation (including enhanced production of juvenile crust) for the first well-documented supercontinent at 1.9 Ga was at least 500 My, decreasing to about 300 My for Rodinia and then to 100 to 200 My for Gondwana and Pangea. Paralleling this decrease is a decrease in the volume of juvenile continental crust produced during each supercontinent cycle. Approximately 40% of the present continental crust may have been produced during the formation of supercratons or a supercontinent at 2.7 Ga, 30% during the formation of the 1.9 Ga supercontinent, and only about 12% during the formation of Rodinia (Plate 5). The total amount of juvenile continental crust produced during the Phanerozoic appears to have been about 18% of the present volume. Only the first two supercontinent events at 2.7 and 1.9 Ga, both of which may be associated with mantle-plume events, contained significant volumes of juvenile crust.

What may have been responsible for decreases in the length of supercontinent cycles and in the volume of juvenile crust produced during each supercontinent cycle? If related to mantle-plume events, the major cause may have been the decreasing "strength" of the 660-km discontinuity with time. As the 660-km discontinuity became more permeable to descending slabs with falling mantle temperature, slabs would begin to steadily sink through the boundary, and thus fewer slabs would accumulate at the boundary. Hence, when an avalanche occurred, it would be relatively small. The small volume of plates in an avalanche may have had two effects:

1. The duration of a mantle-plume event should have been shorter and perhaps more limited in geographic extent.
2. Because of the smaller mass of sinking slabs, the number of mantle plumes generated would have decreased; thus, the volume of juvenile crust associated with these plumes would have decreased.

If the proposed mantle-plume model is viable, the decreasing aggregation time of supercontinents with time may reflect cooling of the mantle as radiogenic heat decreases with time.

One aspect of the catastrophic crustal growth model that is not well understood is that of whether the supercontinent cycle can operate independently of mantle-plume events (Condie, 2000). Peltier et al. (1997) suggested that the supercontinent cycle is caused by slab avalanche events in the mantle. In their model, the avalanches produce mantle downwellings directly over the avalanches, which act as "catchment basins" for an aggregating supercontinent. However, if supercontinents accumulate over geoid lows and breakup over mantle upwellings (Anderson, 1982; Gurnis, 1988; Lowman and Jarvis, 1996), both of which are a consequence of the supercontinent cycle, slab avalanches in the mantle may not be a necessary part of supercontinent formation. As described earlier, mantle upwellings that eventually break supercontinents result from thermal shielding of a large volume of mantle from subduction and cooling. If so, how are mantle-plume events related to the supercontinent cycle? Perhaps slab avalanches can be considered an "add-on" to the supercontinent cycle (Fig. 9.18). A slab avalanche, which initiates plume bombardment of the lithosphere and consequent production of juvenile crust, may occur any time during the supercontinent cycle. In my model (Condie, 1998), supercontinent breakup involving increased rates of subduction may trigger a slab avalanche, and this may be the only common denominator to both cycles.

As the Earth gets older, mantle-plume events should fade out before the supercontinent cycle comes to an end because the 660-km seismic discontinuity becomes more permeable to descending slabs as the mantle cools and the Rayleigh number drops. Except for three possible small mantle-plume events in the Ordovician, late Paleozoic, and Cretaceous, mantle-plume events appear to have ended after 1.9 Ga. The supercontinent cycle, however, may continue as long as convection continues in the mantle in response to the shielding of parts of the mantle by large lithospheric plates.

Two Types of Mantle-Plume Events

From precise U-Pb zircon ages it is clear that an increased production rate of juvenile crust correlates with the formation, *not* the breakup, phases of supercontinents (Fig. 9.4). Also, at least in the 1.9 and 2.7 Ga crustal events, juvenile crust peaks correlated with alleged mantle-plume events. However, there appears to be two types of mantle-plume events: one associated with supercontinent formation and one associated with supercontinent breakup (Condie, 1998; Condie, 2000). As described previously, it is commonly believed that mantle plumes associated with mantle upwellings are responsible for fragmenting supercontinents. If a supercontinent (or supercontinents) is not large enough to provide sufficient mantle shielding to produce an upwelling, it may not completely fragment, as appears to be the case with the 2.7 and 1.9 Ga supercontinents (Condie, 2002a). There is no evidence for large volumes of juvenile crust associated with this type of mantle-plume event, referred to as a *shielding mantle-plume event*. The juvenile crust produced in association with a shielding event is chiefly flood basalts and associated mafic underplating. Although plume heads can reach diameters up to about 2500 km, the volume of juvenile mafic crust associated with a given plume, as evidenced by the volume of Phanerozoic flood basalts and mafic underplates (from reflection seismology), is probably relatively small. For instance, if the entire high seismic-velocity layer at the base of Proterozoic continental crust (Durrheim and Mooney, 1991) was composed of later plume underplate, which is unlikely, it comprised only 10 to 25% of the Proterozoic crust. However, there are two important uncertainties: the volume of oceanic plateaus associated with a shielding mantle-plume event and the fraction of these plateaus that eventually collides and accretes to the continents. If zircon ages from the continental crust are representative, however, it would appear that the volume of oceanic plateaus accreted to the continents is relatively small, at least since the Archean (Condie, 2001).

So what is different about the mantle-plume events that may be associated with peaks in juvenile crust production at 1.9 and 2.7 Ga? These events, called *catastrophic mantle-plume events,* must be triggered by some process other than plate shielding. They also differ from shielding mantle-plume events in that they are short lived, a less than 100-My duration, in contrast to shielding events that last more than 200 My. Because large volumes of continental crust are associated with these mantle-plume events, they must be more intense and perhaps more widespread than shielding events. As explained earlier, the breakup of supercontinents may trigger slab avalanches at the 660-km discontinuity, resulting in catastrophic mantle-plume events. Although such a model may work for the 1.9 Ga event (and perhaps the three events in the Phanerozoic), it will not work for the 2.7 Ga event because there is no evidence for an earlier supercontinent fragmenting.

Crustal growth associated with catastrophic mantle-plume events chiefly occurs by the addition of arc components to the continents either along continental margins or by collisions of oceanic arcs and continents (Condie, 2001). If geologic sampling is representative of continental accretion in the last 2.5 Gy, oceanic-plateau accretion plays a relatively minor role in continental growth. The fact that supercontinent formation may occur simultaneously with a 1.9 Ga mantle-plume event may not be coincidental. Perhaps the breakup of the late Archean supercratons from 2.3 to 2.2 Ga triggered the 1.9 Ga mantle-plume event. In this sense, a catastrophic event provides positive feedback for crustal growth that began during a shielding mantle-plume event. Also, a growing supercontinent may contribute to the preservation of juvenile crust by trapping it in collisional and accretionary orogens.

What about the other numerous mantle-plume events recognized by Abbott and Isley (2002) using plume proxies (Fig. 9.9)? Most of these are not associated with increased rates of juvenile crust production, and many are not associated with supercontinent formation or breakup. Possibly many of the plume proxy events are not global mantle-plume events but more localized events affecting only certain parts of the crust. Also, as suggested by Abbott and Isley (2002), there appears to be two or three scales of periodicity in mantle-plume events, some of which may be related to asteroid or cometary impacts on the Earth.

And finally, assuming that both mantle-plume and supercontinent events are recorded in the geologic record, how can the two events be distinguished? Timing is important; hence, precise U-Pb zircon dating of events is critical. Catastrophic mantle-plume events are short lived (<100 My), whereas supercontinent events last more than 200 My. However, shielding mantle-plume events are comparable in length to supercontinent events. As investigators obtain more precise ages, we can look for rapid decay of catastrophic events with time. For instance, rapid input of CO_2 into the atmosphere associated with a short-lived mantle-plume event may show an abrupt increase in black shale deposition and global warming followed by a gradual decline as the mantle plumes cool with time. In contrast, supercontinent events and shielding mantle-plume events should show a gradual onset and last more than 200 My. As exemplified by the opposing effects of supercontinent formation and a mantle-plume event on black shale deposition and sea level, a global effect may be minor or even missing if the two events cancel each other. In contrast, positive feedback, such as the effects of supercontinent formation and a mantle-plume event on juvenile crust production rate, should enhance the magnitude of excursions of global change.

Further Reading

Condie, K. C., 2001. Mantle Plumes and Their Record in Earth History. Cambridge University Press, Cambridge, UK, 306 pp.

Davies, G. F., 1999. Dynamic Earth Plates, Plumes, and Mantle Convection. Cambridge University Press, Cambridge, UK, 458 pp.

Ernst, R. E., and Buchan, K. L. (eds.), 2001. Mantle Plumes: Their Identification Through Time. Geological Society of America, Spec. Paper 352.

Yoshida, M., Windley, B. F., and Dasgupta, S., 2003. Proterozoic East Gondwana: Supercontinent Assembly and Breakup. Geological Society of London, Special Publication No. 206, 480 pp.

Comparative Planetary Evolution

10

Introduction

Most data favor an origin for the planets by condensation and accretion of a gaseous solar nebula in which the Sun forms at the center (Boss et al., 1989; Taylor, 1992). Considering that the age of the universe is on the order of 15 Ga, the formation of the solar system about 4.6 Ga is a relatively recent event. Because it is not yet possible to observe planetary formation in other gaseous nebulae, investigators depend upon a variety of indirect evidence to reconstruct the conditions under which the planets in the solar system formed. Geophysical and geochemical data provide the most important constraints. Also, because the interiors of planets are not accessible for sampling, investigators often rely on meteorites to learn more about planetary interiors. Most scientists now agree that the solar system formed from a gaseous dust cloud, known as the solar nebula, which will be described later in the chapter.

In this the final chapter, I look at the Earth as a member of the solar system by comparing it to other planets. As you have seen in earlier chapters, much is known about the structure and the history of the Earth, and it is important to emphasize the uniqueness of the Earth in comparison with the other planets. First, plate tectonics and continents seem to be unique to the Earth. Why should only one of the terrestrial planets have plate tectonics? Although scientists do not have an answer to this question yet, I shall explore some possible reasons, especially in comparison with our sister planet Venus. The oceans and the presence of free oxygen in the terrestrial atmosphere are other unique features in the solar system, as I explained in Chapter 6. In addition, the Earth appears to be the only member of the solar system with life, although this may not have always been so. As I review other bodies in our solar system, and especially as I described the origin of the solar system, remember these unique terrestrial features that somehow must be accommodated in any model of planetary origin and evolution.

Impact Chronology in the Inner Solar System

It is well known that impacts on planetary surfaces play an important role in planetary evolution (Glikson, 1993). Impact effects are known on varying scales from dust size to planetary size and have occurred throughout the history of the solar system. A considerable amount of effort has gone into studying the impact record of the inner solar system as recorded on the lunar surface, and this record is now used to estimate the ages of craters on other planets. The sequence of events (volcanism, rifting, erosion, etc.) on a planetary or satellite surface can be deduced from the crosscutting relationships of craters, and if investigators can date important events, this relative history can be tied to absolute ages (Price et al., 1996). This, of course, assumes investigators can estimate the impact flux rate with time. It is well known that most of the large impacts on the lunar surface were early, terminating about 3.9 Ga with the large Imbrium impact. Although the termination of major impacts in the inner solar system was likely early, it may not have been synchronous throughout the system. However, the striking similarity of age–crater density curves on the Moon, Mercury, Venus, Mars, and asteroids strongly suggests a similar impact history for the inner part of the solar system; hence, it is probably valid to use the lunar timescale throughout this region. All of these bodies appear to have been "resurfaced" by an intense period of impact cratering about 3.9 Ga (Kring and Cohen, 2002). The source of the impacting debris was probably the asteroid belt.

In addition, the early impact history of the Earth is at least partly preserved in early Archean sediments in South Africa and Western Australia. At least 10 thin layers of sand-size spherules, interpreted as impact spherules, were deposited in both regions within two relatively narrow time windows: 3.47 to 3.23 and 2.64 to 2.49 Ga (Simonson and Harnik, 2000). Each layer is thought to be the product a large asteroid-size impactor. The oldest recognized impact, dated by precise U-Pb isotopic ages from impact-produced zircons, occurs in both South Africa and Western Australia at 3470 ± 2 Ma (Byerly et al., 2002). Confirming the presence of asteroid material in the impact layers are anomalous Cr isotopic compositions and large Ir anomalies in some of the spherule beds (Kyte et al., 2003).

Members of the Solar System

Planets

Basically, there are three types of planets in the solar system. The largest distinction is the break between the inner silicate planets, known as the **terrestrial planets,** and the outer **giant planets.** The giant planets can further be divided into the gas giants, Jupiter and Saturn, and the ice giants, Uranus and Neptune. In this respect, the planets reflect the three main components in the solar nebula: gas, ice, and rock. In this section, I review the chief characteristics of the planets, emphasizing the terrestrial planets.

Mercury

Mercury, the closest planet to the Sun, is a peculiar planet in several respects. First, the orbital eccentricity and the inclination of the orbit to the ecliptic are greater than those any other planet except Pluto (Table 10.1). Also, the high mean density (5.43 g/cm^3) of Mercury implies an Fe-Ni core that comprises about 66% of the planet's mass, thus having a core/mantle ratio greater than that of any other planet (Fig. 10.1). Although Mercury's magnetic field is weak, like the Earth, it has a dipole field probably generated by an active dynamo in a liquid outer core. Results from Mariner 10 show that the surface of Mercury is similar to that of the Moon in terms of crater distributions and probable age. Spectral studies of the Mercurian surface are consistent with a crust rich in calcic plagioclase and low-Fe pyroxenes, perhaps much like the lunar highlands' crust (Tyler et al., 1988; Solomon, 2003). The occurrence of a weak sodium cloud around the planet supports this idea. There are two types of plains on Mercury's surface: the early intercrater plains formed before 4 Ga and the smooth plains formed about 3.9 Ga. These plains have been attributed both to debris sheets from large impacts and to fluid basaltic lavas, and their origin continues to be a subject of controversy (Taylor, 1992; Solomon, 2003). For instance, microwave and infrared radiation reflected from the Mercurian surface indicate high albedo inconsistent with the presence of basalt (Jeanloz et al., 1995). These data would seem to favor a crust on Mercury, including the plains, composed almost entirely of anorthosite.

Mercury also displays a global network of large lobate scarps up to 1 km in height that may be high-angle reverse faults. They indicate a contraction in planetary radius of 2 to 4 km, probably because of rapid cooling of the mantle. The faulting that produced the lobate scarps occurred both before and after major volcanism. Employing the lunar

Table 10.1 Properties of the Planets									
	Mean distance to Sun (AU)	Orbital period (d, days; y, years)	Mean orbital velocity (km/sec)	Mass (Earth = 1)	Equatorial radius (km)	Mean density (g/cm^3)	Zero pressure density (g/cm^3)	Area/ mass (Earth = 1)	Core/ mantle (Earth = 1)
Mercury	0.387	88d	48	0.056	2439	5.43	5.3	2.5	12
Venus	0.723	225d	35	0.82	6051	5.24	3.95	1.1	0.9
Earth	1	365d	30	1	6378	5.52	4	1	1
Moon	1	27.3d*	1*	0.012	1738	3.34	3.3	6.1	0.12
Mars	1.52	687d	24	0.11	3398	3.94	3.75	2.5	0.8
Jupiter	5.2	11.9y	13	318	71,600	1.32			
Saturn	9.2	29.5y	9.7	95.2	60,000	0.69			
Uranus	19.1	84y	6.8	14.5	25,600	1.27			
Neptune	30	164y	5.4	17.2	24,760	1.64			
Pluto	39.44	248y	4.7	0.0022	1150	2.03			

* Period and velocity about the Earth.

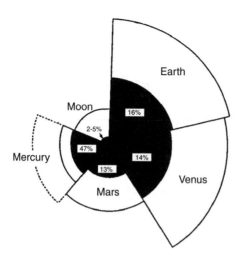

impact timescale, crater distributions indicate that the contraction occurred before massive bombardment of the planet at 3.9 Ga and probably before 4 Ga.

Two ideas have been suggested to explain the relatively high density of Mercury: (1) high-temperature evaporation of the silicate mantle, thus enhancing the proportion of iron in the planet, and (2) removal of part of the mantle by collision with another planet during or soon after the accretion of Mercury (Vilas et al., 1988). The reflectance data suggesting a plagioclase-rich crust and the presence of a sodium cloud around Mercury do not favor the evaporation theory, because volatile elements such as sodium should have been removed and lost during the high-temperature evaporation event. The impactor idea has received increased support from stochastic modeling of planetary accretion in the inner solar system, which indicates that numerous potential impactors (some as large as Mars) may have existed in the solar nebula. To have a core/mantle ratio similar to the Earth and Venus, Mercury would have had to lose about half of its original mass during the collision (Fig. 10.1). The fate of this fragmented material, mostly silicates and oxides, is unknown, but it may have been swept away by solar radiation or possibly accreted to Venus or the Earth if it crossed their orbits. The highly eccentric orbit of Mercury also may have developed during the early collision.

Mars

The Mariner and Viking missions indicate that Mars is quite different from the Moon and Mercury (Taylor, 1992; Carr et al., 1993). The Martian surface includes major shield volcanoes, fracture zones, rifts, and large canyons that appear to have been cut by running water. Also, much of the planet is covered with wind-blown dust, and more than half of the surface is covered by a variably cratered terrain similar to the surfaces of the Moon and Mercury. Large near-circular basins are similar to lunar mare basins and probably formed about 3.9 Ga. Much of the northern hemisphere of Mars is comprised of volcanic plains and large stratovolcanoes. In contrast, the southern hemisphere is covered by the ancient cratered terrain that is greater than 4 Gy in age. The Tharsis bulge, which was probably formed by a large mantle upwelling, is the dominant structural feature on the Martian surface (Zuber, 2001). It is about 10 km high at the center and 8000 km across, comprising about 25% of the Martian surface. The Tharsis region includes gigantic shield volcanoes and volcanic plains, and the Valles Marineris represents an enormous rift valley that spans one-quarter of the equator. The size of Martian volcanoes implies a thick lithosphere on Mars. The distribution and surface features of lava flows on Mars indicate low viscosities of eruption and a total volume of lavas much greater than those on the Earth. Topography and gravity measurements by the Mars Global Surveyor show that Mars has two distinct crustal zones: the northern lowlands with uniform crustal thickness and the southern highlands with relatively thin crust (Zuber et al., 2000; Zuber, 2001). In addition, the northern hemisphere crust is relatively smooth and has far fewer craters than the southern hemisphere crust. Because the buried crater distributions in the north (as deduced from rims of buried craters) are similar to those in the south, it is likely that the crust in both hemispheres is similar in age (Zuber, 2001).

A peculiar feature of Mars is the presence of widespread wind-blown dust, perhaps analogous to terrestrial loess deposits. The Martian dust is the cause of the variations in albedo on the planet's surface, once thought to be canals. In Martian bright areas, dust may accumulate to more than 1 m in thickness. The movement of dust on the surface by gigantic dust storms appears to be related to seasonal changes. On the whole, erosion on the Martian surface has been slow, consistent with the unweathered composition of the rocks analyzed by the Viking Landers. The redistribution of sediment by wind occurs at a rate much greater than weathering and erosion, such that much of the dusty sediment on Mars was produced early in Martian history and has been reworked ever since.

Chemical analyses from the Viking landing sites suggest that the dominant volcanic rocks on Mars are Fe-rich basalts and that weathering of the basalts occurs in a hydrous, oxidizing environment. The Russian Phobos mission obtained chemical analyses of a large area also consistent with basaltic rocks (Fig. 10.2). Remote sensing studies show a strong concentration of Fe^{+3} on the surface indicative of hematite, probably in the wind-blown dust. Because Mars has an uncompressed density much less than that of the Earth and Venus (Table 10.1), it must also have a distinctly different composition than these planets. If Mars has a carbonaceous chondrite composition and is completely differentiated, the core mass is about 21% of total mass and the core radius is about 50% of the planetary radius. It would appear that compared with Venus and the Earth, Mars is more volatile, rich by at least a factor of two, and its core probably contains a substantial amount of sulfur (Taylor, 1992). It is also possible to learn about the composition of the Martian crust from meteorites that appear to have been ejected from the planet's surface by one or more impacts in the last 20 My. Chemical analysis of these meteorites shows they are picrites (Mg-rich basalts) and basalts, in agreement with the surface measurements (Zuber, 2001).

One of the most puzzling aspects of Martian geology is the role that water has played on the planetary surface (Carr et al., 1993; Baker, 2001). Although water is frozen on

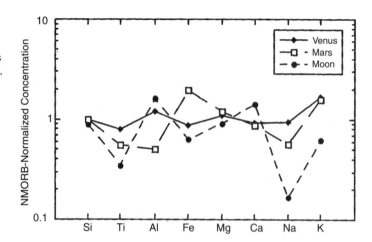

Figure 10.2 Normal midocean-ridge basalt (NMORB), normalized, major element distributions in crustal rocks from Venus, Mars, and the Moon.

Mars today, scientists see evidence for running water in the past, such as dry valleys and canyons (Fig. 10.3). Clearly, climatic conditions on Mars must have been warm enough to permit running water sometime in the past (Jakosky and Phillips, 2001). The most intriguing features are the large, flat valleys up to 150 km wide, some up to 2000 km long, which appear to have been cut by running water. These valleys are similar to Pleistocene glacial features on the Earth caused by cataclysmic flooding, and the Martian analogues are generally interpreted to have resulted from gigantic floods (Baker, 2001). The sizes of the Martian outflow channels imply immense discharges of water, much larger than any floods on the Earth for which there is evidence preserved in the geologic record (Fig. 10.4). Although most Martian floods appear to have occurred early (4–2 Ga), some occurred as recently as 10 Ma. What caused the floods remains problematic, and all may not be of the same origin. One possible cause is volcanism, which could have melted permafrost, suddenly releasing huge volumes of water. Other Martian canyons with numerous tributaries look like terrestrial canyons and appear to have been cut more slowly by rivers or spring sapping by ground water (Fig. 10.3). Most of these canyons are in the ancient cratered terrain, the oldest part of the Martian crust, indicating the presence of transient warm climates on Mars more than 4 Ga. The presence of large, buried channels in the northern lowlands is consistent with northward transport of water and sediment before the end of northern hemisphere resurfacing (Zuber et al., 2000).

Unlike the Earth, Mars has no global dipole magnetic field, as confirmed by the Mars Global Surveyor mission. However, the Surveyor mission found strong, spatially variable magnetizations in the Martian crust. These crustal magnetizations, an order of magnitude larger than the strongest magnetizations found in terrestrial rocks and minerals, are probably thermal remanent magnetization. Hence, the magnetizations were produced by cooling events, which occurred when a global magnetic field was present in the planet (Stevenson, 2001). This implies that early in Martian history the planet had an active geodynamo in the core as the Earth does today.

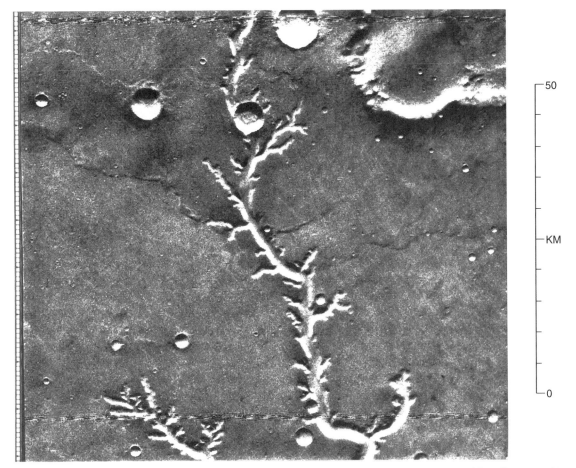

Figure 10.3   Nirgal Vallis, a branching canyon on the Martian surface thought to have been cut by running water. The valley is incised into crust probably older than 3.5 Ga. Viking Orbiter frame 466A54. Courtesy of NASA.

The geologic history of Mars was probably similar to that of the Moon and Mercury for the first few hundred million years. Core formation was early, probably coinciding with a transient magma ocean. Part of the early crust survived major impacting that terminated about 3.9 Ga and occurs as the ancient cratered terrain in the southern hemisphere. This early crust may have dated to 4.5 Ga based on the age of Shergotty, Nakhla, and Chassigny (SNC) meteorites probably derived from the Martian surface. After rapid crystallization of the magma ocean before 4.4 Ga, extensive melting in the upper mantle formed a thick basaltic crust. Heat loss was chiefly by mantle plumes. The Tharsis bulge developed from about 3.9 to 3.8 Ga, perhaps in response to a gigantic mantle plume (Mutch et al., 1976). With a drop in surface temperature, permafrost formed, which later locally and perhaps catastrophically melted and caused massive floods that cut many of the large flat-bottom valleys. The major erosional events probably occurred during or just

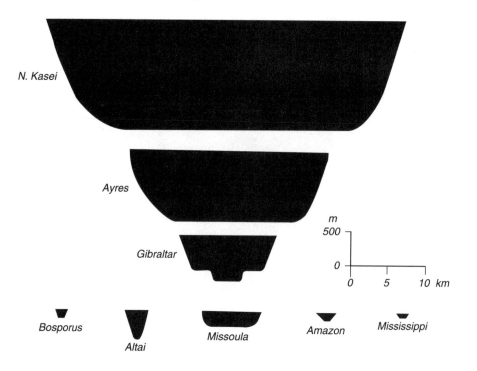

N. Kasei

Ayres

Gibraltar

m
500

0

0   *5*   *10 km*

Bosporus   *Altai*   *Missoula*   *Amazon*   *Mississippi*

before the terminal large impact event about 3.9 Ga. Continued fracturing and volcanism on Mars extended to at least 1000 Ma and perhaps 100 Ma.

Venus

Comparison with the Earth. Unlike the other terrestrial planets, Venus is similar to the Earth both in size and mean density (5.24 and 5.52 g/cm^3, respectively) (Table 10.1). After correcting for pressure differences, the uncompressed density of Venus is within 1% of that of the Earth, indicating that both planets are similar in composition, with Venus having a somewhat smaller core/mantle ratio. Although both planets have similar amounts of N_2 and CO_2, most of the Earth's CO_2 is not in the atmosphere but in carbonates. Venus also differs from the Earth by the near absence of water and the high density and temperature of its atmosphere. As described later, Venus may at one time have lost massive amounts of water by the loss of hydrogen from the upper atmosphere. Unlike the Earth, Venus lacks a satellite, has a slow retrograde rotation (244 Earth days for one rotation), and does not have a measurable magnetic field. Because Venus orbits the Sun in only 225 days, the day on Venus (244 Earth days) is longer than the year. The absence of a magnetic field in Venus may be caused by the absence of a solid inner core because, as described in Chapter 5, crystallization of an inner core may be required for a dynamo to operate in the outer core of a planet. Of the total Venusian surface, 84% is flat rolling plains, some of which are more than 1 km above the average plain elevation. Only 8% of the surface is true highlands; the remainder (16%) lies below the average radius, forming broad, shallow basins. This is unlike the topographic distribution on the Earth, which is bimodal because of plate tectonics (Fig. 10.5). The unimodal distribution of elevation on Venus does not support the existence of plate tectonics on Venus today.

The spectacular Magellan imagery indicates that unlike the Earth, deformation on Venus is distributed over thousands of kilometers rather than occurring in narrow orogenic belts (Solomon et al., 1992). There are numerous examples of compressional tectonic features on Venus, such as Maxwell Montes deformational belt in the western part of Ishtar Terra (Fig. 10.6). Ishtar Terra is a highland about 3 km above the mean planetary radius surrounded by compressional features suggestive of tectonic convergence resulting in crustal thickening. Maxwell Montes stands 11 km above the surrounding plains and shows a wrinkle-like pattern suggestive of compressional deformation (Kreep and Hansen, 1994). The deformation in this belt appears to have occurred passively in response to horizontal stresses from below.

Coronae are large circular features (60–2600 km in diameter with most 100–300 km) with a great diversity of morphologies (Stefan et al., 2001). Almost all coronae occur between 80° N and 80° S latitude and show a high concentration in equatorial areas. Venus is the only planet known to have coronae. An approximately inverse correlation between crater and corona density suggests that the volcano–tectonic process that forms coronae may be the same process that destroys craters (Stefanick and Jurdy, 1996). The most widely accepted models for the origin of coronae are those involving mantle plumes. A rising plume creates a region of uplift accompanied by radial deformation and dyke emplacement (Copp et al., 1998). Volcanism may also accompany this stage. As the plume head spreads at the base of the lithosphere, it elevates the surface, producing annuli in some coronae. This is generally followed by collapse as the plume head cools.

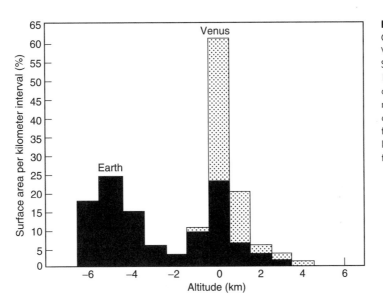

Figure 10.5

Comparison of relief on Venus and the Earth. Surface height is plotted in 1-km intervals as a function of surface area. Height is measured from the sphere of average planetary radius for Venus and from the sea level for the Earth. Modified from Pettengill et al. (1980).

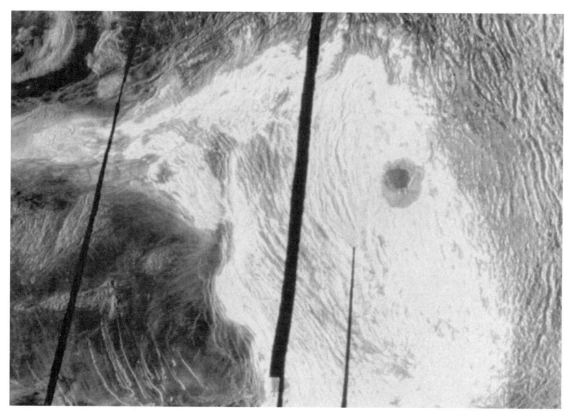

Figure 10.6   Magellan image of Maxwell Montes, the highest mountain range on Venus, which stands 11 km above the average diameter of the planet. The complex pattern of intersecting ridges and valleys reflects intense folding and shearing of the crust. Courtesy of U. S. Geological Survey.

Another unique and peculiar feature of the Venusian surface is the closely packed sets of grooves and ridges known as *tesserae,* which appear to result from compression. A combination of structural, mechanical, topographic, and geologic evidence suggests that tesserae record interaction of deep mantle plumes with an ancient, globally thin lithosphere, resulting in regions of thickened crust (Hansen et al., 1999).

Perhaps the most important data from the Magellan mission are those related to impact craters (Kaula, 1995). Unlike the Moon, Mars, and Mercury, Venus does not preserve a record of heavy bombardment from the early history of the solar system (Price and Suppe, 1994). Crater size–age distribution shows an average age of the Venusian surface of only 600 to 400 Ma, indicating extensive resurfacing of the planetary surface at this time. Most of this resurfacing is with low-viscosity lavas, presumably mostly basalts as inferred from the Venera geochemical data. Crater distribution also indicates a rapid decline in the resurfacing rate within the last tens of millions of years. However, results suggest that some large volcanoes (72 Ma), some basalt flows (128 Ma), some rifts (130 Ma), and many coronae (120 Ma), are much younger than the average age the resurfaced plains and probably represent ongoing volcanic and tectonic activity (Price et al., 1996).

The differences between Venus and the Earth, with the lower bulk density of Venus, affect the nature and rates of surface processes (weathering, erosion, and deposition), tectonic processes, and volcanic processes. Because a planet's thermal and tectonic history depends on its size and the area/mass ratio as described later, Venus and the Earth are expected to have similar histories. However, the surface features of Venus are quite different from those of the Earth, raising questions about how Venus transfers heat to the surface and whether plate tectonics has ever been active. The chief differences between the Earth and Venus appear to have two underlying causes: (1) small differences in planetary mass leading to different cooling, degassing, and tectonic histories, and (2) differences in distance from the Sun, resulting in different atmospheric histories.

Surface Composition. Much has been learned about the surface of Venus from scientific missions by the United States and Russia. The Russian Venera landings on the Venusian surface have provided a large amount of data on the structure and composition of the crust. Results suggest that most of the Venusian surface is composed of blocky bedrock surfaces and that less than one-fourth contains porous, soil-like material (McGill et al., 1983). The Venera Landers have also revealed the presence of abundant volcanic features, complex tectonic deformation, and unusual ovoid features of probable volcanic–tectonic origin. Reflectance studies of the Venusian surface suggest that iron oxides may be important components. Partial chemical analyses made by the Venera Landers indicate that basalt is the most important rock type. The high K_2O recorded by Venera 8 and 13 is suggestive of alkali basalt, and the results from the other Venera landings clearly indicate tholeiitic basalt, perhaps with geochemical affinities to terrestrial ocean-ridge tholeiites (Fig. 10.2). A Venusian crust composed chiefly of basalt is consistent with the presence of thousands of small shield volcanoes that occur on the volcanic plains, typically 1 to 10 km in diameter and with slopes of about 5 degrees. The size and distribution of these volcanoes resembles terrestrial oceanic-island and seamount volcanoes.

Venusian Core. Venus has no global magnetic field, although it likely has a molten outer core with or without an inner core (Stevenson, 2003). The absence of a dynamo in the outer core probably reflects the lack of convection caused either because an inner core is absent or because the outer core is not cooling. If the inside of Venus is hotter than the corresponding depth in the Earth, which seems likely, an inner core is not expected. Alternatively, or in addition, the Venusian core may not be cooling at present because it is still recovering from heat loss associated with a resurfacing event some 500 Ma.

Cooling and Tectonics. To understand the tectonic and volcanic processes on Venus, it is first necessary to understand how heat is lost from the mantle. Four sources of information are important in this regard: the amount of ^{40}Ar in the Venusian atmosphere, lithosphere thickness, topography, and gravity anomalies. The amount of ^{40}Ar in planetary atmospheres can be used as a rough index of past tectonic and volcanic activity because it is produced in planetary interiors by radioactive decay and requires tectonic–volcanic processes to escape. Venus has about one-third as much ^{40}Ar in its atmosphere as does the Earth, which implies less tectonic and volcanic activity for comparable ^{40}K contents.

In contrast to the Earth, where at least 90% of the heat is lost by the production and subduction of oceanic lithosphere, there is no evidence for plate tectonics on Venus. The difficulty of initiating and sustaining subduction on Venus is probably because of a combination of high mantle viscosity; high fault strength; and thick, relatively buoyant basaltic crust. Thus, it would appear that Venus, like the Moon, Mercury, and Mars, must lose its heat through conduction from the lithosphere, perhaps transmitted upward chiefly by mantle plumes. The base of the thermal lithosphere in terrestrial ocean basins is about 150 km deep, where the average geotherm intersects the wet mantle solidus. On Venus, however, where the mantle is likely dry, an average geotherm does not intersect the dry solidus, indicating the absence of a distinct boundary between the lithosphere and the mantle (Fig. 10.7). The base of the elastic lithosphere in ocean basins is at the 500° C isotherm, or about 50 km deep. Because 500° C is near the average surface temperature of Venus, there is no elastic lithosphere on Venus. Another important difference between Venus and Earth is the strong positive correlation between gravity and topography on Venus, implying compensation depths in the Venusian mantle of 100 to 1000 km. This requires strong coupling of the mantle and lithosphere, and hence the absence of an asthenosphere, agreeing with the thermal arguments presented previously. This situation may have arisen from a lack of water in Venus. One of the important consequences of a stiff mantle is the inability to recycle lithosphere through the mantle, again showing that plate tectonics cannot occur on Venus. The deformed plateaus and the lack of features characteristic of brittle deformation, such as long faults, suggest the Venusian lithosphere behaves more like a viscous fluid than a brittle solid. The steep-sided high-elevation plateaus on Venus, however, attest to the strength of the Venusian lithosphere.

Figure 10.7

Comparison of geotherms from an average terrestrial ocean basin and Venus. Conduction is assumed to be the only mode of lithospheric heat transfer on Venus. Also shown are wet (0.1% water) and dry mantle solidi.

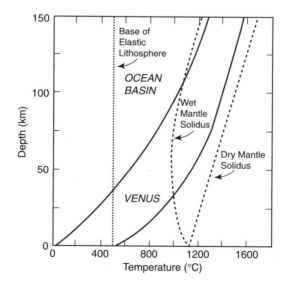

Two thermal-tectonic models have been proposed for Venus: the conduction and the mantle-plume models (Bindschadler et al., 1992). In the conduction model, Venus loses heat by simple conduction through the lithosphere, and tectonics is a result of compression and tension in the lithosphere in response to the changing thermal state of the planet. It is likely that such a model describes the Moon, Mercury, and Mars at present. Not favoring a conduction model for Venus, however, is the implication that the topography is young because it cannot be supported for long with warm, thin lithosphere.

In the mantle-plume model, which is preferred by most investigators, Venus is assumed to lose heat from large mantle plumes coupled with delamination and sinking of the lithosphere (Turcotte, 1995; Turcotte, 1996) (Fig. 10.8). Also consistent with plumes are the deep levels of isostatic gravity compensation beneath large topographic features, suggesting the existence of plumes beneath these features. Unlike the Earth, most of the topographic and structural features on Venus can be accounted for by mantle plumes, with compressional forces over mantle downwellings responsible for the compressional features on the surface. On the whole, the geophysical observations from Venus support the idea that mantle downwelling is the dominant driving force for deformation of the surface of Venus. The return flow in the mantle would also occur in downwellings and would undoubtedly involve delamination and sinking of significant volumes of the lithosphere. This is a striking contrast to the way the Earth cools, as shown in Figure 10.8.

Although the age of the Venusian surface is likely variable, studies of crater distributions indicate that more than 80% of the surface had all of its craters removed in a short period, probably between 10 and 100 My (Nimmo and McKenzie, 1998). It is still debated

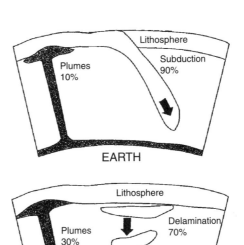

Figure 10.8
Comparison of the Earth's cooling mechanisms with possible cooling mechanisms on Venus. Estimates of magnitudes given in percentages. Modified from Turcotte (1995).

whether this resurfacing of the planet from 600 to 400 Ma was caused by a catastrophic planetwide mantle-plume event, a short-lived plate tectonic event in the waning stages today, or simply ongoing resurfacing whose mean age is 600 to 400 Ma (Strom et al., 1994; Nimmo and McKenzie, 1998).

Giant Planets

Jupiter and Saturn, the two largest planets, have densities indicating that they are composed chiefly of hydrogen and helium (Table 10.1). In the outer parts of the planets, these elements occur as ices and gases and at greater depths as fluids. The cores of the giant planets include a mixture of high-density ices and silicates. Relative to the Sun, the giant planets are enriched in elements heavier than He. Magnetic fields of these planets vary significantly in orientation or magnitude, and the origin of these fields is poorly understood. They are not, however, produced by dynamo action in a liquid Fe core, as is the case in the terrestrial planets. Unlike Jupiter and Saturn, the densities of Uranus and Neptune require a greater silicate fraction in their interiors. Models for Uranus, for instance, suggest a silicate core and icy inner mantle composed chiefly of water, CH_4, and NH_3 and a gaseous and icy outer mantle composed chiefly of H_2 and He. Neptune must have an even greater proportion of silicate and ice. Except for Jupiter, with a 3-degree inclination to the ecliptic, the outer planets are highly tilted in their orbits (Saturn 26.7 degrees, Uranus 98 degrees, and Neptune 29 degrees). Such large tilts probably result from collisions with other planets early in the history of the solar system. Whatever hit Uranus to knock it completely over must have had a mass similar to that of the Earth.

Satellites and Planetary Rings

General Features

There are about 60 satellites in the solar system. Although there is great diversity in the satellites and no two are alike, three general classes of planetary satellites are recognized:

1. *Regular satellites,* which include most of the larger satellites and many of the smaller satellites, are those that revolve in or near the plane of the planetary equator and revolve in the direction in which the parent planet moves about the Sun.
2. *Irregular satellites* have highly inclined, often retrograde, and eccentric orbits, and many are far from the planet. Many of Jupiter's satellites belong to this category as do the outermost satellites of Saturn and Neptune (Phoebe and Nereid, respectively). Most, if not all, of these satellites were captured by the parent planet.
3. *Collisional shards* are small, often irregular-shaped satellites that appear to have been continually eroded by ongoing collisions with smaller bodies. Many of the satellites of Saturn and Uranus are of this type. Phobos and Deimos, the tiny satellites of Mars, may be captured asteroids.

There are regularities in satellite systems that are important in constraining satellite origin. For instance, the large, regular satellites of Jupiter, Saturn, and Uranus have low inclination, prograde orbits indicative of formation from an equatorial disk (Stevenson, 1986). Although regular satellites extend to 20 to 50 planet radii, they do not form a scale model of the solar system. Although the large satellites are mostly rocky or rock–ice mixtures, small satellites tend to be more ice rich, suggesting that some of the larger satellites may have lost ice or accreted rock. Volatile ices, such as CH_4 and N_2, appear only on satellites distant from both the Sun and the parent planet, reflecting the cold temperatures necessary for their formation. One thing that emerges from an attempt to classify satellites is that no general theory of satellite formation is possible.

Planetary Rings

Since the Voyager photos of planetary rings in the outer planets, the origin of planetary rings has taken on new significance. Some have suggested that the rings of Saturn can be used as an analogue for the solar nebula from which the solar system formed. Although Jupiter, Saturn, Uranus, and Neptune are now all known to have ring systems, they are all different, and no common theory can explain all of them. Although the rings of Saturn are large in diameter, the thickness of the rings is probably less than 50 m. The average particle size in the rings is only a few meters, and single particles orbit the planet in about 1 day. Three models have attracted most attention for the origin of planetary rings. In the first two models, rings are formed with the parent planet as remnants of an accretionary disk or of broken pieces of satellites. Neither of these origins is likely, however, because rings formed in such a manner should not have survived beyond a few million years. Alternatively, the rings may be debris from the disruption of captured comets such as Chiron. In this model, the small particles become rings and the larger fragments may become satellites. If the rings around the giant planets are the remains of captured comets, they are latecomers to the solar system because, as you shall see later in this chapter, comets are among the youngest members of the solar system.

The Moon

As a planetary satellite, there are many unique features about the Moon. Among the more important are the following, all of which must be accommodated by any acceptable model for lunar origin:

1. The orbit of the Moon about the Earth is neither in the equatorial plane of the Earth nor in the ecliptic; it is inclined 6.7 degrees to the ecliptic (Fig. 10.9).
2. Except for the Pluto–Charon pair, the Moon has the largest mass of any satellite–planetary system.
3. The Moon has a low density compared with that of the terrestrial planets, implying a relatively low iron content.
4. The Moon is strongly depleted in volatile elements and enriched in some refractory elements such as Ti, Al, and U.

Figure 10.9 Orbital relations of the Earth–Moon system.

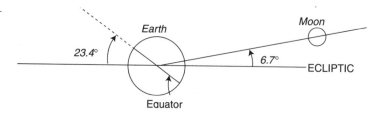

5. The angular momentum of the Earth–Moon system is anomalously high compared to other planet–satellite systems.
6. The Moon rotates in the same direction as does the Earth.

A great deal has been learned about the geochemistry and geophysics of the Moon from the Apollo landings (Taylor, 1982; Taylor, 1992). Although average lunar density is much less than the average Earth density (Table 10.1), its uncompressed density is about the same as the Earth's mantle, implying that the Moon is composed largely of Fe and Mg silicates. Unlike most other satellites, which are mixtures of silicates and water ice, the Moon must have formed in the inner part of the solar system.

From seismometers placed on the Moon by astronauts, scientists can deduce the broad structure of the lunar interior. The Moon has a thick crust (60–100 km) comprising about 12% of the lunar volume, and it appears to have formed soon after planetary accretion, about 4.5 to 4.4 Ga (Taylor, 1992). From the limited sampling of the lunar crust by the astronauts, scientists have learned that it is composed chiefly of anorthosites and gabbroic anorthosites as represented by exposures in the lunar highlands. These rocks typically have cumulus igneous textures, although they have been modified by impact brecciation. Sm-Nd isotopic dating indicates that this plagioclase-rich crust formed about 4.45 Ga. As you shall see later, it appears to have formed by crystallization of an extensive magma ocean. Also characteristic of the lunar surface are the mare basins, large impact basins formed before 3.9 Ga, covering about 17% of the lunar surface (Fig. 10.10). These basins are flooded with basalt flows only 1 to 2 km thick and probably erupted chiefly from fissures. Isotopically dated mare basalts range from 3.9 to 2.5 Ga. The impacts that formed the mare basins did not initiate the melting that produced the basalts, which were erupted up to hundreds of millions of years later and thus represent a secondary crust on the Moon. The youngest basaltic eruptions may be as young as 1 Ga. The lunar crust overlies a mantle composed of two layers. The upper layer or lithosphere extends to a depth of 400 to 500 km and is probably composed of cumulate ultramafic rocks. The second layer extends to about 1100 km, where a sharp break in seismic velocity occurs. Although evidence is still not definitive, it appears that the Moon has a small metallic core (300–500 km in diameter), comprising 2 to 5% of the lunar volume (Fig. 10.1).

Although the Moon does not have a magnetic field, remnant magnetization in lunar rocks suggests a lunarwide magnetic field at least between 3.9 and 3.6 Ga (Fuller and Cisowski, 1987). The maximum strength of this field was probably only about one-half that of the present Earth's field. It is likely this field was generated by fluid motions in the lunar core, much like the present Earth's field is produced. A steady decrease in the

Figure 10.10 Oblique view of the southern part of the Imbrium basin, one of the large mare basins on the Moon. Courtesy of the Lunar and Planetary Institute.

magnetic field after 3.9 Ga reflects cooling and complete solidification of the lunar core by no later than 3 Ga.

The most popular model for lunar evolution involves the production of an ultramafic magma ocean that covered the entire Moon to a depth of 500 km or more and crystallized in less than 100 My beginning about 4.45 Ga (Fig. 10.11). Plagioclase floats, producing an anorthositic crust; pyroxenes; and olivine largely sink, producing an incompatible-element depleted upper mantle. Later partial melting of this mantle produces the mare basalts. Detailed models for crystallization of the magma ocean indicate the process was complex, involving floating "rockbergs" and cycles of assimilation, mixing, and trapping of residual liquids. The quenched surface and anorthositic rafts were also continually broken up by impact.

Satellite Origin

Regular satellites in the outer solar system are commonly thought to form in disks around their parent planets. The disks may have formed directly from the solar nebula as the planets accreted, and the satellites may have grown because of the collision of small bodies within the disks. Alternatively, the disks could form by the breakup of planetesimals (small planets) when they came within the **Roche limit** (the distance at which tidal forces of the planet fragment a satellite). Still other possible origins for planetary disks include spin-off because of contraction of a planet, outward transfer of angular momentum, and massive collisions between accreting planets. The collisional scenario is particularly interesting in that it forms the basis for the most widely accepted model for the origin of the Moon. Regardless of the way it originates, once a disk is formed, computer modeling indicates that satellites will accrete in short periods of 1 My. The irregular satellites and

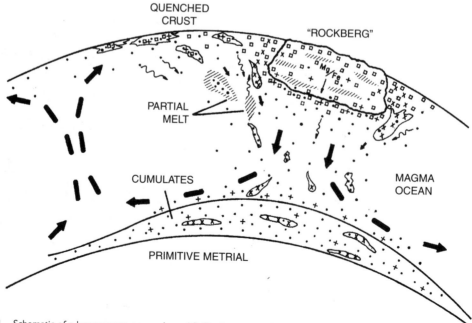

Figure 10.11 Schematic of a lunar magma ocean about 4.5 Ga. Arrows are flow patterns, and rockbergs are principally anorthosites. Modified from Longhi (1978).

collisional shards, however, cannot be readily explained by the disk accretion model. This has led to the idea that many satellites have been captured by the gravity field of their parent planet during a near collision of the two bodies. As dictated by their compositions, some of the rocky satellites in the outer solar system may have accreted in the inner solar system, and their orbits were perturbed in a manner that took them into the outer solar system, where they were captured by one of the Jovian planets.

Comets and Other Icy Bodies

Comets are important probes of the early history of the solar system because, compared with other bodies in the solar system, they appear to have been least affected by thermal and collisional events (Wyckoff, 1991). Comet heads are small (radii of 1–10 km) and have a low density ($0.1–1$ gm/cm^3). Because of their highly elliptical orbits, comets reside in a "dormant state" most of the time in the outer reaches of the solar system at temperatures of no more than $180°$ C in what are known as the Edgeworth-Kuiper belt (30–1000 AU from the Sun, in which 1 astronomical unit = Earth–Sun distance) and the Oort cloud (1000–50,000 AU) (Stern, 2003). Perturbations of cometary orbits by passing stars have randomized them, making it difficult to determine where comets originally formed in the solar nebula. However, the presence of CO_2 and sulfur in comets suggests they formed in the outer, cold regions of the nebula. Only for a few months do most comets come close enough to the Sun (0.5–1.5 AU) to form vaporized tails. Short-period comets, which orbit the Sun in less than 200 years, appear to come from just beyond Neptune from 35 to 50 AU in the Edgeworth-Kuiper belt.

Closely related to comets are Pluto, Triton (the large satellite of Neptune), and related icy bodies in the outer solar system. Pluto, which is about 0.2 times the mass of the Moon, has a highly inclined and eccentric orbit (Table 10.1). It has one satellite, Charon, whose orbit is inclined 90 degrees to the ecliptic. Charon is much less dense than Pluto and contains a large volume of ice, whereas Pluto has greater silicate content. It is generally thought that Charon was produced by a collision with Pluto, which stripped some of the ice from Pluto and reaccreted it into the satellite. Scientists now realize that Pluto is really not a planet but is more closely related to comets. However, it has too great a density to be a typical comet (Table 10.1), perhaps increasing its density during the collision that formed Charon.

Knowledge of cometary composition was greatly increased from the Giotto mission in 1986 during a close approach to Halley's comet (Jessberger et al., 1989). The nucleus of Halley's comet is irregular in shape, and its surface is covered with craters and a layer of dark dust up to 1 m thick. How much dust (silicates and oxides) resides inside this or other comets is still unknown. Model calculations, however, indicate that the dust/ice ratio in comets is 0.5 to 0.9. Data show the gaseous component in Halley's comet is composed chiefly of water vapor with only traces of CO_2, CO, CH_4, and NH_3 (Nuth, 2001). Because the ratios of these gases are dissimilar from the Sun, it appears that the material composing Halley's comet is not primitive but has been fractionated. Also, compared with solar abundances, hydrogen is strongly depleted in Halley's comet.

So what is the origin of the great clouds of comets in the solar system? Because the original solar nebula probably did not have enough mass to extend to the Oort cloud, comets must have been added later to form this cloud. The giant planets appear to have ejected icy bodies out of the inner solar system to more than 50 AU, and in this manner the outer part of the solar system became populated with great clouds of comets. In this respect, comets are some of the youngest members of the solar system.

Asteroids

Asteroids are small planetary bodies, most of which revolve about the Sun in an orbit between Mars and Jupiter (Lebofsky et al., 1989; Taylor, 1992). Of the 10,000 or so known asteroids, most occur between 2 and 3 AU from the Sun (Fig. 10.12). The total mass of the asteroid belt is only about 5% of that of the Moon. Only a few large asteroids are recognized, the largest of which is Ceres with a diameter of 933 km. Most asteroids are less than 100 km in diameter, and there is high frequency with diameters of 20 to 30 km. Three main groups of asteroids are recognized: (1) the near-Earth asteroids (Apollo, Aten, and Amor classes), some of which have orbits that cross that of the Earth; (2) the main belt asteroids; and (3) the Trojans revolving in the orbit of Jupiter (Fig. 10.12). Most meteorites arriving on the Earth are coming from the Apollo asteroids. The orbital gaps in which no asteroids occur in the asteroid belt (for instance, at 3.8 and 2.1 AU; Fig. 10.12) appear to reflect orbital perturbations caused by resonances in the gravity field of Jupiter. In terms of spectral studies, asteroids vary significantly in composition (Table 10.2), and some can be matched to meteorite groups. Within the asteroid belt, there is a zonal arrangement that reflects chemical composition. S, C, P, and D asteroid classes occupy successive rings outward in the belt, whereas M types predominate near the middle and B and F types near the outer edge. The broad pattern is that fractionated asteroids dominate in the inner part of the belt and that low-albedo primitive types (class C) occur only in the outer portions of the belt. Thus, asteroids inward of 2 AU are igneous asteroids, and the proportion of igneous to primitive asteroids decreases outward such that by 3.5 AU there are no igneous types represented. Although asteroids are continually colliding with each other as indicated by the angular and irregular shapes of most, the remarkable compositional zonation in the asteroid belt indicates that mixing and stirring in the belt must be relatively minor.

The existence of the asteroid belt raises some interesting questions about the origin of solar system. Why is there such a depletion in mass in this belt in comparison to that predicted by the interpolation of planetary masses? Was there ever a single, small planet in the asteroid belt, and if not, why? Cooling rate data from iron meteorites that come from the asteroids, as well as an estimate of the tidal forces of Jupiter, indicate that a single planet never existed in the asteroid belt. The tidal forces of Jupiter would fragment the planet before it grew to planetary size. Hence, it appears that the asteroids accreted directly from the solar nebula as small bodies and have subsequently been broken into even smaller fragments by continuing collisions. The depletion of mass in the asteroid belt may also be caused by Jupiter. Because of its large gravitation field, it is likely that

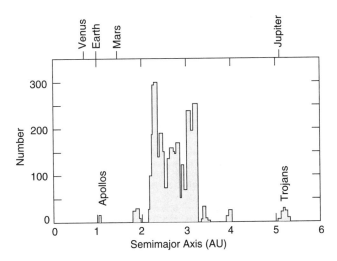

Figure 10.12   Histogram of the frequency of asteroid distances from the Sun. Distances from the Sun are in astronomical units (1 AU = Earth–Sun distance).

Jupiter swept up most of the mass in this part of the solar system and ejected it out of the solar system. It would appear that asteroid growth stopped in most bodies when they reached about 100 km in diameter as the belt ran out of material. Furthermore, the preservation of what appears to be basaltic crust on some asteroids suggests that they have survived for 4.55 Gy, when melting occurred in many asteroids as determined by dating fragments of these bodies that arrived on the Earth as meteorites.

Table 10.2   Characteristics of Major Asteroid Classes

Classes	Characteristics
Low-Albedo Classes	
C	Common in the outer part of the asteroid belt at 3 AU; perhaps parental to carbonaceous chondrites
D	Found close to the orbit of Jupiter (5 AU); no meteorite analogues
P	Common near 4 AU; no meteorite analogues
Moderate-Albedo Classes	
M	Common in main asteroid belt (3 AU); similar to Fe meteorites
Q	Apollo and related asteroids that cross the Earth's orbit; parental to major chondrite meteorite groups
S	Common in inner parts of asteroid belt (2 AU); some in Earth-crossing orbits; may be parental to pallasites and some Fe meteorites
V	4 Vesta and related asteroids about 2 AU; parental to some basaltic achondrites

Meteorites

Meteorites are small extraterrestrial bodies that have fallen on the Earth. Most meteorites fall as showers of many fragments, and more than 3000 individual meteorites have been described. Meteorites range in size from dust particles to bodies hundreds of meters across. Those with masses greater than 500 gm fall on the Earth at a rate of about one per 10^6 km^2 per year. Meteorites have also been found on the Moon's surface and presumably occur on other planetary surfaces. One of the best preserved and largest suites of meteorites is found within the ice sheets of Antarctica. To avoid contamination, when they are chopped out of the ice, they are given special care and documentation similar to the samples collected on the Moon's surface. Trajectories of meteorites entering the Earth's atmosphere have been measured and indicate that most come from the asteroid belt. Others are fragments ejected off the lunar surface or other planetary surfaces by impact, and some may be remnants of comets. Most meteorites appear to have been produced during collisions between asteroids. Meteorites date from 4.56 to 4.55 Ga, supporting an origin as fragments of asteroids formed during the early stages of accretion in the solar nebula. Many meteorites are breccias in that they are composed of an amalgamation of angular rock fragments tightly welded together. These breccias formed during collisions on the surfaces of asteroids. In some cases, melting occurred around fragment boundaries as reflected by the presence of glass.

Meteorites are classified as *stones* (including chondrites and achondrites), *stony-irons,* and *irons,* depending on the relative amounts of silicates and Fe-Ni metal phases present. *Chondrites,* the most widespread meteorites, are composed partly of small silicate spheroids known as *chondrules* (Fig. 10.13) and have chemical compositions similar to the Sun. *Achondrites,* which lack chondrules, commonly have igneous textures, appear to have crystallized from magmas, and thus preserve the earliest record of magmatism in the solar system. Some achondrites are breccias that probably formed on asteroid surfaces by impact. Stony-irons and irons have textures and chemical compositions that suggest they formed in asteroid interiors by fractionation and the segregation of melts. The metal in meteorites is composed of two phases of Fe-Ni, which are intergrown and produce the Widmanstatten structure visible on polished surfaces of iron meteorites.

Chondrites

Chondrites are the most common meteorites and are composed of two components: chondrules and matrix (Fig. 10.13). Chondrules have a restricted size range of 0.2 to 4 mm in diameter, with most less than 1 mm. Despite considerable diversity in the composition of individual chondrules, mean compositions of chondrules from various groups of meteorites are similar, suggesting that they were well mixed before accreting into parent bodies. One group of chondrites, the **carbonaceous chondrites,** is of special interest. These meteorites are hydrated, contain carbonaceous matter, and have not been subjected to temperatures greater than 200° C, or carbonaceous compounds would not survive. They consist of a matrix of hydrated Mg silicates (principally chlorite and

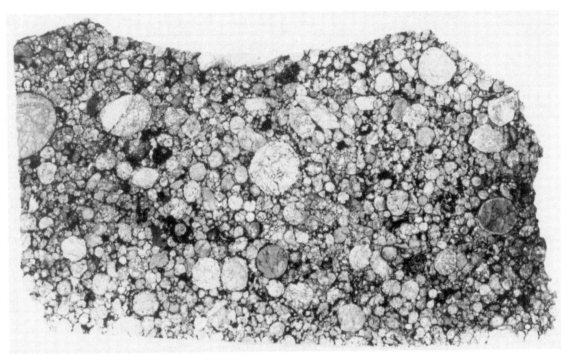

Figure 10.13   Thin section of the Inman chondrite. The photomicrograph shows chondrules (circular bodies), which are about 1 mm in diameter. The chondrules, which formed by rapid cooling of liquid droplets, are composed of olivine and pyroxene crystals surrounded by metallic iron (black). The long dimension is about 2.5 cm. Courtesy of Rhian Jones.

serpentine) enclosing chondrules of olivine and pyroxene. An important chemical feature of carbonaceous chondrites is that they contain elements in approximately solar ratios, suggesting that they are primitive, and many investigators think that one class of carbonaceous chondrites contains samples of the primitive solar nebula from which the solar system formed. Matrices typically show wide variations in chemical and mineralogical composition that are thought to reflect differences in chemical composition within the solar nebula.

Although it is clear that chondrules are the products of rapid cooling of liquid droplets, it is not yet agreed how and where in the solar nebula this occurred. Relative to primitive carbonaceous chondrites, chondrules are enriched in lithophile elements and depleted in siderophile and chalcophile elements. This provides an important boundary condition for chondrule origin in that the material from which they formed must have undergone earlier melting to fractionate elements before chondrule formation. This would seem to eliminate an origin for chondrules by direct condensation from the solar nebula. An alternative to nebular condensation is that chondrules formed by the melting of preexisting solids in the solar nebula, the only mechanism that can explain all of the physical and chemical constraints. It would appear that metal, sulfide, and silicate phases must have been in the nebula before chondrule formation and that chondrules formed by rapid melting and cooling of these substances. What caused the melting? Perhaps there were nebular flares, analogous to modern solar flares, that released sudden bursts of energy into the nebular cloud and instantly melted local clumps of dust, which chilled and formed chondrules. Later, these chondrules were accreted into asteroids.

Shergotty, Nakhla, and Chassigny Meteorites

The SNC meteorites have distinct chemical compositions requiring that they come from a rather evolved planet (Marti et al., 1995). SNC meteorites are fine-grained, igneous cumulates of mafic or komatiitic composition. The most convincing evidence that they have come from Mars is the presence of a trapped atmospheric component similar to the composition of the Martian atmosphere, as determined from spectral studies. Also, the major element composition of shergottites is similar to the compositions measured by the Viking Lander on the Martian surface. Because of the problems of ejecting material from the Martian surface, it is possible that the SNC meteorites were all derived from a single, large impact that occurred about 200 Ma. Sm-Nd isotopic ages suggest that most of these meteorites crystallized from magmas about 1.3 Ga at shallow depths. At least one Martian meteorite from Antarctica, however, yields an age of about 4.5 Ga and appears to represent a fragment of the Martian ancient cratered terrain, possibly some of the oldest crust preserved in the solar system.

Refractory Inclusions

Meteorite breccias contain a variety of components, which have been subjected to detailed geochemical studies. Among these are inclusions rich in refractory elements (such as Ca, Al, Ti, and Zr) known as Ca- and Al-rich inclusions (CAIs), which range in size from dust to a few centimeters across. Stable isotopic compositions of these inclusions indicate that they are foreign to the solar system (Taylor, 1992). Because the sequence of mineral appearance in CAIs does not follow that predicted by condensation in a progressively cooling solar nebula and differs from inclusion to inclusion, it is probable that local rather than widespread heating occurred in their nebular sources. Some appear to be direct condensates from a nebular cloud; others show evidence favoring an origin as a residue from evaporation. The ages of CAIs indicate that they became incorporated into the solar nebula during the early stages of condensation and accretion. Where they came from and how they became incorporated in the solar nebula, however, remain mysteries.

Iron Meteorites and Parent Body Cooling Rates

It is likely that most iron meteorites come from the cores of asteroids. For such cores to form, the parent bodies must have melted soon after or during accretion, and molten Fe-Ni must have settled to the center of the bodies. It is possible to constrain the cooling rates of iron meteorites from the thickness and Ni content of kamacite bands. This, in turn, provides a means of estimating the size of the parent body, because smaller parent bodies cool faster than larger ones. Calculated cooling rates are generally in the range of 1 to 10° C/My, which indicates an upper limit for the radius of parent bodies of 300 km, with most lying between 100 and 200 km. Such results clearly eliminate the possibility that a single planet was the parent body for meteorites and asteroids. This conclusion is consistent with that deduced from estimates of Jupiter's tidal forces, which indicate that a single planet could not form in the asteroid belt.

Asteroid Sources

From a combination of mineralogical and spectral studies of meteorites and from spectral studies of asteroids, it has been possible to assign the possible parent bodies of some meteorites to specific asteroid groups (Febofsky et al., 1989). The most remarkable spectral match is between the visible spectrum of the third-largest asteroid, 4 Vesta, and a group of meteorites known as basaltic achondrites (Fig. 10.14). Supporting this source, 4 Vesta occurs in an orbit with a 3:1 resonance, which is an "escape hatch" for material knocked off 4 Vesta to enter the inner solar system.

Most groups of meteorites do not seem to have spectral matches among the asteroids, including the most common meteorites, the chondrites. This may be because of what is generally referred to as *space weathering* of asteroid surfaces, which changes their spectral characteristics (Pieters and McFadden, 1994). However, no evidence to support this idea has yet been recovered. Another possibility is that the asteroid parents of most chondrites are smaller than investigators can resolve with remote sensing on the Earth.

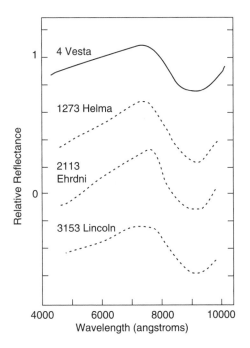

Figure 10.14
Comparison of the visible reflectance of asteroid 4 Vesta with three basaltic achondrite meteorites that may have been derived from Vesta.

Meteorite Chronology

Sm-Nd, Rb-Sr, and Re-Os isochron ages record the times of accretion and partial melting in meteorite parent bodies. These ages cluster between 4.56 and 4.53 Ga, which is currently the best estimate for the age of the solar system (Fig. 10.15). The oldest reliably dated objects in the solar system are the CAIs in the Allende meteorite at 4566 Ma. The oldest Sm-Nd ages that reflect melting of asteroids are from basaltic achondrites at 4539 ± 4 Ma. Iron meteorite ages range from 4.56 to 4.46 Ga. K-Ar dates from meteorites reflect cooling ages and generally fall in the range of 4.4 to 4.0 Ga.

Fragmentation of asteroids by continual collisions exposes new surfaces to bombardment with cosmic rays. Interactions of cosmic rays with elements in the outer meter of meteorites produces radioactive isotopes that can be used to date major times of parent body breakup (Marti and Graf, 1992). Stone meteorites have cosmic-ray exposure ages of 100 to 5 Ma, as illustrated by L-chondrites (Fig. 10.16), and irons are chiefly 1000 to 200 Ma. The peak about 40 Ma in the L-chondrite exposure ages is interpreted as a major collisional event between asteroids at this time. The differences in exposure ages between

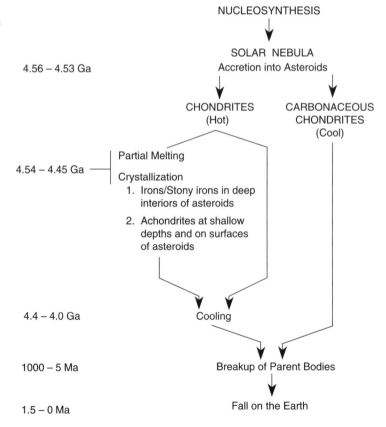

Figure 10.15 Summary of the history of meteorites and asteroids.

NUCLEOSYNTHESIS

SOLAR NEBULA
Accretion into Asteroids

4.56 – 4.53 Ga

CHONDRITES (Hot)   CARBONACEOUS CHONDRITES (Cool)

4.54 – 4.45 Ga

Partial Melting

Crystallization
1. Irons/Stony irons in deep interiors of asteroids
2. Achondrites at shallow depths and on surfaces of asteroids

4.4 – 4.0 Ga   Cooling

1000 – 5 Ma   Breakup of Parent Bodies

1.5 – 0 Ma   Fall on the Earth

Figure 10.16 Histogram of cosmic-ray exposure ages of L-chondrites. Modified from Marti and Graf (1992).

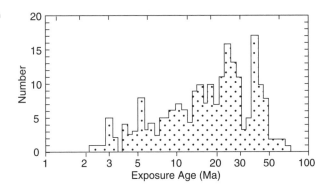

stones and irons reflect chiefly that irons are more resistant to collisional destruction than stones. After a meteorite falls on the Earth, it is shielded from cosmic rays. The amount of parent isotope remaining can be used to calculate a **terrestrial age,** the time at which the meteorite fell on the Earth's surface and became effectively cut off from a high cosmic-ray flux. Although most terrestrial ages are less than 100 years, some as old as 1.5 Ma have been reported.

Chemical Composition of the Earth and the Moon

Because it is not possible to sample the interior of the Earth and the Moon, indirect methods must be used to estimate their composition (Hart and Zindler, 1986; McDonough and Sun, 1995). It is generally agreed that the Earth and other bodies in the solar system formed by condensation and accretion from a solar nebula and that the composition of the Sun roughly reflects the composition of this nebula. Nucleosynthesis models for the origin of the elements also provide limiting conditions on the composition of the planets. As you saw in Chapter 4, it is possible to estimate the composition of the Earth's upper mantle from the analysis of mantle xenoliths and basalts, both of which transmit information about the composition of their mantle sources to the surface. Meteorite compositions and high-pressure experimental data also provide important input on the overall composition of the Earth.

Shock-wave experimental results indicate a mean atomic weight for the Earth of about 27 (mantle = 22.4 and core = 47.0) and show that it is composed chiefly of iron, silicon, magnesium, and oxygen. When meteorite classes are mixed to give the correct core/mantle mass ratio (32:68) and mean atomic weight of the Earth, results indicate that iron and oxygen are the most abundant elements followed by silicon and magnesium (Table 10.3). Almost 94% of the Earth is composed of these four elements.

From lunar heat-flow results, correlations among refractory elements, and density and moment of inertia considerations, it is possible to estimate bulk lunar composition

Table 10.3 Major Element Composition of the Earth and the Moon

	Earth[1]	Earth[2]	Moon[3]
Fe	29.9	28.2	8.3
O	30.9	32.4	44.7
Si	17.4	17.2	20.3
Mg	15.9	15.9	19.3
Ca	1.9	1.6	3.2
Al	1.4	1.5	3.2
Ni	1.7	1.6	0.6
Na	0.9	0.25	0.06
K	0.02	0.02	0.01
Ti	0.05	0.07	0.2

Values in weight percentages.

[1] Nonvolatile portion of Type-I carbonaceous chondrites with FeO/FeO+MgO of 0.12 and sufficient SiO_2 reduced to Si to yield a metal/silicate ratio of 32:68 (Ringwood, 1966).

[2] From Allegre et al. (1995b).

[3] Based on Ca, Al, Ti = 5× Type-I carbonaceous chondrites; FeO = 12% to accommodate lunar density; and Si/Mg = chondritic ratio (from Taylor, 1982).

(Table 10.3). Compared to the Earth, the Moon is depleted in Fe, Ni, Na, and S and is enriched in other major elements. The bulk composition of the Moon is commonly likened to the composition of the Earth's mantle because of similar densities. The data indicate, however, that the Moon is enriched in refractory elements such as Ti and Al compared with the Earth's mantle.

To explain planetary formation, elements can be divided into three geochemical groups. **Volatile elements** are those elements that can be volatilized from silicate melts under moderately reducing conditions at temperatures below 1400° C. **Refractory elements** are not volatilized under the same conditions. Refractory elements can be further subdivided into **oxyphile** and **siderophile groups** depending on whether they follow oxygen or iron, respectively, under moderately reducing conditions. Both the Earth and the Moon differ from carbonaceous chondrites in the distribution of elements in these groups (Fig. 10.17). Compared with carbonaceous chondrites, both bodies are depleted in volatile and siderophile elements and enriched in oxyphile refractory elements. Any model for the accretion of the Earth and the Moon from the solar nebula must explain these peculiar element distribution patterns.

Vertical zoning of elements has occurred in both the Earth and the Moon. This zoning is the result of element **fractionation,** the segregation of elements with similar geochemical properties. Fractionation results from physical and chemical processes such as condensation, melting, and fractional crystallization. Large-ion lithophile elements (such as K, Rb, Th, and U) have been strongly enriched in the Earth's upper mantle and even more so in the crust in relation to the mantle and core. In contrast, siderophile refractory elements (such as Mn, Fe, Ni, and Co) are concentrated chiefly in the mantle or core, and oxyphile refractory elements (such as Ti, Zr, and La) are found mainly in the mantle.

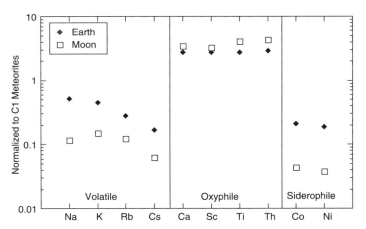

Figure 10.17   Volatile, oxyphile, and siderophile element distributions in the Earth and the Moon normalized to Type-I carbonaceous chondrite (CI) meteorites.

Age and Early Evolution of the Earth

Extinct Radioactivity

Radiogenic isotopes that were present when the solar system formed and that have short half-lives are no longer found in materials accreted into planetary bodies. These isotopes, with half-lives of a few million to tens of millions of years, can be useful in constraining the timing of accretionary events in the earliest stages in the history of the solar system. Their usefulness depends not only on their half-lives but also on how effectively accretion fractionates the parent from the daughter isotope. If this fractionation is strong, anomalous amounts of the daughter isotope will occur in planetary bodies today and will show up in the Earth's mantle and in basalts derived from the mantle. From the amount of daughter isotope present (over background level), ages of early fractionation events can be estimated. The most useful isotopic pairs are ^{92}Nb/^{92}Zr ($t_{1/2}$ = 36 My), ^{182}W/^{182}Hf ($t_{1/2}$ = 9 My), ^{146}Sm/^{142}Nd ($t_{1/2}$ = 103 My), ^{129}I/^{129}Xe ($t_{1/2}$ = 16 My), and ^{244}Pu/^{136}Xe ($t_{1/2}$ = 82 My). An example of using the ^{182}W/^{182}Hf method to constrain the age of the Earth's core was given in Chapter 5.

First 700 Million Years

Because the Earth is mostly inaccessible to sampling and it is a continuously evolving system, it is difficult to date (Zhang, 2002). Contributing to this difficulty, the Earth accreted over some interval; hence, if an age is obtained, what event in this interval does it date? The first isotopic ages of the Earth were model Pb ages of about 4.55 Ga obtained from sediments and oceanic basalts (Patterson et al., 1955). Until recent precise ages were obtained from meteorites, other attempts to date the Earth have yielded ages in the range from 4.55 to 4.45 Ga. Reconsideration of model Pb ages indicates that the first major differentiation event in the Earth occurred about 4.45 Ga. It would thus seem that the Earth is 100 My younger than the most primitive meteorites.

As previously mentioned, refractory inclusions (CAIs) in the Allende meteorite are the oldest dated objects in the solar system at 4566 Ma. The accretion of most chondrites began a maximum of about 3 My later and lasted no more than 8 My. The model Pb ages from the Earth probably represent the mean age of core formation, when iron was separated from the mantle and U and Pb were fractionated from each other. Both of these ages are strikingly similar from 4460 to 4450 Ma, suggesting that both processes went on simultaneously during the terminal stages of planetary accretion (Fig. 10.18). This being the case, the obtained age of the Earth is the age of early differentiation, and how this age is related to the onset of planetary accretion is not well known. Recent high-resolution measurements using the $^{182}Hf/^{182}W$ method suggest that bodies in the inner part of the solar system all formed in the first 30 My after accretion began (Yin et al., 2002). Most of the Earth's growth was in the first 10 My, and it was 80 to 90% complete by 30 My after time zero (about 4525 Ma) (Fig. 10.18).

Although the oldest ages of rocks dated from the Moon are about 4450 Ma, model ages suggest that the Moon accreted 4500 to 4480 Ma. Hf isotopic data, however, suggest that the Moon formed about 30 My after solar system formation, when the Earth was complete or nearly complete (Yin et al., 2002). If the Moon was formed by accretion of material left after a Mars-size body hit the Earth, as described later, it would appear that this impact occurred from 4530 to 4520 Ma. If the Earth began to accrete when the CAIs accreted at 4566 Ma, then the total time between accretion and the first major melting event from 4450 to 4400 Ma is 75 to 100 My (Fig. 10.18).

In its early stages of growth, the Earth probably had a magma ocean sustained by the heat sources described previously and especially from impacts. All or most of the early atmosphere must have been lost at or before 4500 Ma by early T-Tauri events and planetary

Figure 10.18 Isotopic timescale for the accretion of meteorites, the Earth, and the Moon. CIA, Chemical Index of Alteration.

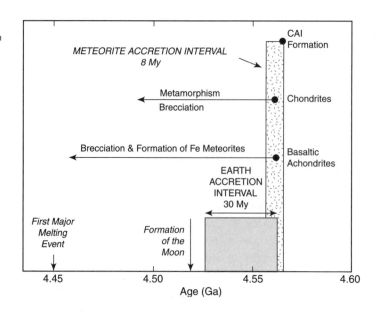

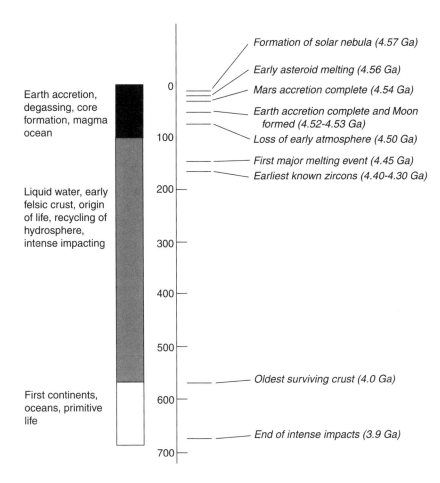

Earth accretion, degassing, core formation, magma ocean

Liquid water, early felsic crust, origin of life, recycling of hydrosphere, intense impacting

First continents, oceans, primitive life

Formation of solar nebula (4.57 Ga)

Early asteroid melting (4.56 Ga)

Mars accretion complete (4.54 Ga)

Earth accretion complete and Moon formed (4.52-4.53 Ga)

Loss of early atmosphere (4.50 Ga)

First major melting event (4.45 Ga)

Earliest known zircons (4.40-4.30 Ga)

Oldest surviving crust (4.0 Ga)

End of intense impacts (3.9 Ga)

Figure 10.19 Summary of the first 700 My of the Earth's history.

collisions (Fig. 10.19). As the magma ocean rapidly cooled, the outer part of the Earth would have solidified and the first primitive crust would have formed. Because of intense bombardment of the inner solar system about 3900 Ma, little of this early crust survives, although detrital zircons with ages of 4.3 to 4.4 Ga are probably derived from this early crust (Fig. 10.19).

Comparative Evolution of the Atmospheres of the Earth, Venus, and Mars

The only three terrestrial planets to have retained atmospheres are the Earth, Venus, and Mars, yet the compositions and densities of their atmospheres differ significantly (Table 6.1). The greenhouse effect, the distance from the Sun, and the planetary mass all may have played roles in producing such different atmospheres (Prinn and Fegley, 1987; Hunten, 1993). Planetary surface temperature is controlled largely by the greenhouse effect; hence, only when significant amounts of CO_2, water, or CH_4 accumulate in the atmosphere will the surface temperature begin to rise. This can be illustrated by the effect of progressively increasing water pressure in an early planetary atmosphere (Walker, 1977). Beginning with estimated surface temperatures as dictated by the distance from the Sun, water vapor will increase in each atmosphere in response to planetary degassing until it intersects the water vapor saturation curve (Fig. 10. 20). For the Earth, this occurs about 0.01 atm when water begins to condense and precipitate on the surface. Further degassing does not increase the water content of the atmosphere or the surface temperature but rains out and begins to form the oceans, a process that occurred on the Earth probably between 4.53 and 4.50 Ga. The growing oceans stabilize the CO_2 content of the atmosphere through the carbon cycle described in Chapter 6. Because Venus is closer to the Sun, it had a higher initial surface temperature than the Earth. This results in the greenhouse effect raising the surface temperature before the vapor saturation curve is intersected. As degassing continues, water and CO_2 may rapidly accumulate in the atmosphere, producing a **runaway greenhouse** in which atmospheric temperature continues to rise

Figure 10.20 Idealized trajectories for increasing atmospheric water vapor pressure on Venus, the Earth, and Mars. Modified from Walker (1977).

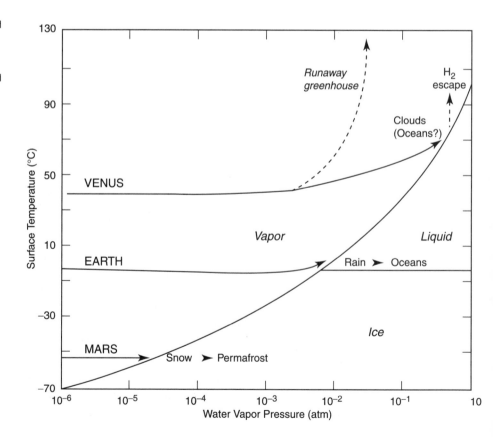

indefinitely (Fig. 10.20). The reason water is virtually absent in the Venusian atmosphere today is because of photolysis in the upper atmosphere that produces H_2, which readily escapes. The remaining O_2 may be incorporated in weathered rocks on the surface and perhaps returned to the Venusian mantle by tectonic processes. It is noteworthy that the amount of CO_2 in the Venusian atmosphere is approximately equivalent to the amount in near-surface reservoirs on the Earth (most of it in marine carbonates). Hence, if it were not for the oceans that provide a vital link in the carbon cycle, the Earth would probably have an uninhabitable atmosphere like Venus.

Mars is farther from the Sun than the Earth and has a colder initial surface temperature. Hence, as water increases in the atmosphere and intersects the vapor saturation curve, snow rather than liquid water precipitates (Fig. 10.20). As with the Earth, neither the surface temperature nor the atmospheric water vapor pressure increases with further degassing. Upon further cooling (because of the lack of a greenhouse effect), CO_2 would also freeze, forming the Martian polar ice caps. The sparsity of light gases such as H_2 and N_2 on Mars probably reflects the relatively small mass of the planet and its inability to retain light gases, which escape from its gravitational field.

Although this model accounts for the general features of the evolution of the atmospheres of the Earth, Venus, and Mars, some aspects require modification. For instance, the model assumes that Mars was cold and frozen from the beginning, an unlikely assumption in terms of the large canyons on the surface that were almost certainly cut by running water before 4 Ga, as previously explained (Carr, 1987). Therefore, it is probable that an early CO_2 greenhouse effect existed on Mars, and this necessitates some kind of tectonic recycling of CO_2. Because Mars is too small to retain oceans and it is unlikely that plate tectonics ever operated, it is unlikely that an Earth-like carbon cycle controlled CO_2 levels in the early atmosphere. Although the nature of the Martian early recycling system is unknown, the planet must have cooled more rapidly than the Earth, perhaps trapping most of the CO_2 and at least some of the water in its interior. Because of insufficient greenhouse heating, the surface cooled and most of the remaining CO_2 condensed as ice on the surface. It is probable that today most of the degassed water and some of the CO_2 on Mars occur as permafrost near the surface.

The runaway greenhouse model for Venus assumes that Venus never had oceans, another questionable assumption in terms of the high deuterium/hydrogen (D/H) ratio in trace amounts of water in the Venusian atmosphere. This ratio, which is 100 times higher than that in the Earth's water, may have developed during preferential loss of H_2 from the Venusian atmosphere, enriching the residual water in the heavier deuterium isotope, which has a much smaller escape probability than hydrogen (Gurwell, 1995). Although this observation does not prove the existence of early oceans on Venus, the planet at one time must have had at least 10^3 times more water than at present, either in oceans or as water vapor in the atmosphere. An alternative and preferred evolutionary path for Venus leads to the condensation of water from 4.5 to 4.4 Ga, the formation of water vapor clouds, and possibly the formation of oceans (Fig. 10.20). On the Earth, water is blocked from entering the stratosphere by a cold trap (a temperature minimum) 10 to 15 km above the surface (Fig. 6.1), and most water vapor condenses when it reaches the cold trap.

However, if water vapor in the lower atmosphere exceeds 20%, as it must have on Venus, it produces greenhouse warming. This moves the cold trap to a high altitude (~100 km), where it no longer is effective in preventing water vapor from rising into the upper atmosphere, where the vapor can undergo photolysis and hydrogen escape. Thus, on Venus, hydrogen escape could have eliminated an oceanic volume of water in less than 30 My. Some of the missing water also may be housed in hydrous minerals in the interior of Venus (Lecuyer et al., 2000). Following the early loss of water, the Venusian atmosphere would rapidly evolve into a runaway greenhouse caused by the remaining CO_2, and this is the situation that exists today on Venus.

Continuously Habitable Zone

Life requires a narrow range in surface temperature and hence a narrow range in the content of greenhouse gases. From knowledge of temperature and gas distributions in the inner solar system, the width of this zone, known as the **continuously habitable zone,** can be estimated. If the Earth, for instance, had accreted in an orbit 5% closer to the Sun, the atmosphere would continue to rise in temperature and the oceans would evaporate leading to the runaway greenhouse described previously. The outer extreme of the continuously habitable zone is less certain but appears to extend somewhat beyond the orbit of Mars. Thus, the continuously habitable zone in the solar system would appear to extend from about 0.95 to 1.5 AU (Kasting et al., 1993b). This observation is important in that the continuously habitable zone is wide enough that habitable planets may exist in other planetary systems.

Although in any planetary system most life, and all higher forms of life, would be limited to the continuously habitable zone, some microbes that can withstand extreme conditions, as described in Chapter 7, may survive outside of this zone; in particular, they may survive on some satellites of the Jovian planets.

Condensation and Accretion of the Planets

Solar Nebula

Although many models have been proposed for the origin of the solar system, only those that begin with a gaseous dust cloud appear to be consistent with data from both astrophysics and cosmochemistry (Taylor, 1999). Stars form by contraction of gaseous nebulae and planets and other bodies form by condensation and accretion as nebulae cool (Hartmann, 1983). Dust around main sequence stars, such as Vega, can be detected by excess infrared radiation above the level expected from the stellar photosphere (Weinberger, 2002).

Just how the solar system formed can be considered in terms of three questions:

1. How did the Sun acquire the gaseous material from which the planets formed?
2. What is the history of condensation of the gaseous material?
3. What are the processes and history of planetary accretion?

Regarding the first question, one viewpoint is that the Sun, already in existence, attracted material into a gaseous nebula about itself. Another proposes that the preexisting Sun captured a solar nebula of appropriate mass and angular momentum to form the solar system. Most theories, however, call upon condensation and accretion of the Sun and planets from the same cloud approximately the same time. All models have in common a gaseous nebula from which the planets form. The minimum mass of such a nebula is about 1% of a solar mass. The mechanisms by which the nebula becomes concentrated into a disk with the Sun at the center are not well understood. One possibility is the transferring of angular momentum from the Sun to the nebula, caused either by hydromagnetic coupling during rotational instability of the Sun or by turbulent convection in the nebula. Another possibility is a nearby supernova, which may have triggered the collapse of the solar nebula. In either case, condensing matter rapidly collapses into a disk about the Sun (Boss et al., 1989; Wetherill, 1994) or into a series of Saturn-like rings, which condense and accrete into the planets (Fig. 10.21). Small **planetesimals** form within the cloud and spiral toward the ecliptic plane, where they begin to collide with one another and grow into "planetary embryos," which range in size from that of the Moon to that of Mercury. It is these embryos that collide and grow into the terrestrial planets and the silicate–ice cores of the giant outer planets. Most of the gas in the nebula was swept away within a few million years by high-energy winds emanating from the Sun. Because the gas giants had to form before this gas left the solar system, Jupiter and Saturn must have formed early. About 5 AU, it was cold enough that water being swept outward froze into ice, piled up, and aggregated to form the core of Jupiter. This embryo grew to about 10 times the size of the Earth in as little as 500,000 years, when it began to capture escaping gases and grew into a giant. Uranus and Neptune accumulated later because they are composed chiefly of ice and silicates with little gas, the latter of which had escaped before their formation.

Two processes are important in planet formation. **Condensation** is the production of solid dust grains as the gases in the solar nebula cool. **Accretion** is the collision of the dust grains to form clumps and progressively larger bodies, some of which grow into planets (Fig. 10.22). Condensation begins during the collapse of the nebula into a disk and leads to the production of silicate and oxide particles as well as to other compounds, all of which are composed chiefly of Mg, Al, Na, O, Si, Fe, Ca, and Ni. In the cooler parts

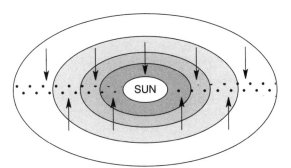

Figure 10.21 Schematic cross-section of the primitive solar nebula. Temperatures in the nebula decrease outward from the Sun from hottest (dark gray) to coolest (white). Arrows indicate the direction of the gravitational sinking of solid particles into the ecliptic.

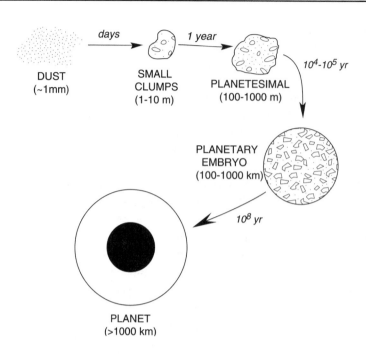

of the nebula, ices of C, N, O, Ne, S, Ar, and halogens form. The remaining gaseous mixture is composed chiefly of H_2 and He. The ices are especially abundant in the outer part of the disk where the giant planets form, and silicate–oxide particles are concentrated in the inner part and produce the terrestrial planets. Thermodynamic considerations indicate that iron should be present in the cloud initially only in an oxidized state. Type-I carbonaceous chondrites may represent a sample of this primitive nebula. These meteorites contain only oxidized iron and large amounts of volatile components of the preceding ices (including organic compounds), and they have not been heated to more than 200° C to preserve their volatile constituents.

Although it is commonly assumed that the solar nebula was well mixed, differences in isotopic abundances and ages of various meteorites suggest that it was not (Wasson, 1985; Taylor, 1992). Variations in the abundance of rare gases and isotopic variations of such elements as Mg, O, Si, Ca, and Ba indicate large mass-dependent fractionation effects. The isotopic anomalies appear to require a sudden injection of neutrons into the early solar nebula. The source of such a neutron burst may have been one or more supernova, which also may have triggered the collapse of the solar nebula.

Planetary accretion models fall into two general categories. **Homogeneous accretion** involves condensation of a nebula followed by the accretion of planets and other bodies as the nebula continues to cool. In contrast, **heterogeneous accretion** is when condensation and accretion occur at the same time. The specific mechanisms by which particulate matter evolves into small, meter-size bodies and by which these bodies, in turn, grow into planetesimals and planets are becoming better understood. Most experimental data suggest that grains stick together during the early stages of accretion by weak electrostatic forces (Van der Waal forces) important in producing centimeter-size objects. Theoretical studies indicate that a large fraction of the gaseous cloud will condense and accrete rapidly into several hundred asteroid-size bodies rather than into a swarm of small particles.

Computer models suggest that the first generation of planetesimals will increase in radius to 100 to 1000 m in about 1 year after condensation (Fig. 10.22). Wetherill (1986; 1994) has simulated further planetary growth by stochastic modeling using a high-speed computer. His results indicate that the planetesimals grow into planets in two distinct stages. During the first stage (10^4–10^5 years), planetesimals collide at low velocities and rapidly grow into planetary embryos. These embryos are commonly referred to as runaway embryos because in each local zone of the nebula a single body grows more rapidly than its neighbors within the zone and because before it stops growing it cannibalistically consumes its neighbors. In the second stage, which lasts about 10^8 years, these embryos collide with each other to form the terrestrial planets and cores of the outer giant planets. During this stage, there would be widespread mixing of accreting materials in the inner solar system as well as "giant impacts" in which one embryo collides with another. It is during this stage that a Mars-size body may have collided with the Earth to form the Moon and that another body collided with Mercury, stripping it of much of its mantle. The different tilts and spins of the planets also support massive impacts during the early history of the solar system. The terrestrial planets should have grown to within 50 km of their final radius in the first 25 My, and accretion should have been 98% complete by 10^8 years. These times are consistent with the $^{182}W/^{182}Hf$ isotopic ages for the formation of the Earth (Fig. 10.18).

The Earth and other inner planets accreted long after the outer giant planets from volatile-poor planetesimals that were probably already differentiated into metallic cores and silicate mantles in a gas-free inner part of the nebula (Taylor, 1999). During the time of supercollisions, the Earth was probably struck at a high velocity many times by bodies ranging up to Mercury in size and at least once by a Mars-size body. Much of the kinetic energy associated with these collisions would have been expended in heating the interior of the planet, and calculations show that the Earth should have been largely melted long before it completed its growth. Thus, a widespread magma ocean would exist on the planet. It is during this stage that the Earth's core formed and the planet was extensively degassed to form the atmosphere. Core formation was the first step in producing a zoned planet.

The giant planets, Jupiter and Saturn, consist chiefly of H_2 and He accreted onto rocky cores that also grew by planetesimal collisions. Once a planet reaches 10 to 50 Earth masses, it is able to attract H_2 and He from the nebula and to grow into a giant planet. Two models have been proposed for the origin of the outer planets (Pollack and Bodenheimer, 1989). In the first model, the cores of these planets form by aggregating refractory components with volatile elements accreting to make the mantles. One problem with this model is that the elements in the core should be highly soluble at the high temperatures and pressures in the planetary interiors; hence, a core should not form. The model consistent with most data involves the accretion of silicates and ices to form planetary cores followed by the accretion of gaseous components to about 15 Earth masses and then by rapid accretion of H_2 and He. In the case of Jupiter, a massive core of about 10 Earth masses was able to form in less than a million years. Once formed, Jupiter dominated the later evolution of the solar system because of its tremendous mass and gravity. It depleted both the asteroid belt and the region in which Mars would later form of material, thus accounting for the small amount of mass in this region, including the small size of Mars.

Heterogeneous Accretion Models

Heterogeneous accretion models call upon planetary growth by simultaneous condensation and accretion of various compounds as the temperature falls in an originally hot solar nebula (Ringwood, 1979). The product is a zoned planet (Grossman, 1972). Cameron (1973) has suggested a model whereby the planets develop in a rotating, disk-shaped solar nebula as it cools. Cooling the gas as dissipation proceeds results in condensation over a range of temperatures. The sequence of compounds condensed from the nebula at a pressure of 10^{-4} atm is summarized in Table 10.4. Heterogeneous accretion should produce a planet with high-temperature refractory components in the center overlain sequentially by metal, various silicates, compounds with oxidized iron, and hydrated silicates–oxides and ices. At the surface there should be a deep magma ocean (Righter, 2003).

Although it is appealing that scientists can produce a zoned planet directly, heterogeneous accretion models are faced with many obstacles. For instance, there is a large degree of overlap in the condensation temperatures given in Table 10.4; hence, a zoned planet is not clearly predicted. Also, the model predicts an Earth with a small core of Ca-Al-Ti oxides, which is not consistent with geophysical data. The problem is not alleviated if iron melts and sinks to the center, displacing the oxide layer upward, because there is no seismic evidence for a Ca-Al-Ti layer above the core. These, with several

Table 10.4 Condensation Sequence from the Solar Nebula at 10^{-4} atm	
Phases	**Temperature (°C)**
Ca, Al, Ti and related refractory oxides	1250–1600
Metallic Fe-Ni	1030–1200
Forsterite (Mg_2SiO_4) and enstatite ($MgSiO_3$)	1030–1170
Ca-plagioclase	900–1100
Na-K-feldspars	~730
Troilite (FeS)	430
Fe-Mg pyroxenes and olivine	300
Fe oxides	100–200
Carbonaceous compounds	100–200
Hydrated Mg silicates	0–100
Ices	< 0
Modified from Grossman (1972).	

other significant geochemical and isotopic problems, seem to present insurmountable stumbling blocks for a strictly heterogeneous accretion model.

Homogeneous Accretion Models

In homogeneous accretion models, condensation is essentially complete before accretion begins (Taylor, 1999; Drake, 2000). Compositional zonation in the solar nebula is caused by decreasing temperature outward from the Sun (Fig. 10.21), and this is reflected in planetary compositions. Refractory oxides, metals, and Mg silicates are enriched in the inner part of the cloud where Mercury accretes; Mg-Fe silicates and metal in the region from Venus to the asteroids; and mixed silicates and ices in the outer part of the nebula where the giant planets accrete. Homogeneous accretion models produce an amazingly good match between predicted and observed planetary compositions (Fig. 10.23). If, for instance, volatiles were blown outward when Mercury was accreting refractory oxides, metal, and Mg silicates, Venus and the Earth should be enriched in volatile components compared to Mercury. The lower mean density of Venus and the Earth is consistent with this prediction. Mars has a still lower density, and it is probably enriched in oxidized iron compounds relative to the Earth and Venus. Likewise, the giant outer planets are composed of mixed silicates and ices, reducing their densities dramatically.

If homogeneously accreted, how do the terrestrial planets become zoned? It must be by melting, which segregates by such processes as fractional crystallization and the

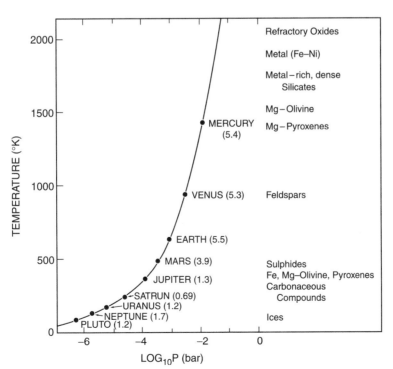

Figure 10.23 Calculated temperature and pressure distribution in the solar nebula late in planetary accretion history. Shown also is the distribution of the planets, their densities, and the predicted composition from condensation experimental data (Table 10.4).

sinking of molten iron to the core. Where does the heat come from to melt the planets and asteroids? Some of the major known heat sources are as follows:

1. *Accretional energy.* This energy depends on the impact velocities of accreting bodies and the amount of input energy retained by the growing planet. Accretional energy alone appears to have been sufficient, if entirely retained in the planet, to largely melt the terrestrial planets while they were accreting.
2. *Gravitational collapse.* As a planet grows, the interior is subjected to higher pressures and minerals undergo changes to phases with more densely packed structures. Most of these changes are exothermic, as described in Chapter 4, and large amounts of energy are liberated into planetary interiors.
3. *Radiogenic heat sources.* Radioactive isotopes liberate significant amounts of heat during decay. Short-lived radioactive isotopes, such as ^{26}Al and ^{244}Pu, may have contributed significant quantities of heat to planets during accretion. Long-lived isotopes, principally ^{40}K, ^{235}U, ^{238}U, and ^{232}Th, are important heat producers throughout planetary history.
4. *Core formation.* Core formation is a strongly exothermic process and appears to have occurred over a relatively short period (≤ 100 Ma) that began during the late stages of planetary accretion.

The order of magnitude estimates indicate that there was enough heat available during accretion of the terrestrial planets and the Moon, if retained, to completely melt these bodies. Because volatiles are retained in the terrestrial planets, however, it is unlikely that they were ever completely molten. Hence, it is necessary to remove the early heat rapidly, and convection seems the only process capable of bringing heat from planetary interiors to the surface in a short enough interval (≤ 100 Ma) to escape complete melting. During the late stages of accretion, collisions deposited enough energy, if fully retained, to partly or fully melt the Earth, perhaps multiple times (Drake, 2000). The result is a magma ocean, perhaps extending nearly to the center of the Earth. Metal would sink through the magma ocean and accumulate at the center as the core grew. The temperature dependence of mantle viscosity appears to be the most important factor controlling planetary thermal history (Carlson, 1994; Schubert et al., 2001). Initially, when a planet is hot and viscosity is low, chaotic mantle convection rapidly cools the planet and crystallizes magma oceans (Drake, 2000). As mantle viscosity increases, beginning 100 My after accretion, convection should cool planets at reduced rates.

Accretion of the Earth

With astrophysical models, the chemical composition of planets provides an important constraint on planetary accretion (Greenberg, 1989; Wetherill, 1990; Taylor, 1992). Any model for the accretion of the Earth must account for a depletion in volatile elements in the silicate portion of the planet (V–K, Fig. 10.24) relative to the solar nebula, whose composition is assumed to be that of Type-I carbonaceous chondrites. This increasing depletion with increasing condensation temperature may reflect depletions in the nebular

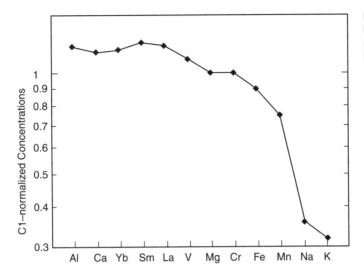

Figure 10.24 Type-I carbonaceous chondrite (CI) normalized element distributions in the Earth.

material in the region from which the Earth accreted. Alternatively, the trend could result from mixing between two components in the nebula, one a "normal" component relatively enriched in refractory elements (Al–La, Fig. 10.24) and one strongly depleted volatiles. In this case, the earliest stage of the Earth's accretion would be dominated by refractory elements and would be followed by a second stage in which mixed refractory and volatile components accrete as the temperature of the nebula continues to decrease (partly heterogeneous accretion). Depletions in volatile elements in portions of the inner solar nebula may have been caused by an intense T-Tauri wind emitted from the Sun during its early history. This wind could have blown volatile constituents from the inner to the outer part of the solar system.

What about the depletions in siderophile elements in the silicate fraction of the Earth (Fig. 10.17)? These can best be explained by core formation during which these elements (such as Co and Ni), with their strong affinities for iron, are purged from the mantle as molten iron sinks to the Earth's center to form the core. Thus, it would appear that the volatile element depletions in the Earth occurred during the early stages of planetary accretion and that the siderophile element depletions occurred during the late stages.

Volatile-depleted primitive materials have usually been assumed to represent the building blocks from which the Earth was accreted. However, there are some unique compositional differences between terrestrial materials (mantle xenoliths, and basalts) and known meteorites. These include significant differences in Mg/Si, Al/Si, D/H, Ar/H_2O, and Kr/Xe ratios and in oxygen and osmium isotopic ratios (Drake and Righter, 2002). Thus, it would appear that the Earth is not made of asteroids with the common chondritic or achondritic compositions that characterize meteorites that hit the Earth. This implies that distinct compositional zones were maintained in the dust cloud from which the solar system accreted. This means that there was not widespread mixing of the material in the inner solar system during accretion and that planets growing in the inner solar system received most of their masses from narrow rings around the Sun (Drake and Righter, 2002).

Extrasolar Planets

As the ability of scientists to detect planets around stars increases with improved instrumentation and data reduction techniques, many new planets have been discovered (more than 100 in 2003) (Bodenheimer and Lin, 2002). The chief method of planet detection is a radial-velocity survey. As a planet orbits a star, its gravity causes the star to wobble, producing a periodic change in the spectrum of light emitted by the star that can be used to infer the planet's presence. Also, the star's wobble can be measured by tracking its position. Planets can also be detected as they pass in front of the star.

The most unexpected aspect of the new planets is their peculiar properties (Taylor, 1999). Most extrasolar planets have masses of 0.75 to 3 Jovian masses, but masses less than 0.5 of a Jovian mass are difficult to detect. Unlike planets in the solar system, most extrasolar planets are much closer to their parent star than Mercury is to the Sun and they have orbital periods of only a few days. Most also have anomalous eccentricities (>0.1) compared to Jupiter (0.048) and Saturn (0.055). These properties are difficult to explain with current models for the origin of planetary systems, which work for the solar system. Clearly, the conditions that existed to make our solar system are not easily reproduced in other planetary systems.

Origin of the Moon

Scientific results from the Apollo missions to the Moon have provided a voluminous amount of data on the structure, composition, and history of the Moon (Taylor, 1982; Taylor, 1992). Seismic data are generally interpreted in terms of five zones within the Moon (from the surface inward): a plagioclase-rich crust (dominantly gabbroic anorthosite) 60 to 100 km thick, an upper mantle composed chiefly of pyroxenes and olivine 300 to 400 km thick, a lower mantle of the same composition (with olivine dominating) about 500 km thick, and 200 to 400 km of iron core.

Among the more important constraints that any model for lunar origin must satisfy are the following:

1. The Moon does not revolve in the equatorial plane of the Earth or in the ecliptic plane. The lunar orbit is inclined 5.1 degrees to the ecliptic, whereas the Earth's equatorial plane is inclined at 23.4 degrees (Fig. 10.9).
2. Tidal dissipation calculations indicate that the Moon is retreating from the Earth, resulting in an increase of 15 sec/My in the length of the day. Orbital calculations and the Roche limit indicate that the Moon has not been closer to the Earth than about 24,000 km.
3. The Moon is enriched in refractory oxyphile elements and depleted in refractory siderophile and volatile elements relative to the Earth (Fig. 10.17).

Particularly important is the low density of the Moon (3.34 gm/cm^3) compared with other terrestrial planets, which indicates the Moon is significantly lower in iron than these planets.

4. The Earth–Moon system has an anomalously large amount of angular momentum (3.45×10^{41} gm/cm^2/sec) compared with the other planets.

5. The oxygen isotopic composition of lunar igneous rocks collected during the Apollo missions is the same as that of mantle-derived rocks from the Earth. Because oxygen isotopic composition seems to vary with position in the solar system, the similarity of oxygen isotopes in lunar and terrestrial igneous rocks suggests that both bodies formed in the same part of the solar system approximately the same distance from the Sun.

6. Isotopic ages from igneous rocks on the lunar surface range from about 4.46 to 3.1 Ga. Model ages indicate that the anorthositic rocks of the lunar highlands crust formed from about 4.46 to 4.45 Ga.

Models for the origin of the Moon generally fall into one of four categories: (1) fission from the Earth, (2) the double-planet scenario in which the Moon accretes from a sediment ring around the Earth, (3) capture by the Earth, and (4) impact on the Earth's surface by a Mars-size body (Fig. 10.25). Any acceptable model must account for the preceding constraints, and thus far none of these models is completely acceptable. Each of the hypotheses will be briefly described, and a summary of how well each model complies with major constraints is given in Table 10.5.

Fission Models

Fission models involve the separation of the Moon from the Earth during an early stage of rapid spinning when tidal forces overcame gravitational forces. One version of the fission hypothesis (Wise, 1963) suggests that the formation of the Earth's core reduced the amount of inertia, increasing the rotational rate and spinning off material to form the Moon. Such a model is attractive in that it accounts for the similarity in density between the Earth's mantle and the Moon and for the absence of a large metallic lunar core. The hypothesis is also consistent with the Moon's revolution in the same direction as the Earth rotates, the circular shape of the lunar orbit, the existence of a lunar bulge facing the Earth, and the similar oxygen isotopic ratios of the two bodies. The model also explains the iron-poor character of the Moon if fission occurred after core formation in the Earth, because the Moon would be formed largely from the Earth's mantle.

Fission models, however, face several major obstacles. For instance, they do not explain the inclination of the lunar orbit, and they require more than four times the total angular momentum available in the present Earth–Moon system. If the Earth–Moon system ever had this excess angular momentum, no acceptable mechanism has been proposed to lose it (Wood, 1986). Also, lunar igneous rocks are more depleted in siderophile and volatile elements than terrestrial igneous rocks, indicating that the lunar interior is not similar in composition to the Earth's mantle. Although some investigators favor a fission

Figure 10.25   Models for the origin of the Moon.

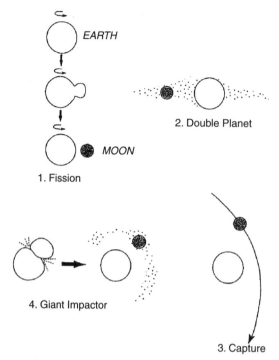

2. Double Planet

1. Fission

4. Giant Impactor

3. Capture

model (Binden, 1986), the preceding problems seem to render the model highly implausible (Table 10.5).

Double-Planet Models

Double-planet models involve an accreting Earth with simultaneous accretion of the Moon from orbiting solid particles (Harris and Kaula, 1975). A major advantage of these models, also known as *coaccretion* or *precipitation models,* is that they do not invoke special, low-probability events. The models assume that as the Earth accreted, solid particles accumulated in orbit about the Earth and accreted to form the Moon. The general scenario is as follows: the Earth accretes first and its core forms during

Table 10.5   Summary of Major Constraints on Models of Lunar Origin

	Fission	Capture	Double Planet	Giant Impactor
Angular momentum	I	C(I)	I	C
Lunar rotation	C	C(C)	C	C
Orbital characteristics	I	C(C)	I	C
Chemical composition	I	I(I)	C	C

Values in parenthesis for are disintegrative capture. C, consistent; I, inconsistent.

accretion; as the Earth heats, material is vaporized from the surface, forming a ring around the Earth from which the Moon accretes. Because core formation extracts siderophile elements from the mantle, the material vaporized from the Earth is depleted in these elements; hence, the Moon, which accretes from this material, is also depleted in these elements. Because volatile elements are largely lost by intense solar radiation from a T-Tauri wind after the Earth accretes (but before the Moon accretes), the material from which the Moon accretes is depleted in volatile elements relative to the Earth. This leaves the material from which the Moon accretes relatively enriched in oxyphile refractory elements.

The most serious problems with the double-planet models are they do not seem capable of explaining the large amount of angular momentum in the Earth–Moon system (Wood, 1986) and they do not readily explain the inclined lunar orbit.

Capture Models

Capture models propose that the Moon and the Earth formed in different parts of the solar nebula and that early in the history of the solar system the Moon or its predecessor approached the Earth and was captured (Taylor, 1992). Both catastrophic and noncatastrophic models of lunar capture have been described involving retrograde and prograde orbits for the Moon before capture.

Capture models fall into two categories. In an **intact capture,** a fully accreted Moon is captured by the Earth. In a **disintegrative capture,** a planetesimal comes within the Earth's Roche limit, it is fragmented by tidal forces with most of the debris captured in orbit about the Earth, and the debris reaccretes to form the Moon (Wood and Mitler, 1974). Although intact capture models may explain the high angular momentum and inclined lunar orbit, they cannot readily account for geochemical differences between the two bodies. The similar oxygen isotopic ratios between lunar and terrestrial igneous rocks suggest that both bodies formed in the same part of the solar system, yet the capture model does not offer a ready explanation for the depletion of siderophile and volatile elements in the Moon. Also, intact lunar capture is improbable because it requires a specific approach velocity and trajectory. Disintegrative capture models cannot account for the high angular momentum in the Earth–Moon system.

Giant Impactor Model

The giant impactor model involves a glancing collision of the Earth with a Mars-size body during which debris from both the Earth and the colliding planet collect and accrete in orbit about the Earth to form the Moon (Hartmann, 1986; Cameron, 1986). Such a model has the potential of eliminating the angular momentum and nonequatorial lunar orbit problems and of explaining the chemical differences between the Earth and the Moon. One of the major factors that led to the giant impactor model is the stochastic models of Wetherill (1985), which indicate that numerous large bodies formed in the inner solar system during the early stages of planetary accretion. Results suggest the existence of at least 10 bodies larger than Mercury and several equal to or larger than Mars. Wetherill (1986) estimates that about one-third of these objects collided with and accreted to the Earth, providing 50 to 75% of the Earth's mass. The obliquities of planets and the slow retrograde motion of Venus are also most reasonably explained by late-stage impact of large planetesimals.

In the giant impactor model, a Mars-size planet collides with the Earth during the late stages of accretion (Taylor, 1993) (Fig. 10.26). Such an impact can easily account for the anomalously high angular momentum in the Earth–Moon system (Newsom and Taylor, 1989). The Moon is derived chiefly from the mantle of the impactor, thus accounting for the geochemical differences between the two bodies. The high energy of the impact results not only in complete disruption of the impacting planet but also in widespread melting of the Earth, producing a magma ocean. Computer simulations of the impact event indicate that the material from which the Moon accretes comes primarily from the mantle of the impactor (Fig. 10.26). Because iron had already segregated into the Earth's core, low iron (and other siderophile element) content of the Moon is predicted by the model. An additional attribute of the model is that the material that escapes from the system after impact is mostly liquid or vapor phases, thus leaving dust enriched in refractory

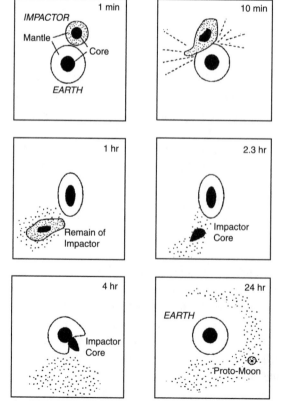

Figure 10.26 Computer simulation of the formation of the Moon by a giant impactor. The model is for an oblique collision with the Earth of an impactor of 0.14 Earth masses at a velocity of 5 km/sec. Time elapsed since the impact is given in each box. Modified from Newsom and Taylor (1989).

elements from which the Moon forms, accounting for the extreme depletion in volatile elements in the Moon. Whether the remaining material immediately accretes into the Moon or forms several small lumps that coalesce to form the Moon on timescales on the order of 100 years is unknown. In either case, the Moon heats rapidly, forming a widespread magma ocean.

Although the giant impactor model needs to be more fully evaluated, it appears to be capable of accommodating more of the constraints related to lunar origin than any of the competing models (Table 10.5).

Earth's Rotational History

Integration of equations of motion of the Moon indicate there has been a minimum in the Earth–Moon distance in the geologic past, that the inclination of the lunar orbit has decreased with time, and that the eccentricity of the lunar orbit has increased as the Earth–Moon distance has increased (Lambeck, 1980). It has long been known that angular momentum is being transferred from the Earth's spin to lunar orbital motion, which moves the Moon from the Earth. The current rate of this retreat is about 5 cm/year. The corresponding rotational rate of the Earth has been decreasing at a rate of about 5×10^{-22} rad/sec. Tidal torques cause the inclination of the Moon's orbit to vary slowly with time as a function of the Earth–Moon distance. When this distance is greater than about 10 Earth radii, the lunar orbital plane moves toward the ecliptic, and when it is less than 10 Earth radii, it moves toward the Earth's equatorial plane. For an initially eccentric orbit, the transfer of angular momentum is greater at perigee than apogee, and hence, the degree of eccentricity increases with time. These evolutionary changes in the angular momentum of the Earth–Moon system change the length of the terrestrial day and month.

Many groups of organisms secrete sequential layers related to cyclical, astronomical phenomena. Such organisms, known as **biologic rhythmites** (Pannella, 1972), provide a means of independently evaluating the calculated orbital retreat of the Moon. The most important groups are corals and bivalves. Daily increments in these organisms are controlled by successive alternation of daylight and darkness. Seasonal increments reflect changes in the length of sunlight per day, seasonal changes in food supply, and in some instances tidal changes. Actual growth patterns are complex, and local environmental factors may cause difficulties in identifying periodic growth patterns. Results from Phanerozoic bivalves and corals, however, are consistent and suggest a decreasing rotational rate of the Earth of about 5.3×10^{-22} rad/sec (Fig. 10.27), which is in agreement with astronomical values. This corresponds to 420 to 430 days per year from 500 to 400 Ma. Stromatolites also deposit regular bands, although a quantitative relationship of banding to astronomical rhythm has not yet been well established (Cao, 1991). The interpretation of stromatolite bands from the Biwabik Iron Formation in Minnesota is that at 2000 Ma there were 800 to 900 days per year (Mohr, 1975) (Fig. 10.27). Another type of geochronometer has been recognized in laminated fine-grained sediments, known as **sedimentologic rhythmites.** Cyclically laminated Neoproterozoic (750–650 Ma) tidal

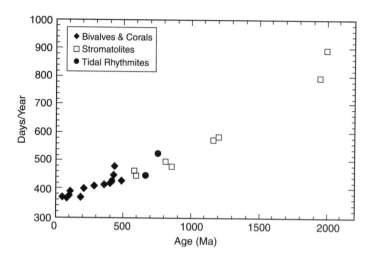

sediments from Australia have been interpreted to record paleotidal periods and paleorotation of the Earth (Williams, 1989). Results suggest that 650 Ma there were 13 months per year and 450 days per year. When both biologic and sedimentary rhythmite data are plotted with time, results suggest a sublinear relationship with the number of days per year decreasing about 0.2 days/My, in agreement with the extrapolation of astronomical calculations.

If the Moon came within the Roche limit of the Earth (~2.9 Earth radii) after 4.5 Ga, a record of such a close encounter should be preserved. Even if the Moon survived this encounter without disintegrating, the energy dissipated in the two bodies would largely melt both bodies and completely disrupt and recycle any crust. The preservation of 4.45 Ga crust on the Moon strongly indicates that the Moon has not been within the Earth's Roche limit since it formed about 4.5 Ga; thus, the relationship in Figure 10.27 cannot be extrapolated much before 2 Ga.

Comparative Planetary Evolution

From this survey of the solar system, it seems clear that no two bodies in the solar system have identical histories. Although the terrestrial planets have many features in common, as do the outer giant planets, it would appear that each planet has its own unique history. At least from my perspective, the Earth seems to be most peculiar. Not only is it the only planet with oceans, an oxygen-bearing atmosphere, and living organisms, but it also is the only planet in which plate-tectonic processes are known to be active. As a final topic in this chapter, I compare the evolutionary histories of the terrestrial planets and in doing so identify some of the important variables that control and direct the paths of planetary evolution.

As you have seen, the terrestrial planets and the Moon have similar densities (Table 10.1) and thus, on the whole, similar bulk compositions. Each of them is evolving toward a stage of thermal and tectonic stability and quiescence as they cool. The rate at which a planet approaches this final stage depends on a variety of factors that directly or indirectly control the loss of heat (Carlson, 1994; Taylor, 1992; Taylor, 1999). First, the position of a planet in the solar system is important because, as you have seen, it reflects the condensation sequence of elements from the cooling solar nebula. Also important are the abundances of radiogenic isotopes that contribute to heating planetary bodies. The Moon, for instance, contains considerably smaller amounts of U, Th, and K than the Earth and therefore will not produce as much radiogenic heat. Analyses of fine-grained materials from the Viking landing sites suggest that Mars is also depleted in radiogenic isotopes compared with the Earth. Planetary mass is important in that the amount of accretional and gravitational energy directly depends on mass. Planetary size is also important in that greater area/mass ratios result in more rapid heat loss from planetary surfaces. For instance, the Moon, Mars, and Mercury should cool much faster than Venus and the Earth because of their higher area/mass ratios (Table 10.1). Also important is the size of the iron core in that much of initial planetary heat is produced during core formation. Except for Mercury, the Earth has the highest core/mantle ratio, followed by slightly lower values for Venus and Mars. As described previously, the very high core/mantle ratio for Mercury is probably a result of the loss of some of the Mercurian mantle by an early giant impact and thus is not indicative of a large contribution of heat from core formation. The volatile contents and especially the water content of planetary mantles and the rate of volatile release are important in controlling atmosphere development, the amount of melting, fractional crystallization trends, and the viscosity of planetary interiors, which, in turn, affects the rate of cooling. Convection, mantle-plume activity, or both appear to be the primary mechanisms by which heat is lost from the terrestrial planets. Only the Earth, however, requires mantle convection at present and supports plate tectonics.

There are three types of planetary crusts. Primary crust forms during or immediately after planetary accretion by cooling at the surface. Secondary crusts arise later from partial melting of recycled primary crust or from partial melting of planetary interiors. By analogy with the preserved primary crust on the Moon and probably on Mercury, primary crusts on the terrestrial planets may have been anorthositic in composition, produced by floatation of plagioclase during rapid crystallization of magma oceans. Secondary crusts, however, are typically basaltic in composition and form only after the crystallization of magma oceans. They are produced by partial melting of ultramafic rocks in planetary mantles and in mantle plumes. Examples include the lunar mare basalts, the Earth's oceanic crust, and perhaps most of the crust preserved on Mars and Venus. Tertiary crust is formed by partial melting and further processing of secondary crust. The Earth's continents may be the only example of tertiary crust in the solar system.

Although every planet has its own unique history, the primary differences in planetary thermal history are controlled chiefly by heat productivities, volatile-element contents, and cooling rates. Distance from the Sun is also an important variable, especially in terms of planetary composition and atmosphere evolution. A qualitative portrayal of planetary thermal histories is illustrated in Figure 10.28. The temperature scale is schematic. All terrestrial planets underwent rapid heating during late stages of accretion, reaching

Figure 10.28 Schematic
thermal evolution of the
terrestrial planets and the
Moon. Lower threshold
temperatures for planetary
processes: Tc, convection;
Tm, magmatism; Tp,
mantle-plume generation at
the core–mantle boundary.

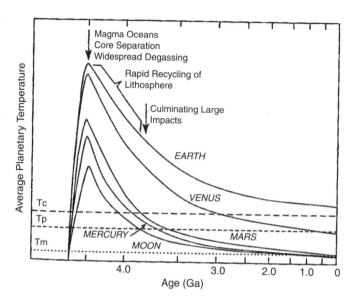

maximum temperatures about 4.5 Ga, at which time widespread magma oceans were
produced. Also, just before this time, molten iron descended to planetary centers, forming metal cores, and planetary mantles rapidly degassed. Rapid, chaotic convection in the
magma oceans resulted in rapid cooling and crystallization, producing a transient
primary crust probably of anorthositic composition. Partial melting of recycled crust,
mantle ultramafic rocks, or both produced widespread secondary basaltic crusts on all of
the terrestrial planets, a process still occurring on the Earth as ocean-ridge basalts and
oceanic plateaus are produced. At least on the Earth and Venus, where substantial heat
was available, early crust and lithosphere were rapidly recycled into the mantle, aided by
intense giant impacts, which continued until about 3.9 Ga (Fig. 10.28). Although continental crust is not preserved on the Earth until 4 Ga, it may have been produced and rapidly recycled into the mantle before this time (Chapter 4). Rapid cooling of the smaller
planets, including Mercury and probably Mars, led to thickened and strong lithospheres
by 4.45 Ga. These planets, as well as the Moon, are known as **one-plate planets,** because
their thick, strong, planetwide lithosphere is basically all one plate. Most of the magmatic
activity on these planets resulted from mantle-plume activity and occurred before 3.9 Ga.
The Moon formed by the accretion of material blasted from the Earth by a giant impact
about 4.53 Ga, and it rapidly heated and melted to form a magma ocean, which crystallized by 4.5 Ga to produce the anorthositic lunar highlands crust. The youngest volcanism
appears to have been about 2.5 Ga on the Moon, perhaps 1.5 Ga on Mercury, and maybe
as recent as 200 Ma on Mars.

Because of their greater initial heat, Venus and the Earth cooled more slowly than the
other terrestrial planets. Venus may have passed through the minimum temperature for
convection some 3 Ga, but the Earth has not yet reached that point in its cooling history
(Fig. 10.28). If the resurfacing of Venus from 600 to 400 Ma was caused by a catastrophic

mantle-plume event, it may have been the last such event, as suggested by the intersection of the Venus cooling curve with Tp, the lower temperature limit for plume production at the core–mantle interface. Volcanism may still be active on Venus, however. The most intriguing question—that of why Venus and the Earth followed such different cooling paths even though both planets are similar in mass and density—remains problematic. One possibility is that although the Earth cooled chiefly by convection and plate tectonics, Venus cooled by mantle plumes and conduction through the lithosphere. If this is the case, investigators are left with the question of why different cooling mechanisms dominated in each planet. Some investigators suggest that it has to do with the absence of water on Venus. They argue that in a dry planet, the lithosphere is thick and strong as it is on Venus, and there is no melting to produce a low-velocity zone; thus plates cannot move about or subduct. Taking this one step further, you might ask why Venus is dry yet the Earth is wet. As suggested previously, Venus may have rapidly lost its water as hydrogen escaped from the atmosphere soon after or during planetary accretion, because the surface of the planet was too hot for oceans to survive. And why was the surface too hot? Perhaps this is because Venus is closer than the Earth to the Sun. If this line of reasoning is correct, it may be that the position that a planet accretes in a gaseous nebula is one of the most important variables controlling its evolution. The reason Mars did not sustain plate tectonics or accumulate an ocean may be because of its small mass and rapid cooling, resulting in a lithosphere too thick and strong to subduct. If there was enough water to form Martian oceans, it either escaped by hydrogen loss or is trapped in the interior of the planet.

If the preceding scenario bears any resemblance to what really happened, it would appear that two important features led to a unique history for the Earth: (1) its position in the solar system and (2) its relatively large mass. Without both of these, the Earth may have evolved into quite a different planet than the one we live on.

Further Reading

Atreya, S. K., Pollack, J. B., and Matthews, M. S. (eds.), 1989. Origin and Evolution of Planetary and Satellite Atmospheres. University of Arizona Press, Tucson, 881 pp.

Beatty, J. K., Petersen, C. C., and Chaikin, A. L., 1999. The New Solar System, Fourth Edition. Cambridge University Press, Cambridge, UK, 430 pp.

de Pater, I., and Lissauer, J., 2001. Planetary Sciences. Cambridge University Press, Cambridge, UK, 544 pp.

Lewis, J. S., 1997. Physics and Chemistry of the Solar System, Revised Edition. Academic Press, San Diego, 591 pp.

McSween, H. Y. Jr., 1999. Meteorites and Their Parent Bodies. Cambridge University Press, Cambridge, UK, 322 pp.

Rosenberg, G. D., and Runcorn, S. K. (eds.), 1975. Growth Rhythms and the History of the Earth's Rotation. J. Wiley & Sons, New York, 559 pp.

Taylor, S. R., 1992. Solar System Evolution: A New Perspective. Cambridge University Press, Cambridge, UK, 307 pp.

References

Abbott, D. H., 1996. Plumes and hotspots as sources of greenstone belts. Lithos, 37: 113–127.

Abbott, D. H., and Hoffman, S. E., 1984. Archean plate tectonics revisited. Tectonics, 3: 429–448.

Abbott, D. H., and Isley, A. E., 2002. Extraterrestrial influences on mantle plume activity. Earth Planet. Sci. Lett., 205: 53–62.

Abbott, D. H., and Mooney, W., 1995. The structural and geochemical evolution of the continental crust: Support for the oceanic plateau model of continental growth. Rev. Geophys., Suppl. 99: 231–242.

Abbott, D. H., Burgess, L., Longhi, J., and Smith, W. H. F., 1994. An empirical thermal history of the Earth's upper mantle. J. Geophys. Res., 99: 13,835–13,850.

Abelson, P. H., 1966. Chemical events on the primitive Earth. Proc. Nat. Acad. Sci., 55: 1365–1372.

Adatte, T., Keller, G., Harting, M., Stueben, D., Kramar, U., and Stinnesbeck, W., 2003. Cretaceous–Tertiary events: A multicausal scenario. In: Mantle Plumes: Physical Processes, Chemical Signatures, and Biological Effects (abstract volume). Cardiff University, Cardiff, UK, p. 1.

Ahall, K. I., Persson, P. O., and Skiold, T., 1995. Westward accretion of the Baltic shield: Implications from the 1.6-Ga Amal-Horred belt, southwestern Sweden. Precamb. Res., 70: 235–251.

Ahrens, T. J., Holland, K. G., and Chen, G. Q., 2002. Phase diagram of iron: Revised core temperatures. Geophys. Res. Lett., 29 (7): 54-1 to 54-4.

Alabaster, T., Pearce, J. A., and Malpas, J., 1982. The volcanic stratigraphy and petrogenesis of the Oman ophiolitic complex. Contrib. Mineral. Petrol., 81: 168–183.

Albarede, F., 1998. Time-dependent models of U-Th-He and K-Ar evolution and the layering of mantle convection. Chem. Geol., 145: 413–429.

Albarede, F., and van der Hilst, R. D., 1999. New mantle convection model may reconcile conflicting evidence. EOS, 80: 535–539.

Alfe, D., Gillan, M. J., and Price, G. D., 2002. Composition and temperature of the Earth's core constrained by combining ab initio calculations and seismic data. Earth Planet. Sci. Lett., 195: 91–98.

Algeo, T. J., and Seslavinsky, K. B., 1995. The Paleozoic world: Continental flooding, hypsometry, and sea level. Am. J. Sci., 295: 787–822.

Allegre, C. J., 1982. Chemical geodynamics. Tectonics, 81: 109–132.

Allegre, C. J., and Minster, J. F., 1978. Quantitative models of trace element behavior in magmatic processes. Earth Planet. Sci. Lett., 38: 1–25.

Allegre, C. J., Brevart, O., Dupre, B., and Minster, J. F., 1980. Isotopic and chemical effects produced in a continuously differentiating convecting earth mantle. Phil. Trans. Roy. Soc. Lond., 297A: 447–477.

Allegre, C. J., Moreira, M., and Staudacher, T., 1995a. ^{4}He/^{3}He dispersion and mantle convection. Geophys. Res. Lett., 22: 2325–2328.

Allegre, C. J., Poirier, J. P., Humler, E., and Hofmann, A. W., 1995b. The chemical composition of the Earth. Earth Planet. Sci. Lett., 134: 515–526.

Allen, P. A., and Homewood, P. (eds.), 1986. Foreland Basins. Blackwell Scientific Publications, Oxford, 453 pp.

Alvarez, W. L., 1986. Toward a theory of impact crises. EOS 67 (35): 649–655.

Alvarez, W. L., Alvarez, L. W., Asaro, F., and Michel, H. V., 1982. Extraterrestrial cause for the K–T extinction. Science, 208: 1095–1108.

Amelin, Y., Lee, D. C., Halliday, A. N., and Pidgeon, R. T., 1999. Nature of the Earth's earliest crust from hafnium isotopes in single detrital zircons. Nature, 399: 252–255.

Anbar, A. D., and Knoll, A. H., 2002. Proterozoic ocean chemistry and evolution: A bioinorganic bridge? Science, 297: 1137–1142.

Anderson, D. L., 1982. Hotspots, polar wander, Mesozoic convection, and the geoid. Nature, 297: 391–393.

Anderson, D. L., Sammis, C., and Jordan, T., 1971. Composition and evolution of the mantle and core. Science, 171: 1103–1112.

Anderson, J. L., 1983. Proterozoic anorogenic plutonism of North America. Geol. Soc. Am., Mem., 161: 133–154.

Anderson, J. L., and Morrison, J., 1992. The role of anorogenic granites in the Proterozoic crustal development of North America. In: K. C. Condie (ed.), Proterozoic Crustal Evolution. Elsevier, Amsterdam, pp. 263–300.

Arculus, R. J., and Johnson, R. W., 1978. Criticism of generalized models for the magmatic evolution of arc–trench systems. Earth Planet. Sci. Lett., 39: 118–126.

Argus, D. F., and Heflin, M. B., 1995. Plate motion and crustal deformation estimated with geodetic data from the GPS. Geophys. Res. Lett., 22: 1973–1976.

Armstrong, R. L., 1981. Radiogenic isotopes: The case for crustal recycling on a near-steady-state no-continental-growth Earth. Phil. Trans. Roy. Soc. Lond., A301: 443–472.

Armstrong, R. L., 1991. The persistent myth of crust growth. Austral. J. Earth Sci., 38: 613–630.

Arndt, N. T., 1983. Role of a thin, komatiite-rich oceanic crust in the Archean plate-tectonic process. Geology, 11: 372–375.

Arndt, N. T., 1991. High Ni in Archean tholeiites. Tectonophysics, 187: 411–420.

Arndt, N. T., 1994. Archean komatiites. In: K. C. Condie (ed.), Archean Crustal Evolution. Elsevier, Amsterdam, pp. 11–44.

Arndt, N. T., 1999. Why was flood volcanism on submerged continental platforms so common in the Precambrian? Precamb. Res., 97: 155–164.

Arndt, N. T., and Goldstein, S. L., 1987. Use and abuse of crust formation ages. Geology, 15: 893–895.

Artemieva, I. M., and Mooney, W. D., 2001. Thermal thickness and evolution of Precambrian lithosphere: A global study. J. Geophys. Res., 106: 16,387–16,414.

Ashwal, L. D., and Myers, J. S., 1994. Archean anorthosites. In: K. C. Condie (ed.), Archean Crustal Evolution. Elsevier, Amsterdam, pp. 315–355.

Asmerom, Y., et al., 1991. Strontium isotopic variations of Neoproterozoic seawater: Implications for crustal evolution. Geochim. Cosmochim. Acta, 55: 2883–2894.

Aspler, L. B., and Chiarenzelli, J. R., 1998. Two Neoarchean supercontinents? Evidence from the Paleoproterozoic. Sediment. Geol. 120: 75–104.

Ave'Lallemant, H. G., and Carter, N. L., 1970. Syntectonic recrystallization of olivine and modes of flow in the upper mantle. Geol. Soc. Am. Bull., 81: 2203–2220.

Ayers, L. D., and Thurston, P. C., 1985. Archean supracrustal sequences in the Canadian shield: An overview. Geol. Assoc. Can., Spec. Paper 28: 343–380.

BABEL Working Group, 1990. Evidence for early Proterozoic plate tectonics from seismic reflection profiles in the Baltic shield. Nature, 348: 34–38.

Baker, V. R., 2001. Water and the Martian landscape. Nature, 412: 228–236.

Baragar, W. R. A., Lamber, M. B., Baglow, N., and Gibson, I. L., 1990. The sheeted dyke zone in the Troodos ophiolite. In: J. Malpas et al. (eds.), Ophiolites: Oceanic Crustal Analogues. Cyprus Geol. Survey, pp. 37–51.

Baragar, W. R. A., Ernst, R. E., Hulbert, L., and Peterson, T., 1996. Longitudinal petrochemical variation in the Mackenzie dyke swarm, northwestern Canadian shield. J. Petrol., 37: 317–359.

Barley, M. E., 1993. Volcanic, sedimentary, and tectonostratigraphic environments of the 3.46-Ga Warrawoona Megasequence: A review. Precamb. Res., 60: 47–67.

Barley, M. E., Pickard, A. L., and Sylvester, P. J., 1997. Emplacement of a large igneous province as a possible cause of BIF 2.45 Ga. Nature, 385: 55–58.

Barley, M. E., Krapez, B., Groves, D. I., and Kerrich, R., 1998. The late Archean bonanza: Metallogenic and environmental consequences of the interaction between mantle plumes, lithospheric tectonics, and global cyclicity. Precamb. Res., 91: 65–90.

Barnes, C. R., 1999. Paleoceanography and paleoclimatology: An Earth system perspective. Chem. Geol., 161: 17–35.

Barron, E. J., Fawcett, P. J., Peterson, W. H., Pollard, D., and Thompson, S. L., 1995. A "simulation" of mid-Cretaceous climate. Paleoceanography, 10: 953–962.

Bartley, J. K., et al., 2001. Global events across the Mesoproterozoic–Neoproterozoic boundary: Ca and Sr isotopic evidence from Siberia. Precamb. Res., 111: 165–202.

Barton, J. M. Jr., et al., 1994. Discrete metamorphic events in the Limpopo belt, southern Africa: Implications for the application of P–T paths in complex metamorphic terrains. Geology, 22: 1035–1038.

Beck, R. A., Burbank, D. W., Sercombe, W. J., Olson, T. L., and Khan, A. M., 1995. Organic carbon exhumation and global warming during the early Himalayan collision. Geology, 23: 387–390.

Becker, L., Poreda, R. J., Hunt, A. G., Bunch, T. E., and Rampino, M., 2001. Impact event at the Permian–Triassic boundary: Evidence from extraterrestrial noble gases in fullerenes. Science, 291: 1530–1533.

Bednarz, U., and Schmincke, H., 1994. Petrological and chemical evolution of the northeastern Troodos extrusive series, Cyprus. J. Petrol., 35: 489–523.

Bekker, A., Holland, H. D., Wang, P. L., Rumble, D. III, Stein, H. J., Hannah, J. L., Coetzee, L. L., and Beukes, N. J., 2004. Dating the rise of atmospheric oxygen. Nature, 427: 117–120.

Belcher, C. M., Collinson, M. R., Sweet, A. R., Hildebrand, A. R., and Scott, A. C., 2003. Fireball passes and nothing burns: The role of thermal radiation in the Cretaceous–Tertiary event—Evidence from the charcoal record of North America. Geology, 31: 1061–1064.

Belonoshko, A. B., Ahuja, R., and Johansson, B., 2003. Stability of the body-centered cubic phase of iron in the Earth's inner core. Nature, 424: 1032–1034.

Bengtson, S. (ed.), 1994. Early Life on Earth. Columbia University Press, New York, 630 pp.

Benton, M. J., 1995. Diversification and extinction in the history of life. Science, 268: 52–58.

Benz, W., and Cameron, A. G. W., 1990. Terrestrial effects of the giant impact. In: H. E. Newsom and J. H. Jones (eds.), Origin of the Earth. Oxford University Press, New York, pp. 61–67.

Bergman, M. I., 1997. Measurements of elastic anisotropy due to solidification texturing and the implications for the Earth's inner core. Nature, 489: 60–63.

Berner, R. A., 1983. Burial of organic carbon and pyrite sulfur in sediments over Phanerozoic time: A new theory. Geochim. Cosmochim. Acta, 47: 855–862.

Berner, R. A., 1987. Models for carbon and sulfur cycles and atmospheric oxygen: Application to Paleozoic history. Am. J. Sci., 287: 177–196.

Berner, R. A., 1994. 3Geocarb II: A revised model of atmospheric CO_2 over Phanerozoic time. Am. J. Sci., 294: 56–91.

Berner, R. A., 1999. A new look at the long-term carbon cycle. GSA Today, 9 (11): 1–6.

Berner, R. A., 2001. Modeling atmospheric O_2 over Phaerozoic time. Geochim. Cosmochim. Acta, 65: 685–694.

Berner, R. A., and Canfield, D. E., 1989. A new model for atmospheric oxygen over Phanerozoic time. Am. J. Sci., 289: 333–361.

Beukes, J. J., Klein, C., Kaufman, A. J., and Hayes, J. M., 1990. Carbonate petrography, kerogen distribution, and carbon and oxygen isotope variations in an early Proterozoic transition from limestone to iron-formation deposition, Transvaal Supergroup, South Africa. Bull. Soc. Econ. Geol., 85: 663– 690.

Bickle, M. J., 1986. Implications of melting for stabilization of the lithosphere and heat loss in the Archean. Earth Planet. Sci. Lett., 80: 314–324.

Bickle, M. J., 1996. Metamorphic decarbonation, silicate weathering, and the long-term carbon cycle. Terra Nova, 8: 270–276.

Bickle, M. J., Nisbet, E. G., and Martin, A., 1994. Archean greenstone belts are not oceanic crust. J. Geol., 102: 121–137.

Binden, A. B., 1986. The binary fission origin of the moon. In: W. K. Hartmann, R. J. Phillips, and G. J. Taylor (eds.), Origin of the Moon. Lunar and Planetary Institute, Houston, pp. 519–550.

Bindschadler, D. L., Schubert, G., and Kaula, W. M., 1992. Coldspots and hotspots: Global tectonics and mantle dynamics of Venus. J. Geophys. Res., 97: 13,495–13,532.

Bleeker, W., 2002. Archean tectonics: A review, with illustrations from the Slave craton. Geol. Soc. Lond., Spec. Publ., 199: 151–181.

Bleeker, W., 2003. The late Archean record: A puzzle in ca. 35 pieces. Lithos, 71: 99–134.

Blichert-Toft, J., and Albarede, F., 1994. Short-lived chemical heterogeneities in the Archean mantle with implications for mantle convection. Science, 263: 1593–1596.

Blichert-Toft, J., Frey, F. A., and Albarede, F., 1999. Hf isotope evidence for pelagic sediments in the source of Hawaiian basalts. Science, 285: 879–882.

Bloxham, J., 1992. The steady part of the secular variation of Earth's magnetic field. J. Geophys. Res., 97: 19,565–19,579.

Bodenheimer, P., and Lin, D. N. C., 2002. Implications of extrasolar planets for understanding planet formation. Ann. Rev. Earth Planet. Sci., 30: 113–148.

Boehler, R., 1996. Experimental constraints on melting conditions relevant to core formation. Geochim. Cosmochim. Acta, 60: 1109–1112.

Bogue, S. W., and Merrill, R. T., 1992. The character of the field during geomagnetic reversals. Ann. Rev. Earth Planet. Sci., 20: 181–219.

Boher, M., Abouchami, W., Michard, A., Albarede, F., and Arndt, N. T., 1992. Crustal growth in West Africa at 2.1 Ga. J. Geophys. Res., 97: 345–369.

Bohlen, S. R., 1991. On the formation of granulites. J. Met. Geol., 9: 223–229.

Bohor, B. F., Modreski, P. J., and Foord, E. E., 1987. Shocked quartz in the K–T boundary clays: Evidence for a global distribution. Science, 236: 705–709.

Bois, C., Bouche, P., and Pelet, R., 1980. Historie géologique et répartition des réserves d'hydrocarbures dans le monde [Global geologic history and distribution of hydrocarbon reserves]. Revue de l'Institut Français du Pétrole, 35: 237–298.

Bonati, E., 1967. Mechanisms of deep-sea volcanism in the South Pacific. In: P. H. Abelson (ed.), Researches in Geochemistry. Wiley, New York, pp. 453–491.

Bortolotti, V., Marroni, M., Pandolfi, L., Principi, G., and Saccani, E., 2002. Interaction between midocean-ridge and subduction magmatism in Albanian ophiolites. J. Geol., 110: 561–576.

Boryta, M., and Condie, K. C., 1990. Geochemistry and origin of the Archean Beit Bridge Complex, Limpopo belt, South Africa. J. Geol. Soc. Lond., 147: 229–239.

Boschi, L., and Dziewonski, A. M., 1999. High- and low-resolution images of the Earth's mantle: Implications of different approaches to tomographic modeling. J. Geophys. Res., 104: 25,567–25,594.

Boss, A. P., Morfill, G. E., and Tscharnuter, W. M., 1989. Models of the formation and evolution of the solar nebula. In: S. K. Atreya, J. B. Pollack, and M. S. Matthews (eds.), Origin and Evolution of Planetary and Satellite Atmospheres. University Arizona Press, Tucson, pp. 35–77.

Bott, M. H. P., 1979. Subsidence mechanisms at passive continental margins. Am. Assoc. Petrol Geol., Mem., 29: 3–9.

Bottinga, Y., and Allegre, C. J., 1978. Partial melting under spreading ridges. Phil. Trans. Roy. Soc. Lond., 288A: 501–525.

Bowring, S. A., 1990. The Acasta gneisses: Remnant of Earth's early crust. In: H. E. Newsom and J. H. Jones (eds.), Origin of the Earth. Oxford University Press, Oxford, pp. 319–343.

Bowring, S. A., and Housh, T., 1995. The Earth's early evolution. Science, 269: 1535–1540.

Bowring, S. A., Williams, I. S., and Compston, W., 1989. 3.96-Ga gneisses from the Slave province, Northwest Territories, Canada. Geology, 17: 971–975.

Bowring, S. A., Grotzinger, J. P., Isachsen, C. E., Knoll, A. H., Pelechaty, S. M., and Kolosov, P., 1993. Calibrating rates of early Cambrian evolution. Science, 261: 1293–1298.

Bowring, S. A., Erwin, D. H., Jin, Y. G., Martin, M. W., Davidek, K., and Wang, W., 1998. U/Pb zircon geochronology and tempo of the end-Permian mass extinction. Science, 280: 1039–1045.

Boyd, F. R., 1989. Compositional distinction between oceanic and cratonic lithosphere. Earth Planet. Sci. Lett., 96: 15–26.

Bradshaw, T. K., Hawkesworth, C. J., and Gallagher, K., 1993. Basaltic volcanism in the southern Basin and Range: No role for a mantle plume. Earth Planet. Sci. Lett., 116: 45–62.

Brady, P. V., 1991. The effect of silicate weathering on global temperature and atmospheric CO_2. J. Geophys. Res., 96: 18,101–18,106.

Brasier, M. D., and Lindsey, J. F., 1998. A billion years of environmental stability and emergence of eukaryotes: New data from northern Australia. Geology, 26: 555–558.

Bratton, J. F., 1999. Clathrate eustasy: Methane hydrate melting as a mechanism for geologically rapid sea-level fall. Geology, 27: 915–918.

Breuer, D., and Spohn, T., 1995. Possible flush instability in mantle convection at the Archean–Proterozoic transition. Nature, 378: 608–610.

Brocks, J. J., Logan, G. A., Buick, R., and Summons, R. E., 1999. Archean molecular fossils and the early rise of eukaryotes. Science, 285: 1033–1036.

Brooks, C., Hart, S. R., Hofman, A., and Jarnes, D. E., 1976. Rb/Sr mantle isochrons from oceanic areas. Earth Planet. Sci. Lett., 32: 51–61.

Brown, M., 1993. P–T–t evolution of orogenic belts and the causes of regional metamorphism. J. Geol. Soc. Lond., 150: 227–241.

Brown, M., Rushmer, T., and Sawyer, E. W., 1995. Introduction to special section: Mechanisms and consequences of melt segregation from crustal protoliths. J. Geophys. Res., 100: 15,551–15,563.

Bruhn, D., Groebner, N., and Kohlstedt, D. L., 2000. An interconnected network of core-forming melts produced by shear deformation. Nature, 403: 883–886.

Bryan, P., and Gordon, R. G., 1986. Errors in minimum plate velocity determined from paleomagnetic data. J. Geophys. Res., 91: 462–470.

Buchan, K. L., Halls, H. C., and Mortensen, J. K., 1996. Paleomagnetism, U/Pb geochronology, and geochemistry of Marathon dykes, Superior Province, and comparison with the Fort Frances swarm. Can. J. Earth. Sci., 33: 1583–1595.

Buffett, B. A., 2000. Earth's core and the geodynamo. Science, 288: 2007–2012.

Buffett, B. A., et al., 1996. On the thermal evolution of the Earth's core. J. Geophys. Res., 101: 7989–8006.

Buick, R., and Dunlop, J. S. R., 1990. Evaporitic sediments of early Archean age from the Warrawoona Group, Western Australia. Sedimentology, 37: 247–278.

Buick, R., Groves, D. I., and Dunlop, J. S. R., 1995a. Geological origin of described stromatolites older than 3.2 Ga: Comment and reply. Geology, 23: 191–192.

Buick, R., Thornett, J R., McNaughton, N. J., Smith, J. B., Barley, M. E., and Savage, M., 1995b. Record of emergent continental crust 3.5 billion years ago in the Pilbara craton of Australia. Nature, 375: 574–577.

Burke, K., Dewey, J. F., and Kidd, W. S. F., 1976. Precambrian paleomagnetic results compatible with contemporary operation of the Wilson cycle. Tectonophysics, 33: 287–299.

Byerly, G. R., Lowe, D. R., Wooden, J. L., and Xie, X., 2002. An Archean impact layer from the Pilbara and Kaapvaal cratons. Science, 297: 1325–1327.

Cairns-Smith, A. G., 1982. Genetic Takeover and Mineral Origins of Life. Cambridge University Press, Cambridge, UK, 477 pp.

Caldeira, K., and Rampino, M. R., 1991. The mid-Cretaceous superplume, carbon dioxide, and global warming. Geophys. Res. Lett., 18: 987–990.

Calvert, A. J., Sawyer, E. W., Davis, W. J., and Ludden, J. N., 1995. Archean subduction inferred from seismic images of a mantle suture in the Superior province. Nature, 375: 670–674.

Cameron, A. G. W., 1973. Accumulation processes in the primitive solar nebula. Icarus, 18: 407–450.

Cameron, A. G. W., 1986. The impact theory for origin of the Moon. In: W. K. Hartmann, R. J. Phillips, and G. J. Taylor (eds.), Origin of the Moon. Lunar and Planetary Institute, Houston, pp. 609–616.

Campbell, I. H., 2002. Implications of Nb/U, Th/U, and Sm/Nd in plume magmas for the relationship between continental and oceanic crust formation and the development of the depleted mantle. Geochim. Cosmochim. Acta, 66: 1651–1661.

Campbell, I. H., and Griffiths, R. W., 1992a. The changing nature of mantle hotspots through time: Implications for the chemical evolution of the mantle. J. Geol., 92: 497–523.

Campbell, I. H., and Griffiths, R. W., 1992b. The evolution of the mantle's chemical structure. Lithos, 30: 389–399.

Campbell, I. H., Griffiths, R. W., and Hill, R. I., 1989. Melting in an Archean mantle plume: Heads it's basalts, tails it's komatiites. Nature, 339: 697–699.

Cande, S. C., Leslie, R. B., Parra, J. C., and Hobart, M., 1987. Interaction between the Chile ridge and the Chile trench: Geophysical and geothermal evidence. J. Geophys. Res., 92: 495–520.

Canfield, D. E., 1998. A new model for Proterozoic ocean chemistry. Nature, 396: 450–453.

Canfield, D. E., and Teske, A., 1996. Late Proterozoic rise in atmospheric oxygen from phylogenetic and stable isotopic studies. Nature, 382: 127–132.

Canfield, D. E., Habicht, K. S., and Thamdrup, B., 2000. The Archean sulfur cycle and the early history of atmospheric oxygen. Science, 288: 658–661.

Cao, R., 1991. Origin and order of cyclic growth patterns in mat-ministromatolite biotherms from the Proterozoic Wumishan Formation, north China. Precamb. Res., 52: 167–178.

Caputo, M. V., and Crowell, J. C., 1985. Migration of glacial centers across Gondwana during Paleozoic era. Geol. Soc. Am. Bull., 96: 1020–1036.

Card, K. D., and Ciesielski, A., 1986. Subdivisions of the Superior province of the Canadian shield. Geosci. Can., 13: 5–13.

Carey, S., and Sigurdsson, H. A., 1984. A model of volcanigenic sedimentation in marginal basins. Geol. Soc. Lond., Spec. Publ., 16: 37–58.

Carlson, R. L., 1995a. A plate cooling model relating rates of plate motion to the age of the lithosphere at trenches. Geophys. Res. Lett., 22: 1977–1980.

Carlson, R. W., 1994. Mechanisms of Earth differentiation: Consequences for the chemical structure of the mantle. Rev. Geophys., 32: 337–361.

Carlson, R. W., 1995b. Physical and chemical evidence on the cause and source characteristics of flood basalt volcanism. Austral. J. Earth Sci., 38: 525–544.

Carlson, R. W., Shirey, S. B., Pearson, D. G., and Boyd, F. R., 1994. The mantle beneath continents. Carnegie Inst. Wash. Year Book 93: 109–117.

Carr, M. H., 1996. Water on Mars. Oxford University Press, Oxford.

Carr, M. H., Kuzmin, R. O., and Masson, P. L., 1993. Geology of Mars. Episodes, 16: 307–315.

Carr, M. H., et al., 1998. Evidence for a subsurface ocean on Europa. Nature, 391: 363–365.

Catling, D. C., Zahnle, K. J., and McKay, C. P., 2001. Biogenic methane, hydrogen escape, and the irreversible oxidation of early Earth. Science, 293: 839–843.

Cavalier-Smith, T., 1987. The simultaneous symbiotic origin of mitochondria, chloroplasts, and microbodies. N. Y. Acad. Sci., Annals, 503: 55–71.

Cawood, P. A., and Suhr, G., 1992. Generation and obduction of ophiolites: Constraints from the Bay of Islands Complex, western Newfoundland. Tectonics, 11: 884–897.

Chandler, F. W., 1988. Quartz arenites: Review and interpretation. Sed. Geol., 58: 105–126.

Chang, S., 1994. The planetary setting of prebiotic evolution. In: S. Bengtson (ed.), Early Life on Earth. Columbia University Press, New York, pp. 10–23.

Chapman, D. S., 1986. Thermal gradients in the continental crust. In: J. B. Dawson, D. A. Carswell, J. Hall, and K. H. Wedepohl (eds.), The Nature of the Lower Continental Crust. Geological Society, London, Special Publication No. 24, pp. 63–70.

Chapman, D. S., and Furlong, K. P., 1992. Thermal state of the continental lower crust. In: D. M. Fountain, R. Arculus, and R. W. Kay (eds.), Continental Lower Crust. Elsevier, Amsterdam, pp. 179–199.

Charlou, J. L., Fouquet, Y., Bougalt, H., Donval, J. P., Etoubleau, J., Jean-Baptiste, P., Dapoigney, A., Appriou, P., and Rona, P. S., 1998. Intense CH_4 plumes generated by serpentinization of ultramafic rocks at the intersection of the 15°20' N fracture zone and the mid-Atlantic ridge. Geochim. Cosmochim. Acta, 62: 2323–2333.

Christensen, N. I., 1966. Elasticity of ultrabasic rocks. J. Geophys. Res., 71: 5921–5931.

Christensen, N. I., and Mooney, W. D., 1995. Seismic velocity structure and composition of the continental crust: A global view. J. Geophys. Res., 100: 9761–9788.

Christensen, U. R., 1995. Effects of phase transitions on mantle convection. Ann. Rev. Earth Planet. Sci., 23: 65–87.

Christensen, U. R., and Yuen, D. A., 1985. Layered convection induced by phase transitions. J. Geophys. Res., 90: 10,291–10,300.

Chyba, C. F., 1993. The violent environment of the origin of life: Progress and uncertainties. Geochim. Cosmochim. Acta, 57: 3351–3358.

Chyba, C. F., 2000. Energy for microbial life on Europa. Nature, 403: 381–382.

Cloos, M., 1993. Lithospheric buoyancy and collisional orogenesis: Subduction of oceanic plateaus, continental margins, island arcs, spreading ridges, and seamounts. Geol. Soc. Am. Bull., 105: 715–737.

Cloud, P., 1968a. Atmospheric and hydrospheric evolution on the primitive Earth. Science, 160: 729–736.

Cloud, P., 1968b. Pre-Metazoan evolution and the origins of the Metazoa. In: E. T. Drake (ed.), Evolution and Environment. Yale University Press, New Haven, pp. 1–72.

Cloud, P., 1973. Paleoecological significance of the banded iron formation. Econ. Geol., 68: 1135–1143.

Coffin, M. F., and Eldholm, O., 1994. Large igneous provinces: Crustal structure, dimensions, and external consequences. Rev. Geophys., 32: 1–36.

Cogley, J. G., 1984. Continental margins and the extent and number of continents. Rev. Geophys., 22: 101–122.

Cogley, J. G., and Henderson-Sellers, A., 1984. The origin and earliest state of the Earth's hydrosphere. Rev. Geophys. Space Phys., 22: 131–175.

Coleman, R. G., 1977. Ophiolites. Springer-Verlag, Berlin, 229 pp.

Collerson, K. D., and Kamber, B. S., 1999. Evolution of the continents and the atmosphere inferred from Th-U-Nb systematics of the depleted mantle. Science, 283: 1519–1522.

Condie, K. C., 1981. Archean Greenstone Belts. Elsevier, Amsterdam, 434 pp.

Condie, K. C., 1989. Geochemical changes in basalts and andesites across the Archean–Proterozoic boundary: Identification and significance. Lithos, 23: 1–18.

Condie, K. C., 1990a. Geochemical characteristics of Precambrian basaltic greenstones. In: R. P. Hall and D. J. Hughes (eds.), Early Precambrian Basic Magmatism. Blackie & Son, Glasgow, UK, pp. 40–55.

Condie, K. C., 1990b. Growth and accretion of continental crust: Inferences based on Laurentia. Chem. Geol., 83: 183–194.

Condie, K. C., 1992a. Evolutionary changes at the Archean–Proterozoic boundary. In: J. E. Glover and S. E. Ho (eds.), The Archean: Terrains, Processes, and Metallogeny (proceeding volume of the Third International

Archean Symposium 1990, Perth, Australia). University of Western Australia, Crawley, Western Australia, Publication No. 22, pp. 177–189.

Condie, K. C., 1992b. Proterozoic terranes and continental accretion in southwestern North America. In: K. C. Condie (ed.), Proterozoic Crustal Evolution. Elsevier, Amsterdam, pp. 447–480.

Condie, K. C., 1993. Chemical composition and evolution of the upper continental crust: Contrasting results from surface samples and shales. Chem. Geol., 104: 1–37.

Condie, K. C., 1994. Greenstones through time. In: K. C. Condie (ed.), Archean Crustal Evolution. Elsevier, Amsterdam, pp. 85–120.

Condie, K. C., 1995. Episodic ages of greenstones: A key to mantle dynamics. Geophys. Res. Lett., 22: 2215–2218.

Condie, K. C., 1997. Contrasting sources for upper and lower continental crust: The greenstone connection. J. Geol., 105: 729–736.

Condie, K. C., 1998. Episodic continental growth and super-continents: A mantle avalanche connection? Earth Planet. Sci. Lett., 163: 97–108.

Condie, K. C., 1999. Mafic crustal xenoliths and the origin of the lower continental crust. Lithos, 46: 95–101.

Condie, K. C., 2000. Episodic continental growth models: Afterthoughts and extensions. Tectonophysics, 322: 153–162.

Condie, K. C., 2001. Mantle Plumes and Their Record in Earth History. Cambridge University Press, Cambridge, UK, 306 pp.

Condie, K. C., 2002a. The supercontinent cycle: Are there two patterns of cyclicity? J. African Earth Sci., 35 (2): 179–183.

Condie, K. C., 2002b. Breakup of a Paleoproterozoic super-continent. Gondwana Res., 5 (1): 41–43.

Condie, K. C., 2003a. Incompatible element ratios in oceanic basalts and komatiites: Tracking deep mantle sources and continental growth rates with time. Geochem. Geophys. Geosys., 4 (1), pp. 1.

Condie, K. C., 2003b. Supercontinents, superplumes, and continental growth: The Neoproterozoic record. Geol. Soc. Lond., Spec. Publ., 206: 1–21.

Condie, K. C., 2004. High field strength/element ratios in Archean basalts: A window to evolving sources of mantle plumes. Lithos (In Press).

Condie, K. C., and Chomiak, B., 1996. Continental accretion: Contrasting Mesozoic and early Proterozoic tectonic regimes in the North America. Tectonophysics, 265: 101–126.

Condie, K. C., and Rosen, O. M., 1994. Laurentia–Siberia connection revisited. Geology, 22: 168–170.

Condie, K. C., and Wronkiewicz, D. J., 1990. The Cr/Th ratio in Precambrian pelites from the Kaapvaal craton as an index of craton evolution. Earth Planet. Sci. Lett., 97: 256–267.

Condie, K. C., Des Marais, D. J., and Abbott, D., 2000. Geologic evidence for a mantle superplume event at 1.9 Ga. Geochem. Geophys. Geosys., 1, doi: 2000GC000095 (online).

Conrad, C. P., and Lithgow-Bertelloni, C., 2002. How mantle slabs drive plate tectonics. Science, 298: 207–209.

Cook, F. A., Brown, L. D., Kaufman, S., Oliver, J. E., and Petersen, J., 1981. COCORP seismic profiling of the Appalachian orogen beneath the coastal plain of Georgia. Geol. Soc. Am. Bull., 92: 738–748.

Cook, P. J., 1992. Phosphogenesis around the Proterozoic–Phanerozoic transition. J. Geol. Soc. Lond., 149: 615–620.

Cook, P. J., and McElhinny, M. W., 1979. A reevaluation of the spatial and temporal distribution of sedimentary phosphate deposits in the light of plate tectonics. Econ. Geol., 74: 315–330.

Cooper, G., Klimmich, N., Belisle, W., Sarinana, J., Brabham, K., and Garrel, L., 2001. Carbonaceous meteorites as a source of sugar-related organic compounds for the early Earth. Nature, 414: 879–883.

Copp, D. L., Guest, J. E., and Stofan, E. R., 1998. New insights into coronae evolution: Mapping on Venus. J. Geophys. Res., 103: 19,401–19,417.

Corliss, J. B., Baross, J. A., and Hoffman, S. E., 1981. A hypothesis concerning the relationship between submarine hot springs and the origin of life on Earth. Ocenaol. Acta, Spec. Publ.: 59–69.

Coulton, A. J., Harper, G. D., and O'Hanley, D. S., 1995. Oceanic versus emplacement age serpentinization in the Josephine ophiolite: Implications for the nature of the Moho at intermediate- and slow-spreading ridges. J. Geophys. Res., 100: 22,245–22,260.

Courtillot, V., and Besse, J., 1987. Magnetic field reversals, polar wander, and core–mantle coupling. Science, 237: 1140–1147.

Courtillot, V., Jaupart, C., Manighetti, I., Tapponnier, P., and Besse, J., 1999. On causal links between flood basalts and continental breakup. Earth Planet. Sci. Lett., 166: 177–195.

Courtillot, V. E., and Cisowski, S., 1987. The Cretaceous-Tertiary boundary events: External or internal causes? EOS, 68: 193.

Cowan, D. S., 1986. The origin of some common types of melange in the western Cordillera of North America. In: N. Nasu, K. Kobayashi, S. Nyeda, I. Kushiro, and H. Kagami (eds.), Formation of Active Ocean Margins. D. Reidel, Dordrecht, pp. 257–272.

Coward, M. P., Jan, M. Q., Rex, O., Tarney, J., Thirwall, M., and Windley, B. F., 1982. Geotectonic framework of the Himalaya of northern Pakistan. J. Geol. Soc. Lond., 139: 299–308.

Coward, M. P., Broughton, R. D., Luff, I. W., Peterson, M. G., Pudsey, C. J., Rex, D. C., and Asifkhan, M., 1986. Collision tectonics in the northwestern Himalayas. Geol. Soc. Lond., Spec. Publ., 19: 203–220.

Cox, A., 1969. Geomagnetic reversals. Science, 163: 237–245.

Cronin, V. S., 1991. The cycloid relative-motion model and the kinematics of transform faulting. Tectonophysics, 187: 215–249.

Crough, S. T., 1983. Hotspot swells. Ann. Rev. Earth Planet. Sci., 11: 165–193.

Crowell, J. C., 1999. Pre-Mesozoic ice ages: Their bearing on understanding the climate system. Geol. Soc. Am., Mem., 192: 1–112.

Crowley, K. D., and Kuhlman, S. L., 1988. Apatite thermo-chronometry of western Canadian shield: Implications for origin of the Williston basin. Geophys. Res. Lett., 15: 221–224.

Crowley, T. J., 1983. The geologic record of climatic change. Rev. Geophys. Space Phys., 21: 828–877.

Crowley, T. J., and Berner, R. A., 2001. CO_2 and climate change. Science, 292: 870–872.

Culotta, R. C., Pratt, T., and Oliver, J., 1990. A tale of two sutures: COCORP's deep seismic surveys of the Grenville province in the eastern US midcontinent. Geology, 18: 646–649.

Dahlen, F. A., and Suppe, J., 1988. Mechanics, growth, and erosion of mountain belts. Geol. Soc. Am., Spec. Paper, 218: 161–178.

Dallmeyer, R. D., and Brown, M., 1992. Rapid Variscan exhumation of Eo-Variscan metamorphic rocks from southern Brittany, France (program with abstracts). Geol. Soc. Am., 24: A236.

Dalziel, I. W. D., 1997. Neoproterozoic–Paleozoic geography and tectonics: Review, hypothesis, environmental speculation. Geol. Soc. Am. Bull., 109: 16–42.

Dalziel, I. W. D., Dalla Salda, L. H., and Gahagan, L. M., 1994. Paleozoic Laurentia–Gondwana interaction and the origin of the Appalachian–Andean mountain system. Geol. Soc. Am. Bull., 106: 243–252.

Dalziel, I. W. D., Lawver, L. A., and Murphy, J. B., 2000. Plumes, orogenesis, and supercontinental fragmentation. Earth Planet. Sci. Lett., 178: 1–11.

Danielson, A., Moller, P., and Dulski, P., 1992. The Eu anomalies in BIFs and the thermal history of the oceanic crust. Chem. Geol., 97: 89–100.

Dauphas, N., 2003. The dual origin of the terrestrial atmosphere. Icarus, 165: 326–339.

Davaille, A., 1999. Simultaneous generation of hotspots and superswells by convection in a heterogeneous planetary mantle. Nature, 402: 756–760.

Davies, G. F., 1980. Review of oceanic and global heat flow estimates. Rev. Geophys., 18: 718–722.

Davies, G. F., 1992. On the emergence of plate tectonics. Geology, 20: 963–966.

Davies, G. F., 1993. Conjectures on the thermal and tectonic evolution of the Earth. Lithos, 30: 281–289.

Davies, G. F., 1995. Penetration of plates and plumes trough the mantle transition zone. Earth Planet. Sci. Lett., 133: 507–516.

Davies, G. F., 1996. Punctuated tectonic evolution of the Earth. Earth Planet. Sci. Lett., 136: 363–379.

Davies, G. F., 1999. Dynamic Earth Plates, Plumes, and Mantle Convection. Cambridge University Press, Cambridge, UK, 458 pp.

Davies, G. F., 2002. Stirring geochemistry in mantle convection models with stiff plates and slabs. Geochim. Cosmochim. Acta, 66: 3125–3142.

Davies, G. F., and Richards, M. A., 1992. Mantle convection. J. Geol., 100: 151–206.

DeBari, S. M., and Sleep, N. H., 1991. High-Mg, low-Al bulk composition of the Talkeetna island arc, Alaska: Implications for primary magmas and the nature of the arc crust. Geol. Soc. Am. Bull., 103: 37–47.

de Duve, C., 1995. The beginnings of life on Earth. Am. Scient., 83: 428–437.

DeLong, S. E., and Fox, P. J., 1977. Geological consequence of ridge subduction. In: M. Talwani and W. C. Pittman III (eds.), Island Arcs, Deep-Sea Trenches, and Back-Arc Basins. Am. Geophys. Union, Maurice Ewing Series 1, pp. 221–228.

DeMets, C., Gordon, R. G., Stein, S., and Argus, D. F., 1990. Current plate motions. Geophys. Res. Lett., 14: 911–914.

Deming, D., 1999. On the possible influence of extraterrestrial volatiles on Earth's climate and the origin of the oceans. Paleogeog. Paleoclimat. Paleoecol., 146: 33–51.

Derry, L. A., and Jacobsen, S. R., 1990. The chemical evolution of Precambrian seawater: Evidence from REEs in BIF. Geochim. Cosmochim. Acta, 54: 2965–2977.

Derry, L. A., Kaufman, A. J., and Jacobsen, S. B., 1992. Sedimentary cycling and environmental change in the late Proterozoic: Evidence from stable and radiogenic isotopes. Geochim. Cosmochim. Acta., 56: 1317–1329.

Des Marais, D. J., 1994. The Archean atmosphere: Its composition and fate. In: K. C. Condie (ed.), Archean Crustal Evolution. Elsevier, Amsterdam, pp. 505–523.

Des Marais, D. J., 1997. Long-term evolution of the bio–geochemical carbon cycle. In: J. F. Banfield and K. H. Nielsen (eds.), Geomicrobiology: Interactions Between Microbes and Minerals. Reviews in Mineralogy, Vol. 35. Mineralogical Society of America, Washington, DC, pp. 429–448.

Des Marais, D. J., Strauss, H., Summons, R. E., and Hayes, J. M., 1992. Carbon isotope evidence for the stepwise oxidation of the Proterozoic environment. Nature, 359: 605–609.

Desrochers, J. P., Hubert, C., Ludden, J. N., and Pilote, P., 1993. Accretion of Archean oceanic plateau fragments in the Abitibi greenstone belt, Canada. Geology, 21: 451–454.

Dewey, J. F., 1988. Extensional collapse of orogens. Tectonics, 7: 1123–1139.

Dewey, J. F., and Kidd, W. S. F., 1977. Geometry of plate accretion. Geol. Soc. Am. Bull., 88: 960–968.

Dewey, J. F., Hempton, M. R., Kidd, W. S. F., Saroglu, F., and Sengor, A. M. C., 1986. Shortening of continental lithosphere: The neotectonics of eastern Anatolia—A young collision zone. Geol. Soc. Lond., Spec. Publ., 19: 3–36.

de Wit, M. J., 1998. On Archean granites, greenstones, cratons, and tectonics: Does the evidence demand a verdict? Precamb. Res., 91: 181–226.

de Wit, M. J., and Ashwal, L. D., 1995. Greenstone belts: What are they? S. African J. Geol., 98: 505–520.

de Wit, M. J., et al., 1992. Formation of an Archean continent. Nature, 357: 553–562.

Dick, H. J. B., Lin, J., and Schouten, H., 2003. An ultra-slow-spreading class of ocean ridge. Nature: 426: 405–412.

Dickinson, W. R., 1970. Relations of andesites, granites, and derivative sandstones to arc–trench tectonics. Rev. Geophys., 8: 813–860.

Dickinson, W. R., 1973. Widths of modern arc–trench gaps proportional to past duration of igneous activity in associated magmatic arcs. J. Geophys. Res., 78: 3376–3389.

Dickinson, W. R., and Seely, D. R., 1986. Structure and stratigraphy of forearc regions. Am. Assoc. Petrol. Geol. Bull., 63: 2–31.

Dilek, Y., and Eddy, E. A., 1992. The Troodos and Kizaldag ophiolites as structural models for slow-spreading ridge segments. J. Geol., 100: 305–322.

Donnadieu, Y., Ramstein, G., Fluteau, F., Besse, J., and Meert, J., 2002. Is high obliquity a plausible cause for Neoproterozoic glaciations? Geophys. Res. Lett., 29, doi: 10.1029/2002GL015902 (online).

Doolittle, W. F., 1999. Phylogenetic classification and the universal tree. Science, 284: 2124–2128.

Dott, R. H. Jr., 2003. The importance of eolian abrasion in supermature quartz sandstones and the paradox of weathering on vegetation-free landscapes. J. Geol., 111: 387–405.

Douglass, J., Schilling, J. G., and Fontignie, D., 1999. Plume-ridge interactions of the Discovery and Shona mantle plumes with the southern mid-Atlantic ridge (40-55° S). J. Geophys. Res., 104: 2941–2962.

Downes, H., 1993. The nature of the lower continental crust of Europe: Petrological and geochemical evidence from xenoliths. Phys. Earth Planet. Int., 79: 195–218.

Drake, M. J., 2000. Accretion and primary differentiation of the Earth: A personal journey. Geochim. Cosmochim. Acta, 64: 2363–2370.

Drake, M. J., and Righter, K., 2002. Determining the composition of the Earth. Nature 416: 39–44.

Duba, A., 1992. Earth's core not so hot. Nature, 359: 197–198.

Dubrovinsky, L., et al., 2003. Iron–silica interaction at extreme conditions and the electrically conducting layer at the base of Earth's mantle. Nature, 422: 58–61.

Duffy, T. S., Zha, C., Downs, R. T., Mao, H., and Hemley, R. J., 1995. Elasticity of forsterite to 16 GPa and the composition of the upper mantle. Nature, 378: 170–173.

Dumberry, M., and Bloxham, J., 2003. Torque balance, Taylor's constraint, and torsional oscillations in the numerical model of the geodynamo. Phys. Earth Planet. Interiors, 140: 29–51.

Duncan, R. A., 1991. Ocean drilling and the volcanic record of hotspots. GSA Today, 1 (10): 213–219.

Duncan, R. A., and Erba, E., 2003. Thermal and compositional effects of oceanic LIPs: Consequences for Cretaceous ocean anoxic events. In: Mantle Plumes: Physical Processes, Chemical Signatures, and Biological Effects (abstract volume). Cardiff University, Cardiff, UK, p. 10.

Duncan, R. A., and Richards, M. A., 1991. Hotspots, mantle plumes, flood basalts, and true polar wander. Rev. Geophys., 29: 31–50.

Dunlop, D. J., 1995. Magnetism in rocks. J. Geophys. Res., 100: 2161–2174.

Durrheim, R. J., and Mooney, W. D., 1991. Archean and Proterozoic crustal evolution: Evidence from crustal seismology. Geology, 19: 606–609.

Durrheim, R. J., and Mooney, W. D., 1994. Evolution of the Precambrian lithosphere: Seismological and geochemical constraints. J. Geophys. Res., 99: 15,359–15,374.

DuToit, A., 1937. Our Wandering Continents. Oliver and Boyd, London, 366 pp.

Elthon, D., and Scarfe, C. M., 1984. High pressure phase equilibria of a high-MgO basalt and the genesis of primary oceanic basalts. Am. Mineral., 69: 1–15.

Emslie, R. F., 1978. Anorthosite massifs, rapakivi granites, and late Proterozoic rifting of North America. Precamb. Res., 7: 61–98.

England, P. C., and Richardson, S. W., 1977. The influence of erosion upon the mineral facies of rocks from different metamorphic environments. J. Geol. Soc. Lond., 134: 201–213.

Eriksson, K. A., and Fedo, C. M., 1994. Archean synrift and stable-shelf sedimentary successions. In: K. C. Condie (ed.), Archean Crustal Evolution. Elsevier, Amsterdam, pp. 171–203.

Eriksson, P. G., 1999. Sea level changes and the continental freeboard concept: General principles and application to the Precambrian. Precamb. Res., 97: 143–154.

Eriksson, P. G., and Cheney, E. S., 1992. Evidence for the transition to an oxygen-rich atmosphere during the evolution of redbeds in the lower Proterozoic sequences of southern Africa. Precamb. Res., 54: 257–269.

Eriksson, P. G., Mazumder, R., Sarkar, S., Bose, P. K., Altermann, W., and van der Merwe, R., 1999. The 2.7–2.0-Ga volcano–sedimentary record of Africa, India, and Australia: Evidence for global and local changes in sea level and continental freeboard. Precamb. Res. 97: 269–302.

Eriksson, P. G., et al., 2002. Late Archean superplume events: A Kaapvaal-Pilbara perspective. J. Geodyn., 34: 207–247.

Ernst, R. E., and Buchan, K. L., 2001. The use of mafic dike swarms in identifying and locating mantle plumes. Geol. Soc. Am., Spec. Paper, 352: 247–266.

Ernst, R. E., and Buchan, K. L., 2003. Recognizing mantle plumes in the geological record. Ann. Rev. Earth Planet. Sci., 31: 469–523.

Ernst, R. E., Head, J. W., Parfitt, E., Grosfils, E., and Wilson, L., 1995. Giant radiating dyke swarms on Earth and Venus. Earth Sci. Rev., 39: 1–58.

Ernst, W. G., 1972. Occurrence and mineralogic evolution of blueschist belts with time. Am. J. Sci., 272: 657–668.

Erwin, D. H., 1993. The Great Paleozoic Crisis. Columbia University Press, New York.

Erwin, D. H., 1994. The Permo–Triassic extinction. Nature, 367: 231–235.

Erwin, D. H., 2003. Impact at the Permo–Triassic boundary: A critical evaluation. Astrobiology, 3: 67–74.

Estey, L. H., and Douglas, B. J., 1986. Upper mantle anisotropy: A preliminary model. J. Geophys. Res., 91: 11,393–11,406.

Evans, D. A. D., Li, Z. X., Kirschvinik, J. L., and Wingate, M. T. D., 2000. A high-quality mid-Neoproterozoic paleomagnetic pole from south China, with implications for ice ages and the breakup configuration of Rodinia. Precamb. Res., 100: 313–334.

Eyles, N., 1993. Earth's glacial record and its tectonic setting. Earth Sci. Rev., 35: 1–248.

Fahrig, W. F., 1987. The tectonic settings of continental mafic dyke swarms: Failed arm and early passive margin. Geol. Assoc. Can., Spec. Paper, 34: 331–348.

Fanale, F. P., 1971. A case for catastrophic early degassing of the Earth. Chem. Geol., 8: 79–105.

Farmer, J., 1998. Thermophiles, early biosphere evolution, and the origin of life on Earth: Implications for the exobiological exploration of Mars. J. Geophys. Res., 103: 28,457–28,461.

Farquhar, J., and Wing, B. A., 2003. Multiple sulfur isotopes and the evolution of the atmosphere. Earth Planet. Sci. Lett., 213: 1–13.

Farquhar, J., Bao, H. M., and Thiemens, M., 2000. Atmospheric influence of Earth's earliest sulfur cycle. Science, 289: 756–758.

Faure, K., de Wit, M. J., and Willis, J. P., 1995. Late Permian global coal hiatus linked to ^{13}C-depleted CO_2 flux into the atmosphere during the final consolidation of Pangaea. Geology, 23: 507–510.

Fedo, C. M., Myers, J. S., and Appel, P. W. U., 2001. Depositional setting and paleogeographic implications of Earth's oldest supracrustal rocks, the >3.7-Ga Isua greenstone belt, West Greenland. Sediment. Geol., 141: 61–77.

Fischer, A. G., 1984. Biological innovations and the sedimentary record. In: H. D. Holland and A. F. Trendall (eds.), Patterns of Change in Earth Evolution. Springer-Verlag, Berlin, pp. 145–157.

Fisher, D. E., 1985. Radiogenic rare gases and the evolutionary history of the depleted mantle. J. Geophys. Res., 90: 1801–1807.

Fisher, R. V., 1984. Submarine volcaniclastic rocks. Geol. Soc. Lond., Spec. Publ., 16: 5–28.

Fitton, J. G., Saunders, A. D., Larsen, L. M., Hardarson, B. S., and Norry, M. J., 1998. Volcanic rocks from the southeastern Greenland margin at 63° N: Composition, petrogenesis, and mantle sources. Proc. Ocean Drill. Prog., Sci. Res., 152: 331–350.

Folk, R. L., 1976. Reddening of desert sands: Simpson Desert, NT, Australia. J. Sed. Petrol., 46: 604–615.

Follmi, K. B., 1996. The phosphorus cycle, phosphogenesis, and marine phosphate-rich deposits. Earth Sci. Rev., 40: 55–124.

Foote, M., 2003. Origination and extinction through the Phanerozoic: A new approach. J. Geol., 111: 125–148.

Forsyth, D. W., 1996. Partial melting beneath a mid-Atlantic segment detected by teleseismic PKP delays. Geophys. Res. Lett., 23: 463–466.

Fountain, D. M., and Salisbury, M. H., 1981. Exposed crustal sections through the continental crust: Implications for crustal structure, petrology, and evolution. Earth Planet. Sci. Lett., 56: 263–277.

Fountain, D. M., Furlong, K. P., and Salisbury, M. H., 1987. A heat production model of a shield area and its implications for the heat flow–heat production relationship. Geophys. Res. Lett., 14: 283–286.

Frakes, L. A., Francis, J. E., and Syktus, J. L., 1992. Climate Modes of the Phanerozoic. Cambridge University Press, New York, 274 pp.

Friend, C. R. L., Nutman, A. P., and McGregor, V. R., 1988. Late Archean terrane accretion in the Godthab region, southern West Greenland. Nature, 335: 535–538.

Frost, C. D., and Frost, B. R., 1997. Reduced rapakivi-type granites: The tholeiite connection. Geology, 25: 647–650.

Froude, D. O., Ireland, T. R., Kinny, P. O., Williams, I. S., and Compston, W., 1983. Ion microprobe identification of 4100–4200-Ma-old terrestrial zircons. Nature, 304: 616–618.

Fryer, P., 1996. Evolution of the Mariana convergent plate margin system. Rev. Geophys., 34: 89–125.

Fuller, M., and Cisowski, S. M., 1987. Review of lunar magnetism. In: J. A. Jacobs (ed.), Geomagnetism. Academic Press, New York, pp. 307-455.

Fyfe, W. S., 1978. The evolution of the Earth's crust: Modern plate tectonics to ancient hot spot tectonics? Chem. Geol., 23: 89–114.

Gaffin, S., 1987. Ridge-volume dependence on seafloor generation rate and inversion using long-term sea level change. Am. J. Sci., 287: 596–611.

Gaherty, J. B., and Jordan, T. H., 1995. Lehmann discontinuity as the base of an anisotropic layer beneath continents. Science, 268: 1468–1471.

Galer, S. J. G., 1991. Interrelationships between continental freeboard, tectonics, and mantle temperature. Earth Planet. Sci. Lett., 105: 214–228.

Galer, S. J. G., and Mezger, K., 1998. Metamorphism, denudation, and sea level in the Archean and cooling of the Earth. Precamb. Res., 92: 389–412.

Gans, P. B., 1987. An open-system two-layer crustal stretching model for the eastern Great Basin. Tectonics, 6: 1–12.

Gao, S., and Wedepohl, K. H., 1995. The negative Eu anomaly in Archean sedimentary rocks: Implications for decomposition, age, and importance of their granitic sources. Earth Planet. Sci. Lett., 133: 81–94.

Garfunkel, Z., 1986. Review of oceanic transform activity and development. J. Geol. Soc. Lond., 143: 775–784.

Gastil, G., 1960. The distribution of mineral dates in time and space. Am. J. Sci., 258: 1–35.

Genda, H., and Abe, Y., 2003. Survival of a protoatmosphere through the stage of giant impacts: The mechanical aspect. Icarus, 164: 149–162.

Gibbs, A. K., Montgomery, C. W., O'Day, P. A., and Erslev, E. A., 1986. The Archean–Proterozoic transition: Evidence form the geochemistry of metasedimentary rocks of Guyana and Montana. Geochim. Cosmochim. Acta, 50: 2125–2141.

Gilbert, W., 1986. The RNA world. Nature, 319: 618.

Glatzmaier, G. A., 2002. Geodynamo simulations: How realistic are they? Ann. Rev. Earth Planet. Sci., 30: 237–257.

Glikson, A. Y., 1993. Asteroids and early Precambrian crustal evolution. Earth Sci. Rev., 35: 285–319.

Goff, J. A., Holliger, K., and Levander, A., 1994. Modal fields: A new method for characterization of random velocity heterogeneity. Geophys. Res. Lett., 21: 493–496.

Goldfarb, R. J., Groves, D. I., and Gardoll, S., 2001. Orogenic gold and geologic time: A global synthesis. Ore Geol. Rev., 18: 1–75.

Goldstein, S. J., and Jacobsen, S. B., 1988. Nd and Sr isotopic systematics of river water suspended material: Implications for crustal evolution. Earth Planet. Sci. Lett., 87: 249–265.

Gordon, R. G., and Stein, S., 1992. Global tectonics and space geodesy. Science, 256: 333–342.

Goto, K., Hamaguchi, H., and Suzuki, Z., 1985. Earthquake-generating stresses in a descending slab. Tectonophysics, 112: 111–128.

Graham, J. B., Dudley, R., Aguilar, N. M., and Gans, C., 1995. Implications of the late Paleozoic oxygen pulse for physiology and evolution. Nature, 375: 117–120.

Graham, S. A., et al., 1986. Provenance modeling as a technique for analyzing source terrane evolution and controls on foreland sedimentation. In: A. Allen and P. Homewood (eds.), Foreland Basins. Blackwell Scientific Publications, Oxford, pp. 425–436.

Grand, S. P., 1987. Tomographic inversion for shear velocity beneath the North American plate. J. Geophys. Res., 92: 14,065–14,090.

Grand, S. P., and Helmbuger, D. V., 1984. Upper mantle shear structure beneath the northwest Atlantic Ocean. J. Geophys. Res., 89: 11,465–11,475.

Grand, S. P., van der Hilst, R. D., and Widiyantoro, S., 1997. Global seismic tomography: A snapshot of convection in the Earth. GSA Today, 7 (4): 1–7.

Grattan, J., 2003. Pollution and paradigms: Lessons from the Laki fissure eruption. In: Mantle Plumes: Physical Processes, Chemical Signatures, and Biological Effects (abstract volume). Cardiff University, Cardiff, UK, p. 19.

Greenberg, R., 1989. Planetary accretion. In: S. K. Atreya, J. B. Pollack, and M. S. Matthews (eds.), Origin and Evolution of Planetary and Satellite Atmospheres. University Arizona Press, Tucson, pp. 137–165.

Greff-Lefftz, M., and Legros, H., 1999. Core rotational dynamics and geological events. Science, 286: 1707–1709.

Gregory, R. T., 1991. Oxygen isotope history of seawater revisited: Timescales for boundary event changes in the oxygen isotope composition of seawater. In: H. P. Taylor Jr. et al. (eds.), Stable Isotope Geochemistry. Geochemical Society, Special Publication No. 3, pp. 65–76.

Grey, K., Walter, M. R., and Calver, C. R., 2003. Neoproterozoic biotic diversification: Snowball Earth or aftermath of the Acraman impact? Geology, 31: 459–462.

Grieve, R. A. F., 1980. Impact bombardment and its role in protocontinental growth on the early Earth. Precamb. Res., 10: 217–247.

Griffin, W. L., and O'Reilly, S. Y., 1987. Is the continental Moho the crust–mantle boundary? Geology, 15: 241–244.

Griffin, W. L., O'Reilly, S. Y., Ryan, C. G., Gaul, O., and Ionov, D. A., 1998. Secular variation in the composition of subcontinental lithospheric mantle: Geophysical and geodynamic implications. Am. Geophys. Union–Geol. Soc. Am., Geodynam. Series, 26: 1–25.

Griffin, W. L., O'Reilly, S. Y., and Ryan, C. G., 1999. The composition and origin of subcontinental lithospheric mantle. In: Y. Fei, C. M. Berka, and B. O. Mysen (eds.), Mantle Petrology: Field Observations and High Pressure Experimentation. Geochemical Society, Special Publication No. 6, pp. 13–45.

Griffin, W. L., O'Reilly, S. Y., Abe, N., Aulbach, S., Davies, R. M., Pearson, N. J., Doyle, B. J., and Kivi, K., 2003. The origin and evolution of Archean lithospheric mantle. Precamb. Res., 127: 19–41.

Griffiths, R. W., and Campbell, I. H., 1990. Stirring and structure in mantle plumes. Earth Planet. Sci. Lett., 99: 66–78.

Grossman, L., 1972. Condensation in the primitive solar nebula. Geochim. Cosmochim. Acta, 36: 597–619.

Grotzinger, J. P., 1990. Geochemical model for Proterozoic stromatolite decline. Am. J. Sci., 290A: 80–103.

Grotzinger, J. P., and Kasting, J. F., 1993. New constraints on Precambrian ocean composition. J. Geol., 101: 235–243.

Grotzinger, J. P., and Knoll, A. H., 1999. Stomatolites in Precambrian carbonates: Evolutionary mileposts or environmental dipsticks? Ann. Rev. Earth Planet. Sci., 27: 313–358.

Grotzinger, J. P., Bowring, S. A., Saylor, B. Z., and Kaufman, A. J., 1995. Biostratigraphic and geochronologic constraints on early animal evolution. Science, 270: 589–604.

Groves, D. I., and Barley, M. E., 1994. Archean mineralization. In: K. C. Condie (ed.), Archean Crustal Evolution. Elsevier, Amsterdam, pp. 461–504.

Gubbins, D., 1994. Geomagnetic polarity reversals: A connection with secular variation and core–mantle interaction? Rev. Geophys., 32: 61–83.

Guillou, L., and Jaupart, C., 1995. On the effect of continents on mantle convection. J. Geophys. Res., 100: 24,217–24,238.

Gurnis, M., 1988. Large-scale mantle convection and the aggregation and dispersal of supercontinents. Nature, 332: 695–699.

Gurnis, M., 1993. Phanerozoic marine inundation of continents driven by dynamic topography above subductiing slabs. Nature, 364: 589–593.

Gurnis, M., and Davies, G. F., 1986. Apparent episodic crustal growth arising from a smoothly evolving mantle. Geology, 14: 396–399.

Gurwell, M. A., 1995. Evolution of deuterium on Venus. Nature, 378: 22–23.

Haapala, I., and Ramo, T., 1990. Petrogenesis of the Proterozoic rapakivi granites of Finland. Geol. Soc. Am., Spec. Paper, 246: 275–286.

Hager, B. H., and Clayton, R. W., 1989. Constraints on the structure of mantle convection using seismic observations, flow models, and the geoid. In: W. R. Peltier (ed.), Mantle Convection. Gordon & Breach, New York, pp. 657–763.

Hager, B. H., Clayton, R. W., Richards, M. A., Comer, R. P., and Dziewonski, A. M., 1985. Lower mantle heterogeneity, dynamic topography, and the geoid. Nature, 313: 541–545.

Haggerty, S. E., and Sautter, V., 1990. Ultradeep ultramafic upper mantle xenoliths. Science, 248: 993–996.

Hale, C. J., 1987. Paleomagnetic data suggest link between the Archean–Proterozoic boundary and inner core nucleation. Nature, 329: 233–236.

Hallam, A., 1973. Provinciality, diversity, and extinction of Mesozoic marine invertebrates in relation to plate movements. In: D. H. Tarling and S. K. Runcorn (eds.), Implications of Continental Drift to the Earth Sciences, Vol. 1. Academic Press, London, pp. 287–294.

Hallam, A., 1974. Changing patterns of provinciality and diversity of fossil animals in relation to plate tectonics. J. Biogeo., 1: 213–225.

Hallam, A., 1987. End-Cretaceous mass extinction event: Argument for terrestrial causation. Science, 29: 1237–1242.

Hallam, A., and Wignall, P. B., 1999. Mass extinctions and sea-level changes. Earth Sci. Rev., 48: 217–250.

Hamilton, W. B., 1988. Plate tectonics and island arcs. Geol. Soc. Am. Bull., 100: 1503–1527.

Hamilton, W. B., 1998. Archean magmatism and deformation were not products of plate tectonics. Precamb. Res., 91: 143–179.

Hanan, B. B., and Graham, D. W., 1996. Lead and helium isotope evidence from oceanic basalts for a common deep source of mantle plumes. Science, 272: 991–995.

Hansen, V. L., Banks, B. K., and Ghent, R. R., 1999. Tessera terrain and crustal plateaus, Venus. Geology, 27: 1071–1074.

Haq, B. U., 1998. Gas hydrates: Greenhouse nightmare? Energy panacea or pipe dream? GSA Today, 8 (11): 1–6.

Hardebeck, J., and Anderson, D. L., 1996. Eustasy as a test of a Cretaceous superplume hypothesis. Earth Planet. Sci. Lett., 137: 101–108.

Hardie, L. A., 1996. Secular variation in seawater chemistry: An explanation for the coupled secular variation in the mineralogies of marine limestones and potash evaporites over the past 600 My. Geology, 24: 279–283.

Hardie, L. A., 2003. Secular variations in Precambrian seawater chemistry and the timing of Precambrian aragonite seas and calcite seas. Geology, 31: 785–788.

Hargraves, R. B., 1986. Faster-spreading or greater ridge length in the Archean? Geology, 14: 750–752.

Harlan, S. S., Heaman, L., LeCheminant, A. N., and Premo, W. R., 2003. Gunbarrel mafic magmatic event: A key 780-Ma time marker for Rodinia plate reconstructions. Geology, 31: 1053–1056.

Harley, S. L., and Black, L. P., 1987. The Archean geological evolution of Enderby Land, Antarctica. Geol. Soc. Lond., Spec. Publ., 27: 285–296.

Harper, C. L., and Jacobsen, S. B., 1996. Evidence for ^{182}Hf in the early solar system and constraints on the timescale of terrestrial accretion and core formation. Geochim. Cosmochim. Acta, 60: 1131–1153.

Harris, A. W., and Kaula, W. M., 1975. A coaccretional model of satellite formation. Icarus, 24: 516–524.

Harris, N. B. W., 1995. Significance of weathering Himalayan metasedimentary rocks and leucogranites for the Sr isotope evolution of seawater during the early Miocene. Geology, 23: 795–798.

Harris, N. B. W., Pearce, J. A., and Tindle, A. G., 1986. Geochemical characteristics of collision-zone magmatism. In: M. P. Coward and A. C. Ries (eds.), Collision Tectonics. Geol. Soc. Lond., Spec. Publ., 19: 67–81.

Harrison, C. G. A., 1987. Marine magnetic anomalies: The origin of the stripes. Ann. Rev. Earth Planet. Sci., 15: 505–534.

Harrison, T. M., Copeland, P., Kidd, W. S. F., and Yin, A., 1992. Raising Tibet. Science, 255: 1663–1670.

Hart, S. R., 1984. A large-scale anomaly in the Southern Hemisphere mantle. Nature, 309: 753–757.

Hart, S. R., 1986. In search of a bulk-Earth composition. Chem. Geol., 57: 247–267.

Hart, S. R., 1988. Heterogeneous mantle domains: Signatures, genesis, and mixing chronologies. Earth Planet. Sci. Lett., 90: 273–296.

Hart, S. R., Hauri, E. H., Oschmann, L. A., and Whitehead, J. A., 1992. Mantle plumes and entrainment: Isotopic evidence. Science, 256: 517–519.

Hartmann, W. K., 1983. Moons and Planets. Wadsworth Publishing, Belmont, CA, 509 pp.

Hartmann, W. K., 1986. Moon origin: The impact–trigger hypothesis. In: W. K. Hartmann, R. J. Phillips, and G. J. Taylor (eds.), Origin of the Moon. Lunar and Planetary Institute, Houston, pp. 579–608.

Hartnady, C. J. H., 1991. About turn for supercontinents. Nature, 352: 476–478.

Hartnady, C. J. H., and le Roex, A. P., 1985. Southern ocean hotspot tracks and the Cenozoic absolute motion of the African, Antarctic, and South American plates. Earth Planet. Sci. Lett., 75: 245–257.

Hauri, E. H., and Hart. S. R., 1993. Re/Os isotope systematics of HIMU and EMII oceanic island basalts from the south Pacific Ocean. Earth Planet. Sci. Lett., 114: 353–371.

Hauri, E. H., Whitehead, J. A., and Hart, S. R., 1994. Fluid dynamic and geochemical aspects of entrainment in mantle plumes. J. Geophys. Res., 99: 24,275–24,300.

Hawkesworth, C. J., Erlank, A. J., Kempton, P. D., and Waters, F. G., 1990. Mantle metasomatism: Isotope and trace-element trends in xenoliths from Kimberley, South Africa. Chem. Geol., 85: 19–34.

Hawkesworth, C. J., Gallahger, K., Hergt, J. M., and McDermott, F., 1994. Destructive plate margin magmatism: Geochemistry and melt generation. Lithos, 33: 169–188.

Hedges, S. B., Parker, P. H., Sibley, C. G., and Kumar, S., 1996. Continental breakup and the ordinal diversification of birds and mammals. Nature, 381: 226–229.

Heirtzler, J. R., et al., 1968. Marine magnetic anomalies, geomagnetic field reversals, and motions of the ocean floor and continents. J. Geophys. Res., 73: 2119–2135.

Heizler, M. T., 1993. Thermal history of the continental lithosphere using multiple diffusion domain thermochronometry (PhD thesis). University of California, Los Angeles, p. 295.

Helffrich, G. R., and Wood, B. J., 2001. The Earth's mantle. Nature, 412: 501–507.

Helmstaedt, H. H., and Scott, D. J., 1992. The Proterozoic ophiolite problem. In: K. C. Condie (ed.), Proterozoic Crustal Evolution. Elsevier, Amsterdam, pp. 55–95.

Herzberg, C., 1995. Generation of plume magmas through time: An experimental perspective. Chem. Geol., 126: 1–16.

Hey, R., Duennebier, F. K., and Morgan, W. J., 1980. Propagating rifts on midocean ridges. J. Geophys. Res., 85: 3647–3658.

Hildebrand, A. R., et al., 1991. Chicxulub crater: A possible K–T boundary impact crater on the Yucatan Peninsula, Mexico. Geology, 19: 867–871.

Hinrichs, K. U., 2002. Microbial fixation of methane carbon at 2.7 Ga: Was an anaerobic mechanism possible? Geochem. Geophys. Geosys., 3 (7), doi: 10.1029/2001GC000286 (online).

Hirn, A., Nercessian, A., Sapin, M., Jobert, G., Xu, Z. X., 1984. Lhasa block and bordering sutures: A continuation of the 500-km Moho traverse through Tibet. Nature, 307: 25–27.

Hirose, K., 2002. Phase transitions in pyrolitic mantle around 670-km depth: Implications for upwelling of plumes from the lower mantle. J. Geophys. Res., 107 (B4), doi: 10.1029/2001JB000597 (online).

Hirose, K., Fei, Y., Ma, Y., and Mao, H. K., 1999. The fate of subducted basaltic crust in the Earth's lower mantle. Nature, 397: 53–56.

Hoffman, P. F., 1988. United plates of America, the birth of a craton. Ann. Rev. Earth Planet. Sci., 16: 543–603.

Hoffman, P. F., 1989. Speculations on Laurentia's first gigayear. Geology, 17: 135–138.

Hoffman, P. F., 1990. Geological constraints on the origin of the mantle root beneath the Canadian shield. Phil. Trans. Roy. Soc. Lond., 331A: 523–532.

Hoffman, P. F., 1991. Did the breakout of Laurentia turn Gondwanaland inside-out? Science, 252: 1409–1412.

Hoffman, P. F., Kaufman, A. J., Halverson, G. P., and Schrag, D. P., 1998. A Neoproterozoic snowball Earth. Science, 281: 1342–1346.

Hoffman, S. E., Wilson, M., and Stakes, D. S., 1986. Inferred oxygen isotope profile of Archean oceanic crust, Onverwacht Group, South Africa. Nature, 321: 55–58.

Hofmann, A. W., 1988. Chemical differentiation of the Earth: The relationship between mantle, continental crust, and oceanic crust. Earth Planet. Sci. Lett., 90: 297–314.

Hofmann, A. W., 1997. Mantle geochemistry: The message from oceanic volcanism. Nature, 385: 219–229.

Hofmann, A. W., Jochum, K. P., Seufert, M., and White, W. M., 1986. Nb and Pb in oceanic basalts: New constraints on mantle evolution. Earth Planet. Sci. Lett., 79: 33–45.

Hofmann, H. J., Grey, K., Hickman, A. H., and Thorpe, R. I., 1999. Origin of 3.45 Ga coniform stromatolites in Warrawoona Group, Western Australia. Geol. Soc. Am. Bull., 111: 1256–1262.

Holbrook, W. S., Mooney, W. D., and Christensen, N. J., 1992. The seismic velocity structure of the deep continental crust. In: D. M. Fountain, R. Arculus, and R. W. Kay (eds.), Continental Lower Crust. Elsevier, Amsterdam, pp. 1–43.

Holland, H. D., 1984. The Chemical Evolution of the Atmosphere and Oceans. Princeton University Press, Princeton, 582 pp.

Holland, H. D., 1992. Distribution and paleoenvironmental interpretation of Proterozoic paleosols. In: J. W. Schopf and C. Klein (eds.), The Proterozoic Bisophere: A Multidisciplinary Study. Cambridge University Press, New York, pp. 153–155.

Holland, H. D., 1994. Early Proterozoic atmospheric change. In: S. Bergtson (ed.), Early Life on Earth. Columbia University Press, New York, pp. 237–244.

Holland, H. D., 2002. Volcanic gases, black smokers, and the great oxidation event. Geochim. Cosmochim. Acta, 66: 3811–3826.

Holland, H. D., and Beukes, N. J., 1990. A paleoweathering profile from Griqualand West, South Africa: Evidence for a dramatic rise in atmospheric oxygen between 2.2 and 1.9 Ga. Am. J. Sci., 290A: 1–34.

Holland, H. D., and Zimmermann, H., 2000. The dolomite problem revisited. Intern. Geol. Rev., 42: 481–490.

Holland, H. D., Lazar, B., and McGaffrey, M., 1986. Evolution of the atmosphere and oceans. Nature, 320: 27–33.

Hollerbach, R., and Jones, C. A., 1993. Influence of the Earth's inner core on geomagnetic fluctuations and reversals. Nature, 365: 541–543.

Holliger, K. Levander, A. R., and Goff, J. A., 1993. Stochastic modeling of the reflective lower crust: Petrophysical and geological evidence from the Ivrea zone. J. Geophys. Res., 98: 11,967–11,980.

Hollings, P., and Kerrich, R., 1999. Trace element systematics of ultramafic and mafic volcanic rocks from the 3-Ga North Caribou greenstone belt, northwestern Superior province. Precamb. Res., 93: 257–279.

Holmden, C., and Muehlenbachs, K., 1993. The $^{18}O/^{16}O$ ratio of 2-Ga seawater inferred from ancient oceanic crust. Science, 259: 1733–1736.

Holser, W. T., 1984. Gradual and abrupt shifts in ocean chemistry during Phanerozoic time. In: H. D. Holland and A. F. Trendall (eds.), Patterns of Change in Earth Evolution. Springer-Verlag, Berlin, pp. 123–143.

Holser, W. T., Schidlowski, M., Mackenzie, F. T., and Maynard, J. B., 1988. Geochemical cycles of carbon and sulfur. In: C. B. Gregor, R. M. Garrels, F. T. Mackenzie, and J. B. Maynard (eds.), Chemical Cycles in the Evolution of the Earth. John Wiley, New York, pp. 105–173.

Homewood, P., Allen, P. A., and Williams, G. D., 1986. Dynamics of the Molasse basin of western Switzerland. In: A. Allen and P. Homewood (eds.), Foreland Basins. Blackwell Scientific Publications, Oxford, pp. 199–218.

Horita, J., Zimmermann, H., and Holland, H. D., 2002. Chemical evolution of seawater during the Phanerozoic: Implications from the record of marine evaporites. Geochim. Cosmochim. Acta, 66: 3733–3756.

Humphris, S. E., et al., 1995. The internal structure of an active seafloor massive sulphide deposit. Nature, 377: 713–716.

Hunten, D. M., 1993. Atmospheric evolution of the terrestrial planets. Science, 259: 915–920.

Hunter, D. R., 1975. The regional geological setting of the Bushveld Complex. University Witwatersrand, Economic Geology Research Unit, Johannesburg, 18 pp.

Hurley, P. M., and Rand, J. R., 1969. Predrift continental nuclei. Science, 164: 1229–1242.

Hut, P., et al., 1987. Comet showers as a cause of mass extinctions. Nature, 329: 118–125.

Hyde, W. T., Crowley, T. J., Baum, S. K., and Peltier, R., 2000. Neoproterozoic "snowball Earth" simulations with a coupled climate–ice sheet model. Nature, 405: 425–429.

Imbrie, J., 1985. A theoretical framework for the Pleistocene ice ages. Quart. J. Geol. Lond., 142: 417–432.

Ingersoll, R. V., 1988. Tectonics of sedimentary basins. Geol. Soc. Am. Bull., 100: 1704–1719.

Ingram, B. L., Coccioni, R., Montana, A., and Richter, F. M., 1994. Sr isotopic composition of mid-Cretaceous seawater. Science, 264: 546–549.

Irifune, T., and Ringwood, A. E., 1993. Phase transformations in subducted oceanic crust and buoyancy relationships at depths of 600–800 km in the mantle. Earth Planet. Sci. Lett., 117: 101–110.

Irving, E., North, F. K., and Couillard, R., 1974. Oil, climate, and tectonics. Can. J. Earth Sci., 11: 1–17.

Irwin, L. N., and Schulze-Makuch, D., 2001. Assessing the plausibility of life on other worlds. Astrobiology, 1: 143–160.

Ishii, M., and Dziewonski, A. M., 2003. Distinct seismic anisotropy at the center of the Earth. Phys. Earth Planet. Interiors, 140: 203–217.

Isley, A. E., 1995. Hydrothermal plumes and the delivery of iron to BIF. J. Geol., 103: 169–185.

Isley, A. E., and Abbott, D. H., 1999. Plume-related mafic volcanism and the deposition of banded iron formation. J. Geophys. Res., 104: 15,461–15,477.

Isley, A. E., and Abbott, D. H., 2002. Implication for the temporal distribution of high-Mg magmas for mantle plume volcanism through time. J. Geol., 110: 141–158.

Isozaki, Y., 1997. Permo–Triassic boundary superanoxia and stratified superocean: Records from lost deep sea. Science, 276: 235–238.

Ita, J., and Stixrude, L., 1992. Petrology, elasticity, and composition of the mantle transition zone. J. Geophys. Res., 97: 6849–6866.

Ito, E., Morooka, K., Ujike, O., and Katsura, T., 1995. Reactions between molten iron and silicate melts at high pressure: Implications for the chemical evolution of Earth's core. J. Geophys. Res., 100: 5901–5910.

Jacobs, J. A., 1992. Deep Interior of the Earth. Chapman & Hill, New York, 167 pp.

Jacobs, J. A., 1995. The Earth's inner core and the geodynamo: Determining their roles in the Earth's history. EOS, 76 (25): 249–253.

Jacobsen, S. B., and Kaufman, A. J., 1999. The Sr, C, and O isotopic evolution of Neoproterozoic seawater. Chem. Geol., 161: 37–57.

Jacobsen, S. B., and Pimentel-Klose, M., 1988. Nd isotopic variations in Precambrian BIF. Geophys. Res. Lett., 15: 393–396.

Jahren, A. H., 2002. The bio–geochemical consequences of the mid-Cretaceous superplume. J. Geodyn., 34: 177–191.

Jakosky, B. M., and Phillips, R. J., 2001. Mars' volatile and climate history. Nature, 412: 237–244.

James, D. E., Niu, F., and Rokosky, J., 2003. Crustal structure of the Kaapvaal craton and its significance for early crustal evolution. Lithos, 71: 423–429.

Jarchow, C. M., and Thompson, G. A., 1989. The nature of the Mohorovicic discontinuity. Ann. Rev. Earth Planet. Sci., 17: 475–506.

Jaupart, C., and Mareschal, J. C., 1999. The thermal structure and thickness of continental roots. Lithos, 48: 93–114.

Jaupart, C., Mareschal, J. C., Guillou-Frottier, L., and Davaille, A., 1998. Heat flow and thickness of the lithosphere in the Canadian shield. J. Geophys. Res., 103: 15,269–15,286.

Jeanloz, R., 1990. The nature of the Earth's core. Ann. Rev. Earth Planet. Sci., 18: 357–386.

Jeanloz, R., Mitchell, D. L., Sprague, A. L., and de Pater, I., 1995. Evidence for a basalt-free surface on Mercury and implications for internal heat. Science, 268: 1455–1457.

Jenkins, G. S., 1995. Early Earth's climate: Cloud feedback from reduced land fraction and ozone concentrations. Geophys. Res. Lett., 22: 1513–1516.

Jenkins, G. S., Marshall, H. G., and Kuhn, W. R., 1993. Precambrian climate: The effects of land area and Earth's rotation rate. J. Geophys. Res., 98: 8785–8791.

Jenkyns, H. C., 1980. Cretaceous anoxic events: From continents to oceans. J. Geol. Soc. Lond., 137: 171–188.

Jephcoat, A., and Olson, P., 1987. Is the inner core of the Earth pure iron? Nature, 325: 332.

Jessberger, E. K., and Kissel, J., 1989. The composition of comets. In: S. K. Atreya, J. B. Pollack, and M. S. Matthews (eds.), Origin and Evolution of Planetary and Satellite Atmospheres. University Arizona Press, Tucson, pp. 167–191.

Jessop, A. M., and Lewis, T., 1978. Heat flow and heat generation in the Superior province of the Canadian shield. Tectonophysics, 50: 55–77.

Jiang, G., Kennedy, M. J., and Christie-Blick, N., 2003. Stable isotopic evidence for methane seeps in Neoproterozoic postglacial cap carbonates. Nature, 426: 822–826.

Johnson, D. M., 1993. Mesozoic and Cenozoic contributions to crustal growth in the southwestern United States. Earth Planet. Sci. Lett., 118: 75–89.

Johnson, H. P., Van Patten, D., Tivey, M., and Sager, W. W., 1995. Geomagnetic polarity reversal rate for the Phanerozoic. Geophys. Res. Lett., 22: 231–234.

Jolivet, L., Huchon, P., and Ranguin, C., 1989. Tectonic setting of western Pacific marginal basins. Tectonophysics, 160: 23–47.

Jurdy, D. M., Stefanick, M., and Scotese, C. R., 1995. Paleozoic plate dynamics. J. Geophys. Res., 100: 17,965–17,975.

Kah, L. C., Lyons, T. W., and Chesley, J. T., 2001. Geochemistry of a 1.2-Ga carbonate–evaporite succession, northern Baffin and Bylot Islands: Implications for Mesoproterozoic marine evolution. Precamb. Res., 111: 203–234.

Kamber, B. S., and Collerson, K. D., 1999. Origin of ocean island basalts: A new model based on lead and helium isotope systematics. J. Geophys. Res., 104: 15,479–15,491.

Kamo, S. L., and Davis, D. W., 1994. Reassessment of Archean crust development in the Barberton Mountain Land, South Africa, based on U/Pb dating. Tectonics, 13: 167–192.

Kamo, S. L., and Krogh, T. E., 1995. Chicxulub crater source for shocked zircon crystals from the Cretaceous–Tertiary boundary layer, Saskatchewan: Evidence from new U/Pb data. Geology, 23: 281–284.

Kamo, S. L., Czamanske, G. K., Amelin, Y., Fedorenko, V. A., Davis, D. W., and Trofimov, V. R., 2003. Rapid eruption of Siberian flood volcanic rocks and evidence for coincidence with Permian–Triassic boundary and mass extinction at 251 Ma. Earth Planet. Sci. Lett., 214: 75–91.

Kandler, O., 1994. The early diversification of life. In: S. Bengtson (ed.), Early Life on Earth. Columbia University Press, New York, pp. 152–160.

Kaneshima, S., and Helffrich, G., 1999. Dipping low-velocity layer in the mid–lower mantle: Evidence for geochemical heterogeneity. Science, 283: 1888–1891.

Kárason, H., and van der Hilst, R. D., 2000. Constraints on mantle convection from seismic tomography. In: M. R. Richards, R. Gordon, and R. D. van der Hilst (eds.), The History and Dynamics of Global Plate Motion. American Geophysical Union, Washington, DC, Geophysical Monograph 121, pp. 277–288.

Karato, S. I., 1999. Seismic anisotropy of the Earth's inner core resulting from flow induced by Maxwell stresses. Nature, 402: 871–873.

Karhu, J. A., and Holland, H. D., 1996. Carbon isotopes and the rise of atmospheric oxygen. Geology, 24: 867–870.

Karlstrom, K. E., and Houston, R. S., 1984. The Cheyenne belt: Analysis of a Proterozoic suture in southern Wyoming. Precamb. Res. 25: 415–446.

Karlstrom, K. E., Ahall, K. I., Harlan, S. S., Williams, M. L., McLelland, J., and Geissman, J. W., 2001. Long-lived convergent orogen in southern Laurentia, its extensions to Australia and Baltica, and implications for refining Rodinia. Precamb. Res., 111: 5–30.

Karsten, J. L., Klein, E. M., and Sherman, S. B., 1996. Subduction-zone geochemical characteristics in ocean-ridge basalts from the southern Chile ridge: Implications of modern ridge subduction systems for the Archean. Lithos, 37: 143–161.

Kasting, J. F., 1987. Theoretical constraints on oxygen and carbon dioxide concentrations in the Precambrian atmosphere. Precamb. Res., 34: 205–229.

Kasting, J. F., 1991. Box models for the evolution of atmospheric oxygen: An update. Paleogeog. Paleoclimat. Paleoecol., 97: 125–131.

Kasting, J. F., 1993. Earth's early atmosphere. Science, 259: 920–926.

Kasting, J. F., Eggler, D. H., and Raeburn, S. P., 1993a. Mantle redox evolution and the oxidation state of the Archean atmosphere. J. Geol., 101: 245–257.

Kasting, J. F., Whitmire, D. P., and Reynolds, R. T., 1993b. Habitable zones around main sequence stars. Icarus, 101: 1460–1464.

Kato, T., and Ringwood, A. E., 1989. Melting relationships in the system Fe-FeO at high pressures: Implications for the composition and formation of the Earth's core. Phys. Chem. Miner., 16: 524–538.

Kaufman, A. J., 1995. At Home in the Universe: The Search for Laws of Self-Organization and Complexity. Oxford University Press, New York.

Kaufman, A. J., 1997. An ice age in the tropics. Nature, 386: 227–228.

Kaufman, A. J., and Knoll, A. H., 1995. Neoproterozoic variations in the C-isotopic composition of seawater: Stratigraphic and biogeochemical implications. Precamb. Res., 73: 27–49.

Kaula, W. M., 1995. Venus reconsidered. Science, 270: 1460–1464.

Kay, R. W., and Kay, S. M., 1981. The nature of the lower continental crust: Inferences from geophysics, surface geology, and crustal xenoliths. Rev. Geophys. Space Phys., 19: 271–297.

Kay, S. M., and Kay, R. W., 1985. Role of crystal cumulates and the oceanic crust in the formation of the lower crust of the Aleutian arc. Geology, 13: 461–464.

Kay, S. M., Ramos, V. A., Mpodozis, C., and Sruoga, P., 1989. Late Paleozoic to Jurassic silicic magmatism at the Gondwana margin: Analogy to the middle Proterozoic in North America? Geology, 17: 324–328.

Keller, G., Meudt, M., Adatte, T., Berner, Z., Stueben, D., and Tantawy, A. A., 2003. Biotic effects of impacts and volcanism. In: Mantle Plumes: Physical Processes, Chemical Signatures, and Biological Effects (abstract volume). Cardiff University, Cardiff, UK, p. 25.

Kellogg, L. H., 1992. Mixing in the mantle. Ann. Rev. Earth Planet. Sci., 20: 365–388.

Kellogg, L. H., Hager, B. H., and van der Hilst, R. D., 1999. Compositional stratification in the deep mantle. Science, 283: 1881–1884.

Kempton, P. D., et al., 2000. The Iceland plume in space and time: An Sr-Nd-Pb-Hf study of the North Atlantic rifted margin. Earth Planet. Sci. Lett., 177: 255–271.

Kendall, J. M., and Silver, P. G., 1996. Constraints from seismic anisotropy on the nature of the lowermost mantle. Nature, 381: 409–412.

Keppler, H., 1996. Constraints from partitioning experiments of the composition of subduction-zone fluids. Nature, 380: 237–240.

Kerr, A. C., 1994. Lithospheric thinning during the evolution of continental, large, igneous provinces. Geology, 22: 1027–1030.

Kerr, A. C., 1998. Oceanic plateau formation: A cause of mass extinction and black shale deposition around the Cenomanian–Turonian boundary. J. Geol. Soc. Lond., 155: 619–626.

Kerrich, R., Polat, A., Wyman, D., and Hollings, P., 1999. Trace element systematics of Mg- to Fe-tholeiitic basalt suites of the Superior province: Implications for Archean mantle reservoirs and greenstone belt genesis. Lithos, 46: 163–187.

Kershaw, S., 1990. Evolution of Earth's atmosphere and its geological impact. Geol. Today, 6 (2): 55–57.

Kilburn, M. R., and Wood, B. J., 1997. Metal–silicate partitioning and the incompatibility of S and Si during core formation. Earth Planet. Sci. Lett., 152: 139–148.

Kimura, H., and Watanabe, Y., 2001. Oceanic anoxia at the Precambrian–Cambrian boundary. Geology, 29: 995–998.

Kingma, K. J., Cohen, R. E., Hemley, R. J., and Mao. H. K., 1995. Transformation of stishovite to a denser phase at lower mantle pressures. Nature, 374: 243–245.

Kirschvinik, J. L., Ripperdan, R. L., and Evans, D. A., 1997. Evidence for a large-scale reorganization of early Cambrian continental masses by inertial interchange true polar wander. Science, 277: 541–545.

Klein, C., and Beukes, N. J., 1992. Proterozoic iron formations. In: K. C. Condie (ed.), Proterozoic Crustal Evolution. Elsevier, Amsterdam, pp. 383–417.

Klein, G., 1982. Probable sequential arrangement of depositional systems on cratons. Geology, 10: 17–22.

Klein, G., 1986. Sedimentation patterns in relation to rifting, arc volcanism, and tectonic uplift in back-arc basins of the western Pacific Ocean. In: N. Nasu, K. Kobayashi, S. Uyeda, I. Kushiro, and H. Kagami (eds.), Formation of Active Continental Margins. D. Reidel, Dordrecht, pp. 517–550.

Kleine, T., Munker, C., Mezger, K., and Palme, H., 2002. Rapid accretion and early core formation on asteroids and the terrestrial planets from Hf-W chronometry. Nature, 418: 952–955.

Klemperer, S. L., 1987. A relation between continental heat flow and the seismic reflectivity of the lower crust. J. Geophys., 61: 1–11.

Knauth, L. P., and Lowe, D. R., 1978. Oxygen isotope geochemistry of cherts from the Onverwacht Group, South Africa. Earth Planet. Sci. Lett., 41: 209–222.

Knauth, L. P., and Lowe, D. R., 2003. High Archean climatic temperature inferred from oxygen isotope geochemistry of cherts in the 3.5-Ga Swaziland Supergroup, South Africa. Geol. Soc. Am. Bull., 115: 566–580.

Knittle, E., and Jeanloz, R., 1991. Earth's core–mantle boundary: Results of experiments at high pressures and high temperatures. Science, 251: 1438–1443.

Knoll, A. H., 2003. The geological consequences of evolution. Geobiology, 1: 3–14.

Knoll, A. H., and Carroll, S. B., 1999. Early animal evolution: Emerging views from comparative biology and geology. Science, 284: 2129–2136.

Knoll, A. H., Bambach, R. K., Canfield, D. E., and Grotzinger, J. P., 1996. Comparative Earth history and late Permian mass extinction. Science, 273: 452–457.

Kokelaar, P., 1986. Magma–water interactions in subaqueous and emergent basaltic volcanism. Bull. Volcanol., 48: 275–289.

Kominz, M. A., and Bond, G. C., 1991. Unusually large subsidence and sea level events during middle Paleozoic time: New evidence supporting mantle convection models for supercontinent assembly. Geology, 19: 56–60.

Kontak, D. J., and Reynolds, P. H., 1994. ^{40}Ar/^{39}Ar dating of metamorphic and igneous rocks of the Liscomb Complex, Meguma terrane, southern Nova Scotia, Canada. Can. J. Earth Sci., 31: 1643–1653.

Krapez, B., 1993. Sequence stratigraphy of the Archean supracrustal belts of the Pilbara block, Western Australia. Precamb. Res., 60: 1–45.

Kreep, M., and Hansen, V. L., 1994. Structural history of Maxwell Montes, Venus: Implications for Venusian mountain belt formation. J. Geophys. Res., 99: 26,015–26,028.

Kring, D. A., and Cohen, B. A., 2002. Cataclysmic bombardment throughout the inner solar system 3.9–4.0 Ga. J. Geophys. Res., 107 (E2), doi: 10.1029/2001JE001529 (online).

Kroner, A., and Cordani, U., 2003. African, southern Indian, and South American cratons were not part of the Rodinia supercontinent: Evidence from field relationships and geochronology. Tectonophysics, 375: 325–352.

Kroner, A., and Layer, P. W., 1992. Crust formation and plate motion in the early Archean. Science, 256: 1405–1411.

Kroner, A., Compston, W., and Williams, I. S., 1989. Growth of early Archean crust in the ancient gneiss complex of Swaziland as revealed by single zircon dating. Tectonophysics, 161: 271–298.

Kroner, A., Hegner, E., Wendt, J. I., and Byerly, G. R., 1996. The oldest part of the Barberton granitoid-greenstone terrain, South Africa: Evidence for crust formation between 3.5 and 3.7 Ga. Precamb. Res., 78: 105–124.

Kuang, W., and Bloxham, J., 1997. An Earth-like numerical dynamo model. Nature, 389: 371–374.

Kumazawa, M., Helstaedt, H., and Masaki, K., 1971. Elastic properties of eclogite xenoliths form diatremes of the east Colorado Plateau and their implication to the upper mantle structure. J. Geophys. Res., 76: 1231–1247.

Kump, L. R., Kasting, J. F., and Crane, R. G., 1999a. The Earth System. Prentice Hall, Upper Saddle River, NJ, 351 pp.

Kump, L. R., Arthur, M. A., Patzkowsky, M. E., Gibbs, M. T., Pinkus, D. S., and Sheehan, P. M., 1999b. A weathering hypothesis for glaciation at high atmospheric pCO_2 during the late Ordovician. Paleogeog. Paleoclimat. Paleoecol., 152: 173–187.

Kushiro, I., 2001. Partial melting experiments on peridotite and origin of midocean-ridge basalt. Ann. Rev. Earth Planet. Sci., 2001, 29: 71–107.

Kusky, T. M., and Kidd, W. S. F., 1992. Remnants of an Archean oceanic plateau, Belingwe greenstone belt, Zimbabwe. Geology, 20: 43–46.

Kusky, T. M., Li, J. H., and Tucker, R. D., 2001. The Archean Dongwanzi ophiolite complex, north China craton: 2.505-billion-year-old oceanic crust and mantle. Science, 292: 1142–1145.

Kutzner, C., and Christensen, U., 2000. Effects of driving mechanisms in geodynamo models. Geophys. Res. Lett., 27: 29–32.

Kvenvolden, K. A., 1974. Natural evidence for chemical and early biologic evolution. Origins Life, 5: 71–86.

Kvenvolden, K. A., 1999. Potential effects of gas hydrate on human welfare. Proc. Nat. Acad. Sci. USA, 96: 3420–3426.

Kyte, F. T., 1998. A meteorite from the Cretaceous–Tertiary boundary. Nature, 396: 237–239.

Kyte, F. T., Shukolyukov, A., Lugmair, G. W., Lowe, D. R., and Byerly, G. R., 2003. Early Archean spherule beds: Chromium isotopes confirm origin through multiple impacts of projectiles of carbonaceous chondrite type. Geology, 31: 283–286.

Labrosse, S., 2003. Thermal and magnetic evolution of the Earth's core. Phys. Earth Planet. Interiors, 140: 127–143.

Labrosse, S., Poirier, J. P., and Le Mouel, J. L., 2001. The age of the inner core. Earth Planet. Sci. Lett., 190: 111–123.

Lambeck, K., 1980. The Earth's Rotation: Geophysical Cause and Consequences. Cambridge University Press, Cambridge, UK, 449 pp.

Larsen, H. C., and Saunders, A. D., 1998. Tectonism and volcanism at the southeastern Greenland rifted margin: A record of plume impact and later continental rupture. Proc. Ocean Drill. Prog., Sci. Res., 152: 503–533.

Larsen, T. B., and Yuen, D. A., 1997. Fast plumeheads: Temperature-dependent versus non-Newtonian rheology. Geophys. Res. Lett., 24: 1995–1998.

Larson, R. L., 1991a. Latest pulse of Earth: Evidence for a mid-Cretaceous superplume. Geology, 19: 547–550.

Larson, R. L., 1991b. Geological consequences of super-plumes. Geology, 19: 963–966.

Larson, R. L., and Kincaid, C., 1996. Onset of mid-Cretaceous volcanism by elevation of the 670-km thermal boundary layer. Geology, 24: 551–554.

Larson, R. L., and Olson, P., 1991. Mantle plumes control magnetic reversal frequency. Earth Planet. Sci. Lett., 107: 437–447.

Laske, G., and Masters, G., 1999. Limits on differential rotation of the inner core from an analysis of the Earth's free oscillations. Nature, 402: 66–69.

Lawton, T. F., 1986. Compositional trends within a clastic wedge adjacent to a fold–thrust belt: Indianola Group, central Utah, Utah, USA. In: A. Allen and P. Homewood (eds.), Foreland Basins. Blackwell Scientific Publications, Oxford, pp. 411–424.

Lebofsky, L. A., Jones, T. D., and Herbert, F., 1989. Asteroid volatile inventories. In: S. K. Atreya, J. B. Pollack, and M. S. Matthews (eds.), Origin and Evolution of Planetary and Satellite Atmospheres. University Arizona Press, Tucson, pp. 192–229.

Lecuyer, C., Gruau, G., Fruh-Green, G. L., and Picard, C., 1996. Hydrogen isotope composition of early Proterozoic seawater. Geology, 24: 291–294.

Lecuyer, C., Simon, L., and Guyot, F., 2000. Comparison of carbon, nitrogen, and water budgets on Venus and the Earth. Earth Planet. Sci. Lett., 181: 33–40.

Lee, D. C., and Halliday, A. N., 1995. Hf-W chonometry and the timing of terrestrial core formation. Nature, 378: 771–774.

Leitch, E. C., 1984. Marginal basins of the southwestern Pacific and the preservation and recognition of their ancient analogues: A review. Geol. Soc. Lond., Spec. Publ., 16: 97–108.

Lenardic, A., 1997. On the heat flow variation from Archean cratons to Proterozoic mobile belts. J. Geophys. Res., 102: 709–721.

Lenardic, A., and Kaula, W. M., 1996. Near-surface thermal–chemical boundary layer convection at infinite Prandtl number: Two-dimensional numerical experiments. Geophys. J. Int., 126: 689–711.

Li, J., and Agee, C. B., 2001. Element partitioning constraints on the light element composition of the Earth's core. Geophys. Res. Lett., 28: 81–84.

Li, X., and Yuan, X., 2003. Receiver functions in northeastern China: Implications for slab penetration into the lower mantle in northwestern Pacific subduction zone. Earth Planet. Sci., Lett., 216: 679–691.

Li, X., Kind, R., Priestley, K., Sobolev, S. V., Tilmann, F., Yuan, X., and Weber, M., 2000. Mapping the Hawaiian plume conduit with converted seismic waves. Nature, 405: 938–941.

Li, Z. X., and Powell, C. M., 2001. An outline of the paleo-geographic evolution of the Australasian region since the beginning of the Neoproterozoic. Earth Sci. Rev., 53: 237–277.

Litak, R. K., and Brown, L. D., 1989. A modern perspective on the Conrad discontinuity. Am. Geophys. Union, EOS, 70: 722–725.

Lithgow-Bertelloni, C., and Richards, M. A., 1995. Cenozoic plate-driving forces. Geophys. Res. Lett., 22: 1317–1320.

Lithgow-Bertelloni, C., and Silver, P. G., 1998. Dynamic topography, plate-driving forces, and the African super-swell. Nature, 395: 269–272.

Liu, M., 1994. Asymmetric phase effects and mantle convection patterns. Science, 264: 1904–1907.

Longhi, J., 1978. Pyroxene stability and composition of the lunar magma ocean. Proceedings of the Ninth Lunar and Planetary Science Conference, pp. 285–306.

Loper, D. E., 1992. On the correlation between mantle plume flux and the frequency of reversals of the geomagnetic field. Geophys. Res. Lett., 19: 25–28.

Loper, D. E., and Lay, T., 1995. The core–mantle boundary region. J. Geophys. Res., 100: 6397–6420.

Lowe, D. R., 1994a. Accretionary history of the Archean Barberton greenstone belt, southern Africa. Geology, 22: 1099–1102.

Lowe, D. R., 1994b. Archean greenstone-related sedimentary rocks. In: K. C. Condie (ed.), Archean Crustal Evolution, Elsevier, Amsterdam, pp. 121–170.

Lowman, J. P., and Gable, C. W., 1999. Thermal evolution of the mantle following continental aggregation in 3D convection models. Geophys. Res. Lett., 26: 2649–2652.

Lowman, J. P., and Jarvis, G. T., 1996. Continental collisions in wide aspect ratio and high Rayleigh number two-dimensional mantle convection models. J. Geophys. Res., 101: 25,485–25,497.

Lowman, J. P., and Jarvis, G. T., 1999. Effects on mantle heat source distribution on supercontinent stability. J. Geophys. Res., 104: 12,733–12,746.

Lucas, S. B., et al., 1993. Deep seismic profile across a Proterozoic collision zone: Surprises at depth. Nature, 363: 339–342.

Lund, S. P., 1996. A comparison of Holocene paleomagnetic secular variation records from North America. J. Geophys. Res., 101: 8007–8024.

MacDonald, K. C., 1982. Midocean ridges: Fine scale tectonic, volcanic, and hydrothermal processes within the plate boundary zone. Ann. Rev. Earth Planet. Sci., 10: 155–190.

Mahoney, J. J., le Roex, A. P., Peng, Z., Fisher, R. L., and Natland, J. H., 1992. Southwestern limits of Indian Ocean ridge mantle and the origin of low-^{206}Pb/^{204}Pb midocean-ridge basalt: Isotope systematics of the central–southwest Indian ridge (17-50° E). J. Geophys. Res., 97: 19,771–19,790.

Mahoney, J. J., Jones, W. B., Frey, F. A., Salters, V. J., Pyle, D. G., and Davies, H. L., 1995. Geochemical characteristics of lavas from Broken ridge, the Naturaliste Plateau, and southernmost Kerguelen Plateau: Cretaceous plateau volcanism in the southeastern Indian Ocean. Chem. Geol., 120: 315–345.

Mahoney, J. J., Frei, R., Tejada, M. L. G., Mo, X. X., Leat, P. T., and Nagler, T. F., 1998. Tracing the Indian Ocean mantle domain through time: Isotopic results from old West Indian, East Tethyan, and South Pacific seafloors. J. Petrol., 39: 1285–1306.

Mambole, A., and Fleitout, L., 2002. Petrological layering induced by an endothermic phase transition in the Earth's mantle. Geophys. Res. Lett., 29 (22), doi: 10.1029/2002GL014674 (online).

Marakami, T., Utsunomiya, S., Imazu, Y., and Prasad, N., 2001. Direct evidence of late Archean to early Proterozoic anoxic atmosphere from a product of 2.5-Ga ole weathering. Earth Planet. Sci. Lett., 184: 523–528.

Mareschal, J. C., Jaupart, C., Cheng, L. Z., Rolandone, F., Gariepy, C., Bienfait, G., Guillou-Frottier, L., and Lapointe, R., 1999. Heat flow in the trans-Hudson orogen of the Canadian shield: Implications for Proterozoic continental growth. J. Geophys. Res., 104: 29,007–29,024.

Marko, G. M., 1980. Velocity and attenuation in partially molten rocks. J. Geophys. Res., 85: 5173–5189.

Marti, K., and Graf, T., 1992. Cosmic-ray exposure history of ordinary chondrites. Ann. Rev. Earth Planet. Sci., 20: 221–243.

Marti, K., Kim, J. S., Thakur, A. N., McCoy, T. J., and Keil, K., 1995. Signatures of the Martian atmosphere in glass of the Zagami meteorite. Science, 267: 1981–1984.

Martignole, J., 1992. Exhumation of high-grade terranes: A review. Can. J. Earth Sci., 29: 737–745.

Martin, H., 1993. The mechanisms of petrogenesis of the Archean continental crust: Comparison with modern processes. Lithos, 30: 373–388.

Martin, H., 1994. Archean gray gneisses and the genesis of continental crust. In: K. C. Condie (ed.), Archean Crustal Evolution, Elsevier, Amsterdam, pp. 205–260.

Marty, B., and Dauphas, N., 2002. Formation and early evolution of the atmosphere. Geol. Soc. Lond., Spec. Publ., 199: 213–229.

Maruyama, S., 1994. Plume tectonics. J. Geol. Soc. Japan, 100: 24–49.

Maruyama, S., 1997. Pacific-type orogen revisited: Mayashiro-type orogen proposed. Island Arc, 6: 91–120.

Massey, N. W. D., 1986. Metchosin igneous complex, southern Vancouver Island: Ophiolite stratigraphy developed in an emergent island setting. Geology, 14: 602–605.

Matsui, T., Imamura, F., Tajika, E., Nakano, Y., and Fujisawa, Y., 1999. K–T Impact Tsunami, Abstract Vol. 30, Abstract No. 1527. Lunar and Planetary Institute, Houston, CD-ROM.

Mattie, P. D., Condie, K. C., Selverstone, J., and Kyle, P. R., 1997. Composition of the lower continental crust in the Colorado Plateau: Geochemical evidence from mafic xenoliths from the Navajo volcanic field, southwestern United States. Geochim. Cosmochim. Acta, 61: 2007–2021.

Maurrasse, F. J. M., and Sen, G., 1991. Impacts, tsunamis, and the Haitian K–T boundary layer. Science, 252: 1690–1693.

McBirney, A. R., 1976. Some geologic constraints on models for magma generation in orogenic environments. Can. Mineral., 14: 245–254.

McCabe, R., 1984. Implications of paleomagnetic data on the collision-related bending of island arcs. Tectonics, 3: 409–428.

McCollom, T. M., Ritter, G., and Simoneit, B. R. T., 1999. Lipid synthesis under hydrothermal conditions by Fischer-Tropsch-type reactions. Origin Life Evol. Biosphere, 29: 153–166.

McCulloch, M. T., and Bennett, V. C., 1994. Progressive growth of the Earth's continental crust and depleted mantle: Geochemical constraints. Geochim. Cosmochim. Acta, 58: 4717–4738.

McCulloch, M. T., and Gamble, J. A., 1991. Geochemical and geodynamical constraints on subduction-zone magmatism. Earth Planet. Sci. Lett., 102: 358–374.

McDonough, W. F., 1990. Constraints on the composition of the continental lithospheric mantle. Earth Planet. Sci. Lett., 101: 1–18.

McDonough, W. F., and Sun, S. S., 1995. The composition of the Earth. Chem. Geol., 120: 223–253.

McDougall, I., and Harrison, T. M., 1988. Geochronology and Thermochronology by the $^{40}Ar/^{39}Ar$ method. Oxford University Press, New York, 212 p.

McFarlane, A. W., Danielson, A., and Holland, H. D., 1994. Geology and major and trace element chemistry of late Archean weathering profiles in the Fortescue Group, Western Australia: Implications for atmospheric PO_2. Precamb. Res., 65: 297–317.

McGeary, S., Nur, A., and Ben-Avraham, Z., 1985. Spatial gaps in arc volcanism: The effect of collision or subduction of oceanic plateaus. Tectonophysics, 119: 195–221.

McGovern, P. J., and Schubert, G., 1989. Thermal evolution of the Earth: Effects of volatile exchange between atmosphere and interior. Earth Planet. Sci. Lett., 96: 27–37.

McHone, J. F., Niema, R. A., Lewis, C. F., and Yates, A. M., 1989. Stishovite at the K–T boundary, Raton, New Mexico. Science, 243: 1182–1184.

McKay, D. S., et al., 1996. Search for past life on Mars: Possible relic biogenic activity in Martian meteorite ALH84001. Science, 273: 924–930.

McKenzie, D. P., and Bickle, M. J., 1988. The volume and composition of melt generated by extension of the lithosphere. J. Petrol., 29: 625–679.

McKenzie, D. P., and Morgan, W. J., 1969. Evolution of triple junctions. Nature, 224: 125–133.

McLaren, D. J., and Goodfellow, W. D., 1990. Geological and biological consequences of giant impacts. Ann. Rev. Earth Planet. Sci., 18: 123–171.

McLennan, S. M., 1988. Recycling of the continental crust. Pure Appl. Geophys., 128: 683–898.

McLennan, S. M., and Hemming, S., 1992. Sm/Nd elemental and isotopic systematics in sedimentary rocks. Geochim. Cosmochim. Acta, 56: 887–898.

McNamara, A. K., and van Keken, P. E., 2000. Cooling of the Earth: A parameterized convection study of whole versus layered modes. Geochem. Geophys. Geosys., 1, 2000GC000045.

Meert, J. G., and Torsvik, T. H., 2003. The making and unmaking of a supercontinent: Rodinia revisited. Tectonophysics, 375: 261–288.

Meert, J. G., and Van der Voo, R., 1994. The Neoproterozoic glacial intervals: No more snowball Earth? Earth Planet. Sci. Lett., 123: 1–13.

Meissner, R., and Mooney, W., 1998. Weakness of the lower continental crust: A condition for delamination, uplift, and escape. Tectonophysics, 296: 47–60.

Melezhik, V. A., Fallick, A. E., Medvedev, P. V., and Makarikhin, V. V., 1999. Extreme $^{13}C_{carb}$ enrichment in ca. 2.0-Ga magnesite-stromatolite-dolomite-"redbeds" association in a global context: A case for the worldwide signal enhanced by a local environment. Earth Sci. Rev., 48: 71–120.

Melville, R., 1973. Continental drift and plant distribution. In: D. H. Tarling and S. K. Runcorn (eds.), Implications of Continental Drift to the Earth Sciences, Vol. 1. Academic Press, London, pp. 439–446.

Menard, H. W., 1967. Transitional types of crust under small ocean basins. J. Geophys. Res., 72: 3061–3073.

Menzies, M. A., 1990. Archean, Proterozoic, and Phanerozoic lithospheres. In: M. A. Menzies (ed.), Continental Mantle. Clarendon Press, Oxford, pp. 67–85.

Mezger, K., Rawnsley, C. M., Bohlen, S. R., and Hanson, G. N., 1991. U/Pb garnet, sphene, monazite, and rutile ages: Implications for the duration of high-grade metamorphism and cooling histories, Adirondack Mountains, New York. J. Geol., 99: 415–428.

Miller, S. L., 1953. A production of amino acids under possible primitive Earth conditions. Science, 117: 528–529.

Miller, S. L., and Bada, J. L., 1988. Submarine hot springs and the origin of life. Nature, 334: 609–611.

Mohr, P., 1982. Musings on continental rifts. Am. Geophys. Union–Geol. Soc. Am., Geodynam. Series, 8: 293–309.

Mohr, R. E., 1975. Measured periodicities of the Biwabik stromatolites and their geophysical significance. In: G. D. Rosenberg and S. K. Runcorn (eds.), Growth Rhythms and the History of the Earth's Rotation. J. Wiley, New York, pp. 43–55.

Mojzsis, S. J., and Harrison, T. M., 2000. Vestiges of a beginning: Clues to the emergent bidsphee recorded in the oldest known sedimentary rocks. GSA Today, 10 (4): 1–6.

Mojzsis, S. J., Arrhenius, G., McKeegan, K. D., Harrison, T. M., Nutman, A. P., and Friend, C. R. L., 1996. Evidence for life on Earth before 3800 million years ago. Nature, 384: 55–59.

Mojzsis, S. J., Harrison, T. M., and Pidgeon, R. T., 2001. Oxygen isotope evidence from ancient zircons for liquid water at the Earth's surface 4300 Myr ago. Nature, 409: 178–180.

Molnar, P., and Stock, J., 1987. Relative motions of hotspots in the Pacific, Atlantic, and Indian Oceans since late Cretaceous time. Nature, 327: 587–591.

Mooney, W. D., and Meissner, R., 1992. Multigenetic origin of crustal reflectivity: A review of seismic reflection profiling of the continental lower crust and Moho. In: D. M. Fountain, R. Arculus, and R. W. Kay (eds.), Continental Lower Crust. Elsevier, Amsterdam, pp. 45–79.

Mooney, W. D., Laske, G., and Masters, T. G., 1998. CRUST 5.1: A global crustal model at 5 × 5 degrees. J. Geophys. Res., 103: 727–747.

Moore, G. F., Curray, J. R., and Emmel, F. J., 1982. Sedimentation in the Sunda trends and forearc region. Geol. Soc. Lond., Spec. Publ., 10: 245–258.

Moores, E. M., 1982. Origin and emplacement of ophiolites. Rev. Geophys., 20: 735–760.

Moores, E. M., 1991. Southwestern United States–East Antarctic (SWEAT) connection: A hypothesis. Geology, 19: 425–428.

Moores, E. M., 2002. Pre-1-Ga ophiolites: Their tectonic and environmental implications. Geol. Soc. Am. Bull., 114: 80–95.

Moores, E. M., Kellogg, L. H., and Dilek, Y., 2000. Tethyan ophiolites, mantle convection, and tectonic historical contingency: A resolution of the ophiolite conundrum. Geol. Soc. Am., Spec. Paper, 349: 3–12.

Morgan, J. P., Parmentier, E. M., and Lin, J., 1987. Mechanisms for the origin of midocean-ridge axial topography: Implications for the thermal and mechanical structure of accreting plate boundaries. J. Geophys. Res., 92: 12,823–12,836.

Morgan, P., 1985. Crustal radiogenic heat production and the selective survival of ancient continental crust. J. Geophys. Res., 90: C561–C570.

Moyes, A. B., Barton, J. M. Jr., and Groenewald, P. B., 1993. Late Proterozoic to early Paleozoic tectonism in Dronning Maud Land, Antarctica: Supercontinental fragmentation and amalgamation. J. Geol. Soc. Lond., 150: 833–842.

Muehlenbachs, K., and Clayton, R. N., 1976. Oxygen isotope composition of oceanic crust and its bearing on seawater. J. Geophys. Res., 81: 4365–4369.

Mueller, S., and Phillips, R. J., 1991. On the initiation of subduction. J. Geophys. Res., 96: 651–665.

Mueller, W., 1991. Volcanism and related slope to shallow-marine volcaniclastic sedimentation: An Archean example near Chibougamau, Quebec, Canada. Precamb. Res., 49: 1–22.

Muller, R. A., 2002. Avalanches at the core–mantle boundary. Geophys. Res. Lett., 29 (19), doi: 10.1029/2002GL015938 (online).

Mundil, R., Metcalfe, I., Ludwig, K. R., Renne, P. R., Oberli, F., and Nicol, R. S., 2001. Timing of the Permian–Triassic biotic crisis: Implications from new zircon U/Pb age data. Earth Planet. Sci. Lett., 187: 131–145.

Murthy, V. R., 1991. Early differentiation of the Earth and the problem of mantle siderophile elements: A new approach. Science, 253: 303–306.

Nadeau, L., and van Breemen, O., 1994. Do the 1.45–1.39-Ga Montauban Group and the La Bostonnais Complex constitute a Grenvillian accreted terrane? (program with abstracts). Geol. Assoc. Can., 19: A81.

Nance, R. D., Worsley, T. R., and Moody, J. B., 1986. Post-Archean biogeochemical cycles and long-term episodicity in tectonic processes. Geology, 14: 514–518.

Narbonne, G. M., 1998. The Ediacara biota: A terminal Neoproterozoic experiment in the evolution of life. GSA Today, 8 (2): 1–6.

Nelson, B. K., and DePaolo, D. J., 1985. Rapid production of continental crust 1.7 to 1.9 by ago: Nd isotopic evidence from the basement of the North American midcontinent. Geol. Soc. Am. Bull., 96: 746–754.

Nelson, K. D., 1991. A unified view of craton evolution motivated by recent deep seismic reflection and refraction results. Geophys. J. Int., 105: 25–35.

Nesbitt, H. W., and Young, G. M., 1982. Early Proterozoic climates and plate motions inferred from major element chemistry of lutites. Nature, 299: 715–717.

Newsom, H. E., and Sims, K. W. W., 1991. Core formation during early accretion of the Earth. Science, 252: 926–933.

Newsom, H. E., and Taylor, S. R., 1989. Geochemical implications of the formation of the Moon by a single giant impact. Nature, 338: 29–34.

Ni, S., and Helmberger, D. V., 2003. Further constraints on the African superplume structure. Phys. Earth Planet. Interiors, 140: 243–251.

Nicolas, A., 1986. Structure and petrology of peridotites: Clues to their geodynamic environment. Rev. Geophys., 24: 875–895.

Nimmo, F., and McKenzie, D., 1998. Volcanism and tectonics on Venus. Ann. Rev. Earth Planet. Sci., 1998, 26: 23–51.

Nisbet, E. G., 1986. RNA, hydrothermal systems, zeolites, and the origin of life. Episodes, 9: 83–90.

Nisbet, E. G., 1995. Archean ecology: A review of evidence for the early development of bacterial biomes, and speculations on the development of a global-scale biosphere. In: M. P. Coward and A. C. Ries (eds.), Early Precambrian Processes. Geological Society, London, Special Publication No. 95, pp. 27–51.

Nisbet, E. G., 2002. Fermor lecture: The influence of life on the face of the Earth: Garnets and moving continents. Geol. Soc. Lond., Spec. Publ., 199: 275–307.

Nisbet, E. G., and Sleep, N. H., 2001. The habitat and nature of early life. Nature, 409: 1083–1091.

Nisbet, E. G., Cheadle, M. J., Arndt, N. T., and Bickle, M. J., 1993. Constraining the potential temperature of the Archean mantle: A review of the evidence from komatiites. Lithos, 30: 291–307.

Nixon, P. H., Rogers, N. W., Gibson, I. L., and Grey, A., 1981. Depleted and fertile mantle xenoliths from southern African kimberlites. Ann. Rev. Earth Planet. Sci., 9: 285–309.

Noffke, N., Hazen, R., and Nhleko, N., 2003. Earth's earliest microbial mats in a siliciclastic marine environment 2.9-Ga Mozaan Group, South Africa. Geology, 31: 673–676.

Nuth, J. A., 2001. How were the comets made? Am. Scient., 89: 228–235.

Nutman, A. P., 2001. On the scarcity of >3900-Ma detrital zircons in ≥3500-Ma metasediments. Precamb. Res., 105: 93–114.

Nutman, A. P., McGregor, V. R., Friend, C. R. L., Bennett, V. C., and Kinny, P. D., 1996. The Itsaq gneiss complex of southern West Greenland: The world's most extensive record of early crustal evolution. Precamb. Res., 78: 1–39.

Nutman, A. P., Friend, C. R. L., and Bennett, V. C., 2002. Evidence for 3650–3600-Ma assembly of the northern end of the Itsaq gneiss complex, Greenland: Implication for early Archean tectonics. Tectonics, 21 (1), doi: 10.1029/2000TC001203 (online).

Nyblade, A. A., and Pollack, H. N., 1993. A global analysis of heat flow from Precambrian terrains: Implications for the thermal structure of Archean and Proterozoic lithosphere. J. Geophys. Res., 98: 12,207–12,218.

Oberbeck, V. R., Marshall, J. R., and Aggarwal, H., 1993. Impacts, tillites, and the breakup of Gondwanaland. J. Geol., 101: 1–19.

O'Connor, J. M., and Duncan, R. A., 1990. Evolution of the Walvis ridge and Rio Grande rise hotspot system: Implications for African and South American plate motions over plumes. J. Geophys. Res., 95: 17,475–17,502.

Officer, C. B., Hallam, A., Drake, C. L., and Devine, J. D., 1987. Late Cretaceous and paroxysmal Cretaceous–Tertiary extinctions. Nature, 326: 143–149.

Ogasawara, H., Yoshida, A., Imai, E. I., Honda, H., Hatori, K., and Matsuno, K., 2000. Synthesizing oligomers from monomeric nucleotides in simulated hydrothermal environments. Origin Life Evol. Biosphere, 30: 519–526.

Ogawa, M., 2003. Chemical stratification in a two-dimensional convecting mantle with magmatism and moving plates. J. Geophys. Res., 108 (B12), 2561, doi: 10.1029/ 2002JB002205 (online).

Ohmoto, H., 1997. When did the Earth's atmosphere become anoxic? Geochem. News, 93: 12.

Ohmoto, H., Kadegawa, T., and Lowe, D. R., 1993. 3.4-Ga biogenic pyrites from Barberton, South Africa: Sulfur isotope evidence. Science, 262: 555–557.

Ohtani, E., Shibata, T., Kubo, T., and Katao, T., 1995. Stability of hydrous phases in the transition zone and the uppermost part of the lower mantle. Geophys. Res. Lett., 22: 2553–2556.

Olivarez, A. M., and Owen, R. M., 1991. The Eu anomaly of seawater: Implications for fluvial versus hydrothermal REE inputs to the oceans. Chem. Geol., 92: 317–328.

Olson, P., 1997. Probing Earth's dynamo. Nature, 389: 337–338.

Olson, P., and Aurnou, J., 1999. A polar vortex in the Earth's core. Nature, 402: 170–173.

Olson, P., Schuber, G., and Anderson, C., 1987. Plume formation in the D" layer and the roughness of the core– mantle boundary. Nature, 327: 409–413.

Omar, G. I., and Steckler, M. S., 1995. Fission track evidence on the initial rifting of the Red Sea: Two pulses, no propagation. Science, 270: 1341–1344.

O'Nions, R. K., 1987. Relationships between chemical and convective layering in the Earth. J. Geol. Soc. Lond., 144: 259–274.

Oparin, A. I., 1953. The Origin of Life. Dover, New York. 157 pp.

O'Reilly, S. Y., and Griffin, W. L., 1985. A xenolith-derived geotherm for southeastern Australia and its geophysical implications. Tectonophysics, 111: 41–63.

O'Reilly, S. Y., Griffin, W. L., and Gaul, O., 1997. Paleogeotherms in Australia: Basis for 4D lithosphere mapping. Austral. Geol. Survey Org. J., 17: 63–72.

Oro, J., 1994. Early chemical stages in the origin of life. In: S. Bengtson (ed.), Early Life on Earth. Columbia University Press, New York, pp. 48–59.

Orth, C. J., Gilmore, J. S., and Knight, J. D., 1987. Iridium anomaly at the Cretaceous–Tertiary boundary in the Raton basin. In: New Mexico Geological Society Guidebook, Proceedings of the 38th Field Conference. New Mexico Geological Society, Socorro, NM, pp. 265–269.

Pais, M. A., Le Mouel, J. L., Lambeck, K., and Poirier, J. P., 1999. Late Precambrian paradoxical glaciation and obliquity of the Earth: A discussion of dynamical constraints. Earth Planet. Sci. Lett., 174: 155–171.

Pannella, G., 1972. Paleontologic evidence of the Earth's rotational history since early Precambrian. Astrophys. Space Sci., 16: 212–237.

Pasteris, J. D., 1984. Kimberlites: Complex mantle melts. Ann. Rev. Earth Planet. Sci., 12: 133–153.

Patchett, J. P., and Arndt, N. T., 1986. Nd isotopes and tectonics of 1.9–1.7-Ga crusta genesis. Earth Planet. Sci. Lett., 78: 329–338.

Patchett, J. P., and Gehrels, G. E., 1998. Continental influence on Canadian Cordilleran terranes from Nd isotopic study, and significance for crustal growth processes. J. Geol., 106: 269–280.

Patterson, C. C., Tilton, G. R., and Inghram, M. G., 1955. Age of the Earth. Science, 121: 69–75.

Pavlov, A. A., Kasting, J. F., Brown, L. L., Rages, K. A., and Freedman, R., 2000. Greenhouse warming by CH_4 in the atmosphere of early Earth. J. Geophys. Res., 105: 11,981–11,990.

Pavlov, A. A., Hurtgen, M. T., Kasting, J. F., and Arthur, M. A., 2003. Methane-rich Proterozoic atmosphere? Geology, 31: 87–90.

Pearce, J. A., and Peate, D. W., 1995. Tectonic implications of the composition of volcanic arc magmas. Ann. Rev. Earth Planet. Sci., 23: 251–285.

Pearce, J. A., Harris, N. B. W., and Tindle, A. G., 1984. Trace element discrimination diagrams for the tectonic interpretation of granitic rocks. J. Petrol., 25: 956–983.

Pearson, D. G., et al., 1995. Re/Os, Sm/Nd, and Rb/Sr isotope evidence for thick Archean lithospheric mantle beneath the Siberian craton modified by multistage metasomatism. Geochim. Cosmochim. Acta, 59: 959–977.

Pearson, N. J., Alard, O., Griffin, W. L., Jackson, S. E., and O'Reilly, S. Y., 2002. In situ measurement of Re/Os isotopes in mantle sulfides by laser ablation multicollector–inductively coupled plasma mass spectrometry: Analytical methods and preliminary results. Geochim. Cosmochim. Acta, 66: 1037–1050.

Peate, D. W., 1997. The Parana-Etendeka province. Am. Geophys. Union, Mon., 100: 217–245.

Peltier, W. R., Butler, S., and Solheim, L. P., 1997. The influence of phase transformations on mantle mixing and plate tectonics. In: D. J. Crossley (ed.), Earth's Deep Interior. Gordon & Breach, Amsterdam, pp. 405–430.

Pepin, R. O., 1997. Evolution of Earth's noble gases: Consequences of assuming hydrodynamic loss driven by giant impact. Icarus, 126: 148–156.

Percival, J. A., and West, G. F., 1994. The Kapuskasing uplift: A geological and geophysical synthesis. Can. J. Earth Sci., 31: 1256–1286.

Percival, J. A., and Williams, H. R., 1989. Late Archean Quetico accretionary complex, Superior province, Canada. Geology, 17: 23–25.

Percival, J. A., Green, A. G., Milkereit, B., Cook, F. A., Geis, W., and West, G. F., 1989. Seismic reflection profiles across deep continental crust exposed in the Kapuskasing uplift structure. Nature, 342: 416–419.

Percival, J. A., Fountain, D. M., and Salisbury, M. H., 1992. Exposed crustal cross-sections as windows on the lower crust. In: D. M. Fountain, R. Arculus, and R. W. Kay (eds.), Continental Lower Crust. Elsevier, Amsterdam, pp. 317–362.

Perry, E. C. Jr., Ahmad, S. N., and Swulius, T. M., 1978. The oxygen isotope composition of 3800-Ma metamorphosed chert and iron formation from Isukasia, western Greenland. J. Geol., 86: 223–239.

Pesonen, L. J., et al., 2003. Paleomagnetic configuration of continents during the Proterozoic. Tectonophysics, 375: 289–324.

Pettengill, G. H., Campbell, D. B., and Masursky, H., 1980. The surface of Venus. Scient. American, 243: 54–65.

Pfiffner, A., 1992. Alpine orogeny. In: D. Blundell, R. Freeman, and S. Mueller (eds.). A Continent Revealed: The European Geotraverse. Cambridge University Press, Cambridge, UK, pp. 180–190.

Pierson, B. K., 1994. The emergence, diversification, and role of photosynthetic eubacteria. In: S. Bengtson (ed.), Early Life on Earth. Columbia University Press, New York, pp. 161–180.

Pieters, C. M., and McFadden, L. A., 1994. Meteorite and asteroid reflectance spectroscopy: Clues to early solar system processes. Ann. Rev. Earth Planet. Sci., 22: 457–497.

Piper, J. D. A., 1987. Paleomagnetism and the Continental Crust. J. Wiley & Sons, New York, 434 pp.

Pisarevsky, S. A., Wingate, M. T. D., Powell, C. M., Johnson, S., and Evans, D. A. D., 2003. Models of Rodinia assembly and fragmentation. Geol. Soc. Lond., Spec. Publ., 206: 35–55.

Pitman, W. C., and Talwani, M., 1972. Seafloor spreading in the North Atlantic. Geol. Soc. Am. Bull., 83: 619–646.

Plotnick, R. E., 1980. Relationship between biological extinctions and geomagnetic reversals. Geology, 8: 578–581.

Polat, A., Kerrich, R., and Wyman, D. A., 1999. Geochemical diversity in oceanic komatiites and basalts from the late Archean Wawa greenstone belts, Superior province, Canada: Trace element and Nd isotope evidence for a heterogeneous mantle. Precamb. Res., 94: 139–173.

Polet, J., and Anderson, D. L., 1995. Depth extent of cratons as inferred from tomographic studies. Geology, 23: 205–208.

Poli, S., and Schmidt, M. W., 1995. H_2O transport and release in subduction zones: Experimental constraints on basaltic and andesitic systems. J. Geophys. Res., 100: 22,299–22,314.

Pollack, J. B., and Bodenheimer, P., 1989. Theories of the origin and evolution of the giant planets. In: S. K. Atreya, J. B. Pollack, and M. S. Matthews (eds.), Origin and Evolution of Planetary and Satellite Atmospheres. University Arizona Press, Tucson, pp. 564–602.

Poreda, R. J., and Becker, L., 2003. Fullerenes and interplanetary dust at the Permian–Triassic boundary. Astrobiology, 3: 75–90.

Poudjom Djomani, Y. H. P., O'Reilly, S. Y., Griffin, W. L., and Morgan, P., 2001. The density structure of subcontinental lithosphere through time. Earth Planet. Sci. Lett., 184: 605–621.

Powell, C. M., Li, Z. X., McElhinny, M. W., Meert, J. G., and Park, J. K., 1993. Paleomagnetic constraints on timing of the Neoproterozoic breakup of Rodinia and the Cambrian formation of Gondwana. Geology, 21: 889–892.

Pretorius, D. A., 1976. The nature of the Witwatersrand gold–uranium deposits. In: K. H. Wolfe (ed.), Handbook of Stratabound Ore Deposits. Elsevier, Amsterdam, pp. 29–88.

Prevot, M., Mankineu, E. A., Gromme, C. S., and Coe, R. S., 1985. How geomagnetic field vector reverses polarity. Nature, 316: 230–234.

Prevot, M., Mattern, E., Camps, P., and Daignieres, M., 2000. Evidence for a 20-degree tilting of the Earth's rotation axis 110 Ma. Earth Planet. Sci. Lett., 179: 517–528.

Price, M. H., and Suppe, J., 1994. Mean age of rifting and volcanism on Venus deduced from impact crater densities. Nature, 372: 756–759.

Price, M. H., Watson, G., Suppe, J., and Brankman, C., 1996. Dating volcanism and rifting on Venus using impact crater densities. J. Geophys. Res., E2, 101: 4657–4671.

Prokoph, A., Ernst, R. E., and Buchan, K. L., 2004. Time-series analysis of large igneous provinces: 3500 Ma to present. J. Geol., 112: 1–22.

Rampino, M. R., 1994. Tillites, diamictites, and ballistic ejecta of large impacts. J. Geol., 102: 439–456.

Rampino, M. R., and Caldeira, K., 1994. The Goldilocks problem: Climatic evolution and long-term habitability of terrestrial planets. Ann. Rev. Astron. Astrophys., 32: 83–114.

Rapp, R. P., and Watson, E. B., 1995. Dehydration melting of metabasalt at 8-32 kb: Implications for continental growth and crust–mantle recycling. J. Petrol., 36: 891–931.

Rasmussen, B., and Buick, R., 1999. Redox state of the Archean atmosphere: Evidence from detrital heavy minerals in ca. 3250–2750-Ma sandstones from the Pilbara craton, Australia. Geology, 27: 115–118.

Ravizza, G., and Peucker-Ehrenbrink, B., 2003. Chemostratigraphic evidence of Deccan volcanism from the marine osmium isotope record. Science, 302: 1392–1395.

Renne, P. R., Zichao, Z., Richards, M. A., Black, M. T., and Basu, A. R., 1995. Synchrony and causal relations between Permian–Triassic boundary crises and Siberian volcanism. Science, 269: 1413–1416.

Reymer, A., and Schubert, G., 1984. Phanerozoic addition rates to the continental crust and crustal growth. Tectonics, 3: 63–77.

Rhodes, M., and Davies, J. H., 2001. Tomographic imaging of multiple mantle plumes in the uppermost lower mantle. Geophys. J. Intern., 147: 88–92.

Ribe, N. M., and de Valpine, D. P., 1994. The global hotspot distribution and instability of D". Geophys. Res. Lett., 21: 1507–1510.

Richards, M. A., Duncan, R. A., and Courtillot, V. E., 1989. Flood basalts and hotspot tracks: Plume heads and tails. Science, 246: 103–107.

Richardson, S. H., 1990. Age and early evolution of the continental mantel. In: M. A. Menzies (ed.), Continental Mantle. Clarendon Press, Oxford, pp. 55–65.

Richardson, S. H., Harris, J. W., and Gurney. J. J., 1993. Three generations of diamonds from old continental mantle. Nature, 366: 256–258.

Richter, F. M., 1988. A major change in the thermal state of the Earth at the Archean–Proterozoic boundary: Consequences for the nature and preservation of continental lithosphere. J. Petrol., Spec. Lithos. Iss., pp. 39–52.

Richter, F. M., Rowley, D. B., and DePaolo, D. J., 1992. Sr isotope evolution of seawater: The role of tectonics. Earth Planet. Sci. Lett., 109: 11–23.

Righter, K., 2003. Metal–silicate partitioning of siderophile elements and core formation in the early Earth. Ann. Rev. Earth Planet. Sci., 31: 135–174.

Ringwood, A. E., 1979. Origin of the Earth and Moon. Springer-Verlag, New York, 295 pp.

Ritsema, J., and Allen, R. M., 2003. The elusive mantle plume. Earth Planet. Sci. Lett., 207: 1–12.

Rivers, T., 1997. Lithotectonic elements of the Grenville province: Review and tectonic implications. Precamb. Res., 86: 117–154.

Robert, F., 2001. The origin of water on Earth. Science, 293: 1056–1058.

Robertson, D. S., 1991. Geophysical applications of very long baseline interferometry. Rev. Modern Phys., 63: 899–918.

Robinson, P. I., 1973. Paleoclimatology and continental drift. In: D. H. Tarling and S. K. Runcorn (eds.), Implications of Continental Drift to the Earth Sciences, Vol. 1. Academic Press, London, pp. 451–476.

Roering, C., et al., 1992. Tectonic model for the evolution of the Limpopo belt. Precamb. Res., 55: 539–552.

Rogers, J. J. W., 1996. A history of continents in the past 3 billion years. J. Geol., 104: 91–107.

Rona, P. A., 1977. Plate tectonics, energy, and mineral resources: Basic research leading to payoff. EOS 58: 629–639.

Rona, P. A., Klinkhammer, G., Nelson, T. A., Trefry, H., and Elderfield, H., 1986. Black smokers, massive sulfides, and vent biota at the mid-Atlantic ridge. Nature, 321: 33–37.

Ronov, A. B., and Yaroshevsky, A. A., 1969. Chemical composition of the Earth's crust. Am. Geophys. Union, Mon., 13: 37–57.

Ronov, A. B., Bredanova, N. V., and Migdisov, A. A., 1992. General trends in the evolution of the chemical composition of sedimentary and magmatic rocks of the continental Earth crust. Sov. Sci. Rev., Geol. Rev., 1: 1–37.

Rosing, M. T., 1999. ^{13}C-depleted carbon microparticles in >3700-Ma seafloor sedimentary rocks from West Greenland. Science, 283: 674–676.

Rosing, M. T., Rose, N. R., Bridgwater, D., and Thomsen, H. S., 1996. Earliest part of Earth's stratigraphic record: A reappraisal of the >3.7-Ga Isua supracrust sequence. Geology, 24: 43–46.

Ross, M. I., and Scotese, C. R., 1988. A hierarchical tectonic model of the Gulf of Mexico and Caribbean region. Tectonophysics, 155: 139–168.

Rothschild, L. J., and Mancinelli, R. L., 1990. Model of carbon fixation in microbial mats from 3500 Myr ago to the present. Nature, 345: 710–712.

Rowley, D. B., 2002. Rate of plate creation and destruction: 180 Ma to present. Geol. Soc. Am. Bull., 114: 927–933.

Rubey, W. W., 1951. Geologic history of seawater. Geol. Soc. Am. Bull, 62: 1111–1148.

Rudnick, R. L., 1992. Xenoliths: Samples of the lower continental crust. In: D. M. Fountain, R. Arculus, and R. W. Kay (eds.), Continental Lower Crust. Elsevier, Amsterdam, pp. 269–316.

Rudnick, R. L., McDonough, W. F., and O'Connell, R. J., 1998. Thermal structure, thickness, and composition of continental lithosphere. Chem. Geol., 145: 395–411.

Runnegar, B., 1995. Proterozoic eukaryotes: Evidence form biology and geology. In: S. Bengtson (ed.), Early Life on Earth. Columbia University Press, New York, pp. 287–297.

Ruppel, C., 1995. Extensional processes in continental lithosphere. J. Geophys. Res., 100: 24,187–24,215.

Rutter, E. H., and Brodie, K. H., 1992. Rheology of the lower crust. In: D. M. Fountain, R. Arculus, and R. W. Kay (eds.), Continental Lower Crust. Elsevier, Amsterdam, pp. 201–267.

Rutter, E. H., and Neumann, D., 1995. Experimental deformation of partially molten Westerly granite under fluid-absent conditions, with implications for the extraction of granitic magmas. J. Geophys. Res., 100: 15,679–15,716.

Ryan, C. G., Griffin, W. L., and Pearson, N. J., 1996. Garnet geotherms: A technique for derivation of P–T data from Cr-pyrope garnets. J. Geophys. Res., 101: 5611–5625.

Ryder, G., 2002. Mass flux in the ancient Earth–Moon system and benign implications for the origin of life on Earth. J. Geophys. Res., 107, doi: 10.1029/2001JE001583 (online).

Rye, R., and Holland, H. D., 2000. Life associated with a 2.76-Ga ephemeral pond? Evidence from Mt. Roe #2 paleosol. Geology, 28: 483–486.

Rye, R., Kuo, P. H., and Holland, H. D., 1995. Atmospheric CO_2 concentrations before 2.2 Ga. Nature, 378: 603–605.

Ryskin, G., 2003. Methane-driven oceanic eruptions and mass extinctions. Geology, 31: 741–744.

Sacks, I. S., 1983. The subduction of young lithosphere. J. Geophys. Res., 88: 3355–3366.

Sakuyama, M., and Nesbitt, R. W., 1986. Geochemistry of the Quaternary volcanic rocks of the northeast Japan arc. J. Volcanol. Geotherm. Res., 29: 413–450.

Sample, J. C., and Fisher, D. M., 1986. Duplex accretion and underthrusting in an ancient accretionary complex, Kodiak Islands, Alaska. Geology, 14: 160–163.

Samson, S. D., and Patchett, P. J., 1991. The Canadian Cordillera as a modern analogue of Proterozoic crustal growth. Austral. J. Earth Sci., 38: 595–611.

Sanford, A. R., and Einarsson, P., 1982. Magma chambers in rifts. Am. Geophys. Union–Geol. Soc. Am., Geodynam. Series, 8: 147–168.

Sarda, P., Staudacher, T., and Allegre, C. J., 1985. $^{40}Ar/^{39}Ar$ in MORB glasses: Constraints on atmosphere and mantle evolution. Earth Planet. Sci. Lett., 72: 357–375.

Sarda, P., Moreira, M., Staudacher, T., Schilling, J. G., and Allegre, C. J., 2000. Rare gas systematics on the southernmost mid-Atlantic ridge: Constraints on the lower mantle and the Dupal source. J. Geophys. Res., 105: 5973–5996.

Saunders, A. D., Tarney, J., and Weaver, S. D., 1980. Transverse geochemical variations across the Antarctic Peninsula: Implications for the genesis of calc-alkaline magmas. Earth Planet. Sci. Lett., 46: 344–360.

Saunders, A. D., Tarney, J., Kerr, A. C., and Kent, R. W., 1996. The formation and fate of large igneous provinces. Lithos, 37: 81–95.

Sawkins, F. J., 1990. Metal Deposits in Relation to Plate Tectonics. Springer-Verlag, Berlin.

Schermer, E. R., Howell, D. G., and Jones, D. L., 1984. The origin of allochthonous terranes. Ann. Rev. Earth Planet. Sci., 12: 107–131.

Schidlowski, M., Hayes, J. M., and Kaplan, I. R., 1983. Isotopic inferences of ancient biochemistries: Carbon, sulfur, hydrogen, and nitrogen. In: J. W. Schopf (ed.), The Earth's Earliest Biosphere: Its Origin and Evolution. Princeton University Press, Princeton, NJ, pp. 149–186.

Schlanger, S. O., and Jenkyns, H. C., 1976. Cretaceous oceanic anoxic events: Causes and consequences. Geol. Mijnbouw, 55: 179–184.

Scholz, C. H., and Campos, J., 1995. On the mechanism of seismic decoupling and back-arc spreading at subduction zones. J. Geophys. Res., 100: 22,103–22,115.

Schopf, J. W., 1994. The oldest known records of life: Early Archean stromatolites, microfossils, and organic matter. In: S. Bengtson (ed.), Early Life on Earth. Columbia University Press, New York, pp. 193–206.

Schott, B., and Schmeling, H., 1998. Delamination and detachment of a lithospheric root. Tectonophysics, 296: 225–247.

Schreyer, W., 1995. Ultradeep metamorphic rocks: The retrospective viewpoint. J. Geophys. Res., 100: 8353–8366.

Schubert, G., Turcotte, D. L., and Olson, P., 2001. Mantle Convection in the Earth and Planets. Cambridge University Press, Cambridge, UK.

Schwab, F. L., 1978. Secular trends in the composition of sedimentary rocks assemblages: Archean through Pherozoic time. Geology, 6: 532–536.

Sclater, J. G., Jaupart, C., and Galson, D., 1980. The heat flow through oceanic and continental crust and the heat loss of the Earth. Rev. Geophys., 18: 269–311.

Scotese, C. R., 1991. Jurassic and Cretaceous plate tectonic reconstructions. Paleogeog. Paleoclimat. Paleoecol., 87: 493–501.

Scott, D. J., Helmstaedt, H., and Bickle, M. J., 1992. Purtuniq ophiolite, Cape Smith belt, northern Quebec, Canada: A reconstructed section of Early Proterozoic oceanic crust. Geology, 20: 173–176.

Searle, M. P., et al., 1987. The closing of Tethys and the tectonics of the Himalaya. Geol. Soc. Am. Bull., 98: 678–701.

Seber, D., Barazangi, M., Ibenbrahim, A., and Demnati, A., 1996. Geophysical evidence for lithospheric delamination beneath the Alboran Sea and Rit-Betic Mountains. Nature, 379: 785–790.

Seilacher, A., 1994. Early multicellular life: Late Proterozoic fossils and the Cambrian explosion. In: S. Bengtson (ed.), Early Life on Earth. Columbia University Press, New York, pp. 389–400.

Sengor, A. M. C., and Burke, K., 1978. Relative timing of rifting and volcanism on Earth and its tectonic implications. Geophys. Res. Lett., 5: 419–421.

Sengor, A. M. C., Natal'in, B. A., and Burtman, V. S., 1993. Evolution of the Altaid tectonic collage and Paleozoic crustal growth in Eurasia. Nature, 364: 299–307.

Sepkoski, J. J. Jr., 1989. Periodicity in extinction and the problem of catastrophism in the history of life. J. Geol. Soc. Lond., 146: 7–19.

Sepkoski, J. J. Jr., 1996. Patterns of Pohanerozoic extinction: A perspective from global data bases. In: O. H. Walliser (ed.), Global Events and Event Stratigraphy in the Phanerozoic. Springer-Verlag, Berlin, pp. 35–51.

Shaviv, N. J., and Veizer, J., 2003. Celestial driver of Phanerozoic climate? GSA Today, 13 (7): 4–10.

Shaw, D. M., 1976. Development of the early continental crust: Part 2. In: B. F. Windley (ed.), The Early History of the Earth. J. Wiley, New York, pp. 33–54.

Shaw, D. M., Cramer, J. J., Higgins, M. D., and Truscott, M. G., 1986. Composition of the Canadian Precambrian shield and the continental crust of the Earth. Geol. Soc. Lond., Spec. Publ., 24: 275–282.

Sheehan, P. M., 2001. The late Ordovician mass extinction event. Ann. Rev. Earth Planet. Sci., 29: 331–364.

Sherman, D. M., 1995. Stability of possible Fe-FeS and Fe-FeO alloy phases at high pressure and the composition of the Earth's core. Earth Planet. Sci. Lett., 132: 87–98.

Sherman, D. M., 1997. The composition of the Earth's core: Constraints on S and Si versus temperature. Earth Planet. Sci. Lett., 153: 149–155.

Shimamoto, T., 1985. The origin of large or great thrust-type earthquakes along subducting plate boundaries. Tectonophysics, 119: 37–65.

Shixing, Z., and Huineng, C., 1995. Megascopic multicellular organisms form the 1700-My-old Tuanshanzi Formation in the Jixian area, north China. Science, 270: 620–622.

Shock, E. L., 1992. Chemical environment of submarine hydrothermal systems. Origin Life Evol. Biosphere, 22: 67–108.

Shock, E. L., and Schulte, M. D., 1998. Organic synthesis during fluid mixing in hydrothermal systems. J. Geophys. Res., 103: 28,513–28,527.

Shukolyukov, A., and Lugmair, G. W., 1998. Isotopic evidence for the Cretaceous–Tertiary impactor and its type. Science, 282: 927–929.

Sidorin, I., Gurnis, M., and Helmberger, D. V., 1999. Evidence for a ubiquitous seismic discontinuity at the base of the mantle. Science, 286: 1326–1331.

Sillitoe, R. H., 1976. Andean mineralization: A model for the metallogeny of convergent plate margins. Geol. Assoc. Can., Spec. Paper, 14: 59–100.

Silver, P. G., and Chan, W. W., 1991. Shear wave splitting and subcontinental mantle deformation. J. Geophys. Res., 96: 16,429–16,454.

Simonson, B. M., and Harnik, P., 2000. Have distal impact ejecta changed through geologic time? Geology, 28: 975–978.

Sims, K. W. W., Newsom, H. E., and Gladney, E. S., 1990. Chemical fractionation during formation of the Earth's core and continental crust: Clues from As, Sb, W, and Mo. In: H. E. Newsom and J. H. Jones (eds.), Origin of the Earth. Oxford University Press, New York, pp. 291–317.

Sinigoi, S., et al., 1994. Chemical evolution of a large mafic intrusion in the lower crust, Ivrea-Verbano zone, northern Italy. J. Geophys. Res., 99: 21,575–21,590.

Sleep, N. H., 1990. Hotspots and mantle plumes: Some phenomenology. J. Geophys. Res., 95: 6715–6736.

Sleep, N. H., 1992. Archean plate tectonics: What can be learned from continental geology? Can. J. Earth Sci., 29: 2066–2071.

Sleep, N. H., and Windley, B. F., 1982. Archean plate tectonics: Constraints and inferences. J. Geol., 90: 363–379.

Sleep, N. H., Nunn, J. A., and Chou, L., 1980. Platform basins. Ann. Rev. Earth Planet. Sci., 8: 17–34.

Sleep, N. H., Zahnle, K. J., Kasting, J. F., and Morowitz, H. J., 1989. Annihilation of ecosystems by large asteroid impacts on the early Earth. Nature, 342: 139–142.

Sloan, R. E., Rigby, J. K. Jr., Van Valen, L. M., and Gabriel, D., 1986. Gradual dinosaur extinction and simultaneous ungulate radiation in the Hell Creek Formation. Science, 232: 629–633.

Sloss, L. L., 1972. Synchrony of Phanerozoic sedimentary–tectonic events of the North American craton and Russian platform. Inter. Geol. Cong. Rep., Montreal Sec., 6: 24–32.

Small, C., 1995. Observations of ridge–hotspot interactions in the southern ocean. J. Geophys. Res., 100: 17,931–17,946.

Smith, A. G., and Woodcock, N. H., 1982 Tectonic synthesis of the Alpine–Mediterranean region: A review. Am. Geophys. Union–Geol. Soc. Am., Geodynam. Series, 7: 15–38.

Smith, D. E., et al., 1994. Contemporary global horizontal crustal motion. Geophys. J. Inter., 119: 511–520.

Smith, R. B., and Braile, L. W., 1994. The Yellowstone hotspot. J. Volcan. Geotherm. Res., 61: 121–187.

Solomon, S. C., 2003. Mercury: The enigmatic innermost planet. Earth Planet. Sci. Lett., 216: 441–455.

Solomon, S. C., and Toomey, D. R., 1992. The structure of midocean ridges. Ann. Rev. Earth Planet. Sci., 20: 329–364.

Solomon, S. C., et al., 1992. Venus tectonics: An overview of Magellan observations. J. Geophys. Res., 97: 13,199–13,255.

Song, X., and Helmberger, D. V., 1998. Seismic evidence for an inner core transition zone. Science, 282: 924–927.

Spence, W., 1987. Slab pull and the seismotectonics of subducting lithosphere. Rev. Geophys., 25: 55–69.

Spray, J. G., 1984. Possible causes and consequences of upper mantle decoupling and ophiolite displacement. Geol. Soc. Lond., Spec. Publ., 13: 255–267.

Steckler, M., 1984. Changes in sea level. In: H. D. Holland and A. F. Trendall (eds.), Patterns of Change in Earth Evolution. Springer-Verlag, Berlin, pp. 103–121.

Stein, M., and Hofmann, A. W., 1994. Mantle plumes and episodic crustal growth. Nature, 372: 63–68.

Steinberger, B., and O'Connell, R. J., 1998. Advection of plumes in mantle flow: Implications for hotspot motion, mantle viscosity, and plume distribution. Geophys. J. Intern., 132: 412–434.

Stern, S. A., 2003. The evolution of comets in the Oort cloud and Kuiper belt. Nature, 424: 639–642.

Steuber, T., and Veizer, J., 2002. Phanerozoic record of plate tectonic control of seawater chemistry and carbonate sedimentation. Geology, 30: 1123–1126.

Stevens, T. O., and McKinley, J. P., 1995. Lithoautotrophic microbial ecosystems in deep basalt aquifers. Science, 270: 450–454.

Stevenson, D. J., 1990. Fluid dynamics of core formation. In: H. E. Newsom and J. H. Jones (eds.), Origin of the Earth. Oxford University Press, Oxford, pp. 231–249.

Stevenson, D. J., 2001. Mars' core and magnetism. Nature, 412: 214–219.

Stevenson, D. J., 2003. Planetary magnetic fields. Earth Planet. Sci. Lett., 208: 1–11.

Stixrude, L., and Brown, J. M., 1998. The Earth's core. In: R. J. Hemley (ed.), Ultra-High-Pressure Mineralogy: Physics and Chemistry of the Earth's Deep Interior, Vol. 37. Mineralogical Society of America, Washington, DC, pp. 261–282.

Stockwell, C. H., 1965. Structural trends in the Canadian shield. Bull. Am. Assoc. Petrol. Geol., 49: 887–893.

Stockwell, G. S., Beaumont, C., and Boutilier, R., 1986. Geodynamic models of convergent margin tectonics: The transition from rifted margin to overthrust belt and the consequences for foreland basin development. Bull. Am. Assoc. Petrol. Geol., 70: 181–190.

Stoddard, P. R., and Abbott, D., 1996. Influence of the tectosphere upon plate motion. J. Geophys. Res., 101: 5425–5433.

Stofan, E. R., Smrekar, S. E., Tapper, S. W., Guest, J. E., and Grindrod, P. M., 2001. Preliminary analysis of an expanded database for Venus. Geophys. Res. Lett., 28: 4267–4270.

Storey, B. C., 1995. The role of mantle plumes in continental breakup: Case histories from Gondwanaland. Nature, 377: 301–308.

Strik, G., Blake, T. S., Zergers, T. E., White, S. H., and Langereis, C. G., 2003. Paleomagnetism of flood basalts izn the Pilbara craton, Western Australia: Late Archean continental drift and the oldest known reversal of the geomagnetic field. J. Geophys. Res., 108, doi: 10.1029/2003JB002475 (online).

Strom, R. G., Schaber, G. G., and Dawson, D. D., 1994. The global resurfacing of Venus. J. Geophys. Res., 99: 10,899–10,926.

Sun, S. Q., 1994. A reappraisal of dolomite abundance and occurrence in the Phanerozoic. J. Sed. Res., A64: 396–404.

Sun, S., and McDonough, W. F., 1989. Chemical and isotopic systematics of oceanic basalts: Implications for mantle composition and processes. In: A. D. Saunder and J. J. Norry (eds.), Magmatism in the Ocean Basins. Geological Society, London, Special Publication No. 42, pp. 313–345.

Sweet, A. R., 2001. Plants: A yardstick for measuring the environmental consequences of the Cretaceous–Tertiary boundary event. Geosci. Can., 28: 127–138.

Swisher, C. C., et al., 1992. Coeval ^{40}Ar/^{39}Ar ages of 65 Ma from Chicxulub crater melt rock and K–T boundary tektites. Science, 257: 954–958.

Sylvester, P. J., 1994. Archean granite plutons. In: K. C. Condie (ed.), Archean Crustal Evolution, Elsevier, Amsterdam, pp. 261–314.

Sylvester, P. J., Campbell, I. H., and Bowyer, D. A., 1997. Nb/U evidence for early formation of the continental crust. Science, 275: 521–523.

Tackley, P. J., 2000. Mantle convection and plate tectonics: Toward an integrated physical and chemical theory. Science, 288: 2002–2007.

Tackley, P. J., Stevenson, D. J., Glatzmaier, G. A., and Schubert, G., 1994. Effects of multiple phase transitions in a three-dimensional spherical model of convection in the Earth's mantle. J. Geophys. Res., 99: 15,877–15,901.

Takashima, R., Nishi, H., and Yoshida, T., 2002. Geology, petrology, and tectonic setting of the late Jurassic ophiolite in Hokkaido, Japan. J. Asian Earth Sci., 21: 197–215.

Tapponnier, P., Peltzer, G., and Armijo, R., 1986. On the mechanics of the collision between India and Asia. Geol. Soc. Lond., Spec. Publ., 19: 115–158.

Tarduno, J. A., and Gee, J., 1995. Large-scale motion between Pacific and Atlantic hotspots. Nature, 378: 477–479.

Tarduno, J. A., et al., 2003. The Emperor seamounts: Southward motion of the Hawaiian hotspot plume in Earth's mantle. Science, 301: 1064–1069.

Tarney, J., 1992. Geochemistry and significance of mafic dyke swarms in the Proterozoic. In: K. C. Condie (ed.), Proterozoic Crustal Evolution, Elsevier, Amsterdam, pp. 151–179.

Tarney, J., Pickering, K. T., Knipe, R. J., and Dewey. J. F. (eds.), 1991. The behaviour and influence of fluids in subduction zones. Roy. Soc. Lond., Spec. Publ., A335.

Tatsumi, Y., Kani, T., Ishizuka, H., Maruyama, S., and Nishimura, Y., 2000. Activation of Pacific mantle plumes during the Carboniferous: Evidence from accretionary complexes in southwestern Japan. Geology, 28: 580–582.

Taylor, S. R., 1982. Planetary Science: A Lunar Perspective. Lunar and Planetary Institute, Houston, 481 pp.

Taylor, S. R., 1987. The unique lunar composition and its bearing on the origin of the Moon. Geochim. Cosmochim. Acta, 51: 1297–1309.

Taylor, S. R., 1992. Solar System Evolution. Cambridge University Press, Cambridge, UK, 307 pp.

Taylor, S. R., 1993. Early accretional history of the Earth and the Moon-forming event. Lithos, 30: 207–221.

Taylor, S. R., 1999. On the difficulties of making Earth-like planets. Meteor. Planet. Sci., 34: 317–329.

Taylor, S. R., and McLennan, S. M., 1995. The geochemical evolution of the continental crust. Rev. Geophys., 33: 241–265.

Thompson, R. N., and Gibson, S. A., 1994. Magmatic expression of lithospheric thinning across continental rifts. Tectonophysics, 233: 41–68.

Thompson, A. B., and Ridley, J. R., 1987. P–T–t histories of orogenic belts. Phil. Trans. Roy. Soc. Lond., A321: 27–45.

Thurston, P. C., 1994. Archean volcanic patterns. In: K. C. Condie (ed.), Archean Crustal Evolution, Elsevier, Amsterdam, pp. 45–84.

Thurston, P. C., and Chivers, K. M., 1990. Secular variation in greenstone sequence development emphasizing Superior province, Canada. Precamb. Res., 46: 21–58.

Thybo, H., Ross, A. R., and Egorkin, A. V., 2003. Explosion seismic reflections from the Earth's core. Earth Planet. Sci. Lett., 216: 693–702.

Tohver, E., van der Pluijm, B. A., Van der Voo, R., Rizzotto, G., and Scandolara, J. E., 2002. Paleogeography of the Amazon craton at 1.2 Ga: Early Grenvillian collision with the Llano segment of Laurentia. Earth Planet. Sci. Lett., 199: 185–200.

Toksoz, N. M., Minear, J. W., and Julian, B. R., 1971. Temperature field and geophysical effects of the downgoing slab. J. Geophys. Res., 76: 1113–1137.

Tomlinson, K. Y., and Condie, K. C., 2001. Archean mantle plumes: Evidence from greenstone belt geochemistry. Geol. Soc. Am., Mem., 352: 341–357.

Tomlinson, K. Y., Hughes, D. J., Thurston, P. C., and Hall, R. P., 1999. Plume magmatism and crustal growth at 2.9 to 3.0 Ga in the Steep Rock and Lumby Lake area, western Superior province. Lithos, 46: 103–136.

Toon, O. B., 1984. Sudden changes in atmospheric composition and climate. In: H. D. Holland and A. F. Trendall (eds.), Patterns of Change in Earth Evolution. Springer-Verlag, Berlin, pp. 41–61.

Toon, O. B., Turco, R. P., and Covey, C., 1997. Environmental perturbations caused by the impacts of asteroids and comets. Rev. Geophys., 35: 41–78.

Trendall, A. F., 1983. Precambrian iron formation of Australia. Econ. Geol., 68: 1023–1034.

Tromp, J., 2001. Inner-core anisotropy and rotation. Ann. Rev. Earth Planet. Sci., 29: 47–69.

Tunnicliffe, V., and Fowler, M. R., 1996. Influence of seafloor spreading on the global hydrothermal vent fauna. Nature, 379: 531–533.

Turcotte, D. L., 1995. How does Venus lose heat? J. Geophys. Res., 100: 16,931–16,940.

Turcotte, D. L., 1996. Magellan and comparative planetology. J. Geophys. Res., 101: 4765–4773.

Turcotte, D. L., and White, W. M., 2001. Thorium–uranium systematics requires layered mantle convection. J. Geophys. Res., 106: 4265–4276.

Tyler, A. L., Kozlowski, R. W. H., and Lebofsky, L. A., 1988. Determination of rock type on Mercury and the Moon through remote sensing in the thermal infrared. Geophys. Res. Lett., 15: 808–811.

Unrug, R., 1993. The supercontinent cycle and Gondwanaland assembly: Composition of cratons and the timing of suturing events. J. Geodynam., 16: 215–240.

Uyeda, S., 1983. Comparative subductology. Episodes, 2: 19–24.

Uyeda, S., and Miyashiro, A., 1974. Plate tectonics and the Japanese islands: A synthesis. Geol. Soc. Am. Bull., 85: 1159–1170.

Vail, P. R., and Mitchum, R. M. Jr., 1979. Global cycles of relative changes of sea level from seismic stratigraphy. Am. Assoc. Petrol. Geol., Mem., 29: 469–472.

Valet, J. P., and Meynadier, L., 1993. Geomagnetic field intensity and reversals during the past 4 My. Nature, 366: 234–238.

Valet, J. P., Laj, C., and Tucholka, P., 1986. High-resolution sedimentary record of a geomagnetic reversal. Nature, 322: 27–32.

Valley, J. W., Peck, W. H., King, E. M., and Wilde, S. A., 2002. A cool early Earth. Geology, 30: 351–354.

van der Hilst, R. D., 1995. Complex morphology of subducted lithosphere in the mantle beneath the Tonga trench. Nature, 374: 154–157.

van der Hilst, R. D., and Karason, H., 1999. Compositional heterogeneity in the bottom 1000 kilometers of Earth's mantle: Toward a hybrid convection model. Science, 283: 1885–1888.

van der Hilst, R. D., Engdahl, R., Spakman, W., and Nolet, G., 1991. Tomographic imaging of subducted lithosphere below northwestern Pacific island arcs. Nature, 353: 37–43.

van der Hilst, R. D., Widiyantoro, S., and Engdahl, E. R., 1997. Evidence for deep mantle circulation from global tomography. Nature, 386: 578–584.

van der Meijde, M., Marone, F., Giardini, D., and van der Lee, S., 2003. Seismic evidence for water deep in Earth's upper mantle. Science, 300: 1556–1558.

Van der Voo, R., Spakman, W., and Bijwaard, H., 1999. Mesozoic subducted slabs under Siberia. Nature, 397: 246–249.

Van Fossen, M. C., and Kent, D. V., 1992. Paleomagnetism of 122-Ma plutons in New England and the mid-Cretaceous paleomagnetic field in North America: True polar wander or large-scale differential mantle motion? J. Geophys. Res., 97: 19,651–19,661.

Van Kranendonk, M. J., Hickman, A. H., Smithies, R. H., and Nelson, D. R., 2002. Geology and tectonic evolution of the Archean North Pilbara terrain, Pilbara craton, Western Australia. Econ. Geol., 97: 695–732.

Van Kranendonk, M. J., Webb, G. E., and Kamber, B. S., 2003. Geological and trace element evidence for a marine sedimentary environment of deposition and biogenicity of 3.45-Ga strromatolitic carbonates in the Pilbara craton, and support for a reducing Archean ocean. Geobiology, 1: 91–108.

Varsek, J. L., et al., 1993. Lithoprobe crustal reflection structure of the southern Canadian Cordillera 2: Coast Mountains transect. Tectonics, 12: 334–360.

Veevers, J. J., 1990. Tectonic–climatic supercycle in the billion-year plate-tectonic eon: Permian Pangaean icehouse alternates with Cretaceous-dispersed continents greenhouse. Sed. Geol., 68: 1–16.

Veizer, J., 1979. Secular variations in chemical composition of sediments: A review. Phys. Chem. Earth, 11: 269–278.

Veizer, J., 1989. Strontium isotopes in seawater through time. Ann. Rec. Earth Planet. Sci., 17: 141–187.

Veizer, J., and Compston, W., 1976. $^{87}Sr/^{86}Sr$ in Precambrian carbonates as an index of crustal evolution. Geochim. Cosmochim. Acta, 40: 905–914.

Veizer, J., and Jansen, S. L., 1985. Basement and sedimentary recycling: 2—Time dimension to global tectonics. J. Geol., 93: 625–643.

Veizer, J., Compston, W., Clauer, N., and Schidlowski, M., 1983. $^{87}Sr/^{86}Sr$ in late Proterozoic carbonates: Evidence for a mantle event at 900 Ma. Geochim. Cosmochim. Acta, 47: 295–302.

Veizer, J., Hoefs, J., Lowe, D. R., and Thurston, P. C., 1989. Geochemistry of Precambrian carbonates: II—Archean greenstone belts and Archean seawater. Geochim. Cosmochim. Acta, 53: 859–871.

Veizer, J., Bell, K., and Jansen, S. L., 1992. Temporal distribution of carbonatites. Geology, 20: 1147–1149.

Veizer, J., et al., 1999. $^{87}Sr/^{86}Sr$, $\delta13C$, and $\delta^{18}O$ evolution of Phanerozoic seawater. Chem. Geol., 161: 59–88.

Veizer, J., Godderis, Y., and Francois, L. M., 2000. Evidence for decoupling of atmospheric CO_2 and global climate during the Phanerozoic eon. Nature, 408: 698–701.

Vidal, P., Bernard-Griffiths, J., Cocherie, A., Le Fort, P., Peucat, J. J., and Sheppard, S. M., 1984. Geochemical comparison between Himalayan and Hercynian leucogranites. Phys. Earth Planet. Interiors, 35: 179–190.

Vidale, J. E., Ding, X. Y., and Grand, S. P., 1995. The 410-km-depth discontinuity: A sharpness estimate from near- critical reflections. Geophys. Res. Lett., 22: 2557–2560.

Vidale, J. E., Dodge, D. A., and Earle, P. S., 2000. Slow differential rotation of the Earth's inner core indicated by temporal changes in scattering. Nature, 405: 445–447.

Vielzeuf, D., Clemens, J. D., Pin, C., and Moinet, E., 1990. Granites, granulites, and crustal differentiation. In: D. Vielzeuf and P. H. Vidal (eds.), Granulites and Crustal Evolution. Kluwer Academic Press, Norwell, MA, pp. 59–85.

Vigny, C., Ricard, Y., and Froidevaux, C., 1991. The driving mechanism of plate tectonics. Tectonophysics, 187: 345–360.

Vilas, F., et al., 1988. Mercury. University Arizona Press, Tucson.

Vine, F., 1966. Spreading of the ocean floor: Evidence. Science, 154: 1405–1415.

Vine, F., and Matthews, D. H., 1963. Magnetic anomalies over oceanic ridges. Nature, 199: 947–949.

Vinnik, L. P., Green, R. W. E., and Nicolaysen, L. O., 1995. Recent deformations of the deep continental root beneath southern Africa. Nature, 375: 50–52.

Vitorello, I., and Pollack, H. N., 1980. On the variation of continental heat flow with age and the thermal evolution of continents. J. Geophys. Res., 85: 983–995.

Vlastelic, L., Aslanian, D., Dosso, L., Bougault, H., Olivet, J. L., and Geli, L., 1999. Large-scale chemical and thermal division of the Pacific mantle. Nature, 399: 345–350.

Vocadlo, L., Alfe, D., Gillan, M. J., Wood, I. G., Brodholt, J. P., and Price, G. D., 2003. Possible thermal and chemical stabilization of body-centered cubic iron in the Earth's core. Nature, 424: 536–539.

von Brunn, V., and Gold, D. J. D., 1993. Diamictite in the Archean Pongola sequence of southern Africa. J. African Earth Sci., 16: 367–374.

Von Damm, K. L., 1990. Seafloor hydrothermal activity: Black smoker chemistry and chimneys. Ann. Rev. Earth Planet. Sci., 18: 173–204.

von Huene, R., and Scholl, D. W., 1991. Observations at convergent margins concerning sediment subduction, subduction erosion, and the growth of continental crust. Rev. Geophys., 29: 279–316.

von Raumer, J. F., Stampfli, G. M., and Bussy, F., 2003. Gondwana-derived microcontinents: The constituents of the Variscan and Alpine collisional orogens. Tectonophysics, 365: 7–22.

Walker, J. C. G., 1977. Evolution of the Atmosphere. Macmillan, New York, 318 pp.

Walker, J. C. G., 1990. Precambrian evolution of the climate system. Paleogeog. Paleoclimat. Paleoecol., 82: 261–289.

Walker, J. C. G., Klein, C., Schidlowski, M., Schopf, J. W., Stevenson, D. J., and Walter, M. R., 1983. Environmental evolution of the Archean–early Proterozoic Earth. In: J. W. Schopf (ed.), The Earth's Earliest Biosphere: Its Origin and Evolution. Princeton University Press, Princeton, NJ, pp. 260–290.

Walker, R. J., Morgan, J. W., and Horan, M. F., 1995. Osmium-187 enrichment in some plumes: Evidence for core–mantle interaction? Science, 269: 819–821.

Walter, M. R., 1994. Stromatolites: The main geological source of information on the evolution of the early benthos. In: S. Bengtson (ed.), Early Life on Earth. Columbia University Press, New York, pp. 270–286.

Walter, M. R., Veevers, J. J., Claver, C. R., Gorjan, P., and Hill, A. C., 2000. Dating the 840–544-Ma Neoproterozoic interval by isotopes of Sr, C, and S in seawater, and some interpretative models. Precamb. Res., 100: 371–433.

Wang, Y., Weidner, D. J., Liebermann, R. C., and Zhao, Y., 1994. P–V–T equation of state of perovskite: Constraints on composition of the lower mantle. Phys. Earth Planet. Inter., 83: 13–40.

Warner, M., et al., 1996. Seismic reflections from the mantle represent relict subduction zones within the continental lithosphere. Geology, 24: 39–42.

Watanabe, Y., Martini, J. E. J., and Ohmoto, H., 2000. Geochemical evidence for terrestrial ecosystems 2.6 billion years ago. Nature, 408: 574–577.

Waters, F. G., and Erlank, A. J., 1988. Assessment of the vertical extent and distribution of mantle metasomatism below Kimberley, South Africa. J. Petrol., Spec. Lithos. Iss., pp. 185–204.

Weaver, B. L., Wood, D. A., Tarney, J., and Joron, J. L., 1987. Geochemistry of ocean island basalts from the South Atlantic: Ascension, Bouvet, St. Helena, Gough, and Trista da Cunha. Geol. Soc. Lond., Spec. Publ., 30: 253–267.

Wedepohl, K. H., 1995. The composition of the continental crust. Geochim. Cosmochim. Acta, 59: 1217–1232.

Wegener, A., 1912. Die Entstehung der Kontinente. Geol. Rund., 3: 276–292.

Weiguo, S., 1994. Early multicellular fossils. In: S. Bengtson (ed.), Early Life on Earth. Columbia University Press, New York, pp. 358–369.

Weinberger, A. J., 2002. A dusty business. Science, 295: 2027–2028.

Wendlandt, E., DePaolo, D. J., and Baldridge, W. S., 1993. Nd and Sr isotope chronostratigraphy of Colorado Plateau lithosphere: Implications for magmatic and tectonic underplating of the continental crust. Earth Planet. Sci. Lett., 116: 23–43.

Wetherill, G. W., 1990. Formation of the Earth. Ann. Rev. Earth Planet. Sci., 18: 205–256.

Wetherill, G. W., 1994. Provenance of the terrestrial planets. Geochim. Cosmochim. Acta, 58: 4513–4520.

White, R. S., 1988. The Earth's crust and lithosphere. J. Petrol., Spec. Lithos. Iss., pp. 1–10.

White, R. S., and McKenzie, D., 1995. Mantle plumes and flood basalts. J. Geophys. Res., 100: 17,543–17,585.

White, R. S., Spence, G. D., Fowler, S. R., McKenzie, D. P., Westbrook, G. K., and Bowen, A. N., 1987. Magmatism at rifted continental margins. Nature, 330: 439–444.

Wickham, S. M., 1992. Fluids in the deep crust: Petrological and isotopic evidence. In: D. M. Fountain, R. Arculus, and R. W. Kay (eds.), Continental Lower Crust. Elsevier, Amsterdam, pp. 391–421.

Widdel, F., Schnell, S., Heising, S., Ehrenreich, A., Assmus, B., and Schink, B., 1993. Ferrous iron oxidation by anoxygenic phototrophic bacteria. Nature, 362: 834–836.

Wiebe, R. A., 1992. Proterozoic anorthosite complexes. In: K. C. Condie (ed.), Proterozoic Crustal Evolution. Elsevier, Amsterdam, pp. 215–261.

Wiens, D. A., and Stein, S., 1985. Implications of oceanic intraplate seismicity for plate stresses, driving forces, and rheology. Tectonophysics, 116: 143–162.

Wignall, P., 2001. Large igneous provinces and mass extinctions. Earth Sci. Rev., 53: 1–33.

Wignall, P., 2003. The nature of the end-Permian mass extinction and its bearing on volcanic extinction mechanisms. In: Mantle Plumes: Physical Processes, Chemical Signatures, and Biological Effects (abstract volume). Cardiff University, Cardiff, UK, p. 45.

Wilde, P., Quinby-Hunt, M. S., and Berry, B. N., 1990. Vertical advection from oxic or anoxic water from the pycnocline as a cause of rapid extinction or rapid radiations. In: E. G. Kauffman and O. H. Walliser (eds.), Extinction Events in Earth History: Lecture Notes on Earth Science, No. 30. Springer, Heidelberg, pp. 85–97.

Wilde, S. A., Valley, J. W., Peck, W. H., and Graham, C. M., 2001. Evidence from detrital zircon from the existence of continental crust and oceans on the Earth 4.4 Ga. Nature, 409: 175–178.

Williams, G. E., 1975. Late Precambrian glacial climate and the Earth's obliquity. Geol. Mag., 112: 441–465.

Williams, G. E., 1989. Precambrian tidal sedimentary cycles and Earth's paleorotation. EOS, 70: 40–41.

Williams, L. A. J., 1982. Physical aspects of magmatism in continental rifts. Am. Geophys. Union–Geol. Soc. Am., Geodynam. Series, 8: 193–219.

Wilson, J. T., 1963. Evidence from islands on the spreading of the ocean floor. Nature, 197: 536–538.

Wilson, J. T., 1965. Transform faults, oceanic ridges, and magnetic anomalies southwest of Vancouver Island. Science, 150: 482–485.

Wilson, M., 1993. Magmatism and the geodynamics of basin formation. Sediment. Geol., 86: 5–29.

Wilson, P. A., and Norris, R. D., 2001. Warm tropical ocean surface and global anoxia during the mid-Cretaceous period. Nature, 412: 425–428.

Windley, B. F., 1992. Proterozoic collision and accretionary orogens. In: K. C. Condie (ed.), Proterozoic Crustal Evolution, Elsevier, Amsterdam, pp. 419–446.

Windley, B. F., 1993. Proterozoic anorogenic magmatism and its orogenic connections. J. Geol. Soc. Lond., 150: 39–50.

Wise, D. U., 1963. An origin of the Moon by rotational fission during formation of the Earth's core. J. Geophys. Res., 68: 1547–1554.

Wise, D. U., 1973. Freeboard of continents through time. Geol. Soc. Am., Mem., 132: 87–100.

Wolbach, W. S., Lewis, R. S., and Anders, E., 1985. Cretaceous extinctions: Evidence for wildfires and search for meteoric material. Science, 230: 167–170.

Wood, B. J., 1995. The effect of water on the 410-km seismic discontinuity. Science, 268: 74–76.

Wood, D. A., 1979. A variably veined suboceanic upper mantle: Genetic significance for midocean-ridge basalts from geochemical evidence. Geology, 7: 499–503.

Wood, J. A., 1986. Moon over Mauna Loa: A review of hypotheses of formation of the Earth's moon. In: W. K. Hartmann, R. J. Phillips, and G. J. Taylor (eds.), Origin of the Moon. Lunar and Planetary Institute, Houston, pp. 17–56.

Wood, J. A., and Mitler, H. E., 1974. Origin of the Moon by a modified capture mechanism, or half a loaf is better than a whole one. Lunar Sci., 5: 851–853.

Woollard, G. P., 1968. The interrelationship of the crust, upper mantle, and isostatic gravity anomalies in the United States. Am. Geophys. Union, Mon., 12: 312–341.

Workman, R. K., Hart, S. R., Blusztajn, J., Jackson, M., Kurz, M., and Staudigel, H., 2003. Enriched mantle II: A new view from the Samoan hotspot. Geophys. Res. Abstracts, 5: 13,656.

Worsley, T. R., and Nance, R. D., 1989. Carbon redox and climate control through Earth history: A speculative reconstruction. Paleogeog. Paleoclimat. Paleoecol., 75: 259–282.

Worsley, T. R., Nance, R. D., and Moody, J. B., 1984. Global tectonics and eustasy for the past 2 billion years. Marine Geol., 58: 373–400.

Worsley, T. R., Nance, R. D., and Moody, J. B., 1986. Tectonic cycles and the history of the Earth's biogeochemical and paleoceanographic record. Paleoceanography, 1: 233–263.

Wronkiewicz, D. J., and Condie, K. C., 1989. Geochemistry and provenance of sediments from the Pongola Supergroup, South Africa: Evidence for a 3.0-Ga-old continental craton. Geochim. Cosmochim. Acta, 53: 1537–1549.

Wyckoff, S., 1991. Comets: Clues to the early history of the solar system. Earth Sci. Rev., 30: 125–174.

Wyllie, P. J., 1971. The role of water in magma generation and initiation of diapiric uprise in the mantle. J. Geophys. Res., 76: 1328–1338.

Wyman, D. A., and Kerrich, R., 2002. Formation of Archean continental lithospheric roots: The role of mantle plumes. Geology, 30: 543–546.

Xie, Q., and Kerrich, R., 1994. Silicate-perovskite and majorite signature komatiites from the Archean Abitibi greenstone belt: Implications for early mantle differentiation and stratification. J. Geophys. Res., 99: 15,799–15,812.

Xie, Q., Kerrich, R., and Fan, J., 1993. HFSE/REE fractionations recorded in three komatiite–basalt sequences, Archean Abitibi greenstone belt: Implications for multiple plume sources and depths. Geochim. Cosmochim. Acta, 57: 4111–4118.

Xie, Q., McCuaig, T. C., and Kerrich, R., 1995. Secular trends in the melting depths of mantle plumes: Evidence from HFSE/REE systematics of Archean high-Mg lavas and modern oceanic basalts. Chem. Geol., 126: 29–42.

Yang, W., Holland, H. D., and Rye, R., 2002. Evidence for low or no oxygen in the late Archean atmosphere from the 2.76-Ga Mt. Roe #2 paleosol, Western Australia: Part 3. Geochim. Cosmochim. Acta, 66: 3707–3718.

Yin, Q., Jacobsen, S. B., Yamashita, K., Blichert-Toft, J., Telouk, P., and Albarede, F., 2002. A short timescale for terrestrial planet formation from Hf-W chronometry of meteorites. Nature, 418: 949–952.

Yoshida, S. I., Sumita, I., and Kumazawa, M., 1996. Growth model of the inner core coupled with the outer core dynamics and the resulting elastic anisotropy. J. Geophys. Res., 101: 28,085–28,103.

Young, C. J., and Lay, T., 1987. The core–mantle boundary. Ann. Rev. Earth Planet. Sci., 15: 25–46.

Young, G. M., 1991. The geologic record of glaciation: Relevance to the climatic history of Earth. Geosci. Can., 18: 100–108.

Yuen, D. A., Hansen, U., Zhao, W., Vincent, A. P., and Malevsky, A. V., 1993. Hard turbulent thermal convection and thermal evolution of the mantle. J. Geophys. Res., 98: 5355–5373.

Yukutake, T., 1998. Implausibility of thermal convection in the Earth's solid inner core. Phys. Earth Planet. Inter., 108: 1–13.

Zahnle, K. J., and Sleep, N. H., 1997. Impacts and the early evolution of life. In: P. J. Thomas, C. F. Chyba, and C. P. McKay (eds.), Comets and the Origin and Evolution of Life. Springer, New York, pp. 175–208.

Zandt, G., and Ammon, C. J., 1995. Continental crust composition constrained by measurements of crustal Poisson's ratio. Nature, 374: 152–154.

Zaug, A. J., and Cech, T. R., 1986. The interviewing sequence RNA of Tetrahymena is an enzyme. Science, 231: 470–475.

Zegers, T. E., and van Keken, P. E., 2001. Middle Archean continent formation by crustal delamination. Geology, 29: 1083–1086.

Zerr, A., Diegeler, A., and Boehler, R., 1998. Solidus of Earth's deep mantle. Science, 281: 243–245.

Zhai, Y., and Halls, H. C., 1994. Multiple episodes of dike emplacement along the northwest margin of the Superior province, Manitoba. J. Geophys. Res. 99: 21,717–21,732.

Zhang, Y., 2002. The age and accretion of the Earth. Earth Sci. Rev., 59: 235–263.

Zhang, Y. S., and Tanimoto, T., 1992. Ridges, hotspots, and their interaction as observed in seismic velocity maps. Nature, 355: 45–49.

Zhang, Y. S., and Tanimoto, T., 1993. High-resolution global upper mantle structure and plate tectonics. J. Geophys. Res., 98: 9793–9823.

Zhao, D., Akira, H., and Horiuchi, S., 1992. Tomographic imaging of P- and S-wave velocity structure beneath northeastern Japan. J. Geophys. Res., 97: 19,909–19,928.

Zhong, S., and Gurnis, M., 1994. Role of plates and temperature-dependent viscosity in phase change dynamics. J. Geophys. Res., 99: 903–917.

Zhong, S., and Gurnis, M., 1995. Mantle convection with plates and mobile, faulted plate margins. Science, 267: 838–843.

Zindler, A., and Hart, S. R., 1986. Chemical geodynamics. Ann. Rev. Earth Planet. Sci., 14: 493–571.

Zoback, M. L., and Zoback, M., 1980. State of stress in the coterminous United States. J. Geophys. Res., 85: 6113–6145.

Zuber, M. T., 2001. The crust and mantle of Mars. Nature, 412: 220–227.

Zuber, M. T., et al., 2000. Internal structure and early thermal evolution of Mars from Mars Global Surveyor topography and gravity. Science, 287: 1788–1793.

Index